PURE MATHEMATICS

A Second Course

SI Edition

BY

J. K. BACKHOUSE, M.A.

Tutor at the Department of Educational Studies in the University of Oxford and formerly Head of the Mathematics Department at Hampton Grammar School

S. P. T. HOULDSWORTH, M.A.

Lately Headmaster of Sydney Grammar School
Formerly Assistant Master at Harrow School

AND

B. E. D. COOPER, M.A.

Headmaster of St. Bartholomew's Grammar School,
Newbury

LONGMAN

LONGMAN GROUP LIMITED
London
Associated companies, branches and representatives
throughout the world

© J. K. Backhouse, S. P. T. Houldsworth, B. E. D. Cooper, 1963
© SI Edition Longman Group Ltd 1971

First published 1963
SI Edition 1971
Seventh Impression 1978
ISBN 0 582 31798 3

Printed in Singapore by
Kyodo-Shing Loong Printing Industries Pte Ltd

PREFACE

This volume is a sequel to *Pure Mathematics—A First Course*, by J. K. Backhouse and S. P. T. Houldsworth, published by Longman. The two volumes bring the work up to the standard normally required by future scientists and engineers at Advanced Level of the G.C.E. It is hoped that the faster student will be stimulated by the extension of various topics beyond examination requirements, and by the chapters on Complex Numbers and Numerical Methods.

It has been represented to the authors that their course of pure mathematics should have included a chapter on variation, including experimental laws, and they are intending to add such a chapter to a later edition of the First Course.†

The individual reader has been kept in mind, and he is advised to work through questions in the text marked '**Qu.**' as he reads; the class teacher will find that many of these are suitable for oral work.

Certain questions have been marked with an asterisk *. This indicates that they contain a useful result or method for which room has not been found in the text.

We wish to express our thanks to the following Examining Bodies for permission to include questions from past G.C.E. papers in which they own the copyright. The authorities concerned bear no responsibility, of course, for the solutions to their questions, where these are given. Such questions have been marked with the abbreviations shown below:

The Joint Matriculation Board of the Universities of
    Manchester, Liverpool, Leeds, Sheffield, and Birmingham,   N.
The University of Cambridge Local Examinations Syndicate,   C.
The University of London,   L.
The Oxford and Cambridge Schools Examination Board,   O.C.
The Oxford Delegacy of Local Examinations,   O.

In September 1960, Mr B. E. D. Cooper took up his appointment as Headmaster of St Bartholomew's Grammar School, Newbury, and his duties prevented him from continuing to work on this volume. Mr

† This chapter is included in the Second Edition published 1965.

v

J. A. Strover has given us some valuable assistance since that date, and we are pleased to record our appreciation of his help. We also wish to thank Mrs S. L. Parsonson, Mr L. G. W. P. Jones, and Mr M. A. Bloxham for their assistance in working through the answers.

Finally, we wish to express our thanks to all those at Longman, especially Mr E. W. Parker, and William Clowes and Sons Ltd who have helped in the production of the book.

J. K. B.
S. P. T. H.
B. E. D. C.

Harrow-on-the-Hill
November 1962

PREFACE TO THE SECOND EDITION

Pure Mathematics—A First Course and A Second Course have both been modified to employ SI units (Système International d'Unités), the internationally agreed system of units.

For the most part this has involved simply a change of units but on occasions data has had to be altered. In a very few cases completely new questions have been substituted, and there have also been minor amendments to the text. However, apart from the units, the changes are of a minor nature and this edition is compatible with the previous one.

J. K. B.
S. P. T. H.
B. E. D. C.

Oxford
October 1970

CONTENTS

CONTENTS

Chapter 5
PROBABILITY

Chapter 6
THREE-DIMENSIONAL GEOMETRY AND TRIGONOMETRY

Chapter 7
SOME INEQUALITIES AND GRAPHS

Chapter 8
FURTHER EQUATIONS AND FACTORS

Chapter 9
COORDINATE GEOMETRY—I

Chapter 10
SERIES FOR e^z AND $\log_e(1+x)$

Chapter 11
FURTHER DIFFERENTIATION

Chapter 12
FURTHER TRIGONOMETRY

Chapter 13
FURTHER INTEGRATION

Chapter 14
PROJECTION

Chapter 15
MENSURATION AND MOMENTS OF INERTIA

CHAPTER 1

INTEGRATION

1.1. In *A First Course*† we dealt with the differentiation of powers of *x*, polynomials, products and quotients, functions of a function, trigonometrical functions, and we also discussed implicit functions and parameters.

Now that we come to extend the scope of integration we find that it is not, unfortunately, merely a matter of putting into reverse the techniques for differentiation; we have learned a technique for differentiating $(3x^2+2)^4$ as it stands, but can we integrate this function without first expanding it? Even consider the simple function x^n; we can differentiate this for all positive and negative rational values of *n*, but we must bear in mind the gap which remains to be filled later in this book when we discover how to deal with $\int x^{-1}\,dx$.

Integration is, in fact, less susceptible than differentiation to concise systematic treatment. It presents a broad front, and the reader's experience of it will gradually expand, so that by quick recognition of an increasing number of forms of *integrand* (i.e. the function to be integrated) there is developed the power to discriminate between the many possible lines of attack.

Recognizing the presence of a function and its derivative

1.2. The very first thing to search for in any but the simplest integrands is the presence of a function and its derivative; with this, we may often guess the integral to be a certain function of a function, check by differentiation, and adjust the numerical factor. Two examples follow to illustrate this method.

Example 1. *Find* $\int x(3x^2+2)^4\,dx$.

[We note that the *x* outside the bracket is a constant × the derivative of the expression inside the bracket. We deduce that the integral is a function of $(3x^2+2)$.]

† *Pure Mathematics—A First Course*, by J. K. Backhouse and S. P. T. Houldsworth, published by Longmans, Green & Co., 1957, and hereafter referred to as P.M.I.

$$\frac{d}{dx}\{(3x^2+2)^5\} = 5(3x^2+2)^4.6x = 30x(3x^2+2)^4,$$

$$\therefore \frac{d}{dx}\left\{\frac{1}{30}(3x^2+2)^5\right\} = x(3x^2+2)^4.$$

Hence $$\int x(3x^2+2)^4\,dx = \frac{1}{30}(3x^2+2)^5+c.$$

Example 2. *Find* $\int \sin^2 4x \cos 4x\,dx.$

[We note that $\cos 4x$ is a constant $\times$ the derivative of $\sin 4x$, and we deduce that the integral is a function of $\sin 4x$.]

$$\frac{d}{dx}\{\sin^3 4x\} = 3(\sin 4x)^2.\cos 4x.4 = 12\sin^2 4x \cos 4x.$$

Hence $$\int \sin^2 4x \cos 4x\,dx = \frac{1}{12}\sin^3 4x+c.$$

Qu. 1. Differentiate:

(i) $(2x^2+3)^4$,   (ii) $\sqrt{(x^2-2x+1)}$,   (iii) $\dfrac{1}{(2x-1)^2}$,

(iv) $\sin(4x-7)$,   (v) $\tan^3 x$,   (vi) $\cos^2 3x$.

Qu. 2. Find the following integrals, and check by differentiation:

(i) $\int x(x^2+1)^2\,dx$,   (ii) $\int(2x+1)^4\,dx$,   (iii) $\int(x^2+1)^3\,dx$,

(iv) $\int \frac{1}{2}\sin 3x\,dx$,   (v) $\int x^2\sqrt{(x^3+1)}\,dx$,   (vi) $\int \sec^2 x \tan x\,dx$.

Pythagoras' theorem.   Odd powers of sin x, cos x, etc.

1.3. Pythagoras' theorem in the forms $\cos^2 x+\sin^2 x = 1$, $\cot^2 x+1 = \operatorname{cosec}^2 x$, and $1+\tan^2 x = \sec^2 x$, may be used to change some integrands to a form susceptible to the method of §1.2. In particular, it enables us to integrate odd powers of $\sin x$ and $\cos x$.

Example 3. *Find* $\int \sin^5 x\,dx.$

$$\int \sin^5 x\,dx = \int \sin^4 x \sin x\,dx,$$

$$= \int(1-\cos^2 x)^2 \sin x\,dx,$$

$$= \int(\sin x-2\cos^2 x \sin x+\cos^4 x \sin x)\,dx.$$

$$\therefore \int \sin^5 x\,dx = -\cos x+\tfrac{2}{3}\cos^3 x-\tfrac{1}{5}\cos^5 x+c.$$

Qu. 3. Find: (i) $\int \sin^3 x \, dx$, (ii) $\int \cos^5 x \, dx$.

Qu. 4. Find: (i) $\int \cos^3 x \sin^2 x \, dx$, [Write $\cos^3 x$ as $\cos x(1 - \sin^2 x)$.]

(ii) $\int \sin^8 x \cos^2 x \, dx$.

Qu. 5. Find $\int \sec x \tan^3 x \, dx$. [Remember $\dfrac{d}{dx}(\sec x) = \sec x \tan x$.]

Even powers of sin x, cos x†

1.4. Two very important formulae derived from the double-angle formulae are $\cos^2 x = \frac{1}{2}(1 + \cos 2x)$ and $\sin^2 x = \frac{1}{2}(1 - \cos 2x)$. (See P.M.I., p. 285.)

Their use in integrating even powers of $\sin x$ and $\cos x$ is illustrated in the latter part of Exercise 1a, which also gives practice in the use of other formulae, including the factor formulae. (See P.M.I., p. 299.)

Exercise 1a

1. Differentiate:

(i) $(5x^2 - 1)^3$, (ii) $\dfrac{1}{(2x^2 - x + 3)^2}$, (iii) $\sqrt[3]{(x^2 + 4)}$,

(iv) $\cot 5x$, (v) $\cos(5x - 1)$, (vi) $\sin^2 \dfrac{x}{3}$,

(vii) $\tan \sqrt{x}$, (viii) $\sec^2 2x$, (ix) $\sqrt{\operatorname{cosec} x}$.

Find the following integrals in Nos. 2 to 4:

2. (i) $\int x(x^2 - 3)^5 \, dx$, (ii) $\int (3x - 1)^5 \, dx$, (iii) $\int x(x + 2)^2 \, dx$,

(iv) $\int \dfrac{x}{(x^2 + 1)^2} \, dx$, (v) $\int \dfrac{x + 1}{(x^2 + 2x - 5)^3} \, dx$,

(vi) $\int (2x - 3)(x^2 - 3x + 7)^2 \, dx$, (vii) $\int \dfrac{2x}{(4x^2 - 7)^2} \, dx$,

(viii) $\int 2x \sqrt{(3x^2 - 5)} \, dx$, (ix) $\int (x^3 + 1)^2 \, dx$,

(x) $\int \dfrac{x^2 - 1}{\sqrt{(x^3 - 3x)}} \, dx$, (xi) $\int \dfrac{x - 1}{(2x^2 - 4x + 1)^{3/2}} \, dx$,

(xii) $\int (2x^2 - 1)^3 \, dx$.

---

† This section and the latter part of Exercise 1a may with advantage be delayed and done in conjunction with later parts of the chapter.

3. (i) $\int 3 \cos 3x \, dx,$　　　　　　　(ii) $\int \sin (2x + 3) \, dx,$

　　(iii) $\int \cos x \sin x \, dx,$　　　　　(iv) $\int \frac{1}{3} \cos 2x \, dx,$

　　(v) $\int \sin 3x \cos^2 3x \, dx,$　　　(vi) $\int \sec^2 x \tan^2 x \, dx,$

　　(vii) $\int \sec^5 x \tan x \, dx,$　　　(viii) $\int \cos x \sqrt{\sin x} \, dx,$

　　(ix) $\int x \operatorname{cosec}^2 x^2 \, dx,$　　　　(x) $\int \frac{\cos \sqrt{x}}{\sqrt{x}} \, dx,$

　　(xi) $\int \operatorname{cosec}^3 x \cot x \, dx.$

4. (i) $\int \cos^3 x \, dx,$　　　　　　(ii) $\int \cos^5 \frac{x}{2} \, dx,$

　　(iii) $\int \sin^3 2x \, dx,$　　　　　(iv) $\int \cos^3 (2x + 1) \, dx,$

　　(v) $\int \sin^5 x \cos^2 x \, dx,$　　　(vi) $\int \cos^3 x \sin^3 x \, dx,$

　　(vii) $\int \sec^4 x \, dx,$　　　　　(viii) $\int \operatorname{cosec} x \cot^3 x \, dx,$

　　(ix) $\int \tan^5 x \sec x \, dx.$

5. Find $\int \tan x \sec^4 x \, dx$, (i) as a function of $\sec x$, (ii) as a function of $\tan x$, and show that they are the same.

6. Show that the integral given in No. 3(iii) may be obtained in three different forms.

———————

Nos. 7 onwards may be delayed. See footnote to §1.4.

7. Express (i) $\sin^2 \frac{x}{2}$ in terms of $\cos x$,　　(ii) $\cos^2 3x$ in terms of $\cos 6x$

8. Find (i) $\int \cos^2 x \, dx,$　(ii) $\int \sin^2 \frac{x}{2} \, dx,$　(iii) $\int \cos^2 3x \, dx.$

9. Express $\sin^4 x$ in terms of $\cos 2x$, and $\cos^2 2x$ in terms of $\cos 4x$. Show that $\int \sin^4 x \, dx = \frac{3}{8}x - \frac{1}{4} \sin 2x + \frac{1}{32} \sin 4x + c.$

10. Find $\int \cos^4 x \, dx$

11. Find the following integrals:

(i) $\int \sin^2 x \, dx$, (ii) $\int \cos^2 \frac{x}{3} \, dx$, (iii) $\int \sin^4 2x \, dx$, (iv) $\int \cos^4 \frac{x}{2} \, dx$.

12. Write down a formula for $\cos x$ in terms of $\cos \frac{x}{2}$, and show that

$$\int \frac{1}{1+\cos x} \, dx = \tan \frac{x}{2} + c.$$

13. Find the following integrals:

(i) $\int \surd(1+\cos x) \, dx$, (ii) $\int \frac{\cot x}{\surd(1-\cos 2x)} \, dx$,

(iii) $\int \sin 2x \sin^2 x \, dx$, (iv) $\int 2 \sin x \cos \frac{x}{2} \, dx$.

14. (i) Factorize $\sin 3x + \sin x$. (See P.M.I., p. 299.)

(ii) Express $2 \sin 3x \cos 2x$ as the sum of two terms.

(iii) Find $\int \sin 3x \cos 2x \, dx$.

15. Find the following integrals:

(i) $\int \sin x \cos 3x \, dx$, (ii) $\int 2 \cos \frac{3x}{2} \cos \frac{x}{2} \, dx$, (iii) $\int \sin 4x \sin x \, dx$.

Changing the variable

1.5. In Example 1 we found that

$$\int x(3x^2+2)^4 \, dx = \frac{1}{30} (3x^2+2)^5 + c.$$

The integral is a function of $(3x^2+2)$. If we write $3x^2+2$ as u, then the integral is a function of u; this suggests that we might make the substitution $u=3x^2+2$ in the integrand, and *integrate with respect to u.* Let us see how this can be done.

Let $$y = \int x(3x^2+2)^4 \, dx.$$

Then $$\frac{dy}{dx} = x(3x^2+2)^4.$$

If $u=3x^2+2$, x may be expressed as a function of u. Thus y is a function of a function of u, and

$$\frac{dy}{du} = \frac{dy}{dx} \cdot \frac{dx}{du},$$

$$\therefore \frac{dy}{du} = x(3x^2+2)^4 \frac{dx}{du}.$$

Integrating with respect to u,

$$y = \int x(3x^2+2)^4 \frac{dx}{du} \, du.$$

But $u = 3x^2 + 2$,

$$\therefore \frac{du}{dx} = 6x \quad \text{and} \quad \frac{dx}{du} = \frac{1}{6x}.$$

$$\therefore \int x(3x^2+2)^4 \, dx = \int x(3x^2+2)^4 \frac{1}{6x} \, du,$$
$$= \int \tfrac{1}{6}u^4 \, du,$$
$$= \tfrac{1}{30}u^5 + c,$$
$$= \tfrac{1}{30}(3x^2+2)^5 + c.$$

Qu. 6. Find $\int \sin^2 4x \cos 4x \, dx$; put $u = \sin 4x$.

Qu. 7. Find $\int \sin^5 x \, dx$; put $u = \cos x$.

Comparing the foregoing text and questions with the solutions of Examples 1, 2, and 3, it might appear that we have merely introduced a more cumbersome technique; however, the power of changing the variable lies in its application to a wide class of integrals not susceptible to the method of §§1.2, 1.3.

In general, let $f(x)$ be a function of x, and let

$$y = \int f(x) \, dx.$$

Then

$$\frac{dy}{dx} = f(x).$$

If u is a function of x, then y is a function of a function of u, and

$$\frac{dy}{du} = \frac{dy}{dx} \cdot \frac{dx}{du}.$$
$$\therefore \frac{dy}{du} = f(x) \frac{dx}{du},$$
$$\therefore y = \int f(x) \frac{dx}{du} \, du,$$
$$\therefore \int f(x) \, dx = \int f(x) \frac{dx}{du} \, du.$$

Thus an integral with respect to x may be transformed into an integral with respect to a related variable u, by using the above result, and substituting for f(x) and $\frac{dx}{du}$ in terms of u.

Example 4. *Find* $\int x\sqrt{(3x-1)}\,dx$.

$$\int x\sqrt{(3x-1)}\,\frac{dx}{du}\,du = \int \tfrac{1}{3}(u^2+1)u\,\frac{2u}{3}\,du,$$

$$= \int (\tfrac{2}{9}u^4+\tfrac{2}{9}u^2)\,du,$$

$$= \tfrac{2}{45}u^5+\tfrac{2}{27}u^3+c.$$

$$= \tfrac{2}{135}u^3(3u^2+5)+c,$$

$$\therefore \int x\sqrt{(3x-1)}\,dx = \tfrac{2}{135}(3x-1)^{3/2}(9x+2)+c.$$

Side work:
Let $\sqrt{(3x-1)}=u$.

$$x = \tfrac{1}{3}(u^2+1).$$

$$\frac{dx}{du} = \frac{2u}{3}.$$

Qu. 8. Find the following integrals, using the given change of variable:

(i) $\int x\sqrt{(2x+1)}\,dx,\quad \sqrt{(2x+1)} = u,$

(ii) $\int x\sqrt{(2x+1)}\,dx,\qquad 2x+1 = u,$

(iii) $\int x(3x-2)^6\,dx,\qquad 3x-2 = u.$

Exercise 1b

1. Find the following integrals, using the given change of variable:

(i) $\int 3x\sqrt{(4x-1)}\,dx,\qquad \sqrt{(4x-1)} = u,$

(ii) $\int x\sqrt{(5x+2)}\,dx,\qquad 5x+2 = u,$

(iii) $\int x(2x-1)^6\,dx,\qquad 2x-1 = u,$

(iv) $\int \frac{x}{\sqrt{(x-2)}}\,dx,\qquad \sqrt{(x-2)} = u,$

(v) $\int (x+2)(x-1)^4\,dx,\qquad x-1 = u,$

(vi) $\int (x-2)^5(x+3)^2\,dx,\qquad x-2 = u,$

(vii) $\int \frac{x(x-4)}{(x-2)^2}\,dx,\qquad x-2 = u,$

(viii) $\int \frac{x-1}{\sqrt{(2x+3)}}\,dx,\qquad \sqrt{(2x+3)} = u.$

2. Repeat Nos. 1(i) and 1(iv) using a different change of variable in each case.

3. Of each of the following pairs of integrals, one should be found by a suitable change of variable, the other written down at once as a function of a function of x, as in Example 1 on p. 1:

(i) $\int x\sqrt{(3x-4)}\,dx$ and $\int x\sqrt{(3x^2-4)}\,dx$,

(ii) $\int x(x^2+5)^6\,dx$ and $\int x(x+5)^6\,dx$,

(iii) $\int \dfrac{x}{\sqrt{(x-1)}}\,dx$ and $\int \dfrac{x}{\sqrt{(x^2-1)}}\,dx$.

4. Find the following integrals, using a suitable change of variable only when necessary:

(i) $\int x\sqrt{(2x^2+1)}\,dx$,

(ii) $\int \dfrac{3x^2-1}{(x^3-x+4)^3}\,dx$,

(iii) $\int 2x\sqrt{(2x-1)}\,dx$,

(iv) $\int \cos^3 2x\,dx$,

(v) $\int \sin x\sqrt{\cos x}\,dx$,

(vi) $\int \cot^2 x \operatorname{cosec}^2 x\,dx$,

(vii) $\int 2x(4x^2-1)^3\,dx$,

(viii) $\int \dfrac{x}{\sqrt{(2x^2-5)}}\,dx$,

(ix) $\int \dfrac{3x}{\sqrt{(4-x)}}\,dx$,

(x) $\int \dfrac{\sin\sqrt{x}}{\sqrt{x}}\,dx$,

(xi) $\int \tan x\,\sqrt{\sec x}\,dx$,

(xii) $\int \sin^3 \dfrac{x}{2}\cos^2 \dfrac{x}{2}\,dx$,

(xiii) $\int \sin x \sec^2 x\,dx$,

(xiv) $\int \dfrac{\tan x}{\cos x\,\sqrt{(1+\sec x)}}\,dx$,

(xv) $\int \dfrac{\sqrt{(1+\sqrt{x})}}{\sqrt{x}}\,dx$.

Definite integrals and changing the limits

1.6. The method of changing the variable is also applicable to definite integrals. It is usually more convenient to change the limits to those of the new variable at the same time.

As a reminder that one must be ever watchful for the presence of a function and its derivative in an integrand, two examples of this type are also given here.

Example 5. *Evaluate* $\displaystyle\int_{\frac{1}{2}}^{3} x\sqrt{(2x+3)}\ \mathrm{d}x.$

† $\displaystyle\int_{x=\frac{1}{2}}^{x=3} x\sqrt{(2x+3)}\,\frac{\mathrm{d}x}{\mathrm{d}u}\,\mathbf{d}u = \int_{4}^{9} \tfrac{1}{2}(u-3)u^{\frac{1}{2}}\,\tfrac{1}{2}\,\mathrm{d}u,$ Let $2x+3 = u.$

$\hspace{5.5cm} x = \tfrac{1}{2}(u-3).$

$\displaystyle\hspace{2.5cm} = \int_{4}^{9} (\tfrac{1}{4}u^{\frac{3}{2}} - \tfrac{3}{4}u^{\frac{1}{2}})\,\mathrm{d}u, \hspace{1.5cm} \frac{\mathrm{d}x}{\mathrm{d}u} = \tfrac{1}{2}.$

$\displaystyle\hspace{2.5cm} = \left[\tfrac{1}{10}u^{\frac{5}{2}} - \tfrac{1}{2}u^{\frac{3}{2}}\right]_{4}^{9},$

$\hspace{2.5cm} = (24\cdot3 - 13\cdot5) - (3\cdot2 - 4),$

$\hspace{2.5cm} = 11\cdot6.$

x	u
3	9
$\frac{1}{2}$	4

Example 6. *Evaluate* (i) ‡ $\displaystyle\int_{2}^{3} \frac{x}{\sqrt{(x^2-3)}}\ \mathrm{d}x,$ (ii) $\displaystyle\int_{0}^{\frac{\pi}{4}} \cos^3 x \sin x\ \mathrm{d}x.$

(i) $\displaystyle\hspace{1.5cm}\int_{2}^{3} \frac{x}{\sqrt{(x^2-3)}}\ \mathrm{d}x = \left[(x^2-3)^{\frac{1}{2}}\right]_{2}^{3},$

$\hspace{4cm} = (9-3)^{\frac{1}{2}} - (4-3)^{\frac{1}{2}},$

$\hspace{4cm} = \sqrt{6} - 1.$

(ii) $\displaystyle\hspace{1.5cm}\int_{0}^{\frac{\pi}{4}} \cos^3 x \sin x\ \mathrm{d}x = \left[-\tfrac{1}{4}\cos^4 x\right]_{0}^{\frac{\pi}{4}},$

$\hspace{4.5cm} = (-\tfrac{1}{4}\cdot\tfrac{1}{4}) - (-\tfrac{1}{4}),$

$\hspace{4.5cm} = \tfrac{3}{16}.$

Exercise 1c

1. Evaluate the following definite integrals by changing the variable and the limits:

(i) $\displaystyle\int_{2}^{3} x\sqrt{(x-2)}\ \mathrm{d}x,$ (ii) $\displaystyle\int_{0}^{1} x(x-1)^4\ \mathrm{d}x,$

(iii) $\displaystyle\int_{1}^{2} \frac{x}{\sqrt{(2x-1)}}\ \mathrm{d}x,$ (iv) $\displaystyle\int_{1}^{2} (2x-1)(x-2)^3\ \mathrm{d}x,$

(v) $\displaystyle\int_{-\frac{3}{8}}^{0} \frac{x+3}{\sqrt{(2x+1)}}\ \mathrm{d}x.$

---

† The reader should note that, in practice, this integral will of course first be written down as given (i.e. as an integral with respect to x). When it is decided to change the variable, $\mathrm{d}x$ is changed to $\dfrac{\mathrm{d}x}{\mathrm{d}u}\,\mathrm{d}u$; it is then necessary to specify that the limits are still those of x.

‡ Could the integral in Example 6(i) be evaluated between the limits -2 and $+3$? (See P.M.I., p. 104.)

2. Evaluate the following definite integrals either by writing down the integral as a function of x, or by using the given change of variable:

(i) $\int_0^{\frac{\pi}{6}} \sec^4 x \tan x \, dx$, $\quad (\sec x = u)$,

(ii) $\int_0^{\frac{\pi}{2}} \sin^5 x \, dx$, $\qquad (\cos x = u)$,

(iii) $\int_{\frac{\pi}{6}}^{\frac{\pi}{2}} \frac{\cot x}{\sqrt{\operatorname{cosec}^3 x}} \, dx$, $\quad (\operatorname{cosec} x = u)$.

3. Evaluate:

(i) $\int_0^{\frac{1}{2}} \frac{x}{\sqrt{(1-x^2)}} \, dx$, $\qquad$ (ii) $\int_0^4 2x\sqrt{(4-x)} \, dx$,

(iii) $\int_{-1}^0 x(x^2-1)^4 \, dx$, $\qquad$ (iv) $\int_0^{\frac{\pi}{4}} \sec^4 x \, dx$,

(v) $\int_{\frac{1}{2}}^1 \frac{x-2}{(x+2)^3(x-6)^3} \, dx$, $\quad$ (vi) $\int_{-1}^2 (x+1)(2-x)^4 \, dx$,

(vii) $\int_{-\frac{\pi}{2}}^{\frac{\pi}{2}} \cos^3 x \, dx$, $\qquad$ (viii) $\int_{\frac{5}{3}}^{\frac{8}{3}} \frac{x+2}{\sqrt{(3x-4)}} \, dx$,

(ix) $\int_0^{\frac{\pi}{2}} \sin x \sqrt{\cos x} \, dx$.

4. Calculate the area enclosed by the curve $y = \dfrac{x}{\sqrt{(x^2-1)}}$, the x-axis, $x=2$, and $x=3$.

5. Calculate the area under $y = \sin^3 x$ from $x=0$ to $x=\dfrac{2\pi}{3}$.

6. Calculate the volume of the solid generated when the area under $y = \cos x$, from $x=0$ to $x=\dfrac{\pi}{2}$, is rotated through four right angles about the x-axis. (See §1.4.)

7. The area of a uniform lamina is that enclosed by the curve $y = \sin x$, the x-axis, and the line $x=\dfrac{\pi}{2}$. Find the distance from the x-axis of the centre of gravity of the lamina. (See §1.4.)

Introducing inverse trigonometrical functions†

1.7. The expression $\sqrt{(1-x^2)}$ may be reduced to a rational form by changing the variable to u, where $x = \sin u$; thus

$$\sqrt{(1-x^2)} = \sqrt{(1-\sin^2 u)} = \sqrt{\cos^2 u} = \cos u.$$

This is used in the following example.

† A fuller treatment of these functions is given in Chapter 12.

Example 7. *Find* $\int \dfrac{1}{\sqrt{(1-x^2)}}\, dx.$

$$\int \frac{1}{\sqrt{(1-x^2)}} \frac{dx}{du}\, du = \int \frac{1}{\sqrt{(1-\sin^2 u)}} \cos u\, du, \qquad \text{Let } x = \sin u.$$

$$= \int \frac{1}{\cos u} \cos u\, du, \qquad \frac{dx}{du} = \cos u.$$

$$= u + c,$$

$$= \sin^{-1} x + c.$$

This is the first time we have used the notation 'sin^{-1}x'. It is read as 'sine minus one x' (*not* sine *to the* minus one x, since -1 is not an index), and it means 'the angle whose sine is x'; furthermore we must remember that in Example 7 the change of variable involved differentiating sin u, therefore in this case sin^{-1}x must be measured in *radians*. To give some numerical examples, sin$^{-1}1$ is 90° or $\dfrac{\pi}{2}$ radians, and cos$^{-1}\dfrac{1}{\sqrt{2}}$ is 45° or $\dfrac{\pi}{4}$ radians.

Qu. 9. The following angles lie between 0 and 90° inclusive. Express them in degrees, and in radians in terms of π:

(i) $\tan^{-1}1$,   (ii) $\sin^{-1}\frac{1}{2}$,   (iii) $\frac{1}{2}\sin^{-1}1$,

(iv) $\cos^{-1}\frac{1}{2}$,   (v) $\frac{1}{2}\cos^{-1}\frac{1}{2}$,   (vi) $\cos^{-1}1$,

(vii) $2\cos^{-1}\dfrac{\sqrt{3}}{2}$,   (viii) $\frac{1}{3}\cos^{-1}0$,   (ix) $\frac{2}{3}\cot^{-1}1$,

(x) $\sec^{-1}2$,   (xi) $2\operatorname{cosec}^{-1}\sqrt{2}$.

Qu. 10. Use a conversion table† to express the following angles in radians:

(i) 20°,   (ii) 70°,   (iii) 10° 20′,   (iv) 51° 35′,   (v) 86° 3′.

Qu. 11. Express the following angles in degrees and minutes:

(i) 1 radian,   (ii) 0·03 radian,   (iii) 1·25 radians,

(iv) 0·715 radian,   (v) $\dfrac{\pi}{5}$ radian.

Qu. 12. Express the following (acute) angles in radians:

(i) $2\sin^{-1}0\cdot6$,   (ii) $\tan^{-1}1\cdot333$,   (iii) $\frac{2}{3}\cos^{-1}0\cdot3846$.

† See *Four Figure Tables*, by C. Godfrey and A. W. Siddons, Cambridge University Press, 1958.

$$\int \frac{1}{\sqrt{(a^2 - b^2 x^2)}} \, dx$$

1.8. The reader should check that the integral found in §1.7 is not susceptible to the change of variable $\sqrt{(1-x^2)} = u$; $\int \frac{1}{\sqrt{(1-x^2)}} \, dx$ merely becomes $\int \frac{-1}{\sqrt{(1-u^2)}} \, du$. However, changes of variable involving a trigonometrical substitution, such as was successfully applied in this case, open the way to finding a very important group of integrals. Here are two examples of the type of substitution we shall be using.

If $x = 5 \sin u$,

$$\sqrt{(25 - x^2)} = \sqrt{(25 - 25 \sin^2 u)} = \sqrt{\{25(1 - \sin^2 u)\}} = 5 \cos u.$$

If $x = \frac{\sqrt{3}}{2} \sin u$,

$$\sqrt{(3 - 4x^2)} = \sqrt{(3 - 4 \cdot \tfrac{3}{4} \sin^2 u)} = \sqrt{\{3(1 - \sin^2 u)\}} = \sqrt{3} \cos u.$$

Qu. 13. Reduce each of the following to the form $k \cos u$, and give u in terms of x in each case:

(i) $\sqrt{(9 - x^2)}$,   (ii) $\sqrt{(1 - 25x^2)}$,   (iii) $\sqrt{(4 - 9x^2)}$,
(iv) $\sqrt{(7 - x^2)}$,   (v) $\sqrt{(1 - 3x^2)}$,   (vi) $\sqrt{(3 - 2x^2)}$.

We see that to deal with $\sqrt{(a^2 - b^2 x^2)}$ we write

$$a^2 - b^2 x^2 \quad \text{as} \quad a^2 - a^2 \sin^2 u;$$

thus $b^2 x^2 = a^2 \sin^2 u$, and $x = \frac{a}{b} \sin u$. Note that $u = \sin^{-1} \frac{bx}{a}$ and, for the substitution to be valid, and of use, u must be real and not $\frac{\pi}{2}$, so $bx < a$; this condition is implicit in $\sqrt{(a^2 - b^2 x^2)}$ being real and not zero.

Example 8. *Find* $\int \frac{1}{\sqrt{(9 - 4x^2)}} \, dx$.

$$\int \frac{1}{\sqrt{(9 - 4x^2)}} \frac{dx}{du} \, du = \int \frac{1}{\sqrt{(9 - 9 \sin^2 u)}} \cdot \frac{3}{2} \cos u \, du, \qquad \begin{aligned} & 9 - 4x^2. \\ & 9 - 9 \sin^2 u. \end{aligned}$$

$$= \int \frac{1}{3 \cos u} \cdot \frac{3}{2} \cos u \, du, \qquad \begin{aligned} & \text{Let } x = \tfrac{3}{2} \sin u. \\ & \frac{dx}{du} = \frac{3}{2} \cos u. \end{aligned}$$

$$= \int \tfrac{1}{2} \, du,$$

$$= \tfrac{1}{2} u + c,$$

$$= \tfrac{1}{2} \sin^{-1} \frac{2x}{3} + c.$$

Qu. 14. Find the following integrals:

(i) $\int \frac{1}{\sqrt{(4-x^2)}}\, dx$, (ii) $\int \frac{1}{\sqrt{(1-3x^2)}}\, dx$, (iii) $\int \frac{1}{\sqrt{(16-9x^2)}}\, dx$.

Qu. 15. Repeat Qu. 14(i) using the change of variable $x = 2 \cos u$.

$$\int \frac{1}{a^2+b^2x^2}\, dx$$

1.9. In §1.8 we made use of Pythagoras' theorem in the form $\cos^2 u + \sin^2 u = 1$; we shall now find that an alternative form, $1 + \tan^2 u = \sec^2 u$, helps to effect other useful changes of variable.

Qu. 16. Find $\int \frac{1}{1+x^2}\, dx$ by taking x as $\tan u$.

Qu. 17. Reduce each of the following to the form $k \sec^2 u$, and give u in terms of x in each case:

(i) $9 + x^2$,   (ii) $1 + 4x^2$,   (iii) $25 + 9x^2$,
(iv) $3 + x^2$,   (v) $1 + 5x^2$,   (vi) $7 + 3x^2$.

Qu. 18. Find the following integrals:

(i) $\int \frac{1}{4+x^2}\, dx$, (ii) $\int \frac{1}{1+16x^2}\, dx$, (iii) $\int \frac{1}{3+4x^2}\, dx$.

Example 9. *Evaluate* $\int_{\frac{\sqrt3}{2}}^{\frac{3}{2}} \frac{1}{3+4x^2}\, dx$.

$\int_{x=\frac{\sqrt3}{2}}^{x=\frac{3}{2}} \frac{1}{3+4x^2} \frac{dx}{du}\, du$ $\qquad\qquad 3 + 4x^2$.

$\qquad\qquad\qquad\qquad\qquad\qquad\qquad\qquad\qquad 3 + 3\tan^2 u$.

$\displaystyle = \int_{\frac{\pi}{4}}^{\frac{\pi}{3}} \frac{1}{3(1+\tan^2 u)} \frac{\sqrt3}{2}\sec^2 u\, du,$ $\qquad$ Let $x = \frac{\sqrt3}{2} \tan u$.

$\displaystyle = \int_{\frac{\pi}{4}}^{\frac{\pi}{3}} \frac{\sqrt3}{6}\, du,$ $\qquad\qquad\qquad\qquad \frac{dx}{du} = \frac{\sqrt3}{2} \sec^2 u.$

$\displaystyle = \left[\frac{\sqrt3\, u}{6} \right]_{\frac{\pi}{4}}^{\frac{\pi}{3}},$

$\displaystyle = \frac{\sqrt3}{6} \left(\frac{\pi}{3} - \frac{\pi}{4} \right),$

$\displaystyle = \frac{\sqrt3\,\pi}{72}.$

x	$\tan u$	u
$\frac{3}{2}$	$\sqrt3$	$\frac{\pi}{3}$
$\frac{\sqrt3}{2}$	1	$\frac{\pi}{4}$

Exercise 1d

1. The following angles lie between 0 and 90° inclusive.   Express them in degrees, and in radians in terms of π:

   (i) $\cos^{-1} \dfrac{1}{\sqrt{2}}$,   (ii) $\cot^{-1} 1$,   (iii) $\dfrac{1}{\sqrt{3}} \cot^{-1} \dfrac{1}{\sqrt{3}}$,

   (iv) $\sin^{-1} \dfrac{\sqrt{3}}{2}$,   (v) $\sqrt{3} \sin^{-1} \frac{1}{2}$,   (vi) $\frac{1}{3} \sec^{-1} \sqrt{2}$,

   (vii) $\frac{3}{2} \tan^{-1} 1$,   (viii) $\frac{1}{2} \operatorname{cosec}^{-1} 2$.

2. Express the following angles in radians:

   (i) $32°$,   (ii) $60° 21'$,   (iii) $5° 41'$,   (iv) $235° 16'$.

3. Express the following angles in degrees and minutes:

   (i) 2 radians,   (ii) 0·08 radian,   (iii) 1·362 radians,

   (iv) $\dfrac{\pi}{6}$ radian.

4. Express the following (acute) angles in radians:

   (i) $\sin^{-1} 0·8$,   (ii) $\frac{1}{2} \cos^{-1} \left(\dfrac{5}{13} \right)$,   (iii) $2 \tan^{-1} 0·625$.

5. Express the following in the form $k \cos u$, and give u in terms of x in each case:

   (i) $\sqrt{(16 - x^2)}$,   (ii) $\sqrt{(1 - 9x^2)}$,   (iii) $\sqrt{(9 - 4x^2)}$,
   (iv) $\sqrt{(10 - x^2)}$,   (v) $\sqrt{(1 - 6x^2)}$,   (vi) $\sqrt{(5 - 3x^2)}$.

6. Find the following integrals:

   (i) $\displaystyle\int \dfrac{1}{\sqrt{(25 - x^2)}}\, dx$,   (ii) $\displaystyle\int \dfrac{1}{\sqrt{(1 - 4x^2)}}\, dx$,   (iii) $\displaystyle\int \dfrac{1}{\sqrt{(4 - 9x^2)}}\, dx$,

   (iv) $\displaystyle\int \dfrac{1}{\sqrt{(3 - x^2)}}\, dx$,   (v) $\displaystyle\int \dfrac{1}{\sqrt{(1 - 7x^2)}}\, dx$,   (vi) $\displaystyle\int \dfrac{1}{\sqrt{(2 - 3x^2)}}\, dx$.

7. Express the following in the form $k \sec^2 u$, and give u in terms of x in each case:

   (i) $16 + x^2$,   (ii) $1 + 9x^2$,   (iii) $4 + 3x^2$,
   (iv) $2 + x^2$,   (v) $1 + 3x^2$,   (vi) $5 + 2x^2$.

8. Find the following integrals:

   (i) $\displaystyle\int \dfrac{1}{25 + x^2}\, dx$,   (ii) $\displaystyle\int \dfrac{1}{1 + 36x^2}\, dx$,   (iii) $\displaystyle\int \dfrac{1}{16 + 3x^2}\, dx$,

   (iv) $\displaystyle\int \dfrac{1}{5 + x^2}\, dx$,   (v) $\displaystyle\int \dfrac{1}{1 + 6x^2}\, dx$,   (vi) $\displaystyle\int \dfrac{1}{3 + 10x^2}\, dx$.

9. Find the following integrals:

  (i) $\displaystyle\int \frac{1}{9+2x^2} \, dx,$ (ii) $\displaystyle\int \frac{3}{\sqrt{(4-5x^2)}} \, dx,$

  (iii) $\displaystyle\int \frac{1}{\sqrt{(3-2x^2)}} \, dx,$ (iv) $\displaystyle\int \frac{2}{3+5x^2} \, dx.$

10. Evaluate the following integrals, leaving π in your answers:

  (i) $\displaystyle\int_1^{\sqrt{3}} \frac{2}{1+x^2} \, dx,$ (ii) $\displaystyle\int_0^{\sqrt{2}} \frac{1}{\sqrt{(4-x^2)}} \, dx,$ (iii) $\displaystyle\int_{\frac{1}{2}}^1 \frac{3}{\sqrt{(1-x^2)}} \, dx,$

  (iv) $\displaystyle\int_0^3 \frac{1}{9+x^2} \, dx,$ (v) $\displaystyle\int_0^{\frac{1}{6}} \frac{1}{\sqrt{(1-9x^2)}} \, dx,$ (vi) $\displaystyle\int_{-2}^{\sqrt{3}} \frac{1}{5\sqrt{(4-x^2)}} \, dx.$

11a. Find $\displaystyle\int \frac{1}{\sqrt{(9-x^2)}} \, dx$ using (i) $x = 3 \sin u$, (ii) $x = 3 \cos u$.

11b. Evaluate $\displaystyle\int_{\frac{3}{2}}^3 \frac{1}{\sqrt{(9-x^2)}} \, dx$ using (i) $x = 3 \sin u$, (ii) $x = 3 \cos u$.

———————————

12. Find the following integrals, using the given change of variable:

  (i) $\displaystyle\int \frac{1}{\sqrt{\{4-(x+1)^2\}}} \, dx,$ $x+1 = 2 \sin u,$

  (ii) $\displaystyle\int \frac{1}{9+(x-3)^2} \, dx,$ $x-3 = 3 \tan u.$

13. Find the following integrals:

  (i) $\displaystyle\int \frac{1}{(x+3)^2+25} \, dx,$ (ii) $\displaystyle\int \frac{1}{\sqrt{\{4-(x-1)^2\}}} \, dx,$

  (iii) $\displaystyle\int \frac{1}{3(x-2)^2+5} \, dx,$ (iv) $\displaystyle\int \frac{1}{\sqrt{\{9-3(x+1)^2\}}} \, dx.$

*14a. $2x^2-12x+21$ may be written $2(x^2-6x+9)+21-18 = 2(x-3)^2+3.$

  Write the following expressions in the form $a(x+b)^2+c$:

  (i) $x^2-6x+16,$ (ii) $3x^2-12x+14,$ (iii) $2x^2-4x+5.$

14b. Find the following integrals:

  (i) $\displaystyle\int \frac{1}{x^2-2x+5} \, dx,$ (ii) $\displaystyle\int \frac{1}{2x^2+4x+11} \, dx,$

  (iii) $\displaystyle\int \frac{1}{x^2-4x+13} \, dx,$ (iv) $\displaystyle\int \frac{1}{4x^2-8x+7} \, dx.$

15a. $1+6x-3x^2$ may be written $1-3(x^2-2x+1)+3 = 4-3(x-1)^2.$

  Write the following expressions in the form $a-b(x+c)^2$:

  (i) $3-2x-x^2,$ (ii) $5+4x-x^2,$ (iii) $7+2x-2x^2.$

15b. Find the following integrals:

(i) $\int \dfrac{1}{\sqrt{(3-2x-x^2)}}\,dx,$ (ii) $\int \dfrac{1}{\sqrt{(1+8x-4x^2)}}\,dx,$

(iii) $\int \dfrac{1}{\sqrt{(12+4x-x^2)}}\,dx,$ (iv) $\int \dfrac{1}{\sqrt{(-2x^2+12x-9)}}\,dx.$

16. Evaluate:

(i) $\int_{2}^{3} \dfrac{1}{x^2-4x+5}\,dx,$ (ii) $\int_{-1}^{1} \dfrac{1}{\sqrt{(3-2x-x^2)}}\,dx.$

---

*17. Find the following integrals by writing each integrand as two fractions:

(i) $\int \dfrac{3-x}{\sqrt{(1-x^2)}}\,dx,$ (ii) $\int \dfrac{2x+3}{\sqrt{(4-x^2)}}\,dx.$

18. Show that

$$\int \sqrt{(1-x^2)}\,dx = \tfrac{1}{2}\sin^{-1}x + \tfrac{1}{2}x\sqrt{(1-x^2)} + c.$$

Find the following integrals:

(i) $\int \dfrac{x^2}{\sqrt{(1-x^2)}}\,dx,$ (ii) $\int \dfrac{1}{(x^2+9)^2}\,dx,$ (iii) $\int \dfrac{x}{\sqrt{(4-x^4)}}\,dx.$

19. Show that

$$\int \dfrac{1}{(1-x^2)^{3/2}}\,dx = x(1-x^2)^{-\frac{1}{2}} + c.$$

Find the following integrals:

(i) $\int \dfrac{1}{(1-9x^2)\sqrt{(1-9x^2)}}\,dx,$ (ii) $\int \dfrac{1}{x^2\sqrt{(1-x^2)}}\,dx,$

(iii) $\int \dfrac{1}{x\sqrt{(x^2-1)}}\,dx.$

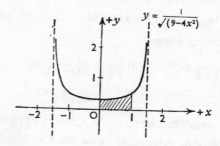

Fig. 1.1

Trigonometrical functions of numbers

1.10. At this stage it is advisable to discuss some of the implications of the definite integrals evaluated in Exercise 1d.

Let us find the area under the curve $y = \dfrac{1}{\sqrt{(9-4x^2)}}$ from $x=0$ to $x=1$. (See Fig. 1.1.) The element of area $= y\,\Delta x = \dfrac{1}{\sqrt{(9-4x^2)}}\,\Delta x$, and the required area is

$$\int_{x=0}^{x=1} \frac{1}{\sqrt{(9-4x^2)}}\,\frac{\mathrm{d}x}{\mathrm{d}u}\,\mathrm{d}u,$$

$$\simeq \int_0^{0\cdot7298} \frac{1}{\sqrt{\{9(1-\sin^2 u)\}}}\,\frac{3}{2}\cos u\,\mathrm{d}u,$$

$$= \left[\tfrac{1}{2}u\right]_0^{0\cdot7298}$$

$$= 0\cdot3649.$$

$9-4x^2.$
$9-9\sin^2 u.$

Let $x = \tfrac{3}{2}\sin u.$
$\dfrac{\mathrm{d}x}{\mathrm{d}u} = \dfrac{3}{2}\cos u.$

x	$\sin u$	u
1	0·6667	41° 49′ = 0·7298 rad.
0	0	0

Now if we retain x as our variable (see Example 8), the integral is evaluated as

$$\left[\tfrac{1}{2}\sin^{-1}\frac{2x}{3}\right]_0^1 = \tfrac{1}{2}\sin^{-1}\tfrac{2}{3}.$$

It may at first sight seem surprising that the number 0·3649 can measure, at one and the same time, units of area, and the angle $\tfrac{1}{2}\sin^{-1}\tfrac{2}{3}$ in radians. These two aspects must be reconciled, and it is apparent that we must extend our definitions of the trigonometrical ratios to include the ratios of *numbers*, as well as of *angles*.

Thus $\sin u$ may be considered as a function of the number u which, as u moves from $-\infty$ to $+\infty$, oscillates between -1 and $+1$ with period 2π. (See P.M.I., p. 260.) The sine of a given *number* is the same as the sine of the *angle* given by that number of *radians*; thus

$$\sin 2 = \sin (2 \text{ radians}) \simeq \sin 114° \ 35' = \sin 65° \ 25' \simeq 0\cdot9093.$$

Qu. 19. Find the values of:

(i) $\cos \dfrac{\pi}{6}$,   (ii) $\sin \tfrac{1}{2}$,   (iii) $\tan 1\cdot2$,   (iv) $\cos 3$,   (v) $\sec 6$.

Qu. 20.   Find the following numbers between $-\frac{\pi}{2}$ and $\frac{\pi}{2}$ inclusive:

(i) $\sin^{-1} 1$,   (ii) $\tan^{-1}(-1)$,   (iii) $\sin^{-1}(\frac{1}{2})$,   (iv) $\tan^{-1} 2$.

Principal values

1.11.   In the course of evaluating the area in §1.10 we established that $\sin^{-1}\frac{2}{3} \simeq 0\cdot 7298$; this value is indicated in Fig. 1.2. However, $Sin^{-1}\frac{2}{3}$ (written with a capital S) has an infinite number of values (four of which are shown in Fig. 1.2); the only one relevant to the integration we performed is that lying between $-\frac{\pi}{2}$ and $\frac{\pi}{2}$, which is written $\sin^{-1}\frac{2}{3}$ (with a small s), and is called the **principal value**.

Similarly, $Tan^{-1} x$ has an infinite number of values, but the principal value $\tan^{-1} x$ lies between $-\frac{\pi}{2}$ and $\frac{\pi}{2}$. However, $\cos^{-1} x$, the principal value of $Cos^{-1} x$, lies between **0 and π**. (See §12.4 for further discussion of principal values.)

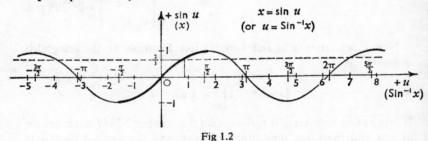

Fig 1.2

Qu. 21.   Sketch the graph of $x = \tan u$ (or $u = Tan^{-1} x$) from $x = -\frac{3\pi}{2}$ to $x = \frac{3\pi}{2}$. Mark clearly that part of the graph giving values of $\tan^{-1} x$.

Qu. 22.   Find the following numbers, in terms of π, or to four significant figures: (i) $\sin^{-1}(\frac{1}{4})$, (ii) $\frac{1}{2}\sin^{-1}(\frac{1}{4})$, (iii) $\operatorname{cosec}^{-1} 6$, (iv) $\cos^{-1}(-\frac{1}{8})$, (v) $\frac{1}{3}\tan^{-1}\sqrt{3}$.

Qu. 23.   Sketch the graph of $y = \dfrac{1}{\sqrt{(1-x^2)}}$, and calculate the area under the curve from $x = \frac{1}{4}$ to $x = \frac{1}{2}$.

Qu. 24.   Find x in terms of y if:

(i) $\sin^{-1} x = \sin^{-1} y + \dfrac{\pi}{3}$,   (ii) $\tan^{-1} x = 2\tan^{-1} y$.

Exercise 1e

1. Find the values of:

(i) $\cos \dfrac{\pi}{12}$,　(ii) $\sin 1 \cdot 5$,　(iii) $\tan 0 \cdot 0806$,　(iv) $\operatorname{cosec} \frac{5}{4}$.

2. Sketch the graph of $x = \cos u$ (or $u = \operatorname{Cos}^{-1} x$) from $x = -\dfrac{\pi}{2}$ to $x = \dfrac{3\pi}{2}$. Mark clearly that part of the graph giving values of $\cos^{-1} x$.

3. Find the following numbers, in terms of π, or to four significant figures:

(i) $\tan^{-1} 3$,　(ii) $\sqrt{3} \tan^{-1} (\sqrt{3})$,　(iii) $\frac{2}{3} \sin^{-1} \dfrac{\sqrt{2}}{2}$,

(iv) $\cos^{-1} (-0 \cdot 375)$,　　　　(v) $\sec^{-1} \pi$.

4. Find x in terms of y if:

(i) $\tan^{-1} x = \tan^{-1} y + \dfrac{\pi}{4}$,　(ii) $\cos^{-1} x = \cos^{-1} y - \dfrac{\pi}{6}$.

5. Sketch the graph of $y = \dfrac{1}{\sqrt{(16 - x^2)}}$, and calculate the area under the curve (i) from $x = 0$ to $x = 2$, (ii) from $x = 2$ to $x = 3$.

6. Sketch the graph of $y = \dfrac{1}{1 + x^2}$ from $x = -3$ to $x = +3$, and calculate the area under the curve (i) from $x = 0$ to $x = 1$, (ii) from $x = 1$ to $x = 2$.

7. A particle moves along a straight line so that t s after starting it is x m from a point O on the straight line, where $x = 10 \cos t$.

(i) How far from O is it after $0, \dfrac{\pi}{2}, \pi, \dfrac{3\pi}{2}, 2\pi$ s?

(ii) When is it first at a distance 5 m from O?

(iii) When is it first 5 m on the negative side of O?

8. A particle moves along a straight line so that t s after the start it is x m from a point O on the line, where $x = 5 \sin \frac{1}{2} t$.

(i) How far from O is it after $0, \pi, 2\pi, 3\pi, 4\pi$ s?

(ii) How long does it take to travel the first 3 m from O?

9. Evaluate the following, correct to three significant figures:

(i) $\displaystyle\int_1^3 \dfrac{1}{1 + x^2}\, dx$,　　　　(ii) $\displaystyle\int_0^{0 \cdot 8} \dfrac{1}{\sqrt{(1 - x^2)}}\, dx$,

(iii) $\displaystyle\int_1^2 \dfrac{1}{4 + 25 x^2}\, dx$,　　　(iv) $\displaystyle\int_1^{\frac{4}{3}} \dfrac{1}{\sqrt{(25 - 9 x^2)}}\, dx$,

(v) $\displaystyle\int_{\frac{1}{2}}^{\sqrt{3}} \dfrac{1}{3 + 4 x^2}\, dx$,　　　(vi) $\displaystyle\int_{-1}^{-\frac{1}{2}} \dfrac{1}{\sqrt{(4 - 2 x^2)}}\, dx$.

Exercise 1f (Miscellaneous)

Find the integrals Nos. 1 to 20, which are arranged by types in the order in which they occur in the chapter:

1. $\int x^2 \sqrt{(x^3 - 1)} \, dx.$

2. $\int \frac{x}{(x^2 - 1)^3} \, dx.$

3. $\int \sin 2x \cos^2 2x \, dx.$

4. $\int \sec^2 x \sqrt{\cot x} \, dx.$

5. $\int \cos^3 4x \, dx.$

6. $\int \sin^2 \frac{x}{3} \, dx.$

7. $\int \cos^4 2x \, dx.$

8. $\int \sqrt{(1 - \cos x)} \, dx.$

9. $\int \sin \frac{x}{3} \cos^2 \frac{x}{6} \, dx.$

10. $\int \cos 3x \cos 2x \, dx.$

11. $\int x(3x - 7)^4 \, dx.$

12. $\int \frac{x}{\sqrt{(5 + x)}} \, dx.$

13. $\int \frac{1}{\sqrt{(6 - 5x^2)}} \, dx.$

14. $\int \frac{1}{1 + 8x^2} \, dx.$

15. $\int \frac{1}{\sqrt{(5 - 4x - x^2)}} \, dx.$

16. $\int \frac{1}{3x^2 + 6x + 5} \, dx.$

17. $\int \frac{x + 1}{\sqrt{(5 - x^2)}} \, dx.$

18. $\int \sqrt{(9 - x^2)} \, dx.$

19. $\int \frac{1}{(4 - x^2)^{\frac{3}{2}}} \, dx.$

20. $\int \frac{1}{x^2 \sqrt{(16 - x^2)}} \, dx.$

21. Find (i) $\int \frac{x}{\sqrt{(1 - x^2)}} \, dx,$ (ii) $\int \frac{2}{1 + x^2} \, dx,$

            (iii) $\int x \sqrt{(1 - x^2)} \, dx,$ (iv) $\int \frac{2}{\sqrt{(1 - x^2)}} \, dx,$

            (v) $\int \frac{2 - x}{\sqrt{(1 - x^2)}} \, dx.$

Find the integrals Nos. 22 to 45:

22. $\int \frac{x + 2}{\sqrt{(3x - 1)}} \, dx.$

23. $\int \frac{1}{2} x \sqrt{(x^2 + 2)} \, dx.$

24. $\int \frac{1}{x^2 \sqrt{(1 - x^2)}} \, dx.$

25. $\int \cos^2 \frac{x}{2} \sin^3 \frac{x}{2} \, dx.$

26. $\int \frac{x}{3 \sqrt{(4 - x^2)}} \, dx.$

27. $\int \frac{3}{\sqrt{(36 - x^2)}} \, dx.$

28. $\int (x - 1)^2 (x + 3)^5 \, dx.$

29. $\int \frac{x^2}{\sqrt{(4 - x^2)}} \, dx.$

30. $\int \frac{2 + x}{\sqrt{(9 - x^2)}} \, dx.$

31. $\int \frac{1}{3x^2 - 12x + 16} \, dx.$

32. $\int \sin^2 \dfrac{x}{5} \, dx.$

33. $\int \sqrt{(\cos 3x + 1)} \, dx.$

34. $\int \sqrt{(4 - x^2)} \, dx.$

35. $\int \sec^5 x \tan x \, dx.$

36 $\int \dfrac{x^2}{\sqrt{(4 - 9x^6)}} \, dx.$

37. $\int \sin^5 2x \, dx.$

38. $\int \dfrac{1}{(1 + x^2)^2} \, dx.$

39. $\int \sin \dfrac{3x}{2} \cos \dfrac{5x}{2} \, dx.$

40. $\int \dfrac{\tan x}{\sqrt{(\cos 2x + 1)}} \, dx.$

41. $\int \dfrac{x}{1 + x^4} \, dx.$

42. $\int \cos x \sqrt{\cos 2x} \, dx.$

43. $\int \dfrac{1}{\cos^2 x + 4 \sin^2 x} \, dx$ (put $\tan x = u$).

44. $\int \dfrac{1}{x\sqrt{(x^2 - 9)}} \, dx.$

45. $\int \dfrac{1}{(1 - x)\sqrt{(1 - x^2)}} \, dx$ (put $x = \cos u$; show integral $= \cot \dfrac{u}{2} + c$).

Evaluate Nos. 46 to 54:

46. $\int_0^3 2x \sqrt{(5x + 1)} \, dx.$

47. $\int_{\frac{\pi}{4}}^{\frac{\pi}{2}} \cos x \, \operatorname{cosec}^3 x \, dx.$

48. $\int_0^\pi \sin \dfrac{x}{2} \cos \dfrac{x}{2} \, dx.$

49. $\int_{-2\pi}^0 \cos^4 \dfrac{x}{4} \, dx.$

50. $\int_0^{\sqrt{3}} \dfrac{1}{(4 - x^2)^{\frac{3}{2}}} \, dx.$

51. $\int_0^{\frac{\pi}{3}} \dfrac{\tan x}{\sqrt{\sec x}} \, dx.$

52. $\int_{2\sqrt{2}}^{2\sqrt{3}} \dfrac{1}{x\sqrt{(x^2 - 4)}} \, dx.$

53. $\int_{-1}^{+1} (x - 3)(x^2 - 6x + 5)^3 \, dx.$

54. $\int_3^4 \dfrac{1}{\sqrt{(-2x^2 + 12x - 16)}} \, dx.$

CHAPTER 2
EXPONENTIAL AND LOGARITHMIC FUNCTIONS

Exponential functions

2.1. The word *exponent* is often used instead of *index*, and functions in which the variable is in the index (such as 2^x, $10^{\sin x}$) are called **exponential functions.**†

The graph of $y = a^x$

2.2. Let us first consider the function 2^x. A table of values follows, and a sketch of $y = 2^x$ is given in Fig. 2.1.

Table of values, $y = 2^x$

x	-3	-2	-1	0	1	2
2^x	$\frac{1}{8}$	$\frac{1}{4}$	$\frac{1}{2}$	1	2	4

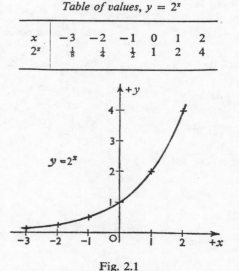

Fig. 2.1

As $x \to -\infty$, $2^x \to 0$, and so the curve approaches the x-axis but does not meet it.

† The graph of $y = 10^x$ is usually encountered during the elementary introduction to logarithms; in P.M.I., §8.5, equations such as $2^x = 3$ are solved. It is probably only in these contexts that the reader has previously met exponential functions.

Qu. 1. Copy and extend the above table to include values of $1 \cdot 5^x$ (from $x = -3$ to $x = +3$), and of $2 \cdot 5^x$, 3^x (both from $x = -2$ to $x = +2$). Sketch, with the same axes, the graphs of $y = 1^x$, $y = 1 \cdot 5^x$, $y = 2^x$, $y = 2 \cdot 5^x$, $y = 3^x$. What do you notice about the gradient of $y = a^x$ at $(0, 1)$ as a takes different values greater than 1?

Qu. 2. How would you deduce the shape of the graph of $y = (\frac{1}{2})^x$ from Fig. 2.1?

The gradient of $y = a^x$ at $(0, 1)$; a limit

2.3. We shall confine our attention for the time being to exponential functions of the form a^x, where a is taken to be a constant greater than 1. Since $a^0 = 1$, the graph of $y = a^x$ (Fig. 2.2) passes through the point A(0, 1), and we let the gradient of the curve at this point be m.

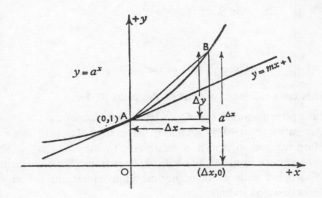

Fig. 2.2

With the usual notation, if B is the point $(\Delta x, a^{\Delta x})$, then the gradient of AB is

$$\frac{\Delta y}{\Delta x} = \frac{a^{\Delta x} - 1}{\Delta x}.$$

Now as $\Delta x \to 0$, the gradient of AB $\to m$.

It follows that the **limit**, as $\Delta x \to 0$, of $\dfrac{a^{\Delta x} - 1}{\Delta x}$ is m, the gradient of $y = a^x$ at $(0, 1)$.

The form of $\dfrac{d}{dx}(a^x)$

2.4. The limit just established enables us to investigate the gradient of $y = a^x$ at any point $P(x, y)$ on the curve. With the usual notation, if Q is the point $(x + \Delta x, y + \Delta y)$,

$$y + \Delta y = a^{x + \Delta x},$$
$$\therefore \Delta y = a^{x + \Delta x} - a^x = a^x(a^{\Delta x} - 1).$$

$\therefore$ the gradient of PQ, $\dfrac{\Delta y}{\Delta x} = a^x \left(\dfrac{a^{\Delta x} - 1}{\Delta x} \right).$ \hfill (1)

Now as $\Delta x \to 0$, the gradient of PQ $\to$ the gradient of the tangent at P; also, since we have shown that

$$\left(\frac{a^{\Delta x} - 1}{\Delta x} \right) \to m,$$

the R.H.S. of (1) $\to ma^x$.

$$\therefore \frac{dy}{dx} = ma^x.$$

Thus $\dfrac{d}{dx}(a^x) = ma^x,$

where m is the gradient of $y = a^x$ at the point (0, 1).

We have already noted (see Qu. 1) that as a increases, the gradients of the curves $y = a^x$ at (0, 1) increase; for every value of a there is an appropriate value of m, and it is reasonable to suppose that we should be able to express m in terms of a. However for the time being we must be satisfied with some numerical approximations for m which we will now proceed to find.

Approximate derivatives of 2^x and 3^x

2.5. The following table was used to draw the graphs of $y = 2^x$ and $y = 3^x$ in Fig. 2.3; $y = m_2 x + 1$ and $y = m_3 x + 1$ are the respective tangents at (0, 1).

Table of values for $y = 2^x$ and $y = 3^x$

x	-2	$-\frac{3}{2}$	-1	$-\frac{1}{2}$	$-\frac{1}{4}$	0	$\frac{1}{4}$	$\frac{1}{2}$	1	$\frac{3}{2}$	2
2^x	0·25	0·35	0·5	0·71	0·84	1	1·19	1·41	2	2·83	4
3^x	0·11	0·19	0·33	0·58	0·76	1	1·32	1·73	3	5·20	9

Qu. 3. (i) Measure the gradients of the two tangents in Fig. 2.3, and deduce approximate expressions for $\frac{d}{dx}(2^x)$ and $\frac{d}{dx}(3^x)$.

(ii) Now calculate the gradient of $y = 2^x$ where $x = 1$, and the gradient of $y = 3^x$ where $x = \frac{1}{2}$. Check from the graph.

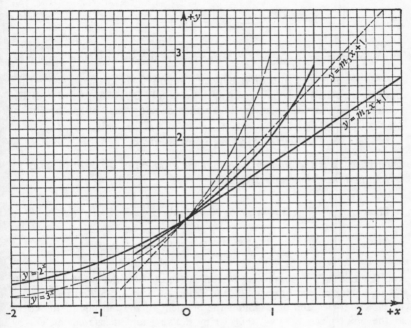

Fig. 2.3

Qu. 4. Tangents were drawn to a graph of $y = 2^x$, and their gradients were measured and entered in the following table:

x	-3	-2	-1	0	1	$\frac{3}{2}$	2	$\frac{5}{2}$
$y = 2^x$	0·125	0·25	0·5	1	2	2·83	4	5·66
$\frac{dy}{dx}$	0·08	0·18	0·37	0·62	1·33	1·90	2·82	3·68

Confirm graphically that these results indicate that $\frac{dy}{dx} \propto y$, and deduce an approximate expression for $\frac{d}{dx}(2^x)$.

The exponential function e^x

2.6. It has been established in the previous section that

$$\frac{d}{dx}(2^x) \simeq 0 \cdot 7 \cdot 2^x,$$

and

$$\frac{d}{dx}(3^x) \simeq 1 \cdot 1 \cdot 3^x.$$

Since, in general, $\frac{d}{dx}(a^x) = ka^x$, these results suggest that for simplicity we should find a value of a between 2 and 3 for which $k = 1$; this number is called e, its value is approximately $2 \cdot 71828$, and it will be found to play a vital part in the further development of mathematics from this point.

Let us now summarize what we know about e.

DEFINITION: e *is the number such that the gradient of* $y = e^x$ *at* $(0, 1)$ *is* 1. e^x *is called the* **exponential function**.

Thus $\qquad \frac{d}{dx}(e^x) = e^x, \quad \text{or if } y = e^x, \quad \frac{dy}{dx} = y.$

Also $\qquad \int e^x \, dx = e^x + c.$

Qu. 5. Letting 1 cm represent $0 \cdot 1$ on each axis, plot the graph of $y = e^x$, taking values of x at intervals of $0 \cdot 05$ from $-0 \cdot 5$ to $+0 \cdot 5$, and making use of tables† for e^x and e^{-x}. Obtain the gradient of the tangent to the curve at the point given by $x = 0 \cdot 08$ (i) by drawing and measurement, (ii) by measuring the ordinate, (iii) by differentiation.

Example 1. *Find* $\frac{dy}{dx}$ *when* $y = e^{3x^2}$.

[Here we have a function of a function of x. This example is written out in full as a reminder of the technique involved, but the reader should be able to differentiate in one step.]

$$y = e^{3x^2}.$$

Let $u = 3x^2$, then $\qquad y = e^u.$

$$\therefore \frac{du}{dx} = 6x \text{ and } \frac{dy}{du} = e^u.$$

† See Godfrey and Siddons, *op. cit.*

Now $$\frac{dy}{dx} = \frac{dy}{du} \cdot \frac{du}{dx} = e^u \cdot 6x.$$

$$\therefore \frac{dy}{dx} = e^{3x^2} \cdot 6x.$$

Example 2. *Find* (*i*) $\int e^{x/2} \, dx$, (*ii*) $\int x^2 \, e^{x^3} \, dx$, (*iii*) $\frac{d}{dx} (e^{3y})$.

(i) Since $\frac{d}{dx} (e^{x/2}) = \frac{1}{2}e^{x/2}$,

$$\int e^{x/2} \, dx = 2e^{x/2} + c.$$

(ii) Since $\frac{d}{dx} (e^{x^3}) = 3x^2 \, e^{x^3}$,

$$\int x^2 \, e^{x^3} \, dx = \frac{1}{3}e^{x^3} + c.$$

(iii) $$\frac{d}{dx} (e^{3y}) = \frac{d}{dy} (e^{3y}) \cdot \frac{dy}{dx} = 3e^{3y} \cdot \frac{dy}{dx}.$$

Qu. 6. Differentiate with respect to x:

(i) $(2x^3 + 1)^5$,   (ii) $\sin (2x^3)$,   (iii) e^{2x^3},   (iv) e^{y^2},

(v) e^{-x^2},   (vi) $e^{\tan x}$,   (vii) $e^{\sqrt{x}}$,   (viii) $e^{\sin y}$.

Qu. 7. Find the following integrals, and check by differentiation:

(i) $\int \frac{x}{(x^2 + 1)^2} \, dx$,   (ii) $\int x \sin (x^2) \, dx$,   (iii) $\int x \, e^{x^2} \, dx$,

(iv) $\int \sin x \, e^{\cos x} \, dx$,   (v) $\int 2e^{x/3} \, dx$,   (vi) $\int 3e^{2x} \, dx$,

(vii) $\int \frac{1}{2}x \, e^{3x^2} \, dx$,   (viii) $\int \operatorname{cosec}^2 2x \, e^{\cot 2x} \, dx$.

Exercise 2a

1. Make rough sketches of the graphs of the following functions:

   (i) e^{2x},   (ii) 2^{-x},   (iii) $2^{1/x}$,   (iv) 2^{x^2},   (v) $3e^x$,   (vi) e^{x+1},
   (vii) e^{-x^2}.

2. With the same axes, sketch the graphs of $y = e^{\sin x}$ and $y = e^{\cos x}$ from $x = 0$ to $x = 4\pi$.

3. With the same axes, sketch the graphs of $y = \tan x$ and $y = e^{\tan x}$ from $x = -\frac{\pi}{2}$ to $x = \frac{3\pi}{2}$.

4. Use the following table of values to draw the graph of $y = 3^x$, taking 2 cm to represent 1 unit on each axis:

x	-2	$-\frac{7}{4}$	$-\frac{3}{2}$	$-\frac{5}{4}$	-1	$-\frac{3}{4}$	$-\frac{1}{2}$	$-\frac{1}{4}$	0
$y = 3^x$	0·11	0·15	0·19	0·25	0·33	0·44	0·58	0·76	1

$\frac{1}{4}$	$\frac{1}{2}$	$\frac{3}{4}$	1	$\frac{5}{4}$	$\frac{3}{2}$	$\frac{7}{4}$	2
1·32	1·73	2·28	3	3·95	5·20	6·84	9

Draw the tangents to the curve at the points given by $x = -\frac{3}{2}, -1, -\frac{1}{2}$, $0, \frac{1}{2}, 1, \frac{3}{2}$, measure their gradients, and confirm graphically that $\frac{dy}{dx} = ky$. Deduce an approximate expression for $\frac{d}{dx}(3^x)$.

Differentiate with respect to x in Nos. 5 to 8.

5. (i) $4e^x$,   (ii) e^{3x},   (iii) e^{2x+1},   (iv) e^{2x^2},

   (v) e^{-2x},   (vi) e^{3y},   (vii) e^{x^2+3},   (viii) e^{x-2},

   (ix) $e^{5/x}$,   (x) $e^{\sqrt[3]{x}}$,   (xi) e^{ax^2+b},   (xii) $e^{\sqrt{t}}$.

6. (i) $e^{\cos x}$,   (ii) $e^{\sec x}$,   (iii) $e^{3\tan y}$,   (iv) $e^{\sin 2x}$,

   (v) $e^{-\cot x}$,   (vi) $e^{\csc^2 x}$,   (vii) $e^{\sqrt{\cos x}}$,   (viii) $e^{a\sin bx}$,

   (ix) $e^{\sin 3t}$,   (x) $e^{\tan x^2}$.

7. (i) $e^{\sqrt{(x^2+1)}}$,   (ii) $e^{(1-x^2)^{-1}}$,   (iii) $e^{\sin^2 4x}$,   (iv) $e^{\tan(x^2+1)}$,

   (v) $e^{\sec^2 3x}$,   (vi) $\dfrac{1}{e^{\csc x}}$,   (vii) $\dfrac{1}{e^{x-2}}$,   (viii) $e^{x\sin x}$,

   (ix) e^{xy},   (x) e^{e^x}.

8. (i) $x^2 e^x$,   (ii) $\dfrac{e^x}{x}$,   (iii) $\dfrac{x}{2}e^{\sin x}$,   (iv) $e^{x^2}\csc x$,

   (v) $\dfrac{e^x}{\sin x}$,   (vi) $\dfrac{\cos x}{x\,e^x}$,   (vii) e^{xe^x},   (viii) $e^{ax}\sec bx$,

   (ix) $\dfrac{e^{ax}}{\sin bx}$,   (x) $\tan^n e^x$,   (xi) $e^x(\cos x + \sin x)$.

9. Find the following integrals:

   (i) $\displaystyle\int 3e^{x/2}\,dx$,   (ii) $\displaystyle\int e^{-x}\,dx$,

   (iii) $\displaystyle\int e^{x/3}\,dx$,   (iv) $\displaystyle\int 2e^{3x-1}\,dx$,

   (v) $\displaystyle\int \frac{x}{2}e^{x^2}\,dx$,   (vi) $\displaystyle\int x^2 e^{-x^3}\,dx$.

(vii) $\int \sin x \, e^{\cos x} \, dx$,

(viii) $\int (1 + \tan^2 x) \, e^{\tan x} \, dx$,

(ix) $\int \dfrac{e^{\cot x}}{\sin^2 x} \, dx$,

(x) $\int x^{-2} \, e^{1/x} \, dx$,

(xi) $\int \dfrac{e^{x^{-2}}}{x^3} \, dx$,

(xii) $\int \dfrac{e^{\sqrt{x}}}{\sqrt{x}} \, dx$,

(xiii) $\int e^{\sin^2 x} \sin 2x \, dx$,

(xiv) $\int e^{\sin 2x} (1 - 2 \sin^2 x) \, dx$,

(xv) $\int \dfrac{x}{e^{x^2}} \, dx$.

10. Find the equation of the tangent to the curve $y = e^x$ at the point given by $x = a$. Deduce the equation of the tangent to the curve which passes through the point $(1, 0)$.

11. Sketch the curve $y = e^{x-2}$, and show that it divides in the ratio $\dfrac{e^2 - 1}{e^2 + 1}$ the area of the rectangle formed by the axes, $x = 2$, $y = 1$.

12. Find the volume of the solid generated by rotation about the x-axis of the area enclosed by $y = e^x$, the axes, $x = 1$.

13. A uniform lamina is bounded by the curve $y = e^{-x^2}$, the axes, $x = 2$. Its area is approximately 0.882; calculate the distance of its centre of gravity from the y-axis, correct to two significant figures.

14. Find $\dfrac{d}{dx}(x \, e^x)$, and deduce $\int x \, e^x \, dx$.

15. Investigate any maximum or minimum values of the function $x \, e^x$, and then sketch the graph of $y = x \, e^x$. Find the equation of the tangent to this curve at the point where $x = -2$.

In Nos. 16 to 19, A and B are constants; in each case show that the differential equation (see P.M.I., p. 88) is satisfied by the given solution:

16. $\dfrac{d^2 y}{dx^2} = 4y$; $y = A \, e^{2x} + B \, e^{-2x}$.

17. $\dfrac{d^2 s}{dt^2} + 4 \dfrac{ds}{dt} = 0$; $s = A + B \, e^{-4t}$.

18. $\dfrac{d^2 y}{dx^2} + 4 \dfrac{dy}{dx} + 3y = 0$; $y = A \, e^{-x} + B \, e^{-3x}$.

19. $\dfrac{d^2 y}{dt^2} - 6 \dfrac{dy}{dt} + 9y = 0$; $y = e^{3t} (A + Bt)$.

20. If $f(x) \equiv e^{4x} \cos 3x$, show that $f'(x) = 5e^{4x} \cos (3x + \alpha)$, where $\tan \alpha = \frac{3}{4}$. Deduce expressions of a similar form for $f''(x)$ and $f'''(x)$.

21. If $f(x) = e^{5x} \sin 12x$, show that $f'(x) = 13e^{5x} \sin (12x + \beta)$, where $\tan \beta = \dfrac{12}{5}$. Write down an expression for $f^n(x)$.

22. Show that the 9th derivative of $e^x \sin x$ is $16\sqrt{2}\,e^x \sin\left(x + \frac{\pi}{4}\right)$.

23. (i) Sketch the curve $x = e^{-t/\pi} \cos t$, taking values of t from 0 to 4π at intervals of $\frac{\pi}{2}$.

   (ii) Find $\frac{dx}{dt}$ in terms of e and π when $t = 0$, π, 2π, 3π, 4π.

   (iii) Find the values of t between 0 and 2π for which x has a minimum and a maximum value.

   (iv) Find the gradient of the curve to four decimal places at the points given by $t = \frac{\pi}{2}$, $\frac{3\pi}{2}$.

Further theory of logarithms

2.7. Since a logarithm is an index (or exponent), the discussion of exponential functions leads naturally to further consideration of logarithms. It is advisable at this stage to restate some of the ideas covered in P.M.I., §§ 8.4, 8.5.

DEFINITION: *The logarithm of b to the base a, written $\log_a b$, is the power to which the base must be raised to equal b.*

Thus, since $\qquad 10^2 = 100, \qquad 2 = \log_{10} 100,$
and if $\qquad a^x = b, \qquad x = \log_a b.$

The reader should already be familiar with the following basic rules:

$$\log_c (ab) = \log_c a + \log_c b.$$
$$\log_c (a/b) = \log_c a - \log_c b.$$
$$\log_c (a^n) = n \log_c a.$$

We shall now show that

$$\log_a b = \frac{\log_c b}{\log_c a}.$$

Let $\qquad x = \log_a b,$
$$\therefore\ a^x = b.$$

Taking logarithms to the base c of each side,

$$\log_c (a^x) = \log_c b,$$
$$\therefore\ x \log_c a = \log_c b,$$
$$\therefore\ x = \frac{\log_c b}{\log_c a}.$$

i.e. $\qquad\qquad\qquad \log_a b = \frac{\log_c b}{\log_c a}.\ \dagger$

† The identity $\frac{b}{a} = \frac{b/c}{a/c}$ provides a mnemonic.

Qu. 8. Express as a single logarithm:
  (i) $2 \log_{10} a - \frac{1}{3} \log_{10} b + 2$,
  (ii) $\log_c (1 + x) - \log_c (1 - x) + A$, where $A = \log_c B$.

Qu. 9. Express in terms of $\log_c a$:

  (i) $\log_c (2a)$,   (ii) $\log_c a^2$,   (iii) $\log_c \frac{1}{a}$,

  (iv) $\log_c \frac{2}{a}$,   (v) $\log_c \sqrt{a}$,   (vi) $\log_c \frac{a}{2}$,

  (vii) $\log_c \frac{1}{a^2}$,   (viii) $\log_c (2a)^{-1}$.

Qu. 10. Solve the equations:
  (i) $3^{2x} = 27$,   (ii) $1 \cdot 2^x = 3$.

Qu. 11.   (i) Prove that $\log_2 10 = \dfrac{1}{\log_{10} 2}$,

          (ii) Evaluate $\log_e 100$ correct to 3 significant figures, taking e as 2·718.

Natural logarithms

2.8. Logarithms to the base e are called *natural logarithms*, or *Naperian logarithms*, in honour of John Napier, a Scotsman, who published the first table of logarithmic sines in 1614. It is not surprising that the idea of logarithms was discovered independently at about the same time by Joost Bürgi, a Swiss; there was a pressing need to reduce labour involved in computation, especially in astronomy and navigation.

Napier's first publication on this topic fired the imagination of Henry Briggs, who visited him to discuss the practical application of the discovery; the fruit of this meeting was the eventual introduction of logarithms to the base 10, or *common logarithms*, for computation. However, it should be remembered that neither Napier nor Bürgi put forward the concept of a *base*, with the logarithm as an index; this idea does not appear to have been fully developed until about the middle of the eighteenth century. The choice of e as the base, though less convenient than 10 for computation, provides a new function $\log_e x$ of fundamental importance.†

Qu. 12. Evaluate $\log_e 2$ correct to three significant figures, taking e as 2·718, and using logarithms to the base 10. Check your answer with a table of natural logarithms.

---

† It has been customary in most mathematical textbooks to denote natural logarithms without the suffix e. However, we shall always specify the base, adopting the alternative recommendation of the British Standards Institution (B.S. 1991: Pt. 1: 1954).

Qu. 13. Find from a table of natural logarithms the values of (i) $\log_e 3$, (ii) $\log_e 15 \cdot 72$, (iii) $\log_e 0 \cdot 02158$, (iv) $\log_e 860 \cdot 9$.

Exercise 2b

1. Express as a single logarithm:
   (i) $2 \log_{10} a - 2 + \log_{10} 2a$,
   (ii) $3 \log_e x + 3 - \log_e 3x$,
   (iii) $4 \log_e (x-3) - 3 \log_e (x-2)$,
   (iv) $\frac{1}{2} \log_e (1+y) + \frac{1}{2} \log_e (1-y) + \log_e k$.

2. Express in terms of $\log_e a$:
   (i) $\log_e 3a$,   (ii) $\log_e a^3$,   (iii) $\log_e (a/3)$,
   (iv) $\log_e (1/a^3)$,   (v) $\log_e (3/a)$,   (vi) $\log_e (\frac{1}{3}a^{-1})$,
   (vii) $\log_e (\sqrt[3]{a})$.

3. Express as the sum or difference of logarithms:
   (i) $\log_e \cot x$,   (ii) $\log_e \tan^2 x$,   (iii) $\log_e (x^2 - 4)$,
   (iv) $\log_e \sqrt{\left(\dfrac{x+1}{x-1}\right)}$,   (v) $\log_e (3 \sin^2 x)$.

4. Solve the equations:
   (i) $\frac{3}{2} \log_{10} a^3 - \log_{10} \sqrt{a} - 2 \log_{10} a = 4$,   (ii) $\log_{10} y - 4 \log_y 10 = 0$.

5. Solve the equations:
   (i) $2^{2/x} = 32$,   (ii) $3^{x+1} = 12$,   (iii) $2^x . 3^{x+1} = 10$,
   (iv) $2^x = \sqrt[x]{10}$.

6. Evaluate $\log_e 3$ correct to three significant figures, taking e as $2 \cdot 718$, and using logarithms to the base 10.

7. Find from a table of natural logarithms the values of:
   (i) $\log_e 5 \cdot 94$,   (ii) $\log_e 25 \cdot 31$,   (iii) $\log_e 0 \cdot 4708$,
   (iv) $\log_e 300 \cdot 2$.

Solve the equations in Nos. 8 to 11.

8. (i) $3^{2(1+x)} - 28.3^x + 3 = 0$,   (ii) $3^{2x} - 4.3^{2+x} + 243 = 0$.

9. (i) $\log_{10} a + \log_a 100 = 3$,   (ii) $\log_e 2 \times \log_e 3 = 5$.

10. (i) $\log_{10} (19x^2 + 4) - 2 \log_{10} x - 2 = 0$,
    (ii) $\log_{10} e . \log_e (x^2 + 1) - 2 \log_{10} e . \log_e x = \log_{10} 5$.

11. (i) $\log_{10} x + \log_{10} y = 1$, $x + y = 11$,
    (ii) $\log_{10} x + \log_{10} y = 0 \cdot 19$, $x + y = 3$ (correct to 2 decimal places).

12. The distance between the 1 and 10 marks on a slide rule is 10 cm. If there are possible setting and reading errors of up to $0 \cdot 01$ cm, find the maximum percentage error in evaluating $3 \cdot 14 \times 4 \cdot 8$.

The derivative of $\log_e x$

2.9. Figure 2.4 shows the graph of $y = \log_e x$, or $x = e^y$; this curve is, of course, the same as the graph of $y = e^x$ with the axes interchanged.

To find $\dfrac{d}{dx} (\log_e x)$, we write $y = \log_e x$ as

$$x = e^y.$$

Differentiating each side with respect to x,†

$$\frac{d}{dx} (x) = \frac{d}{dy} (e^y) \cdot \frac{dy}{dx},$$

$$\therefore 1 = e^y \frac{dy}{dx},$$

$$\therefore \frac{dy}{dx} = \frac{1}{e^y} = \frac{1}{x}.$$

Thus
$$\frac{d}{dx} (\log_e x) = \frac{1}{x}.$$

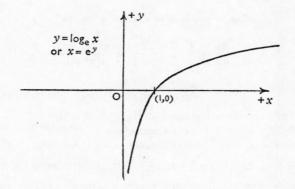

$y = \log_e x$
or $x = e^y$

$(1,0)$

FIG. 2.4

Example 3. *Find* $\dfrac{dy}{dx}$ *if* (i) $y = \log_e 2x$, (ii) $y = \log_e x^2$, (iii) $y = \log_e (x^2 + 2)$,

(iv) $\log_e \dfrac{x}{\sqrt{(x^2 + 1)}}$.

(i) $$y = \log_e 2x = \log_e 2 + \log_e x,$$

$$\therefore \frac{dy}{dx} = \frac{1}{x}.$$

† Or we may differentiate each side with respect to y.

$$\frac{dx}{dy} = e^y = x, \qquad \therefore \frac{dy}{dx} = \frac{1}{\dfrac{dx}{dy}} = \frac{1}{x}$$

(ii) $$y = \log_e x^2 = 2 \log_e x,$$

$$\therefore \frac{dy}{dx} = \frac{2}{x}.$$

(iii) $$y = \log_e (x^2 + 2).$$

Let $u = x^2 + 2$, then $$y = \log_e u.$$

$$\frac{du}{dx} = 2x \quad \text{and} \quad \frac{dy}{du} = \frac{1}{u}.$$

$$\therefore \frac{dy}{dx} = \frac{dy}{du}\frac{du}{dx} = \frac{1}{u} 2x,$$

$$\therefore \frac{dy}{dx} = \frac{2x}{x^2 + 2}.$$

(iv) $$y = \log_e \frac{x}{\sqrt{(x^2 + 1)}} = \log_e x - \tfrac{1}{2} \log_e (x^2 + 1).$$

$$\therefore \frac{dy}{dx} = \frac{1}{x} - \frac{1}{2}\cdot\frac{2x}{x^2 + 1} = \frac{x^2 + 1 - x^2}{x(x^2 + 1)},$$

$$\therefore \frac{dy}{dx} = \frac{1}{x(x^2 + 1)}.$$

Qu. 14. Differentiate with respect to x:

(i) $(x^2 - 2)^5$,   (ii) cosec x^2,   (iii) e^{x^2},

(iv) $\log_e (x^2 - 2)$,   (v) $\log_e \sin^2 x$,   (vi) $\log_e \sin x^2$.

Qu. 15. If $y = \log_e \{x \sqrt{(x + 1)}\}$ find $\dfrac{dy}{dx}$ (i) by differentiating the logarithm of a product as it stands, (ii) by first writing y as the sum of two logarithms.

Qu. 16. Differentiate with respect to x:

(i) $\log_e (3x)$,   (ii) $\log_e (4x)$,   (iii) $\log_e (3x + 1)$,

(iv) $\log_e y$,   (v) $\log_e (2x^3)$,   (vi) $\log_e (x^3 - 2)$,

(vii) $\log_e (x - 1)^3$,   (viii) $\log_e (4t)$,   (ix) $\log_e (3 \sin x)$,

(x) $\log_e \cos 3x$,   (xi) $\log_e (2 \cos^3 x)$,   (xii) $\log_e (4 \sin^2 3x)$,

(xiii) $\log_e \sqrt{(x^2 - 1)}$,   (xiv) $\log_e \dfrac{x}{(x - 1)^2}$.

$$\frac{d}{dx}(a^x) \quad \text{and} \quad \int a^x \, dx$$

2.10. In § 2.4 we found that $\dfrac{d}{dx}(a^x) = ma^x$, m being the gradient of $y = a^x$ at $(0, 1)$. When $a = e$, $m = 1$; we can now find m for other values of a.

Let $\qquad\qquad y = a^x,$

then $\qquad\qquad \log_e y = \log_e a^x = x \log_e a.$

Differentiating with respect to x:

$$\frac{d}{dy}(\log_e y).\frac{dy}{dx} = \frac{d}{dx}(x \log_e a),$$

$$\therefore \frac{1}{y}.\frac{dy}{dx} = \log_e a,$$

$$\therefore \frac{dy}{dx} = y \log_e a,$$

$$\therefore \frac{d}{dx}(a^x) = a^x \log_e a.$$

Qu. 17. Find $\dfrac{d}{dx}(4^x)$ reproducing the above method in full. Find the gradient of the curve $y = 4^x$ at $(2, 16)$.

Qu. 18. Find the gradient at $(0, 1)$ of the following curves, to four decimal places:

(i) $y = 2^x,$ $\qquad$ (ii) $y = 3^x.$

Qu. 19. Differentiate with respect to x:

(i) $10^x,$ $\qquad$ (ii) $2^{3x+1},$ $\qquad$ (iii) $3^{x^2},$ $\qquad$ (iv) $10^{\sin x}.$

Qu. 20. Find $\dfrac{d}{dx}(a^x)$ by writing a as $e^{\log_e a}.$

It follows from the last result that

$$\int a^x \log_e a\, dx = a^x + k,$$

$$\therefore \int a^x\, dx = \frac{a^x}{\log_e a} + c.$$

Qu. 21. Find $\dfrac{d}{dx}(5^x)$ and deduce $\displaystyle\int 5^x\, dx.$

Qu. 22. Find $\dfrac{d}{dx}(2^{x^2})$ and deduce $\displaystyle\int x\, 2^{x^2}\, dx.$

Qu. 23. Find the following integrals:

(i) $\displaystyle\int 3^{2x}\, dx,$ $\qquad$ (ii) $\displaystyle\int x^2\, e^{x^3}\, dx,$ $\qquad$ (iii) $\displaystyle\int 2^{\tan x} \sec^2 x\, dx.$

Exercise 2c

1. Differentiate with respect to x:

(i) $\log_e (4x),$ $\qquad$ (ii) $4 \log_e x,$ $\qquad$ (iii) $\log_e (2x-3),$

(iv) $\log_e (\tfrac{1}{3}y),$ $\qquad$ (v) $\log_e \dfrac{x-1}{2},$ $\qquad$ (vi) $\log_e x^4.$

(vii) $\log_e (x^2 - 1),$　　　　(viii) $\log_e 3x^2,$　　　　· (ix) $3 \log_e x^2,$

(x) $\log_e (x+1)^2,$　　　　(xi) $\log_e (2t^3),$　　　　(xii) $\log_e \dfrac{1}{x},$

(xiii) $\log_e (\tfrac{1}{2}x),$　　　　(xiv) $\log_e \sqrt{x},$　　　　(xv) $\log_e \dfrac{1}{2x},$

(xvi) $\log_e \dfrac{2}{x},$　　　　(xvii) $\log_e x^{-2},$　　　　(xviii) $\log_{10} x,$

(xix) $\log_e \dfrac{1}{t^3},$　　　　(xx) $\log_e \sqrt[3]{x}.$

2. Differentiate with respect to x:
    (i) $\log_e \cos x,$　　　　(ii) $\log_e \sin^2 x,$　　　　(iii) $\log_e \tan 3x,$
    (iv) $\log_e \cos^3 2x,$　　　　(v) $\log_e (2 \cot^2 x),$　　　(vi) $\log_e (3 \cos^2 2x),$
    (vii) $\log_e \tan \dfrac{x}{2},$　　　　(viii) $\log_e \sec x,$　　　　(ix) $\log_e (\sec x + \tan x),$
    (x) $\log_e \operatorname{cosec} x^2,$　　　　　　(xi) $\log_e \dfrac{\sin x + \cos x}{\sin x - \cos x}.$

3. Find:
    (i) $\dfrac{d}{dx} \log_e \sqrt{\dfrac{1-x}{1+x}},$　　　　(ii) $\dfrac{d}{dx} \log_e \{x \sqrt{(x^2-1)}\},$
    (iii) $\dfrac{d}{dx} \log_e \dfrac{(x+1)^2}{\sqrt{(x-1)}},$　　　　(iv) $\dfrac{d}{dx} \log_e \{x + \sqrt{(x^2-1)}\}.$

4. Differentiate with respect to x:
    (i) $\log_e t,$　　　　(ii) $x \log_e x,$　　　　(iii) $x^2 \log_e x,$
    (iv) $\dfrac{\log_e x}{x},$　　　　(v) $x \log_e y,$　　　　(vi) $y \log_e x,$
    (vii) $\dfrac{\log_e x}{x^2},$　　　　(viii) $\dfrac{x}{\log_e x},$　　　　(ix) $(\log_e x)^2,$
    (x) $\log_e (\log_e x^k),$　　　　(xi) $\log_e e^{\sin x}.$

5. Differentiate with respect to x:
    (i) $5^x,$　　　　(ii) $2^{x^2},$　　　　(iii) $3^{2x-1},$
    (iv) $10^{\operatorname{cosec} x},$　　　　(v) $3^{\sqrt{x}},$　　　　(vi) $e^{\log_e x},$
    (vii) $2^{\log_e x},$　　　　(viii) $2^{1/x}.$

6. (i) Find $\dfrac{d}{dx} (3^x)$ and deduce $\displaystyle\int 3^x \, dx.$

    (ii) Find $\dfrac{d}{dx} (2^{x^2})$ and deduce $\displaystyle\int x \, 2^{x^2} \, dx.$

7. Find the following integrals:

    (i) $\displaystyle\int 10^x \, dx,$　　　　(ii) $\displaystyle\int 2^{3x} \, dx,$　　　　(iii) $\displaystyle\int x \, 3^{x^2} \, dx,$

    (iv) $\displaystyle\int 2^{\cos x} \sin x \, dx.$

8. Find $\dfrac{d}{dx} (x \log_e x)$ and deduce $\displaystyle\int \log_e x \, dx.$

9. Find $\dfrac{d}{dx}(x\,2^x)$ and deduce $\int x\,2^x\,dx$.

10. Find (i) $\dfrac{d}{dx}\log_e(x-2)$,   (ii) $\dfrac{d}{dx}\log_e(2-x)$.

Sketch on the same axes the graphs of $\log_e(x-2)$, $\log_e(2-x)$, $y=\dfrac{1}{x-2}$.

11. Sketch on the same axes the following curves:

(i) $y=\log_e x,\quad y=\log_e(-x),\quad y=\dfrac{1}{x}$,

(ii) $y=\log_e\dfrac{1}{x},\quad y=\log_e\left(-\dfrac{1}{x}\right),\quad y=-\dfrac{1}{x}$,

(iii) $y=\log_e(x-3),\quad y=\log_e(3-x),\quad y=\dfrac{1}{x-3}$,

(iv) $y=\log_e\left(\dfrac{1}{x-3}\right),\quad y=\log_e\left(\dfrac{1}{3-x}\right),\quad y=\dfrac{1}{3-x}$.

$$\int \frac{f'(x)}{f(x)}\,dx$$

2.11.   The result $\int x^n\,dx=\dfrac{x^{n+1}}{n+1}+c$ holds for all rational values of n, except $n=-1$; this hitherto puzzling gap may now be filled.   Since we have established that $\dfrac{d}{dx}(\log_e x)=\dfrac{1}{x}$, it follows that

$$\int \frac{1}{x}\,dx = \log_e x+c,$$

or $$\int \frac{1}{x}\,dx = \log_e(kx),\quad \text{where } c=\log_e k.$$

Example 4.   *Find the following integrals*:

$$(i)\ \int \frac{1}{2x}\,dx,\quad (ii)\ \int \frac{1}{2x-1}\,dx.$$

(i)
$$\int \frac{1}{2x}\,dx = \tfrac{1}{2}\int \frac{1}{x}\,dx,$$
$$= \tfrac{1}{2}\log_e x+c,$$
$$\text{or } \log_e(k\sqrt{x}),\ \text{where } c=\log_e k.$$

(ii) $\int \dfrac{1}{2x-1}\,dx$.

This is best tackled in reverse by guessing the form of the integral.

$$\frac{d}{dx}\{\log_e (2x-1)\} = \frac{2}{2x-1}.$$

$$\therefore \int \frac{1}{2x-1}\,dx = \tfrac{1}{2}\log_e (2x-1)+c.$$

Qu. 24. Find the following integrals:

(i) $\int \dfrac{2}{x}\,dx$,   (ii) $\int \dfrac{1}{3x}\,dx$,   (iii) $\int \dfrac{1}{3x-2}\,dx$,   (iv) $\int \dfrac{1}{3x-6}\,dx$.

Qu. 25. Find the following integrals:

(i) $\int \dfrac{1}{2x+3}\,dx$,   using the substitution $u = 2x+3$,

(ii) $\int \dfrac{1}{1-x}\,dx$,   using the substitution $u = 1-x$.

Qu. 26. Evaluate $\int_1^2 \dfrac{3}{x}\,dx$.

Qu. 27. (i) Show that the answer to Example 4(ii) may be written $\log_e \{A(2x-1)^{\frac{1}{2}}\}$, and express c in terms of A.

(ii) If $\tfrac{1}{2}\log_e (x-\tfrac{1}{2})+c$ may be written as $\log_e \{k\sqrt{(2x-1)}\}$, express c in terms of k.

An integral of the form $\displaystyle\int \dfrac{f'(x)}{f(x)}\,dx$ may be reduced to the form $\displaystyle\int \dfrac{1}{u}\,du$, by the substitution $u=f(x)$.

$$\int \frac{f'(x)}{f(x)}\,dx = \int \frac{f'(x)}{f(x)}\cdot\frac{dx}{du}\,du,$$

$$= \int \frac{f'(x)}{u}\cdot\frac{1}{f'(x)}\,du,$$

$$= \int \frac{1}{u}\,du,$$

$$= \log_e u+c,$$

$$= \log_e f(x)+c.$$

Hence $\displaystyle\int \frac{f'(x)}{f(x)}\,dx = \log_e \{k\,f(x)\}.$

From now onwards we must be prepared to recognize, in yet another form, the integrand involving a function of x and its derivative. As before, such an integral may be found by substitution, or usually it may be written down at once.

Example 5. *Find* $\int \dfrac{x}{x^2+1}\,\mathrm{d}x$.

Since $\qquad \dfrac{\mathrm{d}}{\mathrm{d}x} \log_e (x^2+1) = \dfrac{2x}{x^2+1}$,

$$\int \frac{x}{x^2+1}\,\mathrm{d}x = \tfrac{1}{2}\log_e (x^2+1) + c,$$

$$\text{or } \log_e \{k\sqrt{(x^2+1)}\}.$$

Qu. 28. Find the following integrals:

(i) $\int \dfrac{x^2}{(x^3-2)^2}\,\mathrm{d}x$,   (ii) $\int x^2 \cos x^3 \,\mathrm{d}x$,

(iii) $\int x^2 e^{x^3}\,\mathrm{d}x$,   (iv) $\int \dfrac{x^2}{x^3-2}\,\mathrm{d}x$,

(v) $\int \dfrac{x-1}{x^2-2x}\,\mathrm{d}x$ (vi) $\int \dfrac{2x}{3-x^2}\,\mathrm{d}x$,

(vii) $\int \cot x \,\mathrm{d}x$.

Qu. 29. Find $\int \dfrac{x}{x-1}\,\mathrm{d}x$

(i) using the substitution $u = x-1$,

(ii) by first dividing the numerator by the denominator.

$$\int_a^b \frac{1}{x}\,\mathrm{d}x \quad \textbf{when } a, b \textbf{ are negative}$$

†**2.12.** An important point must be cleared up. Reference to Fig. 2.4 on p. 33 reminds us that as the value of x goes from 0 to $+\infty$, the value of $\log_e x$ goes from $-\infty$ to $+\infty$; $\log_e x$ *is not defined for negative values of* x. This presents us with an apparent paradox which may be demonstrated in graphical terms as follows.

Referring to the graph of $y = \dfrac{1}{x}$ in Fig. 2.5 it is apparent that the two shaded areas are equal in magnitude and of opposite sign. However, we soon get into trouble if we seek to evaluate the appropriate integral with the negative limits; thus, is it true to say that

$$\int_{-2}^{-1} \frac{1}{x}\,\mathrm{d}x = \Big[\log_e x\Big]_{-2}^{-1}$$
$$= \text{`}\log_e (-1) - \log_e (-2)\text{'}?$$

† § 2.12 should be delayed until after the reader has answered No. 1 of Exercise 2d.

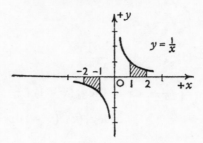

Fig. 2.5

We could now write this as $\log_e\left(\dfrac{-1}{-2}\right) = \log_e \frac{1}{2} = \log_e 2^{-1} = -\log_e 2$, and thus obtain a correct figure for the area, but the working is not valid, since the expression '$\log_e(-1) - \log_e(-2)$' is meaningless.

We surmount this difficulty as soon as we realize that *for negative values of x*, although $\log_e x$ is not defined, $\log_e(-x)$ *does exist*, and

$$\frac{d}{dx}\log_e(-x) = \frac{-1}{-x} = \frac{1}{x}. \quad \text{(See Fig. 2.6.)}$$

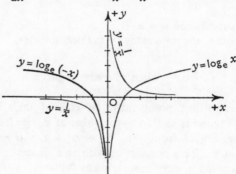

Fig. 2.6

Thus, if a and b are negative, $\displaystyle\int_a^b \frac{1}{x}\,dx = \left[\log_e(-x)\right]_a^b$.

Hence the left hand shaded area in Fig. 2.5 $= \displaystyle\int_{-2}^{-1} \frac{1}{x}\,dx$,

$$= \left[\log_e(-x)\right]_{-2}^{-1},$$
$$= \log_e 1 - \log_e 2,$$
$$= -\log_e 2.$$

Qu. 30.   Evaluate (i) $\int_{-4}^{-3} \dfrac{1}{x}\,dx$,   (ii) $\int_{-1}^{-\frac{1}{2}} \dfrac{1}{x}\,dx$.

Qu. 31.   Evaluate $\int_{-4}^{-2} \dfrac{1}{x}\,dx$, using the change of variable $x = -u$.

Qu. 32.   Can any meaning be assigned to $\int_{-2}^{+2} \dfrac{1}{x}\,dx$?

Example 6.   *Find the area enclosed by the curve* $y = \dfrac{1}{x-2}$ *and*

  (*i*) *the lines* $x=4$, $x=5$, *and the x-axis,*
  (*ii*) *the line* $x=1$, *and the axes.*   (Fig. 2.7.)

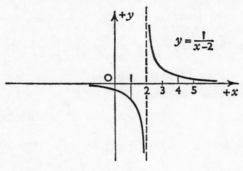

Fig. 2.7

(i)   The required area $= \int_{4}^{5} \dfrac{1}{x-2}\,dx$,

$$= \left[\log_e (x-2)\right]_{4}^{5},$$

$$= \log_e 3 - \log_e 2,$$

$$= \log_e \tfrac{3}{2}.$$

(ii) $\Big[$If we proceed as in (i) but with the new limits, we obtain the meaningless '$\left[\log_e (x-2)\right]_{0}^{1} = \log_e (-1) - \log_e (-2)$.'   We must note that when $x > 2$,

$$\int \frac{1}{x-2}\,dx = \log_e (x-2) + c,$$

but when $x < 2$,

$$\int \frac{1}{x-2}\,dx = \log_e (2-x) + c.\Big]$$

$$\text{The required area} = \int_0^1 \frac{1}{x-2} \, dx,$$

$$= \left[\log_e (2-x) \right]_0^1,$$

$$= \log_e 1 - \log_e 2,$$

$$= -\log_e 2.$$

Qu. 33. Find $\dfrac{d}{dx} \{\log_e (x-3)\}$ and $\dfrac{d}{dx} \{\log_e (3-x)\}$.

Qu. 34. Sketch the curve $y = \dfrac{1}{x-3}$, and evaluate:

$$\text{(i)} \int_5^6 \frac{1}{x-3} \, dx, \qquad \text{(ii)} \int_{-2}^2 \frac{1}{x-3} \, dx.$$

Qu. 35. Sketch the curve $y = \dfrac{1}{2-x}$ and evaluate $\displaystyle\int_3^5 \frac{1}{2-x} \, dx$.

Exercise 2d

1. Find the following integrals:

$$\text{(i)} \int \frac{1}{4x} \, dx, \qquad \text{(ii)} \int \frac{5}{x} \, dx, \qquad \text{(iii)} \int \frac{1}{2x-3} \, dx,$$

$$\text{(iv)} \int \frac{1}{2x+8} \, dx, \qquad \text{(v)} \int \frac{1}{3-2x} \, dx, \qquad \text{(vi)} \int \frac{x}{1-x^2} \, dx,$$

$$\text{(vii)} \int \frac{3x}{x^2-1} \, dx, \qquad \text{(viii)} \int \frac{2x+1}{x^2+x-2} \, dx,$$

$$\text{(ix)} \int \frac{2x-3}{3x^2-9x+4} \, dx, \qquad \text{(x)} \int \frac{x}{x+2} \, dx, \qquad \text{(xi)} \int \frac{3x}{2x+3} \, dx,$$

$$\text{(xii)} \int \frac{2x}{3-x} \, dx, \qquad \text{(xiii)} \int \frac{x-1}{2-x} \, dx, \qquad \text{(xiv)} \int \frac{3-2x}{x-4} \, dx,$$

$$\text{(xv)} \int \tan x \, dx, \qquad \text{(xvi)} \int \cot \frac{x}{2} \, dx, \qquad \text{(xvii)} \int \cot (2x+1) \, dx,$$

$$\text{(xviii)} \int -\tan \frac{x}{3} \, dx, \qquad \text{(xix)} \int \frac{1-\sin 2x}{x-\sin^2 x} \, dx, \qquad \text{(xx)} \int \frac{1-\tan x}{1+\tan x} \, dx,$$

$$\text{(xxi)} \int \frac{2+\tan^2 x}{x+\tan x} \, dx.$$

2. (i) Sketch the curves $y = \log_e (2x-1)$, and $y = \log_e (1-2x)$.

(ii) Find $\dfrac{d}{dx} \log_e (2x-1)$ and $\dfrac{d}{dx} \log_e (1-2x)$.

(iii) Evaluate $\displaystyle\int_1^2 \frac{1}{2x-1} \, dx$ and $\displaystyle\int_{-2}^0 \frac{1}{2x-1} \, dx$.

3. Sketch the curve $y = \dfrac{1}{x-4}$ and evaluate:

$$\text{(i)} \int_1^2 \frac{1}{x-4} \, dx, \qquad \text{(ii)} \int_5^6 \frac{1}{x-4} \, dx.$$

4. (i) Find $\dfrac{d}{dx} \log_e \left(\dfrac{1}{3-x}\right)$ and $\dfrac{d}{dx} \log_e \left(\dfrac{1}{x-3}\right)$.

   (ii) Sketch on the same axes the graphs of $y = -\log_e (3-x)$, $y = -\log_e (x-3)$, $y = \dfrac{1}{3-x}$, and find the area enclosed by the latter, the lines $x = 5$, $x = 6$, and the x-axis.

   (iii) Find the area under $y = \dfrac{1}{3-x}$ from $x = 0$ to $x = 1$.

5. Evaluate the following:

   (i) $\displaystyle\int_2^8 \dfrac{1}{2x}\, dx$,   (ii) $\displaystyle\int_1^{\frac{4}{3}} \dfrac{1}{3x-2}\, dx$,   (iii) $\displaystyle\int_1^3 \dfrac{1}{x-5}\, dx$,

   (iv) $\displaystyle\int_3^5 \dfrac{1}{1-2x}\, dx$,   (v) $\displaystyle\int_{-0.25}^{0.25} \dfrac{1}{2x+1}\, dx$,

   (vi) $\displaystyle\int_{-\sqrt{2}}^0 \dfrac{x}{x^2+2}\, dx$,   (vii) $\displaystyle\int_0^3 \dfrac{2x-1}{x^2-x+1}\, dx$,   (viii) $\displaystyle\int_4^6 \dfrac{x}{x-2}\, dx$,

   (ix) $\displaystyle\int_{-7}^{-5} \dfrac{x+1}{x+3}\, dx$,   (x) $\displaystyle\int_{-0.5}^0 \dfrac{2-x}{x-1}\, dx$,   (xi) $\displaystyle\int_{\frac{\pi}{3}}^{\frac{\pi}{2}} \cot \theta\, d\theta$,

   (xii) $\displaystyle\int_0^{\frac{\pi}{6}} \tan 2x\, dx$,   (xiii) $\displaystyle\int_{\frac{\pi}{6}}^{\frac{\pi}{4}} \dfrac{\sec^2 \theta}{\tan \theta}\, d\theta$.

Exercise 2e (Miscellaneous)

(For integration see Exercises 13e and 13f.)

1. (i) Express $\log_a b$ in terms of logarithms to the base 10. Given that $\log_{10} 2 = 0.3010$, calculate, without using tables, the values to two places of decimals of $\log_2 0.64$ and $\log_5 \sqrt{1.25}$.

   (ii) Solve the equation

   $$\log_2 x + \log_x 2 = 2.5.$$ L.

2. Solve the equation

   $$x^{1.23} = 0.12.$$ L.

3. Solve the equation

   $$\log_a (x^2+3) - \log_a x = 2 \log_a 2.$$ O.C.

4. Solve the equations
   (i) $6^{z+2} = 2(3^{2z})$,   (ii) $\log_x 0.028 = 3$,   (iii) $\log_{10} x = \log_5 (2x)$.   L.

5. If $\log_{10} y = 2 - \log_{10} (x^{2/3})$, express y as a function of x not involving logarithms; hence show that, if $x = 8$, then $y = 25$.   O.C.

6. Prove that

$$\log_a x = \frac{\log_b x}{\log_b a}.$$

Calculate, as accurately as your tables allow, the value of $\log_e 0.3$, given that $\log_{10} e = 0.4343$.　　　　　　O.C.

7. Find to 3 significant figures the possible values of a if

$$\log_a 5 \times \log_a 2 = 10.$$　　　　　　L.

8. Solve the equations:
  (i) $2^{2+2x} + 3 . 2^x - 1 = 0$,
  (ii) $\log_{10} (x^2 + 1) - 2 \log_{10} x = 1$,
  (iii) $\log_{10} x - \log_x 10 = 3$, giving the roots correct to two significant figures.　　　　　　L.

9. Using the same scale and axes, draw the graphs of $y = 2^x$ and $y = 2^{-x}$, between $x = -3$ and $x = +3$. Use your graphs to estimate:
  (i) the root of the equation $2^x - 2^{-x} = 3$,
  (ii) the value of $\sqrt[5]{32} - \sqrt[5]{\frac{1}{32}}$.
Give both results correct to two significant figures.　　　　　　L.

10. (i) Two quantities x and y are connected by the equation $y = a\,e^{bx}$, where a and b are constants. If $y = 1$ when $x = 1$, and $y = 3$ when $x = 4$, find the values of a and b, correct to 3 decimal places.
  (ii) Solve the equation $2^{2x} - 2^{x+2} - 5 = 0$.　　　　　　L.

11. A function y is of the form $y = ax^n + bx$ where a, b, and n are constants. When x is equal successively to 1, 3, and 9, the corresponding values of y are 4, 6, and 15. Find two relations between a and b not involving n, and hence find the numerical values of a, b, and n (the last to three places of decimals).　　　　　　O.C.

12. Prove that $\log_2 e - \log_4 e + \log_8 e - \log_{16} e + \ldots = 1$, where e is the base of natural logarithms. (See p. 216.)　　　　　　N.

13. If $y = A\,e^{-x} \cos(x + \alpha)$, where A and α are constants, prove that:

  (i) $\dfrac{d^2y}{dx^2} + 2\dfrac{dy}{dx} + 2y = 0$,

  (ii) $\dfrac{d^4y}{dx^4} + 4y = 0$.　　　　　　O.C.

14. A particle is moving in a straight line. The displacement x, from an origin O on the line, is given at time t by the equation

$$x = e^{-\frac{1}{2}t}(a \sin t + b \cos t).$$

Initially $t = 0$, $x = 4$, $\dfrac{dx}{dt} = 0$. Find the constants a and b. Determine also (i) the time elapsing from the start before the particle first reaches O, (ii) the time taken from O to attain the greatest displacement on the negative side of the origin.　　　　　　N.

15. Find the maximum and minimum values of the function $(1+2x^2)\,e^{-x^2}$.

O.C.

16. If $y=e^{-x}\cos x$, determine the three values of x between 0 and 3π for which $\dfrac{\mathrm{d}y}{\mathrm{d}x}=0$. Show that the corresponding values of y form a geometric progression with common ratio $-e^{-\pi}$.

N.

17. (i) Find $\dfrac{\mathrm{d}y}{\mathrm{d}x}$ if $y=\log_e\left(\dfrac{3+4\cos x}{4+3\cos x}\right)$.

(ii) If $y=e^{4x}\cos 3x$, prove that

$$\frac{\mathrm{d}^2y}{\mathrm{d}x^2}-8\frac{\mathrm{d}y}{\mathrm{d}x}+25y = 0.$$

O.C.

18. If $y=e^{kt}\cos pt$, prove that

$$\frac{\mathrm{d}^2y}{\mathrm{d}t^2}-2k\frac{\mathrm{d}y}{\mathrm{d}t}+(k^2+p^2)y = 0.$$

If $\dfrac{\mathrm{d}y}{\mathrm{d}t}=2p$ and $\dfrac{\mathrm{d}^2y}{\mathrm{d}t^2}=3p$ when $t=\dfrac{3\pi}{2p}$, calculate k and prove that

$$p = \frac{9\pi}{8\log_e 2}.$$

L.

19. Find the real value of x satisfying the equation

$$e^x-e^{-x} = 4.$$

Show that for this value of x, $e^x+e^{-x}=2\sqrt{5}$.

L.

20. Find the maximum and minimum values of $x^2\,e^{-x}$, and sketch the graph of this function. Find the equation of the tangent to the graph at the point at which $x=1$.

L.

21. A particle moves in a straight line so that, at time t, its distance from the origin is $e^{-\lambda t}\sin(pt+\alpha)$ where λ, p, and α are constants. If it is instantaneously at rest when $t=0$, show that $\lambda=p\cot\alpha$. Show that the time between consecutive positions of instantaneous rest is constant, and that the distances of the particle from the origin when at positions of instantaneous rest diminish in geometrical progression, the common ratio being $e^{-\pi\cot\alpha}$.

L.

CHAPTER 3

PARTIAL FRACTIONS

Introduction

3.1. Early training in algebra teaches us how to 'simplify' an expression such as $\dfrac{1}{x-1} - \dfrac{1}{x+1}$ by reducing it to $\dfrac{2}{x^2-1}$.

We have now reached the stage when the reverse process is of value. Given a fraction such as $\dfrac{5}{x^2+x-6}$ whose denominator factorizes, we may split it up into its component fractions, writing it as $\dfrac{1}{x-2} - \dfrac{1}{x+3}$; it is now said to be in **partial fractions**. Just one example of the several applications of this must suffice for the present. No change of variable yet discussed would enable us to find $\displaystyle\int \dfrac{5}{(x-2)(x+3)}\, dx$ as it stands, but using partial fractions,

$$
\begin{aligned}
\int \frac{5}{(x-2)(x+3)}\, dx &= \int \left\{ \frac{1}{x-2} - \frac{1}{x+3} \right\} dx, \\
&= \log_e (x-2) - \log_e (x+3) + c, \\
&= \log_e \left\{ \frac{k(x-2)}{x+3} \right\}.
\end{aligned}
$$

Qu. 1. Express each of the following as a single fraction:

   (i) $\dfrac{1}{1-x} + \dfrac{2}{1+x}$,   (ii) $\dfrac{2x-1}{x^2+1} - \dfrac{1}{x+1}$,

(iii) $\dfrac{3}{(x-1)^2} + \dfrac{1}{x-1} + \dfrac{2}{x+1}$.

Qu. 2. Express in partial fractions:

(i) $\dfrac{4}{(x-2)(x+2)}$,   (ii) $\dfrac{1}{1-x^2}$,   (iii) $\dfrac{1}{2.3}$,   (iv) $\dfrac{1}{n(n+1)}$.

Unfortunately most partial fractions cannot be obtained by trial and error quite as easily as those in Qu. 2. The reader need only consider attempting Qu. 1 in reverse, to be convinced that we need some technique to find partial fractions; we shall find that this involves us in handling algebraic identities, so we must discuss these briefly.

46

Identities

3.2. Let us first distinguish clearly between an *equation* and an *identity*. $x^2 = 4$ is an *equation*, which is satisfied only by the two values $x = \pm 2$. But

$$x^2 - 4 \equiv (x+2)(x-2),$$

and
$$x^2 + 2x - 2 \equiv (x+1)(x-1) + 2(x+1) - 3,$$

are both **identities**, and for them the L.H.S. = R.H.S. *for any value of x*; moreover, of course, if the R.H.S. is multiplied out, the coefficients of x^2, x and the constant term will be identical on each side.

Example 1. *Find the values of the constants A, B, C such that*
$$5x + 3 \equiv Ax(x+3) + Bx(x-1) + C(x-1)(x+3).$$

First method

Collecting like terms on the R.H.S.,
$$5x + 3 \equiv (A+B+C)x^2 + (3A - B + 2C)x - 3C.$$

Equating coefficients of x^2,
$$0 = A + B + C. \tag{1}$$

Equating coefficients of x,
$$5 = 3A - B + 2C. \tag{2}$$

Equating constant terms,
$$3 = -3C. \tag{3}$$

From (3) $C = -1$, and substituting this value into (1) and (2), and solving these equations simultaneously, we obtain $A = 2$, and $B = -1$.

Second method

$$5x + 3 \equiv Ax(x+3) + Bx(x-1) \quad + C(x-1)(x+3).$$

Putting $x = 0$,
$$3 = \quad 0 \quad + \quad 0 \quad - \quad 3C,\dagger$$
$$\therefore \; C = -1.$$

Putting $x = -3$,
$$-15 + 3 = \quad 0 \quad + B(-3)(-4) + \quad 0,$$
$$\therefore \; -12 = \quad 12B,$$
$$\therefore \; B = -1.$$

Putting $x = 1$,
$$5 + 3 = A.1.4 \quad + \quad 0 \quad + \quad 0,$$
$$\therefore \; A = 2.$$

It should be noted that the identity holds for *any* value of x, but we have chosen those particular values which make all but one term on the R.H.S. vanish each time.

† This should be compared with equation (3) above.

Qu. 3. $2x^2 + 9x - 10 \equiv A(x-3)(x+4) + B(x+2)(x+4) + C(x+2)(x-3)$.

  (i) Obtain three equations in A, B, C by substituting $x = -1, 0, 1$ in this identity.

  (ii) Find the values of A, B, C by substituting more convenient values of x.

Qu. 4.   Find the values of the constants A, B, C in the following identities:

  (i) $22 - 4x - 2x^2 \equiv A(x-1)^2 + B(x-1)(x+3) + C(x+3)$,

                                      using the first method,

  (ii) $5x + 31 \equiv A(x+2)(x-1) + B(x-1)(x-5) + C(x-5)(x+2)$,

                                  using the second method,

  (iii) $13x - 11 \equiv A(3x-2) + B(2x+1)$.

Qu. 5.   Put $x = 1$ to find the value of A in the identity

$$x^2 + x + 7 = A(x^2 + 2) + (Bx + C)(x - 1).$$

Now substitute any other values of x to find B and C.

The substitution method is fast, but often it may be combined with the method of equating coefficients for greater speed and simplicity (e.g. having found A in Qu. 5, equate coefficients of x^2 to find B). The latter method also gives us a deeper insight into the nature of identities. Let us consider the statement

$$\text{`}x^2 - 5x + 8 \equiv A(x+3) + B(x-1)^2.\text{'}$$

Applying the method of substitution we obtain $A = 1$, $B = 2$; however, when A and B are given these values we do *not* have an identity!

This apparently alarming breakdown is readily explained when we apply the method of equating coefficients. This shows that for only *two* unknowns we have the *three* equations, $B = 1$, $A - 2B = -5$, and $3A + B = 8$, which are not consistent. Thus we cannot find values for A and B to form the 'identity' given above.

Since we shall soon be concerned with forming identities, the method of equating coefficients will be a valuable check that the number of unknown constants introduced corresponds to the number of equations to be satisfied.

Qu. 6.   Can values of A, B, C be found which make the following pairs of expressions identical?

  (i) $2x + 3$ and $A(x+1)(x-2) + B(x+1)^2 + C$,

  (ii) $x^2 - 8x + 30$ and $A(x-3)^2 + B(x+2)$.

Exercise 3a

1. Express each of the following as a single fraction:

(i) $\dfrac{3}{x+3} - \dfrac{2}{x-2}$,

(ii) $\dfrac{1}{(x+2)^2} - \dfrac{2}{x+2} + \dfrac{1}{3x-1}$,

(iii) $\dfrac{4}{2+3x^2} - \dfrac{1}{1-x}$,

(iv) $\dfrac{3}{x^2+1} - \dfrac{1}{x-1} + \dfrac{2}{(x-1)^2}$.

2. Express in partial fractions:

(i) $\dfrac{2x}{(3+x)(3-x)}$, (ii) $\dfrac{a}{a^2-b^2}$, (iii) $\dfrac{1}{5.6}$, (iv) $\dfrac{1}{p(1-p)}$.

3. Use the first method of Example 1 to find the values of the constants A, B, C in the following identities:

(i) $31x - 8 \equiv A(x-5) + B(4x+1)$,

(ii) $8 - x \equiv A(x-2)^2 + B(x-2)(x+1) + C(x+1)$,

(iii) $71 + 9x - 2x^2 \equiv A(x+5)(x+2) + B(x+2)(x-3) + C(x-3)(x+5)$,

(iv) $2x^3 - 15x^2 - 10$
$$\equiv A(x-2)(x+1) + B(x+1)(2x^2+1) + C(2x^2+1)(x-2).$$

4. Use the second method of Example 1 to find the values of the constants A, B, C in the following identities:

(i) $2x - 4 \equiv A(3+x) + B(7-x)$,

(ii) $8x + 1 \equiv A(3x-1) + B(2x+3)$,

(iii) $4x^2 + 4x - 26 \equiv A(x+2)(x-4) + B(x-4)(x-1) + C(x-1)(x+2)$,

(iv) $17x^2 - 13x - 16$
$$\equiv A(3x+1)(x-1) + B(x-1)(2x-3) + C(2x-3)(3x+1).$$

5. Can values of A, B, C, D be found which make the following pairs of expressions identical?

(i) $2x^2 - 22x + 53$
and $A(x-5)(x-3) + B(x-3)(x+2) + C(x+2)(x-5)$,

(ii) $x + 7$ and $A(x-2) + B(x+1)^2$,

(iii) $3x^2 + 7x + 11$ and $(Ax+B)(x+2) + C(x^2+5)$,

(iv) $x + 1$ and $A(x-2) + B(x^2+1)$,

(v) $x^3 + 2x^2 - 4x - 2$
and $(Ax+B)(x-2)(x+1) + C(x+1) + D(x-2)$,

(vi) $8 + 3x - x^2$ and $A(x^2+3) + (Bx+C)(2x-5)$,

(vii) $7x^3 + 21x^2 + 12x - 20$
and $A(2x-1)(x+1)^2 + B(x+3)(x+1)^2 + C(x+3)(2x-1)(x+1)$
$+ D(x+3)(2x-1)$.

6. Find the values of A, B, C if $x^3 - 1$ is expressed in the form

$$(x-1)(Ax^2 + Bx + C).$$

Factorize: (i) $x^3 + 1$,   (ii) $x^3 - 8$,   (iii) $x^3 + 27$,
          (iv) $8x^2 - 27$,   (v) $27x^3 + 125$.

7. Express $x^3 + 1$ in the form $x(x-1)(x-2) + Ax(x-1) + Bx + C$.

8. Find the values of a and b if $x^4 + 12x^3 + 46x^2 + ax + b$ is the square of a quadratic expression.

9. Write down the quadratic equation whose roots are α, β. If the same equation may also be written $ax^2 + bx + c = 0$, express $\alpha + \beta$ and $\alpha\beta$ in terms of a, b, c.

10. If α, β, γ are the roots of the equation $px^3 + qx^2 + rx + s = 0$, deduce expressions for $\alpha + \beta + \gamma$, $\beta\gamma + \gamma\alpha + \alpha\beta$, $\alpha\beta\gamma$ in terms of p, q, r, s.

11. Show that the condition for

$$lx + my + n = 0 \quad \text{and} \quad \frac{x}{a}\cos\phi + \frac{y}{b}\sin\phi = 1$$

to represent the same straight line is $a^2 l^2 + b^2 m^2 = n^2$.

Type I—denominator with only linear factors

3.3. We shall find that in the more straightforward cases which we have to deal with at this stage, partial fractions fall into three main types; each will be illustrated by a worked example, and the reader is strongly advised to work through the questions following each of these, before going on to consider the next type.

In practice, a question of considerable length and complexity may depend upon the correct determination of partial fractions in the early stages; to avoid fruitless labour at a later date, the habit of *checking* partial fractions should be firmly established from the start, and they should be thrown back into one fraction mentally, the numerator obtained being checked with the original.

First we deal with a fraction whose denominator consists of only linear factors.

Example 2. *Express* $\dfrac{11x+12}{(2x+3)(x+2)(x-3)}$ *in partial fractions.*

Let $\dfrac{11x+12}{(2x+3)(x+2)(x-3)} \equiv \dfrac{A}{2x+3} + \dfrac{B}{x+2} + \dfrac{C}{x-3}$,

where A, B, C are constants to be found. It follows that

$$\dfrac{11x+12}{(2x+3)(x+2)(x-3)} \equiv \dfrac{A(x+2)(x-3)+B(x-3)(2x+3)+C(2x+3)(x+2)}{(2x+3)(x+2)(x-3)},$$

$$\therefore \ 11x+12 \equiv A(x+2)(x-3)+B(x-3)(2x+3)+C(2x+3)(x+2).$$

Putting $x=3$,

$$33+12 = \qquad 0 \qquad + \qquad 0 \qquad +C.9.5,$$
$$\therefore \ C = 1.$$

Putting $x=-2$,

$$-22+12 = \qquad 0 \qquad + \ B(-5)(-1) + \qquad 0,$$
$$\therefore \ -10 = \qquad 5B,$$
$$\therefore \ B = -2.$$

Putting $x=-\tfrac{3}{2}$,

$$-\tfrac{33}{2}+12 = A\left(\tfrac{1}{2}\right)\left(\dfrac{-9}{2}\right) \ + \qquad 0 \qquad + \qquad 0,$$

$$\therefore \ -\tfrac{9}{2} = \qquad \dfrac{-9A}{4},$$

$$\therefore \ A = 2.$$

$$\therefore \ \dfrac{11x+12}{(2x+3)(x+2)(x-3)} \equiv \dfrac{2}{2x+3} - \dfrac{2}{x+2} + \dfrac{1}{x-3}.$$

$$\left[\text{Since the R.H.S.} \equiv \dfrac{2(x+2)(x-3)-2(x-3)(2x+3)+(2x+3)(x+2)}{(2x+3)(x+2)(x-3)},\right.$$

we check the coefficients in the numerator.$\Big]$

Check: Coefficient of $x^2 = 2-4+2 = 0$.
Coefficient of $x = -2+6+7 = 11$.
Constant term $= -12+18+6 = 12$.

Qu. 7. Express in partial fractions:

(i) $\dfrac{6}{(x+3)(x-3)}$, (ii) $\dfrac{x}{(2+x)(2-x)}$, (iii) $\dfrac{x-1}{3x^2-11x+10}$,

(iv) $\dfrac{3x+1}{(x+2)(x+1)(x-3)}$, (v) $\dfrac{3-4x}{2+3x-2x^2}$.

Type II—denominator with a quadratic factor

3.4. Fractions which can be split solely into partial fractions are necessarily *proper*, by which is meant that *the degree of the numerator is less than the degree of the denominator*.† Moreover, the partial fractions themselves are always proper.

Bearing this in mind we can now discover how to deal with a fraction having in the denominator a quadratic factor which does not factorize

Let $$\frac{3x+1}{(x-1)(x^2+1)} \equiv \frac{A}{x-1} + \frac{\text{`}numerator\text{'}}{x^2+1}.$$

Then $$3x+1 \equiv A(x^2+1) + \text{`}numerator\text{'} \times (x-1).$$

From our previous work on identities, we see, by equating coefficients that there are *three* equations to be satisfied. It follows that there are, *three* constants to determine,‡ and therefore the '*numerator*' must contain two of them; thus the only way to write the second partial fraction, so that it is proper, is in the form $\dfrac{Bx+C}{x^2+1}$.

Example 3. *Express* $\dfrac{3x+1}{(x-1)(x^2+1)}$ *in partial fractions.*

Let $$\frac{3x+1}{(x-1)(x^2+1)} \equiv \frac{A}{x-1} + \frac{Bx+C}{x^2+1}.$$
$$\therefore 3x+1 \equiv A(x^2+1) + (Bx+C)(x-1).$$

Putting $x=1$, $$4 = 2A+0,$$
$$\therefore A = 2.$$

Putting $x=0$, $$1 = A-C,$$
$$\therefore 1 = 2-C,$$
$$\therefore C = 1.$$

---

† With an improper fraction, we divide first, and we obtain a quotient and partial fractions; thus

$$\frac{x^2+x+1}{(x-1)(x+2)} \equiv 1 + \frac{3}{(x-1)(x+2)} \equiv 1 + \frac{1}{x-1} - \frac{1}{x+2},$$

and $$\frac{x^3+2x^2-7x-18}{x^2-9} \equiv x+2 + \frac{2x}{(x+3)(x-3)} \equiv x+2 + \frac{1}{x+3} + \frac{1}{x-3}.$$

‡ In general, the number of constants to be found is the same as the degree of the denominator of the original fraction.

Equating coefficients of x^2,

$$0 = A + B,$$
$$\therefore B = -2.$$

$$\therefore \frac{3x+1}{(x-1)(x^2+1)} \equiv \frac{2}{x-1} + \frac{1-2x}{x^2+1} \equiv \frac{2}{x-1} - \frac{2x-1}{x^2+1}.$$

Check: Coefficient of $x^2 = 2-2 = 0.$
Coefficient of $x = -(-2-1) = +3.$
Constant term $= 2-1 = +1.$

Qu. 8. Express in partial fractions:

(i) $\dfrac{6-x}{(1-x)(4+x^2)}$,

(ii) $\dfrac{4}{(x+1)(2x^2+x+3)}$,

(iii) $\dfrac{5x+2}{(x+1)(x^2-4)}$,

(iv) $\dfrac{3+2x}{(2-x)(3+x^2)}$.

Type III—denominator with a repeated factor

3.5. Here we take as an example $\dfrac{1}{(x+2)(x-1)^2}$. Written as $\dfrac{1}{(x+2)(x^2-2x+1)}$ this suggests Type II and the partial fractions $\dfrac{A}{x+2} + \dfrac{Bx+K}{x^2-2x+1}$. Certainly we have the correct number of constants to be found to identify this expression with the original fraction; however, the denominator of the second partial fraction factorizes, and so we have not gone far enough.

$$\frac{Bx+K}{(x-1)^2} \equiv \frac{B(x-1)+B+K}{(x-1)^2},$$
$$\equiv \frac{B}{x-1} + \frac{B+K}{(x-1)^2}.$$

Writing C for $B+K$, we obtain

$$\frac{Bx+K}{(x-1)^2} \equiv \frac{B}{x-1} + \frac{C}{(x-1)^2}.$$

This indicates the appropriate form when we have a repeated factor. (See also Qu. 10.)

Example 4. *Express* $\dfrac{1}{(x+2)(x-1)^2}$ *in partial fractions.*

Let $\dfrac{1}{(x+2)(x-1)^2} \equiv \dfrac{A}{x+2} + \dfrac{B}{x-1} + \dfrac{C}{(x-1)^2}.$

$\therefore \dfrac{1}{(x+2)(x-1)^2} \equiv \dfrac{A(x-1)^2 + B(x-1)(x+2) + C(x+2)}{(x+2)(x-1)^2}$

$\therefore 1 \equiv A(x-1)^2 + B(x-1)(x+2) + C(x+2).$

Putting $x = -2$,

$$1 = 9A,$$
$$\therefore A = \tfrac{1}{9}.$$

Putting $x = 1$,

$$1 = 3C,$$
$$\therefore C = \tfrac{1}{3}.$$

Equating coefficients of x^2,

$$0 = A + B,$$
$$\therefore B = -\tfrac{1}{9}.$$

$$\therefore \frac{1}{(x+2)(x-1)^2} \equiv \frac{1}{9(x+2)} - \frac{1}{9(x-1)} + \frac{1}{3(x-1)^2}.$$

Check: Expressing the R.H.S. as a single fraction with denominator $(x+2)(x-1)^2$, the numerator is

$$\tfrac{1}{9}\{(x-1)^2 - (x+2)(x-1) + 3(x+2)\}.$$

Coefficient of $x^2 = \tfrac{1}{9}(1-1) = 0.$
Coefficient of $x = \tfrac{1}{9}(-2-1+3) = 0.$
Constant term $= \tfrac{1}{9}(1+2+6) = 1.$

Qu. 9. Express in partial fractions:

(i) $\dfrac{x+1}{(x+3)^2}$,   (ii) $\dfrac{2x^2 - 5x + 7}{(x-2)(x-1)^2}.$

Qu. 10. Find the values of A, B, C, D, if

$$\frac{x^3 - 10x^2 + 26x + 3}{(x+3)(x-1)^3} \equiv \frac{A}{x+3} + \frac{B}{x-1} + \frac{C}{(x-1)^2} + \frac{D}{(x-1)^3}.$$

Improper fractions

3.6. As already implied, an *improper* fraction is one whose *numerator is of degree equal to, or greater than, that of the denominator*. To deal with this we first divide the numerator by the denominator to obtain a quotient and a proper fraction, and then split the latter into partial fractions.   Thus

$$\frac{x^4-2x^3-x^2-4x+4}{(x-3)(x^2+1)} \equiv x+1+\frac{x^2-2x+7}{(x-3)(x^2+1)}, \quad \text{etc.}$$

Often, instead of doing long division, it is quicker to proceed as follows:

$$\frac{2x^2+1}{(x-1)(x+2)} \equiv \frac{2(x^2+x-2)-2x+5}{x^2+x-2} \equiv 2+\frac{5-2x}{(x-1)(x+2)}, \quad \text{etc.}$$

Qu. 11. Express the following in the form of a quotient and a proper fraction:

(i) $\dfrac{x^3+2x^2-2x+2}{(x-1)(x+3)}$ (by long division),

(ii) $\dfrac{3x^2-2x-7}{(x-2)(x+1)}$ (by the short method suggested above).

Qu. 12. Express in partial fractions:

(i) $\dfrac{x^2-7}{(x-2)(x+1)}$,

(ii) $\dfrac{x^3-x^2-4x+1}{x^2-4}$.

Exercise 3b

Express in partial fractions:

1. (i) $\dfrac{x-11}{(x+3)(x-4)}$,

(ii) $\dfrac{x}{25-x^2}$,

(iii) $\dfrac{3x^2-21x+24}{(x+1)(x-2)(x-3)}$,

(iv) $\dfrac{4x^2+x+1}{x(x^2-1)}$,

(v) $\dfrac{8x^2+13x+6}{(x+2)(2x+1)(3x+2)}$,

(vi) $\dfrac{2x^3+x^2-15x-5}{(x+3)(x-2)}$.

2. (i) $\dfrac{5x^2-10x+11}{(x-3)(x^2+4)}$,

(ii) $\dfrac{2x^2-x+3}{(x+1)(x^2+2)}$,

(iii) $\dfrac{3x^2-2x+5}{(x-1)(x^2+5)}$,

(iv) $\dfrac{11x}{(2x-3)(2x^2+1)}$,

(v) $\dfrac{20x+84}{(x+5)(x^2-9)}$,

(vi) $\dfrac{2x^3-x-1}{(x-3)(x^2+1)}$.

3. (i) $\dfrac{x-5}{(x-2)^2}$,

(ii) $\dfrac{5x+4}{(x-1)(x+2)^2}$,

(iii) $\dfrac{5x^2+2}{(3x+1)(x+1)^2}$,

(iv) $\dfrac{x^4+3x-1}{(x+2)(x-1)^2}$.

4. (i) $\dfrac{3x^3+x+1}{(x-2)(x+1)^3}$ (see Qu. 10),

(ii) $\dfrac{3x^2+2x-9}{(x^2-1)^2}$.

5. (i) $\dfrac{x^3 + 2x^2 - 10x - 9}{x^2 - 9}$, (ii) $\dfrac{3(x^2 - 3)}{(x-1)(x+2)}$,

(iii) $\dfrac{2x^4 - 4x^3 - 42}{(x-2)(x^2+3)}$, (iv) $\dfrac{x^4 - 6x^2 + 3}{x(x+1)^2}$.

6. $\dfrac{3x + 7}{x(x+2)(x-1)}$. **7.** $\dfrac{3}{x^2(x+2)}$. **8.** $\dfrac{2x^4 - 17x - 1}{(x-2)(x^2+5)}$.

9. $\dfrac{68 + 11x}{(3+x)(16-x^2)}$. **10.** $\dfrac{2x+1}{x^3-1}$. **11.** $\dfrac{2x^2 + 39x + 12}{(2x+1)^2(x-3)}$.

12. $\dfrac{x+4}{6x^2 - x - 35}$. **13.** $\dfrac{x-2}{x^2(x-1)^2}$. **14.** $\dfrac{7x+2}{125x^3 - 8}$.

15. $\dfrac{x^2 + 2x + 18}{x(x^2+3)^2}$. **16.** $\dfrac{1}{x^4 + 5x^2 + 6}$. **17.** $\dfrac{1}{x^4 - 9}$.

Summation of series

3.7. An introduction to the summation of series was given in P.M.I., Chapter 10. There are some series which may be summed by the use of partial fractions; the method of application is illustrated in the following example.

Example 5. (*i*) *Express* $\dfrac{2}{n(n+1)(n+2)}$ *in partial fractions, and*

(*ii*) *deduce that*

$$\frac{1}{1.2.3} + \frac{1}{2.3.4} + \cdots + \frac{1}{n(n+1)(n+2)} = \frac{1}{4} - \frac{1}{2(n+1)(n+2)}.$$

(i) Let

$$\frac{2}{n(n+1)(n+2)} \equiv \frac{A}{n} + \frac{B}{n+1} + \frac{C}{n+2}.$$
$$\therefore \ 2 \equiv A(n+1)(n+2) + B(n+2)n + Cn(n+1).$$

Putting $n = 0$, $2 = 2A$, $\therefore \ A = 1$.
Putting $n = -1$, $2 = -B$, $\therefore \ B = -2$.
Putting $n = -2$, $2 = 2C$, $\therefore \ C = 1$.

$$\therefore \ \frac{2}{n(n+1)(n+2)} \equiv \frac{1}{n} - \frac{2}{n+1} + \frac{1}{n+2}.$$

Check: Coefficient of $n^2 = 1 - 2 + 1 = 0$.
Coefficient of $n = 3 - 4 + 1 = 0$.
Constant term $= 2$.

(ii) If
$$S \equiv \frac{1}{1.2.3} + \frac{1}{2.3.4} + \ldots + \frac{1}{n(n+1)(n+2)},$$
$$2S \equiv \frac{2}{1.2.3} + \frac{2}{2.3.4} + \ldots + \frac{2}{n(n+1)(n+2)}.$$

[From Part (i) it follows that

$$2S \equiv \left(\frac{1}{1} - \frac{2}{2} + \frac{1}{3}\right) + \left(\frac{1}{2} - \frac{2}{3} + \frac{1}{4}\right) + \left(\frac{1}{3} - \frac{2}{4} + \frac{1}{5}\right) + \ldots + \left(\frac{1}{n} - \frac{2}{n+1} + \frac{1}{n+2}\right).$$

We see that the majority of terms when grouped three together in a different way, such as $\frac{1}{3} - \frac{2}{3} + \frac{1}{3}$, have zero sum. We then have to pick out those terms which remain at the beginning and at the end, and this is most easily done if we set out the working in columns.]

From Part (i)

$$\frac{2}{1.2.3} = \frac{1}{1} - \frac{2}{2} + \frac{1}{3}$$

$$\frac{2}{2.3.4} = \frac{1}{2} - \frac{2}{3} + \frac{1}{4}$$

$$\frac{2}{3.4.5} = \frac{1}{3} - \frac{2}{4} + \frac{1}{5}$$

$$\ldots \ldots \ldots \ldots \ldots \ldots$$

$$\frac{2}{(n-2)(n-1)n} = \frac{1}{n-2} - \frac{2}{n-1} + \frac{1}{n}$$

$$\frac{2}{(n-1)n(n+1)} = \frac{1}{n-1} - \frac{2}{n} + \frac{1}{n+1}$$

$$\frac{2}{n(n+1)(n+2)} = \frac{1}{n} - \frac{2}{n+1} + \frac{1}{n+2}$$

Adding,
$$2S = \frac{1}{2} - \frac{1}{n+1} + \frac{1}{n+2},$$
$$= \frac{1}{2} - \frac{1}{(n+1)(n+2)},$$
$$\therefore S = \frac{1}{4} - \frac{1}{2(n+1)(n+2)}.$$

The reader should also note that as $n \to \infty$, $\frac{1}{2(n+1)(n+2)} \to 0$; thus the infinite series $\frac{1}{1.2.3} + \frac{1}{2.3.4} + \frac{1}{3.4.5} + \ldots$ is *convergent*, and its *sum to infinity* is $\frac{1}{4}$. (See P.M.I., § 11.4.)

Qu. 13. Show that $\dfrac{1}{1.2}+\dfrac{1}{2.3}+\dfrac{1}{3.4}+\ldots+\dfrac{1}{n(n+1)} = 1-\dfrac{1}{n+1}$.

Integration

3.8. We have already shown in § 3.1 how partial fractions may be applied to integration. Two more examples follow.

Example 6. *Find* $\displaystyle\int \dfrac{2x-1}{(x+1)^2}\,dx$.

Let $\dfrac{2x-1}{(x+1)^2} \equiv \dfrac{A}{x+1}+\dfrac{B}{(x+1)^2}$;

we find that $A=2$, $B=-3$.

$$\therefore \int \frac{2x-1}{(x+1)^2}\,dx = \int\left\{\frac{2}{x+1}-\frac{3}{(x+1)^2}\right\}\,dx,$$
$$= 2\log_e(x+1)+3(x+1)^{-1}+c.$$

Qu. 14. Find (i) $\displaystyle\int \dfrac{1}{x^2-9}\,dx$, (ii) $\displaystyle\int \dfrac{2x+2}{(2x-3)^2}\,dx$.

Qu. 15. (i) Find $\displaystyle\int \dfrac{x}{4-x^2}\,dx$ without using partial fractions.

(ii) Find this integral using partial fractions.

Example 7. *Evaluate* $\displaystyle\int_2^3 \dfrac{5+x}{(1-x)(5+x^2)}\,dx$ *correct to three significant figures.*

Let $\dfrac{5+x}{(1-x)(5+x^2)} \equiv \dfrac{A}{1-x}+\dfrac{Bx+C}{5+x^2}$;

we find that $A=1$, $B=1$, $C=0$.

$$\therefore \int_2^3 \frac{5+x}{(1-x)(5+x^2)}\,dx = \int_2^3 \left\{\frac{1}{1-x}+\frac{x}{5+x^2}\right\}\,dx,$$
$$= \left[-\log_e(x-1)+\tfrac{1}{2}\log_e(5+x^2)\right]_2^3,$$
$$= (-\log_e 2+\tfrac{1}{2}\log_e 14)-(-\log_e 1+\tfrac{1}{2}\log_e 9),$$
$$= \tfrac{1}{2}\log_e 14-\log_e 2-\log_e 3,$$
$$= \tfrac{1}{2}(\log_e 10+\log_e 1{\cdot}4)-\log_e 6,$$
$$\simeq \tfrac{1}{2}(2{\cdot}3026+0{\cdot}3365)-1{\cdot}7918,$$
$$\simeq -0{\cdot}4722,$$
$$= -0{\cdot}472 \text{ (correct to three significant figures).}$$

Example 7 also revises an important point. If $x < 1$,

$$\int \frac{1}{1-x}\, dx = -\log_e (1-x) + c;$$

however, if $x > 1$, as the limits show to be the case here,

$$\int \frac{1}{1-x}\, dx = -\log_e (x-1) + c. \qquad \text{(See § 2.12.)}$$

Qu. 16. Can $\int_0^2 \frac{5+x}{(1-x)(5+x^2)}\, dx$ be evaluated?

Qu. 17. Evaluate (i) $\int_2^3 \frac{x-4}{(x+2)(x-1)}\, dx$, (ii) $\int_1^2 \frac{3x^2+2x+2}{(x+1)(x^2+2)}\, dx$.

Exercise 3c

1. Express $\dfrac{2}{n(n+2)}$ in partial fractions, and deduce that

$$\frac{1}{1.3} + \frac{1}{2.4} + \frac{1}{3.5} + \ldots + \frac{1}{n(n+2)} = \frac{3}{4} - \frac{2n+3}{2(n+1)(n+2)}.$$

2. Express $\dfrac{n+3}{(n-1)n(n+1)}$ in partial fractions, and deduce that

$$\frac{5}{1.2.3} + \frac{6}{2.3.4} + \frac{7}{3.4.5} + \ldots + \frac{n+3}{(n-1)n(n+1)} = 1\tfrac{1}{2} - \frac{n+2}{n(n+1)}.$$

3. For the series given in No. 2 write down (i) the nth term, (ii) the sum of the first n terms, (iii) the limit of this sum as $n \to \infty$.

4. Prove that the series $\dfrac{2}{1.2} + \dfrac{2}{2.3} + \dfrac{2}{3.4} + \ldots$ is convergent, and find its sum to infinity.

5. Find the sum of the first n terms of the following series:

(i) $\dfrac{1}{1.4} + \dfrac{1}{2.5} + \dfrac{1}{3.6} + \ldots$,

(ii) $\dfrac{1}{2.4} + \dfrac{1}{4.6} + \dfrac{1}{6.8} + \ldots$,

(iii) $\dfrac{1}{3.6} + \dfrac{1}{6.9} + \dfrac{1}{9.12} + \ldots$,

(iv) $\dfrac{1}{2.6} + \dfrac{1}{4.8} + \dfrac{1}{6.10} + \ldots$,

(v) $\dfrac{1}{1.3.5} + \dfrac{1}{2.4.6} + \dfrac{1}{3.5.7} + \ldots$,

(vi) $\dfrac{1}{3.4.5} + \dfrac{2}{4.5.6} + \dfrac{3}{5.6.7} + \ldots$.

6. Find the sum of the first n terms of the following series, remembering that $2n-1$, $2n+1$, etc. are odd for all integral values of n:

(i) $\dfrac{2}{1.3}+\dfrac{2}{3.5}+\dfrac{2}{5.7}+\cdots,$

(ii) $\dfrac{1}{1.3.5}+\dfrac{1}{3.5.7}+\dfrac{1}{5.7.9}+\cdots,$

(iii) $\dfrac{2}{1.3.5}+\dfrac{3}{3.5.7}+\dfrac{4}{5.7.9}+\cdots.$

7. Find the following integrals:

(i) $\displaystyle\int\dfrac{1}{x(x-2)}\,dx,$ (ii) $\displaystyle\int\dfrac{1}{(x+3)(5x-2)}\,dx,$

(iii) $\displaystyle\int\dfrac{7x+2}{3x^3+x^2}\,dx,$ (iv) $\displaystyle\int\dfrac{x}{16-x^2}\,dx,$

(v) $\displaystyle\int\dfrac{1}{x^2-4x-5}\,dx,$ (vi) $\displaystyle\int\dfrac{x-2}{x^2-4x-5}\,dx,$

(vii) $\displaystyle\int\dfrac{2x^2+2x+3}{(x+2)(x^2+3)}\,dx,$ (viii) $\displaystyle\int\dfrac{22-16x}{(3+x)(2-x)(4-x)}\,dx,$

(ix) $\displaystyle\int\dfrac{4x-33}{(2x+1)(x^2-9)}\,dx,$ (x) $\displaystyle\int\dfrac{5x+2}{(x-2)^2(x+1)}\,dx,$

(xi) $\displaystyle\int\dfrac{x^2-8x+5}{(2x+1)(x^2+9)}\,dx,$ (xii) $\displaystyle\int\dfrac{6-9x}{27x^3+8}\,dx,$

(xiii) $\displaystyle\int\dfrac{x^3-18x-21}{(x+2)(x-5)}\,dx,$ (xiv) $\displaystyle\int\dfrac{37}{4(x-3)(1+4x^2)}\,dx.$

8. (It is intended that all the parts of this question should be answered at one sitting, in order to bring out the comparison between the forms.) Find the following integrals:

(i) $\displaystyle\int\dfrac{1}{1+x^2}\,dx,$ (ii) $\displaystyle\int\dfrac{x}{1+x^2}\,dx,$ (iii) $\displaystyle\int\dfrac{1+x}{1+x^2}\,dx,$

(iv) $\displaystyle\int\dfrac{1}{1-x^2}\,dx,$ (v) $\displaystyle\int\dfrac{x}{1-x^2}\,dx,$ (vi) $\displaystyle\int\dfrac{x}{\sqrt{(1-x^2)}}\,dx,$

(vii) $\displaystyle\int\dfrac{1}{\sqrt{(1-x^2)}}\,dx,$ (viii) $\displaystyle\int\dfrac{1+x}{\sqrt{(1-x^2)}}\,dx,$

(ix) $\displaystyle\int\dfrac{1}{1-x}\,dx,\quad(x<1),$ (x) $\displaystyle\int\dfrac{1}{1-x}\,dx,\quad(x>1),$

(xi) $\displaystyle\int\dfrac{x}{1+x}\,dx,$ (xii) $\displaystyle\int\dfrac{1}{(1-x)^2}\,dx,$ (xiii) $\displaystyle\int\dfrac{x}{(1-x)^2}\,dx.$

9. Evaluate the following, correct to two significant figures:

(i) $\int_3^5 \dfrac{2}{x^2-1}\,dx$,

(ii) $\int_{-1}^0 \dfrac{2}{(1-x)(1+x^2)}\,dx$,

(iii) $\int_2^3 \dfrac{x-9}{x(x-1)(x+3)}\,dx$,

(iv) $\int_0^3 \dfrac{13x+7}{(x-4)(3x^2+2x+3)}\,dx$.

10. Find the volume of the solid generated when the area under $y=\dfrac{1}{x-2}$ from $x=3$ to $x=4$ is rotated through four right angles about the x-axis. If the solid is made of material of uniform density, where is its centre of gravity?

CHAPTER 4

THE BINOMIAL THEOREM

The expansion of $(1+x)^n$ when n is not a positive integer

4.1. The binomial theorem that

$$(1+x)^n = 1+nx+\frac{n(n-1)}{2!}x^2+\frac{n(n-1)(n-2)}{3!}x^3+\dots$$

provided $-1<x<+1$, was used in Chapter 11 of P.M.I., but the general term in the expansion was not discussed for values of n other than positive integers. The term in x^r is found to be

$$\frac{n(n-1)\dots(n-r+1)}{r!}x^r.$$

The proof of this, for values of n other than positive integers, is outside the scope of this book. In the case when n is a positive integer, the term in x^r has been shown (P.M.I., p. 243) to be

$$^nC_rx^r = \frac{n!}{(n-r)!\,r!}x^r.$$

Dividing numerator and denominator by $(n-r)!$, we obtain

$$\frac{n(n-1)\dots(n-r+1)}{r!}x^r.$$

Note that $n(n-1)\dots(n-r+1)$ contains r factors. If the reader can remember that this expression is $n!/(n-r)!$ (when n is a positive integer), it may help him to remember that the last factor, $n-r+1$, is 1 greater than $n-r$.

Example 1. *Find the general terms in the expansions in ascending powers of x of:* (i) $(1+x)^{-1}$, (ii) $(1-2x)^{-3}$.

(i) $(1+x)^{-1}$. The general term is

$$(-1)(-2)\dots(-r)x^r/r!$$
$$= (-1)^r 1.2\dots rx^r/r!$$
$$= (-1)^r x^r.$$

(ii) $(1-2x)^{-3}$. The general term is

$$(-3)(-4)\ldots(-2-r).(-2x)^r/r!$$
$$= (-1)^r 3.4\ldots(r+2).(-1)^r 2^r x^r/r!$$
$$= \frac{r!(r+1)(r+2)}{2}.\frac{2^r x^r}{r!},$$
$$= (r+1)(r+2)2^{r-1}x^r.$$

The expansion obtained in the first part of the last example,

$$(1+x)^{-1} = 1-x+x^2-\ldots+(-1)^r x^r+\ldots,$$

is frequently required and should be memorized. Note that the right-hand side is an infinite geometrical progression with first term 1 and common ratio $-x$, therefore its sum to infinity is $1/(1+x)$, provided $-1<x<+1$. (See P.M.I., p. 232.)

Qu. 1. Write down and simplify the general terms in the expansions of:

(i) $(1+x)^{-2}$,   (ii) $(1-3x)^{-1}$,   (iii) $(1-\frac{1}{2}x)^{-3}$,   (iv) $(1+x)^{-4}$.

It is worth noting that the coefficients in the expansions of $(1-x)^{-1}$, $(1-x)^{-2}$, $(1-x)^{-3},\ldots$ are contained in Pascal's triangle (P.M.I., p. 238).

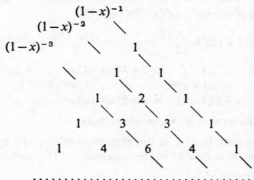

Example 2. *Find the first three terms and the general term in the expansion in ascending powers of x of*

$$\frac{x+5}{(1+3x)(2-x)}.$$

Expressed in partial fractions,

$$\frac{x+5}{(1+3x)(2-x)} = \frac{2}{1+3x}+\frac{1}{2-x}.$$

(The reader should verify that this is so.)

$$2(1+3x)^{-1} = 2\{1-3x+9x^2-\ldots+(-1)^r(3x)^r+\ldots\},$$
$$= 2-6x+18x^2-\ldots+(-1)^r.2(3x)^r+\ldots.$$
$$(2-x)^{-1} = 2^{-1}(1-\tfrac{1}{2}x)^{-1},$$
$$= \tfrac{1}{2}\{1+\tfrac{1}{2}x+\tfrac{1}{4}x^2+\ldots+(\tfrac{1}{2}x)^r+\ldots\},$$
$$= \tfrac{1}{2}+\tfrac{1}{4}x+\tfrac{1}{8}x^2+\ldots+(\tfrac{1}{2})^{r+1}x^r+\ldots.$$

Therefore the sum of the two expansions is

$$2\tfrac{1}{2}-5\tfrac{3}{4}x+18\tfrac{1}{8}x^2+\ldots+\{(-1)^r.2.3^r+(\tfrac{1}{2})^{r+1}\}x^r+\ldots.$$

For the expansion to be valid,

$$-1 < 3x < +1 \quad \text{and} \quad -1 < -\tfrac{1}{2}x < 1.$$

Multiplying the pairs of inequalities by $\tfrac{1}{3}$ and -2 respectively,

$$-\tfrac{1}{3} < x < +\tfrac{1}{3} \quad \text{and} \quad 2 > x > -2.†$$

Therefore the expansion is valid when $-\tfrac{1}{3} < x < +\tfrac{1}{3}$. This is more briefly written $|x| < \tfrac{1}{3}$. $|x|$ is called the *modulus* of x and, in real algebra, is the numerical value of x. Thus $|4| = |-4| = +4$.

Example 3. *Find the first three terms and the term in x^r in the expansion in ascending powers of x of $(x+2)(1+x)^{12}$.*

$$(1+x)^{12} = 1+12x+\frac{12.11}{2!}x^2+\ldots+{}^{12}C_{r-1}x^{r-1}+{}^{12}C_rx^r+\ldots+x^{12}.$$

$$\therefore\ 2(1+x)^{12} = 2+24x+132x^2+\ldots+2x^{12},$$
and $$x(1+x)^{12} = \qquad x+12x^2+66x^3+\ldots+x^{13}.$$
Adding, $$(x+2)(1+x)^{12} = 2+25x+144x^2+\ldots+x^{13}.$$

To find the term in x^r, we must multiply

the term in x^r in the expansion of $(1+x)^{12}$ by 2 and
the term in x^{r-1} in the expansion of $(1+x)^{12}$ by x.

Thus the term in x^r is

$$2.{}^{12}C_rx^r+x.{}^{12}C_{r-1}x^{r-1},$$
$$= \left\{\frac{2.12!}{(12-r)!r!}+\frac{12!}{(13-r)!(r-1)!}\right\}x^r,$$
$$= \frac{12!}{(12-r)!(r-1)!}\left(\frac{2}{r}+\frac{1}{13-r}\right)x^r,$$
$$= \frac{12!(26-r)}{(13-r)!r!}x^r.$$

† When an inequality is multiplied by a negative number, the direction of the inequality sign is reversed. (See § 7.1, p. 122.)

Qu. 2. Find the terms in x^r in the expansions in ascending powers of x of:

(i) $(1-x)(1+x)^{20}$,   (ii) $(2x+3)(1-x)^{10}$.

Qu. 3. Find the general term in the expansion of $(2x-1)/(1+x)^2$ (i) by the method of Example 3, (ii) by expressing the function in partial fractions.

Example 4. *Sum to infinity the series*:

$$(i) \quad 1 - \frac{3}{2}x + \frac{3.9}{2.4}x^2 - \frac{3.9.15}{2.4.6}x^3 + \ldots,$$

$$(ii) \quad \frac{1}{4} - \frac{1.2}{4.8} + \frac{1.2.5}{4.8.12} - \frac{1.2.5.8}{4.8.12.16} + \ldots.$$

(i) Let $\quad S_1 = 1 - \frac{3}{2}x + \frac{3.9}{2.4}x^2 - \frac{3.9.15}{2.4.6}x^3 + \ldots.$

Note that there is a factor of 3^r in the numerator and a factor of 2^r in the denominator of the term in x^r.

$$\therefore S_1 = 1 - \frac{1}{2}(3x) + \frac{1.3}{2^2}\frac{(3x)^2}{2!} - \frac{1.3.5}{2^3}\frac{(3x)^3}{3!} + \ldots,$$

$$= 1 + (-\tfrac{1}{2})(3x) + (-\tfrac{1}{2})(-\tfrac{3}{2})\frac{(3x)^2}{2!} + (-\tfrac{1}{2})(-\tfrac{3}{2})(-\tfrac{5}{2})\frac{(3x)^3}{3!} + \ldots.$$

$$\therefore S_1 = (1+3x)^{-\frac{1}{2}}, \text{ provided } |x| < \tfrac{1}{3}.$$

(ii) Let $\quad S_2 = \frac{1}{4} - \frac{1.2}{4.8} + \frac{1.2.5}{4.8.12} - \frac{1.2.5.8}{4.8.12.16} + \ldots.$

Here the denominators are $4^r . r!$

$$\therefore S_2 = \frac{1}{4} - \frac{1.2}{4^2}\cdot\frac{1}{2!} + \frac{1.2.5}{4^3}\cdot\frac{1}{3!} - \frac{1.2.5.8}{4^4}\cdot\frac{1}{4!} + \ldots.$$

By altering some signs, we can arrange that the factors in the numerators become terms of an arithmetical progression.

$$\therefore S_2 = \frac{1}{4} + \frac{1(-2)}{4^2}\cdot\frac{1}{2!} + \frac{1(-2)(-5)}{4^3}\cdot\frac{1}{3!} + \frac{1(-2)(-5)(-8)}{4^4}\cdot\frac{1}{4!} + \ldots.$$

$$\therefore S_2 = \tfrac{1}{3}\cdot\frac{3}{4} + \frac{\tfrac{1}{3}(-\tfrac{2}{3})}{2!}\cdot\frac{3^2}{4^2} + \frac{\tfrac{1}{3}(-\tfrac{2}{3})(-\tfrac{5}{3})}{3!}\cdot\frac{3^3}{4^3} + \frac{\tfrac{1}{3}(-\tfrac{2}{3})(-\tfrac{5}{3})(-\tfrac{8}{3})}{4!}\cdot\frac{3^4}{4^4} + \ldots.$$

The series has a sum to infinity since $-1 < \tfrac{3}{4} < +1$.

$$\therefore 1 + S_2 = (1 + \tfrac{3}{4})^{\frac{1}{3}}.$$
$$\therefore S_2 = \sqrt[3]{\tfrac{7}{4}} - 1.$$

The expansion of $(1+x)^n$ when $|x|>1$

4.2. It will be recalled that the expansion of $(1+x)^n$ in ascending powers of x is only valid for $|x| < 1$. If, however, $|x| > 1$, the function can be expanded in ascending powers of $1/x$.

$$(1+x)^n = \{x(1+x^{-1})\}^n = x^n(1+x^{-1})^n.$$

When $|x| > 1$, it follows that $|x^{-1}| < 1$, so that we may write

$$(1+x)^n = x^n \left\{ 1 + n\left(\frac{1}{x}\right) + \frac{n(n-1)}{2!}\left(\frac{1}{x}\right)^2 + \dots \right.$$
$$\left. + \frac{n(n-1)\dots(n-r+1)}{r!}\left(\frac{1}{x}\right)^r + \dots \right\}.$$

Qu. 4. Expand the following in ascending powers of $\frac{1}{x}$, giving the ranges of values of x for which the expansions are valid:

  (i) $(1+x)^{-1}$, (ii) $(2+x)^{-2}$, (iii) $(1+3x)^{-2}$,

  (iv) $\dfrac{3}{(x-1)(x-2)}$. (v) $\dfrac{x}{x^2+1}$.

Exercise 4a

1. Find the first three terms and the general terms in the expansions of the following functions in ascending powers of x. State the ranges of values of x for which the expansions are valid.

  (i) $(1+3x)^{-1}$, (ii) $(1-2x)^{-1}$, (iii) $(1+x)^{-2}$,
  (iv) $(1-\frac{1}{2}x)^{-2}$, (v) $(1+x)^{-3}$, (vi) $(2+x)^{-1}$,
  (vii) $\dfrac{1}{(3-x)^2}$, (viii) $\dfrac{1}{(2-3x)^3}$, (ix) $\sqrt{(1+x)}$.

2. Express the following functions in partial fractions and find the first three terms and the general terms in their expansions in ascending powers of x. For what values of x are the expansions valid?

  (i) $\dfrac{3}{(1-x)(1+2x)}$, (ii) $\dfrac{1}{(1+x)(x+2)}$, (iii) $\dfrac{x-1}{x^2+2x+1}$,
  (iv) $\dfrac{5}{1-x-6x^2}$, (v) $\dfrac{x+3}{(x-2)^2}$, (vi) $\dfrac{x+2}{x^2-1}$.

3. Expand the following functions in ascending powers of x, giving the first three terms and the general term, and state the necessary restrictions on the values of x:

  (i) $\dfrac{1}{1+x^2}$, (ii) $\dfrac{x}{1-x^2}$, (iii) $\dfrac{1-x}{1+x}$,

    (iv) $(1+x)(1-x)^{10}$,   (v) $\dfrac{4}{(x+3)(1+x)}$,

    (vi) $\dfrac{x+5}{(3-2x)(x-1)}$,   (vii) $\dfrac{x+7}{(x+1)^2(x-2)}$.

4. Expand the following functions in ascending powers of $1/x$, giving the first three terms and the general terms. State the necessary restrictions on the values of x.

    (i) $(2+x)^{-1}$,   (ii) $(3-x)^{-3}$,   (iii) $(1-2x)^{-2}$,

    (iv) $\dfrac{x+2}{x+1}$,   (v) $\dfrac{x-1}{(x+2)^2}$,   (vi) $\dfrac{1}{x^2-5x+6}$,

    (vii) $\dfrac{2x+4}{(x-1)(x+3)}$,   (viii) $\dfrac{2x}{1-x^2}$,   (ix) $\dfrac{1}{1-x+x^2-x^3}$.

5. Expand $(x-2)^{\frac{1}{2}}$ as a series of descending powers of x as far as the third term. By substituting $x=100$, evaluate $\sqrt{2}$ to five significant figures. [HINT: $\sqrt{98}=7\sqrt{2}$.]

6. Obtain $\sqrt[3]{2}$ to five places of decimals by substituting $x=1000$ in the expansion of $(x+24)^{\frac{1}{3}}$ in descending powers of x.

In Nos. 7 to 10, use the binomial expansion to find the values of:

7. $(16\cdot32)^{\frac{1}{4}}$ to five places of decimals.

8. $\sqrt{9\cdot09}$ to six places of decimals.

9. $\dfrac{1}{(10\cdot04)^2}$ to four significant figures.

10. $\dfrac{1}{\sqrt{17}}$ to four places of decimals.

11. Expand the function $(1+2x)^{\frac{1}{2}}(1-3x)^{-\frac{1}{3}}$ in a series of ascending powers of x as far as the term in x^2. [HINT: multiply the first three terms of the expansion of $(1+2x)^{\frac{1}{2}}$ by those of $(1-3x)^{-\frac{1}{3}}$, ignoring terms in x^3 and higher powers of x.]

In Nos. 12 to 17 expand the functions in series of ascending powers of x as far as the terms indicated.

12. $\dfrac{1}{1+x+2x^2}$, (x^3).   [Write $y=x+2x^2$.]

13. $\dfrac{1}{(1+2x+3x^2)^2}$, (x^3).

14. $\sqrt[3]{(1+3x)}\sqrt{(1+2x)}$, (x^2).

15. $\dfrac{\sqrt[3]{(1-x)^2}}{1+x}$, (x^3).

16. $\dfrac{x}{1 - \sqrt{(1 + 2x)}}$, (x^2). [Expand the denominator; then divide numerator and denominator by x.]

17. $\dfrac{x^2}{1 - \sqrt[3]{(1 - 3x)}}$, (x^3).

18. The field H on the axis of a bar magnet of moment M at a distance d from its centre is approximately $2M/d^3$. Suppose that in calculating the value of H, values of M and d differ by $\pm 2\%$ and $\pm 1\%$ respectively. What is the greatest possible percentage error in calculating the value of H?

19. If a clock with a seconds pendulum registers x s too few per day, what is the time of one beat of the pendulum?

One beat of a seconds pendulum takes $\pi(l/g)^{\frac{1}{2}}$ s, where l is the length of the pendulum and g is a constant. If the length of the pendulum increases by 0.04% owing to expansion, calculate the number of seconds it will have failed to register in a day.

20. If a pendulum beats seconds (see No. 19) at a place where $g = 981$ cm/s^2 and is then removed to a place where g is 0.05% less, how many seconds will it have failed to register in a day?

21. The heat H produced by an electric current flowing through a resistance R with potential difference V for a time t is given by $H = JV^2t/R$, where J is a constant. If V, t, R are given percentage increases x, y, z which are so small that the squares and products of x, y, z may be neglected, find the percentage increase in the value of H.

22. The period of oscillation T of a vibration magnetometer is given by the formula

$$T = 2\pi\sqrt{\dfrac{I}{MH}}.$$

If the quantities I, M, H are estimated with errors of p, q, r per cent, respectively, find the corresponding percentage error in T if the squares and products of p, q, r may be neglected.

Sum to infinity the series in Nos. 23–37 stating the necessary restrictions on the value of x.

23. $1 - \frac{1}{2} + \frac{1}{4} - \frac{1}{8} + \ldots$.

24. $1 - x + \dfrac{1.3}{1.2}x^2 - \dfrac{1.3.5}{1.2.3}x^3 + \dfrac{1.3.5.7}{1.2.3.4}x^4 - \ldots$.

25. $1 + 4x + 12x^2 + \ldots + (n+1)2^n x^n + \ldots$.

26. $1 - \dfrac{2}{3} + \dfrac{2.3}{3.6} - \dfrac{2.3.4}{3.6.9} + \ldots$.

27. $1 + \dfrac{1}{6} - \dfrac{1.2}{6.12} + \dfrac{1.2.5}{6.12.18} - \dfrac{1.2.5.8}{6.12.18.24} + \ldots$

28. $1 - 6x + 24x^2 - \ldots + (-1)^n(n+1)(n+2)2^{n-1}x^n + \ldots$

29. $1 - x - \dfrac{x^2}{2!} - 1.3\,\dfrac{x^3}{3!} - 1.3.5\,\dfrac{x^4}{4!} - \ldots$

30. $1 + \dfrac{1}{4} + \dfrac{1.4}{4.8} + \dfrac{1.4.7}{4.8.12} + \ldots$

31. $1 + 2x - 2.1\,\dfrac{x^2}{2!} + 2.1.4\,\dfrac{x^3}{3!} - 2.1.4.7\,\dfrac{x^4}{4!} + \ldots$

32. $1 + \dfrac{1}{2} - \dfrac{1}{2.4} + \dfrac{1.3}{2.4.6} - \dfrac{1.3.5}{2.4.6.8} + \ldots$

33. $1 - \dfrac{2}{5} + \dfrac{2.5}{5.10} - \dfrac{2.5.8}{5.10.15} + \ldots$

34. $1 + \dfrac{1}{3} - \dfrac{1}{3.6} + \dfrac{1.3}{3.6.9} - \dfrac{1.3.5}{3.6.9.12} + \ldots$

35. $\dfrac{3}{10} + \dfrac{3.5}{10.15} + \dfrac{3.5.7}{10.15.20} + \dfrac{3.5.7.9}{10.15.20.25} + \ldots$

36. $\dfrac{4}{18} + \dfrac{4.7}{18.27} + \dfrac{4.7.10}{18.27.36} + \ldots$

37. $-\dfrac{1}{6} + \dfrac{1.3}{6.9} - \dfrac{1.3.5}{6.9.12} + \dfrac{1.3.5.7}{6.9.12.15} - \ldots$

38. Find the quadratic equation whose roots are

$$-2p + \frac{1}{2}\frac{q}{p} + \frac{1}{8}\frac{q^2}{p^3} + \ldots + \frac{1.3\ldots(2r-3)}{2^r\,r!}\frac{q^r}{p^{2r-1}} + \ldots,$$

$$-\frac{1}{2}\frac{q}{p} - \frac{1}{8}\frac{q^2}{p^3} - \ldots - \frac{1.3\ldots(2r-3)}{2^r\,r!}\frac{q^r}{p^{2r-1}} - \ldots,$$

where $|q| < |p^2|$.

39. Show that

$$1 - \frac{1}{2}\sin^2\theta - \frac{1}{8}\sin^4\theta - \ldots - \frac{1.3\ldots(2r-3)}{2^r\,r!}\sin^{2r}\theta - \ldots,$$

$$= \quad 1 - \frac{1}{2}\tan^2\theta + \frac{3}{8}\tan^4\theta - \ldots + (-1)^r\frac{1.3.5\ldots(2r-1)}{2^r\,r!}\tan^{2r}\theta + \ldots,$$

if $|\theta - n\pi| < \frac{1}{4}\pi$, where n is an integer.

Relations between binomial coefficients

4.3. We first show how the greatest coefficient in a binomial expansion may be found. A similar method may be applied to find the greatest term.

Example 5. *Find the greatest coefficient in the expansion of* $(2x+3)^{12}$.

The coefficient of x^r,

$$u_r = \frac{12!}{(12-r)!\,r!}\, 2^r 3^{12-r},$$

and the coefficient of x^{r+1},

$$u_{r+1} = \frac{12!}{(11-r)!(r+1)!}\, 2^{r+1} 3^{11-r}.$$

The ratio of these coefficients,

$$\frac{u_r}{u_{r+1}} = \frac{r+1}{12-r}\cdot\frac{3}{2}.$$

Therefore $u_r < u_{r+1}$ if

$$\frac{r+1}{12-r}\cdot\frac{3}{2} < 1.$$
$$\therefore\; 3r+3 < 24-2r.\dagger$$
$$\therefore\; 5r < 21.$$
$$\therefore\; r < \tfrac{21}{5} = 4\tfrac{1}{5}.$$

That is, $u_1 < u_2$, $u_2 < u_3$, $u_3 < u_4$, $u_4 < u_5$, but $u_5 \not< u_6$. Therefore the coefficient of x^5 is the largest. Its value is $792.2^5.3^7$.

Qu. 5. Find the greatest *term* in the above expansion when $x = 2$.

Now some series involving the binomial coefficients will be considered. For brevity we shall write

$$(1+x)^n = c_0 + c_1 x + c_2 x^2 + \ldots + c_n x^n.$$

So far we have not assigned any meaning to $^nC_0$. Since, in general, $^nC_r = c_r$, it is most convenient to *define* $^nC_0 = 1$; and, if we *define* $0! = 1$, we can write

† $12-r$ is positive so the inequality sign is unchanged.

$$c_r = {}^nC_r = \frac{n!}{(n-r)!r!}$$

for all values of r from 0 to n.

It should be noted that c_r is only used for ${}^nC_r$. Other coefficients such as ${}^{n-1}C_r$ and ${}^{2n}C_r$ will not be abbreviated in this way.

Example 6. *Find the values of:*

      (i) $c_0 + c_1 + \ldots + c_n$,

      (ii) $c_0 - 2c_1 + 3c_2 - \ldots + (-1)^n(n+1)c_n$,

      (iii) $\frac{1}{2}c_0 + \frac{1}{3}c_1 + \frac{1}{4}c_2 + \ldots + c_n/(n+2)$.

(i) $(1+x)^n = c_0 + c_1 x + \ldots + c_n x^n$.

Substituting $x = 1$,

$$c_0 + c_1 + \ldots + c_n = 2^n.$$

(ii) Remember that $d(x^n)/dx = nx^{n-1}$.

$$x(1+x)^n = c_0 x + c_1 x^2 + c_2 x^3 + \ldots + c_n x^{n+1}.$$

Differentiating with respect to x,

$$(1+x)^n.1 + x.n(1+x)^{n-1} = c_0 + 2c_1 x + 3c_2 x^2 + \ldots + (n+1)c_n x^n.$$

Substituting $x = -1$,

$$c_0 - 2c_1 + 3c_2 - \ldots + (-1)^n(n+1)c_n = 0.$$

(iii) Remember that $\int x^n \, dx = x^{n+1}/(n+1) + k$.

$$(1+x)^n = c_0 + c_1 x + c_2 x^2 + \ldots + c_n x^n.$$
$$\therefore \ x(1+x)^n = c_0 x + c_1 x^2 + c_2 x^3 + \ldots + c_n x^{n+1}.$$

Integrating the right-hand side with respect to x between 0 and 1, we obtain

$$\tfrac{1}{2}c_0 + \tfrac{1}{3}c_1 + \tfrac{1}{4}c_2 + \ldots + c_n/(n+2).$$

For the left-hand side, we write

$$x(1+x)^n = \{(1+x)-1\}(1+x)^n.$$
$$\int_0^1 x(1+x)^n \, dx = \int_0^1 \{(1+x)^{n+1} - (1+x)^n\} \, dx,$$
$$= \left[\frac{(1+x)^{n+2}}{n+2} - \frac{(1+x)^{n+1}}{n+1}\right]_0^1,$$
$$= \frac{2^{n+2}-1}{n+2} - \frac{2^{n+1}-1}{n+1},$$

which is the sum of the series.

Alternatively, the integral could have been evaluated by the substitution $u = 1 + x$.

Certain relations between the binomial coefficients may be obtained by equating coefficients (see p. 47). For example, the identity

$$(1+x)^{n+2} \equiv (1+x)^2(1+x)^n$$

may be expanded in two different ways:

(i) $\qquad 1 + {}^{n+2}C_1 x + \ldots + {}^{n+2}C_r x^r + \ldots + {}^{n+2}C_{n+2} x^{n+2},$

(ii) $\qquad (1 + 2x + x^2)(c_0 + c_1 x + \ldots + c_r x^r + \ldots + c_n x^n).$

The term in x^r in (ii) is obtained by multiplying $c_r x^r$ by 1, $c_{r-1} x^{r-1}$ by $2x$, $c_{r-2} x^{r-2}$ by x^2. Equating coefficients of x^r in the expansions (i) and (ii),

$$^{n+2}C_r = c_r + 2c_{r-1} + c_{r-2} \quad (2 \leqslant r \leqslant n).$$

Qu. 6. What relations are obtained by equating coefficients of

(i) x^{r+1}, $\qquad$ (ii) x^{r+2}?

Example 7. *Prove that*

$$c_0{}^2 + c_1{}^2 + \ldots + c_n{}^2 = {}^{2n}C_n.$$

[The expression ${}^{2n}C_n$ suggests the use of $(1+x)^{2n}$, and the terms $c_r{}^2$ suggest the square of $(1+x)^n$.]

$$(1+x)^n(1+x)^n \equiv (1+x)^{2n}.$$
$$\therefore (c_0 + c_1 x + \ldots + c_n x^n)(c_0 + c_1 x + \ldots + c_n x^n) \equiv (1+x)^{2n}.$$

[${}^{2n}C_n$ is the coefficient of x^n on the right-hand side.]
    Equating coefficients of x^n,

$$c_0 c_n + c_1 c_{n-1} + \ldots + c_{n-1} c_1 + c_n c_0 = {}^{2n}C_n.$$

But $c_r = c_{n-r}$ (see P.M.I., p. 206),

$$\therefore c_0{}^2 + c_1{}^2 + \ldots + c_{n-1}^2 + c_n{}^2 = {}^{2n}C_n.$$

Exercise 4b

1. Find the greatest coefficients in the binomial expansions of the following:

(i) $(x+2)^{10}$,	(ii) $(3x+1)^8$,	(iii) $(4x+3)^{12}$,
(iv) $(2x+5)^{20}$,	(v) $(x+\frac{2}{3})^{11}$,	(vi) $(3x-2)^9$,
(vii) $(12-11x)^{-2}$,	(viii) $(7-5x)^{-3}$.	

2. Find the greatest terms in the binomial expansions of:

(i) $(2x+3y)^{12}$,   when $x = 1$, $y = 3$;

(ii) $(x+2y)^{10}$,   when $x = \frac{1}{2}$, $y = \frac{1}{3}$;

(iii) $(4x+5y)^8$,   when $x = \frac{1}{3}$, $y = \frac{1}{2}$;

(iv) $(3x-5)^{-2}$,   when $x = 1\frac{1}{2}$.

Prove that:

3. $c_0 - c_1 + c_2 - \ldots + (-1)^n c_n = 0$.

4. $c_1 + 2c_2 + 3c_3 + \ldots + nc_n = n \cdot 2^{n-1}$.

5. $c_0 + 2c_1 + 3c_2 + \ldots + (n+1)c_n = 2^{n-1}(n+2)$.

6. $c_1 - 2c_2 + 3c_3 - \ldots + (-1)^{n+1}nc_n = 0$.

7. $2 . 1c_2 + 3 . 2c_3 + \ldots + n(n-1)c_n = n(n-1)2^{n-2}$.

8. $2c_1 - 6c_2 + \ldots + (-1)^{n+1}n(n+1)c_n = 0$.

9. $1^2 c_1 + 2^2 c_2 + \ldots + n^2 c_n = n(n+1)2^{n-2}$.

10. $1^2 c_0 + 2^2 c_1 + \ldots + (n+1)^2 c_n = (n+1)(n+4)2^{n-2}$.

11. $\frac{1}{2}c_0 - \frac{1}{3}c_1 + \ldots + (-1)^n c_n/(n+2) = 1/\{(n+1)(n+2)\}$.

12. $c_0 + \frac{1}{2}c_1 + \frac{1}{3}c_2 + \ldots + c_n/(n+1) = (2^{n+1}-1)/(n+1)$.

13. $\frac{1}{2}c_0 - \frac{1}{6}c_1 + \ldots + (-1)^n c_n/(n+1)(n+2) = 1/(n+2)$.

14. $2c_0 + \frac{3}{2}c_1 + \ldots + \frac{n+2}{n+1}c_n = \frac{2^{n+1}+(n+1)2^n-1}{n+1}$.

15. $^{n+1}C_r = c_r + c_{r-1}, \quad (1 \leqslant r \leqslant n)$.

16. $^{n+3}C_r = c_r + 3c_{r-1} + 3c_{r-2} + c_{r-3}, \quad (3 \leqslant r \leqslant n)$.

17. $c_0{}^2 - c_1{}^2 + \ldots + (-1)^n c_n{}^2 = (-1)^{\frac{1}{2}n}n!/\{(\frac{1}{2}n)!\}^2$ if n is even and is zero if n is odd.

18. $c_0\, {}^{2n}C_0 + c_1\, {}^{2n}C_1 + \ldots + c_n\, {}^{2n}C_n = {}^{3n}C_n$.

19. $c_1{}^2 + 2c_2{}^2 + \ldots + nc_n{}^2 = (2n-1)!/\{(n-1)!\}^2$.

20. $c_0{}^2 + 2c_1{}^2 + \ldots + (n+1)c_n{}^2 = \dfrac{(2n-1)!(n+2)}{n!(n-1)!}$.

21. $c_0{}^2 + \frac{1}{2}c_1{}^2 + \ldots + c_n{}^2/(n+1) = (2n+1)!/\{(n+1)!\}^2$.

Exercise 4c (Miscellaneous)

1. Express the function

$$\frac{1+2x+3x^2}{(1-x)(1+x^2)}$$

in partial fractions.

    If x is so small that powers higher than the third may be neglected, expand the function in the form $A + Bx + Cx^2 + Dx^3$.   N.

2. Find numbers A, B, and C such that the fraction

$$\frac{2x}{(1-x)(1+x^2)} \quad \text{is equal to} \quad \frac{A}{1-x} + \frac{B+Cx}{1+x^2}.$$

Hence obtain the expansion of the fraction in ascending powers of x as far as x^5. Between what values must x lie in order that this expansion may be valid?

N.

3. (i) Without using tables, find the value of
$$\frac{(\sqrt{5}+2)^6-(\sqrt{5}-2)^6}{8\sqrt{5}}.$$

(ii) Expand $(1-3x)^{⅓}$ in ascending powers of x as far as the term in x^3. By taking $x=\frac{1}{8}$, evaluate $\sqrt[3]{5}$ correct to two decimal places. c.

4. Express in partial fractions
$$\frac{x^2+2x+8}{x^3(x+2)}.$$

Hence express in partial fractions
$$\frac{y^2+7}{(y-1)^3(y+1)},$$

and expand this expression in ascending powers of y as far as y^2, stating for what values of y this expansion is valid.

c.

5. (i) Express
$$\frac{7x+3}{(3x-1)(x+1)^2}$$

in partial fractions, and hence find the coefficient of x^n when this expression is expanded in ascending powers of x.

(ii) Write down the first three terms of the binomial expansion of $(1-\frac{1}{1000})^{⅓}$.

Hence evaluate $(37)^{⅓}$ to six decimal places.

o.c.

6. Write down and simplify the first three terms in the binomial expansions of $(1+x)^{½}$ and $(1+x)^{-½}$.

AB is a chord, of length $2ka$, of a circle of radius a. The tangents to the circle at A and B meet in C. Show that, if k is so small compared with unity that k^7 is negligible, the area of the triangle ABC is
$$a^2k^3+\tfrac{1}{2}a^2k^5.$$

L.

7. Use the binomial theorem to evaluate $\dfrac{1}{9\cdot84}-\dfrac{1}{9\cdot85}$ correct to four significant figures.

8. Prove that, if x is so small that its cube and higher powers can be neglected,
$$\sqrt{\frac{1+x}{1-x}} = 1+x+\tfrac{1}{2}x^2.$$

By taking $x=\frac{1}{9}$, prove that $\sqrt{5}$ is approximately equal to $\frac{181}{81}$. c.

9. (i) Find the percentage increase in the value of x^2y^4/z when the percentage increases in x, y, z are p, q, r, respectively, if the squares and higher powers of p, q, r can be neglected.

(ii) Obtain the first four terms in the expansion of

$$\frac{1+x}{(1-x)^3},$$

(a) in a series of ascending powers of x,

(b) in a series of descending powers of x.　　L.

10. (a) By means of the binomial theorem evaluate $(10 \cdot 02)^{10}$ to the nearest thousand.

(b) Write down the expansion of $(1-x)^{-2}$ in ascending powers of x and deduce that

$$\sum_{n=0}^{\infty} (a+bn)x^n = \frac{a+(b-a)x}{(1-x)^2}, \quad \text{when } |x| < 1.$$

Prove that

$$\sum_{n=0}^{\infty} \frac{8n+1}{3^{2n}} = \frac{9}{4}.$$　　N.

11. Use the binomial series to write down the first four terms of the expansion of $(1+y)^{-\frac{1}{2}}$ in a series of ascending powers of y.

Hence find, in terms of $\cos \theta$, the coefficients c_1, c_2, c_3 in the expansion of $(1-2x \cos \theta + x^2)^{-\frac{1}{2}}$ in the form $1 + c_1 x + c_2 x^2 + c_3 x^3 + \ldots$.

Prove that, when $\theta = 0$, every coefficient in the series is equal to $+1$.

[You may assume throughout that the expansions are valid.]　　N.

12. (i) If x is small compared with unity and powers of x higher than x^6 are neglected, show that

$$\frac{1}{1+x^2} = 1 - x^2 + x^4 - x^6.$$

Prove that the error in this approximation is less than x^8.

(ii) Expand $\sqrt{(1+x)}$ as far as the term in x^3.

(iii) Find the coefficients of x^{2n} and x^{2n+1} in the expansion in ascending powers of $(1-x)^2/(1+x)^2$.　　L.

13. Write down the series for $\sqrt{(1+x)}$ in ascending powers of x as far as the term in x^4.

Show also that the error in taking

$$\tfrac{1}{4}(6+x) - \frac{1}{2+x}$$

as an approximation to $\sqrt{(1+x)}$ when x is small is approximately $x^4/128$.　　C.

14. Prove (do not merely verify) that, if E denotes the function

$$\frac{x^2}{2-x+2\sqrt{(1-x)}},$$

then $\qquad\qquad E = 2-x-2\sqrt{(1-x)}.$

Deduce that, if x is small, E is approximately equal to $\frac{1}{4}x^2$.

O.C.

15. Express

$$\frac{1}{(x+2)^2(2x+1)}$$

in partial fractions, and hence expand the expression as a series in ascending powers of x, giving the first four terms and the coefficient of x^n.

Show that, for values of x so small that x^4 may be neglected, the given expression can be represented by

$$\frac{1}{(3x+2)^2}+kx^3$$

for some number k independent of x, and find k.

O.C.

16. Write down, without proof, the binomial expansion for $\sqrt{(1-2x)}$ in ascending powers of x, giving the first three terms and the general term.

Prove that the sum of the first two terms exceeds $\sqrt{(1-2x)}$ by exactly

$$\frac{x^2}{1-x+\sqrt{(1-2x)}}.$$

By putting $x=0.005$, obtain from the first two terms of the expansion an approximation for $\sqrt{11}$, and determine to how many places of decimals your approximation is correct.

C.

17. Show that, if x is so small in comparison with unity that x^3 and higher powers can be neglected,

$$\frac{(1-4x)^{\frac{1}{2}}(1+3x)^{\frac{1}{3}}}{(1+x)^{\frac{1}{2}}} = 1-\tfrac{3}{2}x-\tfrac{33}{8}x^2.$$

L.

18. (a) Find and simplify the term independent of x in the binomial expansion of

$$\left(x^2-\frac{1}{2x}\right)^9.$$

(b) Write down and simplify the first four terms in the expansion in ascending powers of x of $(1+3x)^{\frac{1}{3}}$. Hence evaluate $\sqrt[3]{1.03}$ correct to five places of decimals.

N.

19. (i) Find the values of r and n if, in the expansion of $(1+x)^n$ in ascending powers of x, the $(r+1)$th coefficient is twice the rth and the $(r+10)$th is twice the $(r+11)$th.

(ii) If $\qquad (1+x)^n = c_0 + c_1 x + c_2 x^2 + \ldots + c_n x^n$,

and $\qquad (1+x)^{2n} = C_0 + C_1 x + C_2 x^2 + \ldots + C_{2n} x^{2n}$,

obtain the value of

$$c_0 C_r + c_1 C_{r+1} + c_2 C_{r+2} + \ldots + c_n C_{r+n}$$

where $0 < r \leqslant n$. **L.**

20. Show that for any positive integral value of n there are two values of a such that the coefficients of the powers of x in the three middle terms of the expansion of $(1+ax)^{2n}$ are in arithmetical progression. Show also that the product of these values of a is independent of n. **L.**

21. If c_r is the coefficient of x^r in the binomial expansion of $(1+x)^n$, where n is a positive integer, prove that

(i) $c_0 + c_2 + c_4 + \ldots = c_1 + c_3 + c_5 + \ldots = 2^{n-1}$,

(ii) $c_0^2 + c_1^2 + c_2^2 + \ldots + c_n^2 = (2n)!/(n!)^2$,

(iii) $c_0 + 2c_1 + 3c_2 + \ldots + (n+1)c_n = (n+2)2^{n-1}$. **L.**

22. (i) Find which is the greatest term in the expansion of $(3+2x)^{14}$ in ascending powers of x when $x = 5/2$.

(ii) Prove that three consecutive terms in the expansion of $(1+x)^n$, where n is a positive integer, can never be in geometrical progression. **L.**

23. (a) If the expansion of $(1+x)^n$ in ascending powers of x is denoted by

$$(1+x)^n = c_0 + c_1 x + c_2 x^2 + c_3 x^3 + \ldots + c_r x^r + \ldots,$$

write down the values of c_0, c_1, c_2, c_3 in terms of n, and the value of c_r in terms of n and r. State the range of values of x for which the expansion is valid whatever the value of n.

If $n > -1$, determine the range of values of r for which

$$\left| \frac{c_r}{c_{r-1}} \right| < 1.$$

(b) Use the binomial expansion to calculate the value of

$$(1 + \tfrac{1}{1000})^{200}$$

correct to four places of decimals. **N.**

24. Given that

$$(1 - 2x + 5x^2 - 10x^3)(1+x)^n \equiv 1 + a_1 x + a_2 x^2 + a_3 x^3 + \ldots,$$

and that $a_1^2 = 2a_2$, find the value of n.

What is the value of a_3? **O.C.**

25. (a) Show that, when x is small, the expansion in powers of x of the function

$$(1+x)^p + (1-x)^p - 2(1+x^2)^q$$

is of the form

$$a_2 x^2 + a_4 x^4 + a_6 x^6 + \ldots.$$

If $a_2 = 0$, find q in terms of p.   If, in addition, $a_4 = 0$ and p is not equal to 0, 1, or 2, find the values of p, q, a_6.

(b) Show that, when x is large and positive,

$$(x^2 + x)^{1/2} + (x^2 + 3x)^{1/2} = 2x + 2, \quad \text{approximately.} \qquad \text{N.}$$

CHAPTER 5

PROBABILITY

Introduction

5.1. It is difficult to define exactly what is meant by the word 'probability', but in everyday life we are quite accustomed to hearing someone say that two events are 'equally likely'. By this is usually meant that the speaker does not see any reason to expect one event rather than the other. This may appear to be a negative way of approaching the idea of probability but by starting from the conception of a number of events being equally likely we can obtain some interesting results.

Consider the word BEHAVIOUR. Suppose that the letters are written on identical pieces of card and placed in a box which is shaken to mix them up. Three of the letters are taken from the box by someone who is unable to see which he is taking. What is the probability that there is a vowel among the letters selected?

Three letters can be selected from nine in $^9C_3$ ways. Three consonants can be chosen from the nine letters in $^4C_3$ ways. Therefore there are $^9C_3 - {}^4C_3$ ways in which three letters including at least one vowel may be selected.

Since each of the $^9C_3$ ways of selecting the letters is equally likely, we may say that the probability of having a vowel among the letters chosen is

$$\frac{^9C_3 - {}^4C_3}{^9C_3} = \frac{84-4}{84} = \frac{20}{21}.$$

In general, the probability p of an occurrence or event which may or may not take place in a number of equally likely possibilities may be taken to be

$$p = \frac{\text{number of possibilities in which the event occurs}}{\text{total number of possibilities}}.$$

If q is the probability that the event does not take place, then

$$q = \frac{\text{number of possibilities in which the event does not take place}}{\text{total number of possibilities}}.$$

79

But the event either does or does not take place, so that

$$p+q = 1.$$

Qu. 1. Write down the probability of:
  (i) throwing a six with a die,
  (ii) drawing an Ace from a pack of 52 playing cards,
  (iii) obtaining a prime number when choosing at random a number from
       10, 11, . . . , 19, 20.

Now consider spinning a coin. We assume that, whenever a coin is spun, a 'head' and a 'tail' are equally likely. Since there are two equally likely alternatives, 'head' and 'tail', it follows that the probability of obtaining the event 'head' is $\frac{1}{2}$. It should be stressed that this is so whatever results may have been obtained in previous spins.

In an even number of spins, there is no reason why we should expect more 'heads' than 'tails' and we should expect the proportion of heads to be roughly 0·5.

The author spun some coins when preparing to write this chapter and these results are instructive:

No. of spins	10	100	500
No. of heads	4	46	254
Proportions of heads	0·4	0·46	0·508.

These figures are typical of what may be expected. If 10 spins are made it is quite likely that five heads will be obtained, but it is more likely that there will not be exactly five (see Example 8, p. 86). However, if the number of spins is increased by a significant factor (say 4 or more), the proportion may be expected to come closer to 0·5. Thus, if the reader reported 7 'heads' in 10 spins, the author would not be at all surprised, but a proportion of 0·6 in 500 spins is very unlikely.

At the beginning of this chapter a probability was worked out using combinations. In the two examples which follow, different methods are used, but the reader should note that all three are basically the same in finding the probability from the ratio

$$\frac{\text{number of possibilities in which the event occurs}}{\text{total number of possibilities}}.$$

Example 1. *Two dice are thrown together. What is the probability of a score of eight?*

Each die may fall in six ways, therefore there are $6 \times 6 = 36$ possibilities altogether. Eight is scored in the following 5 ways:

First die	2	3	4	5	6
Second die	6	5	4	3	2

Therefore the probability of scoring eight is 5/36.

Example 2. *If the letters of* DISLOCATE *are arranged at random, what is the probability that the vowels are separated?*

The five consonants DSLCT can be arranged in 5! ways. But for each of these arrangements there are 6 'spaces' in which a vowel may be placed. (4 'spaces' between the letters and one at each end.) These 'spaces' may be filled in $^6P_4$ ways by the 4 vowels. Therefore there are $5! \times {}^6P_4$ arrangements in which the vowels are separated.

But 9 letters may be arranged in 9! ways. Therefore the probability that the vowels are separated is

$$\frac{5! \times {}^6P_4}{9!} = \frac{6.5.4.3}{9.8.7.6} = \frac{5}{42}.$$

Qu. 2. What is the probability that the E's in EMERGENCE are separated when the letters are arranged at random?

Example 3. *In a letter game, the 26 letters of the alphabet are written on separate cards and 10 cards are drawn at random. What is the probability that at least one vowel is included? (Y is not counted as a vowel.)*

Here it is easier to find the probability that there is *no* vowel on the 10 cards drawn.

10 cards with no vowel can be drawn from 21 cards with consonants in $^{21}C_{10}$ ways; but 10 cards can be drawn from 26 in $^{26}C_{10}$ ways. Therefore the probability of drawing 10 cards with no vowel is $^{21}C_{10}/{}^{26}C_{10}$. Hence the probability of drawing at least one vowel is $1 - {}^{21}C_{10}/{}^{26}C_{10}$.

Exercise 5a

1. In a pack of playing cards, what is the probability of drawing: (i) a club, (ii) an Ace, (iii) a court-card (King, Queen, Knave)?
2. With a die, what is the probability of throwing: (i) a six, (ii) an odd number, (iii) not a prime number?

3. When two dice are thrown, what is the probability of a score of: (i) six, (ii) eleven, (iii) two, (iv) not seven?

4. What is the probability of scoring (i) three, (ii) five, (iii) seven with three dice?

5. 'If I take a book and pick a letter at random with a pin, the chance that it is a vowel is 5/26.' Is this statement true or false? Give your reasons.

6. There are four routes between A and B, none of which intersect except at A and B. If one man starts from A and another from B by routes chosen at random, what is the probability that they meet?

7. Thirteen playing cards are dealt to a certain person. What is the probability that all are of the same suit?

8. The digits 1, 2, 3, 4, 5, 6, 7 are written on separate cards. Three are drawn at random, and placed in the order of drawing. What is the probability that the number so formed is greater than 500?

9. In a form list of twenty-five pupils, arranged alphabetically, what is the chance that the two youngest are separated?

10. Four letters are chosen at random from the word WATERING. Find the probability that two or more are vowels.

11. If the letters of TOGETHER are arranged at random, what is the chance that the T's are apart?

12. In a game of bridge, what is the probability of being dealt a hand entirely of honours cards (Ace, King, Queen, Knave, Ten)?

13. Nine people are arranged at random at a round table. What is the probability that two particular people are separated?

14. In a box of children's bricks there are five in each of four colours. If two are picked up at random, what is the chance that their colours are the same?

15. If I play bridge, what is the chance that I am dealt a hand containing cards of not more than three suits at the first deal?

16. In a letter game, cards with letters forming a word are turned face downward and arranged at random in a row. What is the chance that all the E's in PIECEMEAL are separated?

17. A man has three red dahlias, three yellow, two mauve, and two white. When he lifts them he puts them away mixed up together. In the spring he plants four in one bed. What is the probability that they will all be (i) the same colour, (ii) different colours.

18. In a certain country, the numerals in car registration marks range from 1 to 999. What is the probability that the first local car I see on visiting the country should have at least two digits the same in its registration mark?

19. A number is formed by taking three digits at random from the digits 1, 2, 3, 4, 5. What is the probability that the number is a multiple of three (i) if no digit may be repeated, (ii) if repetitions are allowed?

Mutually exclusive events

5.2. Suppose that one letter of the alphabet is chosen at random and consider the possibilities:

(A) that a vowel (a, e, i, o, u) is chosen; and
(B) that a labial (p, b, f, v, m, w) is chosen.

Then the probability of a vowel being chosen $p_A = 5/26$ and the probability that a labial is chosen $p_B = 6/26$. Further, the probability that either a vowel or a labial is chosen $p = 11/26$. Note that (i) a letter cannot be both a vowel and a labial, so that the possibilities A and B are *mutually exclusive*, and (ii) $p = p_A + p_B$.

In general, *if* A *and* B *are two mutually exclusive events with probabilities* p_A *and* p_B, *the probability* p *of either the event* A *or the event* B *is given by*

$$p = p_A + p_B.$$

To prove this, suppose that A and B may arise in n_A and n_B ways out of n equally likely possibilities, then

$$p = \frac{n_A + n_B}{n},$$

$$= \frac{n_A}{n} + \frac{n_B}{n},$$

$$\therefore \ p = p_A + p_B.$$

Qu. 3. Which of the following pairs of events are mutually exclusive?

(i) a number is (a) prime, (b) not prime,
(ii) a playing card is (a) a five, (b) a heart,
(iii) a boy has (a) got influenza, (b) has had influenza,
(iv) a number is (a) odd, (b) even,
(v) a number is (a) a multiple of three, (b) a multiple of five,
(vi) a playing card is (a) a five, (b) a court-card,
(vii) a tree is (a) evergreen, (b) deciduous,
(viii) a woman is (a) British, (b) American,
(ix) a letter is written in (a) the Greek alphabet, (b) the Roman alphabet,
(x) a vehicle (a) is a trailer, (b) has two wheels.

Qu. 4. Find the probability of cutting from a pack of cards, (i) a spade, (ii) a red card. *Deduce* the probability of cutting either a spade or a red card.

Qu. 5. Consider the three conditions:

(A) a number is three or a multiple of three,

(B) a number is five or a multiple of five,

(C) a number is seven or a multiple of seven.

Find the probabilities of a number chosen at random from the numbers 1, 2, ..., 19, 20 satisfying (i) A, (ii) B, (iii) C. Deduce the probability that such a number satisfies (iv) either A or C, (v) either B or C. Why can the probability for a number to satisfy either A or B not be found in this way?

Independent events

5.3. Next consider two events:

(1) drawing an Ace from a pack of playing cards,

(2) throwing a six with a die.

In the first case there are 4 Aces in the 52 cards of the pack, so that the probability of drawing an Ace $p_1 = 4/52$. The probability of throwing a six $p_2 = 1/6$.

Now suppose that a card is drawn and a die is thrown. What is the probability p that both an Ace is drawn and a six is thrown? For each of the 52 cards that may be drawn, the die may fall in 6 different ways, giving 6×52 possibilities. Of these, 4 give both an Ace and a six, so that the probability of this event $p = 4/(6 \times 52)$. Note that (i) the two events are *independent*, i.e. the occurrence of one does not affect the probability of the other event, and (ii) $p = p_1 \times p_2$.

In general, *if* (1) *and* (2) *are two independent events with probabilities* p_1 *and* p_2, *the probability* p *of both events taking place is given by*

$$p = p_1 p_2.$$

To prove this, suppose that (1) and (2) may arise in respectively n_1, n_2 ways out of N_1, N_2 equally likely possibilities. The total number of combinations of possibilities is $N_1 N_2$ and of these the events (1) and (2) both occur $n_1 n_2$ times. Therefore the probability of both events taking place

$$p = \frac{n_1 n_2}{N_1 N_2}.$$

$$\therefore p = p_1 p_2.$$

If p_3 is the probability of another event (3), since we have shown $p_1 p_2$ to be the probability of the event that (1) and (2) should both occur, it follows that the probability that all three events should take place is $(p_1 p_2)p_3 = p_1 p_2 p_3$. Clearly this may be extended to any number of independent events.

Example 4. *Two others and I play a game of pure chance three times. What is the probability of my winning (i) every time, (ii) only the third time?*

My chances of winning and losing any given game are $\frac{1}{3}$ and $\frac{2}{3}$, respectively. Since the events are independent, the probability of my winning every time is $\frac{1}{3} \times \frac{1}{3} \times \frac{1}{3} = \frac{1}{27}$.

Again, the probability that I should lose the first two and win the third is $\frac{2}{3} \times \frac{2}{3} \times \frac{1}{3} = \frac{4}{27}$.

The theorem preceding Example 4 may be extended to the case where p_1 is the probability of the first event and p_2 is the probability of the second *when the first has taken place*. A similar argument shows that the probability of both events taking place is again $p_1 p_2$, and this can clearly be extended to any number of events as before.

Example 5. *Find the probability of drawing first an Ace and then a King from a pack of cards.*

The probability of drawing an Ace is $\frac{4}{52} = \frac{1}{13}$. When an Ace has been drawn, the probability of drawing a King is $\frac{4}{51}$. Therefore the probability of drawing them in this order is $\frac{1}{13} \times \frac{4}{51} = \frac{4}{663}$.

Example 6. *In a large batch of seed, 75% of the plants which could be grown from it could have red flowers and the rest white. Find the probability that two plants grown from it should have flowers of the same colour.*

The probability that the first plant has red flowers is $\frac{3}{4}$. Since the batch of seed is large, the probability that the second plant has red flowers may also be taken to be $\frac{3}{4}$. Therefore the probability that both have red flowers is $\frac{3}{4} \times \frac{3}{4} = \frac{9}{16}$.

The probability that a plant has white flowers is $1 - \frac{3}{4} = \frac{1}{4}$. Therefore the probability that both plants have white flowers is $\frac{1}{4} \times \frac{1}{4} = \frac{1}{16}$.

But the two events above are mutually exclusive, therefore the probability that both plants have flowers the same colour is $\frac{9}{16} + \frac{1}{16} = \frac{5}{8}$.

Qu. 6. What is the probability that the plants in Example 6 have different coloured flowers?

Qu. 7. With the data of Example 6, if the batch was not large, but contained $4n$ seeds, find the probability that both plants should have flowers of the same colour.

Example 7. *If 5% of a consignment of eggs are bad, estimate the chance that half a dozen chosen at random contains at least one bad egg?*

The probability that the first egg is good is $1 - \frac{1}{20} = \frac{19}{20}$. If the consignment is large, the probability that each of the other five eggs is good will be very nearly $\frac{19}{20}$. Therefore the probability that all the eggs are good is $(\frac{19}{20})^6$. Hence the probability that at least one is bad is $1 - (\frac{19}{20})^6 \simeq 0.26$.

Example 8. *Ten coins are spun. What are the probabilities that (i) five heads, (ii) four or six heads are obtained?*

Suppose, for convenience, that the dates on the coins end with the digits $0, 1, \ldots, 9$. We shall refer to the coins by these digits.

(i) The probability that the coins 0, 1, 2, 3, 4 come down heads and 5, 6, 7, 8, 9 come down tails is $(\frac{1}{2})^{10}$.

But the five coins which come down heads may be selected in $^{10}C_5$ ways. Therefore the probability of obtaining five heads is

$$^{10}C_5(\tfrac{1}{2})^{10} = \tfrac{252}{1024} = \tfrac{63}{256}.$$

(ii) Suppose that coins 0, 1, 2, 3 come down heads and the others tails. The probability of this is $(\frac{1}{2})^{10}$. Now the coins that come down heads may be selected in $^{10}C_4$ ways, so that the probability of four heads is

$$^{10}C_4(\tfrac{1}{2})^{10} = \tfrac{210}{1024} = \tfrac{105}{512}.$$

The probability of six heads is the same as the chance of six tails, which has just been found. Therefore the probability of obtaining either four or six heads is $\frac{105}{256}$.

NOTE: It is more likely, in ten spins of a coin, that four or six heads will be obtained than exactly five.

Qu. 8. What is the probability of obtaining seven heads when ten coins are spun?

Example 9. *In a large constituency, 80% of the electorate voted Conservative. If I asked ten people, chosen at random in the constituency, which way they voted, what would be the chance that exactly eight voted Conservative?*

Suppose the people are A, B, C, D, E, F, G, H, I, J. Since the population is large compared with the number asked, it is fair to assume that the probability of a Conservative vote is $\frac{4}{5}$ for each person asked.

The probability that A, B, C, D, E, F, G, H voted conservative is $(\frac{4}{5})^8$, and the probability that I, J did not vote this way is $(\frac{1}{5})^2$. So the probability of both of these events is $(\frac{4}{5})^8 \times (\frac{1}{5})^2$.

Now there are $^{10}C_8$ mutually exclusive ways in which eight conservative and two non-conservative voters could be selected from A, B, C, D, E, F, G, H, I, J. Therefore the probability of exactly eight conservative voters from the ten people is $^{10}C_8(\frac{4}{5})^8(\frac{1}{5})^2 \simeq 0 \cdot 30$.

Exercise 5b

1. If I draw lots with three other people for three tickets, what is the chance that I get (i) all of them, (ii) only the last, (iii) just one ticket?

2. I throw three dice in succession. What is the probability that I obtain a one, a two and a three (i) in that order, (ii) in any order?

3. In a certain batch of seed, the chances of the plants having red, pink, white flowers are in the ratio 4:4:1. When three plants are raised, find the probability that (i) all have white flowers, (ii) none have white flowers, (iii) at least one has white flowers.

4. In the production of a certain precision instrument, 2%, 8%, 15% of three components are found to be defective. What is the probability that at least one of the parts taken for one instrument is defective?

5. In a certain strain of Aster, the chance that a seed produces pink flowers is $\frac{1}{4}$. How many seeds should I sow in order that the probability of obtaining at least one pink flower is more than 99%?

6. In Poker, a player is dealt five cards. What is the probability that the two top cards belong to one suit and the others belong to each of the other three suits?

   What is the probability that he receives the Ace, King, Queen, Knave, and Ten of one suit?

7. In a survey of opinion, people are for, against and non-committal with regard to a certain proposal in the ratio 3:4:2. Find the probability that (i) the first five people interviewed are against the proposal, (ii) the first two answers are different.

8. Smith, Jones, Brown, and Robinson hold amongst themselves a lottery with two prizes and they buy respectively 4, 3, 3, 2 tickets. Find the probability that (i) Smith receives both prizes, (ii) the prizes go to different people.

9. A and B are shooting at a target. The probabilities that they hit the target with a given shot are $\frac{1}{4}$, $\frac{1}{3}$ respectively; so B allows A 5 shots while he has 3. Which is more likely to obtain at least one hit?

10. When propagating a certain bush by cuttings, the chance that a cutting strikes is $\frac{2}{3}$. With eight cuttings, what is the chance of six or more striking?

11. Taking the probability that a child is a boy to be $\frac{1}{2}$, compare the probabilities of finding in a family of four children (i) two boys and two girls, (ii) one boy and three girls or three boys and one girl.

12. If I throw three dice together, what is the chance that I obtain no sixes or just one six?

13. Six people, each with one coin, spin them together and observe the number of heads obtained. Find the probabilities of each number of heads from 0 to 6.

14. With the data of No. 3, find the probability that three plants raised from this seed should have different coloured flowers.

15. With the probabilities of No. 9, C whose chance of a hit is $\frac{1}{2}$ joins in. They all have one shot. What is the probability that the target is hit after C joins?

16. One box contains six black balls and one white, and a second box contains six black balls. If three balls are taken from the first and placed in the second, and then three balls are taken from the second and placed in the first, what is the probability that the white ball is now in the first box?

17. A and B fight a duel with pistols, firing simultaneously. The chance that A kills B with any one shot is $\frac{1}{4}$ and the chance that B kills A is $\frac{2}{5}$. What is the chance that both are alive after the second round? What are their respective chances of survival in a duel to the death?

18. Three musketeers playing with two dice have a stake of three louis d'or on the table to be taken by the first to throw a double six. They are throwing in the order A, B, C when they are interrupted by a call to duty from which they may not all return. If B was just about to throw, how should they divide the stake?

The binomial probability distribution

5.4. Suppose that a man has five packs of cards and draws a card at random from each. What are the probabilities of his drawing 0, 1, 2, 3, 4, 5 court cards (King, Queen, Knave), respectively?

The probability of a success (drawing a court card) is 3/13 and the probability of a failure (not drawing a court card) is 10/13. Denoting a success by S and a failure by F, the probabilities of obtaining successes and failures in the orders indicated are as follows:

FFFFF	FFFFS	FFFSS	FFSSS	FSSSS	SSSSS
$(\frac{10}{13})^5$	$(\frac{10}{13})^4(\frac{3}{13})$	$(\frac{10}{13})^3(\frac{3}{13})^2$	$(\frac{10}{13})^2(\frac{3}{13})^3$	$(\frac{10}{13})(\frac{3}{13})^4$	$(\frac{3}{13})^5$

but the F's and S's may be rearranged in

| 1 | 5 | 10 | 10 | 5 | 1 |

ways, so that the probabilities of respectively

| 0 | 1 | 2 | 3 | 4 | 5 |

successes are

$(\frac{10}{13})^5$ $5(\frac{10}{13})^4(\frac{3}{13})$ $10(\frac{10}{13})^3(\frac{3}{13})^2$ $10(\frac{10}{13})^2(\frac{3}{13})^3$ $5(\frac{10}{13})(\frac{3}{13})^4$ $(\frac{3}{13})^5$

Expressed as decimals correct to 3 significant figures, they are:

| 0·269 | 0·404 | 0·242 | 0·073 | 0·011 | 0·001 |

Note that the probabilities are the terms of the binomial expansion of $(\frac{10}{13} + \frac{3}{13})^5$.

Qu. 9. Show that r S's and $(n-r)$ F's may be arranged in a row in ^nC_r ways.

Now consider the general case. Let the probability of an event occurring on a particular occasion (a success) be p and the probability of the event not taking place (a failure) be q. If n trials are made, the probability of $n-r$ failures followed by r successes is

$$q^{n-r}p^r \quad (r = 0, 1, \ldots, n).$$

But the $n-r$ failures and the r successes may be rearranged in ^nC_r mutually exclusive orders. Therefore the probability of r successes is

$$^nC_r q^{n-r}p^r \quad (r = 0, 1, \ldots, n).$$

Therefore the probabilities of $0, 1, \ldots, n$ successes are given by the terms of the binomial expansion of $(q+p)^n$ in ascending powers of p.

This probability distribution was first investigated in detail by James Bernouilli and given in a posthumous work published in 1713.

Exercise 5c (Miscellaneous)

1. A railway carriage has eight seats, four of them with their backs to the engine. In how many ways can a party of seven people seat themselves in a carriage (i) without restriction, (ii) if a chosen pair must sit together with their backs to the engine? c.

2. A touring cricket team of 15 men contains 5 regular bowlers.
   (i) Show that 450 different elevens can be picked which contain exactly 3 of the five regular bowlers.
   (ii) Find how many different elevens can be picked which contain *at least* 3 of the regular bowlers. c.

3. On each of two parallel lines n points are chosen. Find the number of triangles which have three of these $2n$ points for vertices.
   Find n if half these triangles have a given point for a vertex. L.

4. (i) Show that there are $^{10}C_4$ ways of arranging six similar white and four similar black marbles in line.

Find the number of ways of arranging four white, three black, and two red marbles in line, assuming that marbles of the same colour are indistinguishable. [A numerical answer is required.]

(ii) A bag contains six white and four black marbles. Find the chance that, if two marbles are drawn together, they are both black. c.

5. Find how many three-letter code words can be made with the 26 letters of the alphabet:

(i) when no letter is used more than once in the same word,

(ii) when any letter may be repeated two or three times in the same word,

(iii) when every word contains exactly one vowel (a, e, i, o, u) and no letter more than once. c.

6. (i) A man possesses 24 books, all different. In how many ways can he arrange 12 of them on a shelf? (Leave your answer in terms of factorials.)

(ii) A group of 6 men is chosen from 10 men. Find the probability that a certain 2 men are included in the group. L.

7. Find the number of different arrangements of all the letters of the word ESTEEM taken altogether.

If three letters are chosen at random from the word ESTEEM, find the probability that at least two of them are E's. L.

8. A box of chalks contains 5 white, 4 green, 2 red, 1 yellow. In how many different ways can they be arranged side by side?

Calculate also the probability that the 4 green chalks are together.

(Consider chalks of the same colour identical; leave your answer in factorials.) L.

9. It is required to seat n people at a round table. Show that there are $(n-1)!$ different possible arrangements. How many arrangements are possible if (i) two, (ii) three specified people are to sit together?

Use the results of (i) and (ii) to show that, if n people are given seats at random at a round table, the chance of two particular people sitting together is $\dfrac{2}{n-1}$ and the chance of a group of three particular people sitting together is $\dfrac{6}{(n-1)(n-2)}$. c.

10. Prove that the number of selections of a group of r things that can be made from n things, all of which are different, is

$$\frac{n!}{r!(n-r)!}.$$

A pack of 52 playing cards, all different, contains 20 honours. Show that a selection of three cards from the pack can be made in 22,100 different ways and that 4960 of these contain *no* honour. Deduce the probability that a particular selection of three cards contains *at least* one honour.   c.

11. An urn contains twelve balls, three of each of the colours red, green, blue, yellow. Three balls are drawn from the urn in succession, without replacement. What is the joint probability that the first is red, the second green or blue and the third yellow?

   If this event has happened and two further balls are now drawn, what is the probability that at least one ball of each colour will have been drawn?   N.

12. A cricket team played ten matches of which it lost the first, drew the second and third and won the remaining seven. A supporter tried to forecast *all* ten results before the season began. Find the number of ways:
   (i) of making a forecast of all ten results,
   (ii) of making a correct forecast of the seven wins, it being immaterial whether his other three forecasts turn out to be correct or not,
   (iii) of forecasting correctly the first three results and four of the remaining seven, the other three being incorrectly forecast.   N.

13. Each of six bags contains three coloured balls, one red, one white and one blue. A ball is drawn at random from each bag. Find the probability that two red balls, two white balls and two blue balls are drawn. Find also the probability that exactly four of the six balls are red.   L.

14. One letter is selected from each of the names

   JONES,   THOMSON,   WILKINSON.

   (a) Find the probability that the three letters are the same.
   (b) Show that it is nearly fifteen times as probable that only two of the three letters are the same.   L.

15. (i) If 2 cards are drawn at random from a complete pack, what is the probability that they are both aces?
   (ii) If win, loss, or draw are regarded as equally probable results of any Test Match, what is the probability that one side will win exactly 3 matches in a series of 5?   L.

16. In a large lot of electric light bulbs 5 per cent. of the bulbs are defective. Calculate the probability that a random sample of twenty will contain at most two defective bulbs.

   One-third of the lots presented for inspection have 5 per cent. defective, the rest 10 per cent. defective. If a lot is rejected when a random sample of twenty taken from it contains more than two defective bulbs, find the proportion of lots which are rejected.   N.

17. A box contains red and green sweets mixed together in the ratio 3:2. A handful of twenty sweets is taken at random from the box. Find

    (i) the probability that there will be precisely 15 red sweets in the handful,

    (ii) the probability that the handful of sweets is such that it can be shared equally between five children so that each child receives the same number of red sweets, this number being at least two.   N.

18. An urn contains a large number of black and white marbles in equal proportions and thoroughly mixed. A sample of five marbles is drawn from the urn on two occasions. Show that the probability that the first sample contains precisely one black marble and the second contains precisely three black marbles is 25/512.

Find the probability that the two samples together contain precisely four black marbles.   N.

19. In the seeds from a certain plant lie hidden the potential characteristics, which occur independently, of flower colour and leaf colour. For every three seeds that would, if cultivated to maturity, produce pale-flowered plants, there is one that would produce a dark-flowered plant; and for every three seeds that would produce plants with plain leaves, there is one that would produce a plant with variegated leaves. Calculate the probability that a seed selected at random has the potential characteristics of pale flowers and variegated leaves.

Under controlled conditions of cultivation, the seeds of pale-flowered, variegated-leaved plants have a probability of 0·8 of growth to maturity, the others having a probability of 0·9. If a large random sample of seeds is cultivated, determine the proportion of mature plants which you would expect to have pale flowers and variegated leaves.   N.

20. In a race the odds are 3 to 1 against A being in the first three, 8 to 1 against B being in the first three and 4 to 1 against C being in the first three. What are the odds for or against:

    (a) the first three places being filled in any order by A, B, and C?

    (b) at least one of A, B, C being in the first three?   L.

21. (a) Five per cent. of a large consignment of eggs are bad. Find the probability of getting at least one bad egg in a random sample of a dozen.

    (b) A bag contains b black balls and w white balls where b is greater than w. If they are drawn one by one from the bag, find the probability of drawing first a black, then a white and so on alternately until only black balls remain.   N.

22. (a) Three balls are drawn at random from a bag containing 3 red, 4 white and 5 black balls. Calculate the probabilities that the three balls are (i) all black, (ii) one red, one white, one black.

    (b) The independent probabilities that three components of a television set will need replacing within a year are $\frac{1}{10}$, $\frac{1}{12}$, $\frac{1}{15}$. Calculate the probability that (i) at least one component, (ii) one and only one component, will need replacing.   N.

23. (a) A box contains ten radio valves all apparently sound, although four of them are actually substandard. Find the chance that, if two of the valves are taken from the box together, they are both substandard.

(b) When three marksmen take part in a shooting contest, their chances of hitting the target are $\frac{1}{2}$, $\frac{1}{3}$, $\frac{1}{4}$. Calculate the chance that one, and only one, bullet will hit the target if all three men fire at it simultaneously.　　　　　　　　　　　　　N.

24. Two bags each contain 3 balls, 1 white, and 2 black. A ball is drawn at random from the first bag and placed in the second. A ball is then drawn at random from the second bag and placed in the first. Find the probability that each bag still contains 1 white and 2 black balls.　　L.

25. Find the number of different arrangements in order of p plus signs and m minus signs, all the signs being included in each arrangement.

Find the number of rectangles formed when $n+1$ parallel lines are cut by another set of $n+1$ parallel lines. Show that the number of paths of shortest length by which a point can travel from one corner of the largest rectangle to the opposite corner along segments of these lines is $(2n)!/(n!)^2$.　　　　　　　　　　　　　L.

CHAPTER 6

THREE-DIMENSIONAL GEOMETRY AND TRIGONOMETRY

Drawing a clear figure

6.1. The first part of this chapter will be revision for most readers, but it is likely that in the course of it they will find problems harder than those they have been used to doing.

To begin with, it needs to be emphasized that some of the questions will be very difficult without a clear figure. In general, the four following basic rules should be adopted:

(i) Parallel lines are drawn parallel.

(ii) Vertical lines are drawn parallel to the sides of the paper.

(iii) East–West lines are generally drawn parallel to the bottom of the paper, and North–South lines are drawn at an acute angle to East–West lines.

(iv) All unseen lines should be dotted in.

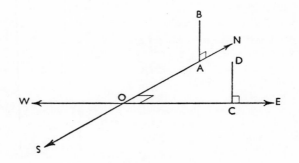

Fig. 6.1

In Fig. 6.1, AB and CD are vertical posts. Notice that the angle NOE is marked a right angle, because this is what it represents.

Qu. 1. Copy Fig. 6.1 and draw AC. Mark in all the right angles at A and C.

The angle between a line and a plane

6.2. When calculating the angle between a line and a plane, we are concerned with calculating the angle between two lines: the given line, and another line lying in the given plane. An infinite number of lines

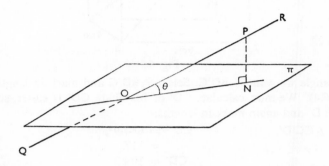

Fig. 6.2

can be drawn, lying in the plane and passing through the point in which the given line meets the plane, and each line will yield a different angle. Which line should we take?

In Fig. 6.2, the line QR meets the plane π in O. In order to find the angle between QR and π, take any point P on QR and drop a perpendicular PN to the plane; now join N to O and θ is the angle required.

We shall see in a later chapter that this is the angle between QR and its *projection* on to π.

Qu. 2. Copy Fig. 6.2 and draw any line lying in π and passing through O. Let M be the foot of the perpendicular from P to this line. Show that angle POM is greater than angle PON.

The result of Qu. 2 shows that θ is the *least* angle between QR and any line which can be drawn in π and passing through O.

Example 1. *Fig. 6.3 represents a rectangular box 9 cm × 6 cm × 6 cm with its lid open at an angle of 30°. Calculate the angle between* BD′ *and the plane* CDD′C′.

BD′ meets the plane CDD′C′ in D′. Take any other point on BD′:B is an obvious point. Drop the perpendicular from B to the plane: BC.

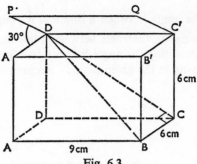

Fig. 6.3

The angle we want is BD'C. Select △BD'C and mark in lengths (see Fig. 6.4). We must calculate CD' or BD' first. CD' is easier, so draw △CC'D' and again mark in lengths.

In △CC'D', $$CD'^2 = 6^2 + 9^2,$$
$$= 117.$$
$$\therefore CD' = 10.82.$$

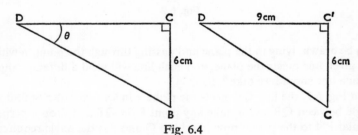

Fig. 6.4

Now mark the length of CD' in △BCD'.
$$\cot \theta = \frac{10.82}{6},$$
$$= 1.803.$$
$$\therefore \theta = 29.0°.$$

Therefore the angle between BD' and the plane CDD'C' is 29.0°.

Qu. 3. Calculate the angle between:
  (i) BD' and the plane BCC'B',
  (ii) AC' and BD'.

Qu. 4. Calculate the angle between BP and the plane ABCD.

The angle between two planes

6.3. When calculating the angle between two planes we are again concerned with calculating the angle between two lines, one in each plane.

Referring to Fig. 6.5, in order to find the angle between two planes π and π', we select a point C on their common line AB and draw lines

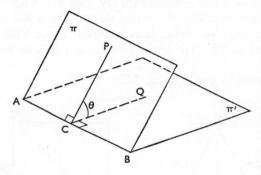

Fig. 6.5

PC and CQ in π and π', respectively, and at right angles to AB. PCQ is the angle we want. This angle is called the *dihedral* angle of the two planes.

Qu. 5. Copy Fig. 6.5 and draw a line PR parallel to AB; join CR. Let P', R' be the feet of the perpendiculars from P, R, respectively, to π'. Show that

angle RCR' < angle PCQ.

Example 2. VABCD *is a right pyramid on a square base* ABCD *of side* 10 *cm. Each sloping edge is* 12 *cm long. Calculate the angle between the faces* VAB *and* VBC.

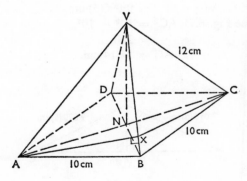

Fig. 6.6

To obtain a good figure first represent the base ABCD as a rhombus, then dot in diagonals to meet at N. Put up the vertical NV and choose V so that AV does not coincide with DV. (See Fig. 6.6.).

VB is the common line of the two planes. Draw AX perpendicular to VB, then, since the figure is symmetrical about VDB, CX is also perpendicular to VB. The angle we want is AXC, so we must work in △AXC.

First we want to find AX and AC.

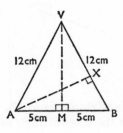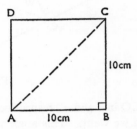

Fig. 6.7

In △VAM, (see Fig. 6.7) $VM^2 = 12^2 - 5^2$,
$$= 119.$$
$$\therefore VM = \sqrt{119} \text{ cm}.$$

Area of △VAB $= 5\sqrt{119} = \frac{1}{2}.AX.12.$

$$\therefore AX = \frac{5\sqrt{119}}{6} \text{cm},$$
$$= 9 \cdot 090(5) \text{ cm}.$$

In △ABC, (see Fig. 6.7) $AC^2 = 10^2 + 10^2$,
$$= 200.$$
$$\therefore AC = 14 \cdot 14 \text{ cm}.$$

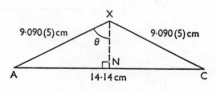

Fig. 6.8

In $\triangle$AXN, (see Fig. 6.8) $\sin \theta = \dfrac{7 \cdot 07}{9 \cdot 091}$,

$$= 0 \cdot 7777.$$
$$\therefore \ \theta = 51 \cdot 05°.$$
$$\therefore \ \angle AXC = 102 \cdot 1°.$$

Therefore the angle between the faces VAB and VBC is $102 \cdot 1°$.

Qu. 6. If, in Example 2, X is any point on VB, prove that $\triangle ABX \equiv \triangle BCX$. Hence prove that if AX is perpendicular to VB, then CX is also perpendicular to VB.

Qu. 7. If, in Example 2, Y is the mid-point of BC, show that VYN is the angle between VBC and ABCD, and calculate it.

Exercise 6a

1. A cuboid is formed by joining the vertices AA′, BB′, CC′, DD′ of two rectangles ABCD and A′B′C′D′. AB = 6 cm, BC = 6 cm, CC′ = 8 cm. X and Y are the mid-points of AD and CD, respectively. Calculate:
   (i) the angle between XB′ and the base ABCD,
   (ii) the angle between the plane XYB′ and the base ABCD,
   (iii) the angle between the plane BB′X and the plane BB′Y.

2. A right pyramid VABCD stands on a rectangular base ABCD. AB = 6 cm, BC = 8 cm, and the height of the pyramid is 12 cm. Calculate:
   (i) the angle which a slant edge makes with the base,
   (ii) the angle which the slant face VAB makes with the base,
   (iii) the angle between the two opposite slant faces VBC and VAD.

3. A hanging lamp is supported by three chains of equal length, fixed to points A, B, C in the ceiling which form an equilateral triangle of side 16 cm, and the lower ends are connected at a point 20 cm below the ceiling. Calculate:
   (i) the length of each chain,
   (ii) the angle which each chain makes with the ceiling.

4. Two equal rectangles 3 m by 4 m are placed so that the longer sides XY coincide. The angle between their planes is 50°. Find the angle between the diagonals which pass through X.

5. A right pyramid stands on a square base of side 8 cm. The height of the pyramid is 10 cm. Calculate the angle between two adjacent faces.

6. Three mutually perpendicular lines meet at O and equal lengths OA, OB, OC are cut off. Find the inclination of ABC to ABO.

7. O is the middle point of the edge AD of a cube, of side 6 cm, whose faces ABCD, A′B′C′D′ are similarly situated. Calculate: (i) the sine of the angle between the plane OCD′ and the face CDD′C′ of the cube, (ii) the sine of the angle between the edge DD′ and the plane OC′D′.

8. Calculate the vertical height and the slope of the slant edges and faces of a regular tetrahedron of side 8 cm, which stands on a horizontal base.

9. The roofs of an L-shaped house slope at 45°. What is the inclination to the horizontal of the line in which the two roofs meet?

10. In Fig. 6.3, taking the same dimensions, calculate the angle between CP and the plane C′D′PQ.

11. A solid is formed by placing a pyramid with square base of side 15 cm and height 20 cm on top of a cuboid with the same dimensions of base and the same height. Calculate the angle which a line drawn from the vertex of the pyramid to a bottom corner of the solid makes with the base.

12. In a regular tetrahedron ABCD, P is the mid-point of AB. Calculate the cosine of the angle between the planes PCD and BCD.

13. In a regular tetrahedron ABCD, Q is the middle point of AD. Find the angle between the line BQ and the plane DBC.

14. In a tetrahedron ABCD, AC = 13 cm, AB = 12 cm, BC = 5 cm, CD = $\sqrt{41}$ cm, BD = 4 cm, and AD = 12 cm. Calculate the cosine of the angle between the planes ABC and BDC.

15. Three adjacent edges of a rectangular box are AB = a cm, AD = b cm, and AF = c cm. Find the angle between the planes BDF and BAD.

16. In a tetrahedron PQRS, P is vertically above Q, one corner of the horizontal base QRS. PS = PR = a, PQ = $2b$, and QR = QS = RS = b. A is the mid-point of PQ. Calculate the sine of the angle between the planes PRS and ARS.

17. A pyramid on a rectangular base has equal slant edges. Prove that a slant edge makes an angle $\cot^{-1}\sqrt{(\cot^2 \alpha + \cot^2 \beta)}$ with the base where α and β are the angles which the slant faces make with the base.

18. In a tetrahedron ABCD, the base ABC is an equilateral triangle of side a cm and the edges DA, DB, DC are all b cm long. X is the centroid of the face ABD. Prove that the angle CX makes with the base is $\tan^{-1}\frac{1}{4} \sqrt{(3b^2/a^2 - 1)}$.

19. OA, OB, OC are unequal mutually perpendicular lines. Prove that cos ∠BAC = cos ∠OAB.cos ∠OAC.

20. The base of a tetrahedron is an equilateral triangle. The slant edges are of equal lengths a and makes angles θ with each other. Prove that the height of the tetrahedron is $\frac{1}{3}a\sqrt{(3 + 6\cos\theta)}$.

Algebraic problems in trigonometry

6.4. Apart from the later questions in Exercise 6a we have so far dealt only with numerical examples. We have obtained solutions, but we shall now want to generalize these solutions in order to understand more clearly how the results depend upon what is given. We shall find, also, that the result often suggests to us the best method of proof.

Two useful hints for solving problems are:

(i) Draw a clear figure marking in the given facts distinctly, and right angles in particular.

(ii) The method is often suggested by the result to be proved.

Example 3. *A man notices two towers, one due North and one in a direction N θ E. If the angle of elevation β of both towers is the same but the height of one is twice the height of the other, prove that*

$$\theta = \cos^{-1} \frac{5 \cot^2 \beta - \cot^2 \alpha}{4 \cot^2 \beta},$$

where α is the angle of elevation of the top of one tower from the top of the other.

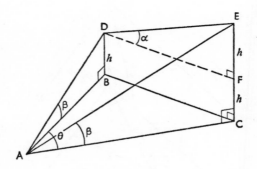

Fig. 6.9

The towers are BD, height h, and CE, height $2h$. The horizontal plane through D cuts CE at F (see Fig. 6.9).

$$\therefore CF = FE = h.$$

[The result shows that we need $\cos \theta$, and this suggests that the cosine formula should be applied to $\triangle ABC$.]

From triangles ABD, DFE, ACE,

$$AB = h \cot \beta, \qquad BC = DF = h \cot \alpha, \qquad AC = 2h \cot \beta.$$

By cosine formula in $\triangle$ABC,

$$\cos \theta = \frac{h^2 \cot^2 \beta + 4h^2 \cot^2 \beta - h^2 \cot^2 \alpha}{2.h \cot \beta.2h \cot \beta},$$

$$= \frac{5h^2 \cot^2 \beta - h^2 \cot^2 \alpha}{4h^2 \cot^2 \beta}.$$

$$\therefore \theta = \cos^{-1} \frac{5 \cot^2 \beta - \cot^2 \alpha}{4 \cot^2 \beta}.$$

Example 4. *Two vertical walls of equal height cast shadows whose widths are b m and c m when the altitude of the sun is θ. If the angle between the walls is α, prove that their height is*

$$\sqrt{\left\{ \frac{b^2 + c^2 + 2bc \cos \alpha}{\cot^2 \theta \sin^2 \alpha} \right\}} \quad \text{m.}$$

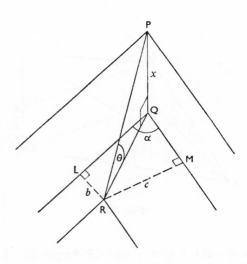

Fig. 6.10

PQ is the line in which the two walls meet, and R is the point shadow of P. L and M are the feet of the perpendiculars from R to the two walls. Let x m be the height of each wall (see Fig. 6.10).

Since $\angle$QLR $= \angle$QMR $= 90°$, RLQM is cyclic, and QR is the diameter of the circumcircle. $\angle$LQM $= \alpha$ therefore $\angle$LRM $= 180° - \alpha$.

By cosine formula in $\triangle$LMR,

$$LM^2 = b^2 + c^2 - 2bc \cos (180° - a),$$
$$= b^2 + c^2 + 2bc \cos a.$$

By sine formula in $\triangle$QLM, $\dfrac{LM}{\sin a} = 2R$,

where R is the circumradius of $\triangle$LMR and therefore of RLQM.

But from $\triangle$PQR, the diameter of the circumcircle QR $= x \cot \theta$,

$$\therefore LM = x \cot \theta . \sin a.$$

$$\therefore x^2 \cot^2 \theta . \sin^2 a = b^2 + c^2 + 2bc \cos a.$$

$$\therefore \text{height of wall} = \sqrt{\left\{ \dfrac{b^2 + c^2 + 2bc \cos a}{\cot^2 \theta \sin^2 a} \right\}} \quad \text{m.}$$

Exercise 6b

1. The angles of elevation of points A, B from a point P are a and β respectively. The bearings of A and B from P are S 20° W and S 40° E, and their distances from P measured on the map are 3 km and 1 km respectively. A is higher than B. Prove that the elevation of A from B is

   $$\tan^{-1} \frac{3 \tan a - \tan \beta}{\sqrt{7}}.$$

2. A pole is set up at C, a point due East of A and due North of B. M is the mid-point of AB. The angles of elevation of the top of the pole from A, M, B are a, γ, β, respectively.

   Prove that $\cot^2 a + \cot^2 \beta = 4 \cot^2 \gamma$. [HINT: This type of question is much simplified by the use of Apollonius' theorem.]

3. A plane slopes down towards the South at an angle a to the horizontal. A road is made up in the plane in the direction ϕ E of N. Prove that the inclination of the road to the horizontal is $\tan^{-1} (\tan a \cos \phi)$.

4. From a point on the ground, two points on the top of a horizontal wall, a m apart, are observed at angles of elevation a and β. The line joining them subtends an angle θ at the point. Prove that the height of the wall is

   $$\frac{a \sin a \sin \beta}{\sqrt{\{\sin^2 a + \sin^2 \beta - 2 \sin a \sin \beta \cos \theta\}}} \quad \text{m .}$$

5. A mast is erected at a point P. At a point B due West, its angle of elevation is a, and at a point C due South, its angle of elevation is β. Prove that its angle of elevation at a point due South of B and due West of C is $\cot^{-1} \sqrt{(\cot^2 a + \cot^2 \beta)}$.

6. A vertical flagstaff of height y m stands on the top A of a tower. The elevation of A from a point due South of it is α, and from a point due East is β. The direct distance between these two points, which are in the horizontal plane through the foot of the tower, is x m and the elevation of the top of the flagstaff from the second point is γ. Prove that

$$y^2 = \frac{x^2(\cot \beta - \cot \gamma)^2}{\cot^2 \gamma (\cot^2 \alpha + \cot^2 \beta)}.$$

7. There are two lights, each l m above level ground, and a m apart. A man, whose height is h m, stands anywhere on the ground. Prove that the line joining the ends of his two shadows cast by the lights is of length $\dfrac{ah}{l-h}$ m.

8. The angles of elevation of the top of a tower measured from three points A, B, C are α, β, γ, respectively. A, B, C are in a straight line such that AB = BC = a, but the line AC does not pass through the base of the tower. Prove that the height of the tower is

$$\frac{a\sqrt{2}}{\{\cot^2 \alpha + \cot^2 \gamma - 2 \cot^2 \beta\}^{\frac{1}{2}}}.$$

9. A man observes a flagpole due North of him. He walks in a direction α N of W for a distance of x m and finds the angle of elevation β is the same. Prove that the angle of elevation when he has walked a further distance x m in the same direction is

$$\tan^{-1} \frac{\tan \beta}{\sqrt{(1 + 8 \sin^2 \alpha)}}.$$

10. From a point A, a lighted window due North of A has an elevation α. From a point B, due West of A, the angle of elevation is β. Prove that the angle of elevation from the mid-point of AB is

$$\tan^{-1} \frac{2}{\sqrt{(3 \cot^2 \alpha + \cot^2 \beta)}}.$$

11. A factory is built within a rectangular plot ABCD. The elevations of the tallest chimney on the building from the three corners A, B, C are α, β, γ respectively. Prove that its elevation from the fourth corner D is

$$\cot^{-1} \sqrt{(\cot^2 \alpha + \cot^2 \gamma - \cot^2 \beta)}.$$

12. A vertical rectangular target faces due South on a horizontal plane. The area of the shadow is $1\frac{1}{4}$ times the area of the target when the sun's altitude is α. Find the bearing of the sun.

13. A vertical tower stands on horizontal ground. From a point P on the ground due South of the tower the angle of elevation of the top of the

tower is α. From a point Q on the ground, South-East of the tower, the angle of elevation is β. Prove that the bearing of Q from P is

$$\tan^{-1} \frac{\cot \beta}{\sqrt{2} \cot \alpha - \cot \beta} \text{ E of N.}$$

14. A triangle ABC is drawn on a plane sloping at θ to the horizontal. A, B are on the same level and C is below them. CA, CB make angles α, β with the horizontal plane through AB.
Prove that

$$\sin \theta = \frac{\sqrt{\{\sin^2 \alpha + \sin^2 \beta - 2 \sin \alpha \sin \beta \cos C\}}}{\sin C}$$

where $\angle ACB = C$.

15. A building at B is x m higher than a building at C, and there is a third building at A, taller than either. The angles of elevation of the tops of the buildings at A, B from the top of that at C are α, β, respectively, and the angle between the vertical planes through CB and CA is θ. The angle between the vertical planes through BA and BC is ϕ. Prove that

the building at A is $\dfrac{x \tan \alpha \sin \phi}{\tan \beta \sin (\theta + \phi)}$ m higher than the building at C.

16. A vertical post of height h m rises from a plane which slopes down towards the South at an angle α to the horizontal. Prove that the length of its shadow when the sun is S θ W at an elevation β is

$$\frac{h \sqrt{(1 + \tan^2 \alpha \cos^2 \theta)}}{\tan \beta + \tan \alpha \cos \theta} \text{ m.}$$

17. A right pyramid of height h stands on a square base of side a. Prove that the angle between adjacent sloping faces is $\cos^{-1} \left(\dfrac{-a^2}{a^2 + 4h^2} \right)$.

18. A right pyramid stands on a regular hexagonal base and its slant faces are equal isosceles triangles of base angle α. Prove that:
   (i) the angle between a slant face and the base is $\cos^{-1} (\sqrt{3} \cot \alpha)$,
   (ii) the angle between adjacent faces is

$$\cos^{-1} \frac{2 \sin^2 \alpha - 3}{2 \sin^2 \alpha}.$$

19. A, B, C are three points on the top of a vertical wall such that $AB = BC = x$, and P is a point in the horizontal plane through the bottom of the wall. The angles of elevation of A, B, C from P are α, β, γ, respectively. If h is the height of the wall, prove that:
   (i) $h^2(\cot^2 \alpha + \cot^2 \gamma - 2 \cot^2 \beta) = 2x^2$,
   (ii) $2h = 3x \sin \alpha$ when $PB = PC = \frac{1}{2}PA$.

L.

20. An aeroplane is flying in a horizontal circle at a uniform speed around a point h m vertically above a point A on horizontal ground. From a point P distance d m due North of A, the greatest and least angles of elevation of the aeroplane are α and β. Show that the angle of elevation of the plane from P at the moment when it is flying due South is

$$\tan^{-1} \frac{2 \sin \alpha \sin \beta}{\sqrt{\{\sin^2 (\alpha - \beta) + \sin^2 (\alpha + \beta)\}}}.$$

21. Two men, A and B, stand a distance a apart on a line running East-West. At noon when the sun's altitude is α, A observes the end of B's shadow and B observes the end of A's shadow. The depression of A's line of sight is β and that of B's is γ. Prove that A's height is

$$a . \sqrt{\frac{\cot^2 \alpha + \cot^2 \gamma}{\cot^2 \beta \cot^2 \gamma - \cot^4 \alpha}}.$$

Three-dimensional geometry

6.5. Applications of calculus and trigonometry are often concerned with three-dimensional figures and so a short introduction to the pure geometry of three dimensions has been included in this volume. It is assumed that the reader is familiar with the abstractions, such as points, lines, planes, and spheres, which form the subject-matter of the geometry of three dimensions. In practice, of course, we are only familiar with approximations to lines, planes, etc., and the initial assumptions, or axioms, of the subject have been chosen as a result of such experience. It is appropriate, therefore, that the reader should verify that the following axioms do, in fact, correspond to his own experience of the geometry of the physical world. Further, he is recommended to make some simple model, with whatever materials are to hand, whenever he has difficulty in visualizing any figure. Pencils, string, books, paper, and card can all be used in this way.

The following list of axioms is not complete but it includes the most important of the properties of incidence of points, lines, and planes.

1. There is one and only one line through two points.
2. There is one and only one plane through three points not in a straight line.
3. A plane contains all the points on a line joining two points in the plane.
†4. Two lines in the same plane have one and only one common point.

---

† For the moment we ignore questions of parallel lines and planes.

†5. Two planes meet in one and only one line.

†6. A plane meets a line in one and only one point.

Qu. 8.   From 1 and 2 above deduce that

   7. There is one and only one plane passing through a given line and a given point not on that line.

Qu. 9.   From 5 and 6 deduce that

†8. Three planes meet in one and only one point, unless they have a common line.

Of the statements above, Nos. 4, 5, 6, 8 need to be modified when parallel lines and planes are considered. This will be done in § 6.8. It is possible to retain the above enunciations by introducing the concepts of points and lines at infinity but this will not be done here.

Skew lines

6.6.   DEFINITION: *Two lines which do not lie in a single plane are called* skew *lines.*

For example, the opposite edges of a tetrahedron are skew and so are AA′, B′C′ in Fig. 6.12, p. 108.

Qu. 10.   How many pairs of skew lines are formed by the twelve edges of a cube?

Example 5.   *Show that one and only one transversal can be drawn to two skew lines through a point not on either.*

Let P be the given point and *l*, *l*′ the skew lines (see Fig. 6.11).

There is one and only one plane π which contains both P and *l*. Similarly there is a unique plane π′ containing both P and *l*′.

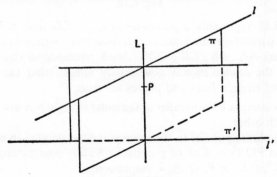

Fig. 6.11

† For the moment we ignore questions of parallel lines and planes.

Planes π, π' meet in a unique line L. Now l, L both lie in plane π, therefore L cuts l. Similarly L cuts l'. But P lies in both planes π, π', so P lies on L.

Hence L is the one and only line through P which cuts l, l'.

Qu. 11. ABC and A'B'C' are two triangles in space such that AA', BB', CC' meet at a point O. Prove that BC, B'C' meet at a point L; CA, C'A' at a point M; AB, A'B' at a point N. Show further that L, M, N are collinear. [This is known as Desargues' theorem. It is also valid for two coplanar triangles. HINT: use Axioms 4 and 5, pp. 106, 107.]

Parallel lines and planes

6.7. DEFINITION: *Two coplanar straight lines are parallel if they never meet however far they are produced in either direction.*

DEFINITION: *Two planes are parallel if they never meet however far they are produced in any direction.*

DEFINITION: *A line and a plane are parallel if they never meet however far either of them is produced in any direction.*

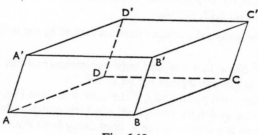

Fig. 6.12

Figure 6.12 represents a *parallelepiped*. ABCD and A'B'C'D' are congruent parallelograms whose corresponding vertices are joined by parallel lines AA', BB', CC', DD'. With reference to this figure, or otherwise, the reader should now satisfy himself that the following properties of straight lines and planes are true.

(i) All straight lines parallel to the same straight line are parallel to each other.

(ii) All planes parallel to the same plane are parallel to each other. [Let PQRS be a plane parallel to ABB'A' and cutting AD, BC, B'C', A'D' in P, Q, R, S, respectively.]

(iii) All lines parallel to a given line and intersecting another line form a plane, and this plane is parallel to the given line.

(iv) In general, a line meets a plane in one point only. What are the two exceptions?

(v) In general, two planes intersect in a straight line. What is the exception?

(vi) A set of parallel planes cuts any other plane in a set of parallel straight lines.

(vii) In general, if two straight lines are parallel, a plane through one of them is parallel to the other.

(viii) In general, three planes meet in a point. What are the three exceptions to this?

Example 6. PQ *and* RS *are skew lines. Prove that one and only one plane can be drawn through* RS *which is parallel to* PQ. (See Fig. 6.13.)

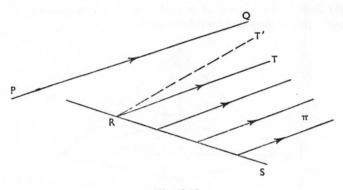

Fig. 6.13

The lines through RS which are parallel to PQ form a plane π, and this plane is parallel to PQ (see (iii) above). Let RT be one such line, and suppose that another plane π' can be drawn through RS which is parallel to PQ. Let π' meet the plane PQTR in RT', say. Since RT' lies in the plane PQTR, either it meets PQ or else it is parallel to it. It cannot meet PQ because it lies in π', which is parallel to PQ. It is not parallel to PQ because π contains all lines through RS which are parallel to PQ. Therefore π' does not exist. Therefore there is one and only one plane π through RS which is parallel to PQ.

Qu. 12. In Fig. 6.12, p. 108, which plane through BD' is parallel to CD?

Qu. 13. Prove that through two skew lines, one and only one pair of parallel planes can be drawn.

DEFINITION: *The angle between two skew lines is the angle between one of them and a line drawn through a point on it and parallel to the other.*

In Fig. 6.12, p. 108, the angle between A'B' and BC is the angle D'A'B'.

Qu. 14. In Fig. 6.12 which is the angle between AD' and CD?

Perpendicular lines and planes

6.8. If the reader has ever had to erect a vertical post (such as a goal post) on level ground, he will know that it is necessary to see that the post appears vertical from two different directions. The next example approaches the same idea more formally.

Example 7. *If a straight line is perpendicular to each of two other straight lines at their point of intersection, it is perpendicular to any, and every, line in the plane containing them.*

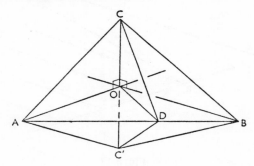

Fig. 6.14

Let OC be perpendicular to OA and OB (see Fig. 6.14).
Produce CO to C' so that CO = OC'. Join A, B to C and C'

$$\triangle AOC \equiv \triangle AOC' \quad \text{(S.A.S.).}$$
$$\therefore \ AC = AC'.$$

Similarly, $BC = BC'.$

$$\therefore \ \triangle ABC \equiv \triangle ABC' \quad \text{(S.S.S.).}$$
$$\therefore \ \angle CAB = \angle C'AB.$$

Now draw any line through O in the plane AOB meeting AB at D.
In triangles CAD and C'AD,

$\angle CAD = \angle C'AD$ (proved), AD is common, CA = C'A (proved).

$$\therefore \quad \triangle CAD \equiv \triangle C'AD \quad (S.A.S.).$$
$$\therefore \quad CD = C'D.$$
$$\therefore \quad \triangle COD \equiv \triangle C'OD \quad (S.S.S.).$$
$$\therefore \quad \angle COD = \angle C'OD = 90°.$$

$\therefore$ CO is perpendicular to any line through O in the plane AOB.

Now OC is skew to any line PQ which lies in the plane AOB and does not pass through O.   However, OC is perpendicular to the line through O which is parallel to PQ and so, by definition, is perpendicular to PQ itself.   Therefore CO is perpendicular to any line in the plane AOB.

The result of this example enables us to see what is meant by a line being perpendicular to a plane.

DEFINITION: *A line is perpendicular to a plane if it is perpendicular to every line lying in the plane.*

The example shows that *a line is perpendicular to a plane if it is perpendicular to any two intersecting lines lying in the plane.*

Qu. 15.   What is the locus in space of points equidistant from two given points?

Qu. 16.   What is the locus in space of points equidistant from three given points?

Qu. 17.   Prove the results obtained in Qu. 15 and 16.   Remember that it is necessary to show:

   (i) that every point which satisfies the condition lies on the locus,

   (ii) that every point on the locus satisfies the condition.

DEFINITION: *Two planes are perpendicular if their dihedral angle* (see p. 97) *is a right angle.*

There now follow some properties of perpendicular lines and planes and the reader should prove these as an exercise.   None of the proofs is long and any difficulty lies in deciding what definitions and results to use.   Class teachers may prefer to use the results below in conjunction with some of the questions of the text as an exercise.

   (i) If a line is perpendicular to a plane, then any plane through that line is also perpendicular to the plane.

   (ii) If two or more straight lines are perpendicular to a plane, they are parallel to each other.

   (iii) If a straight line is perpendicular to a plane, it is perpendicular to all planes which are parallel to the plane.

(iv) All lines which are perpendicular to a given line at the same point on it, lie in a plane.

(v) If two planes are both perpendicular to a third plane, then their intersection is also perpendicular to that plane.

(vi) If two straight lines are drawn from a point outside a given plane, one perpendicular to the plane and the other perpendicular to and intersecting a line in the plane, then the line joining the feet of the perpendiculars is perpendicular to the line in the plane.

Example 8. *One straight line can be drawn perpendicular to both of two skew lines.*

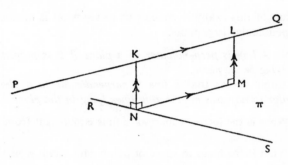

Fig. 6.15

In Fig. 6.15, PQ and RS are two skew lines and L is any point in PQ. Let π be the plane through RS which is parallel to PQ, and let M be the foot of the perpendicular from L on to π. Now draw MN parallel to QP to meet RS in N. PLMN forms a plane (MN∥LP). Draw a line through N parallel to ML; this will meet PQ in a point which we shall call K.

Since ML is perpendicular to π, and NK is parallel to ML, NK is perpendicular to π. Therefore NK is perpendicular to RS.

KLMN is a rectangle (it is a parallelogram and $\angle$ LMN$=90°$). Hence NK is perpendicular to PQ. Therefore NK is perpendicular to both PQ and RS.

Qu. 18. Prove that there is one and only one common perpendicular to two skew lines. [There is a pair of planes through RS and PQ which are parallel to each other (see Example 6). Consider the planes through PQ and RS which are perpendicular to these two planes respectively.]

Qu. 19. Show that the common perpendicular of two skew lines is also the shortest distance between them.

Example 9. *Prove that the lines joining the mid-points of opposite edges of a tetrahedron meet in a point, and bisect each other at that point.*

Let P, Q, R, S, T, U be the mid-points of the edges AB, CD, BC, AD, AC, BD of the tetrahedron ABCD.   Join PS and RQ (see Fig. 6.16).

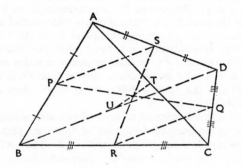

Fig. 6.16

In $\triangle$ABD, PS$\|$BD, PS$=\frac{1}{2}$BD.   (Mid-point theorem.)
In $\triangle$BCD, RQ$\|$BD, RQ$=\frac{1}{2}$BD.   (Mid-point theorem.)

$$\therefore \text{ PS } \| \text{ RQ} \quad \text{and} \quad \text{PS} = \text{RQ}.$$
$$\therefore \text{ PSQR } \text{ is a parallelogram.}$$
$$\therefore \text{ diagonals PQ, RS bisect each other.}$$

Similarly, it can be proved that RUST is a parallelogram and that RS, UT bisect each other.   Therefore PQ, RS, UT all meet, and bisect each other, at the mid-point of RS.

Exercise 6c

1. Prove that the four diagonals of a parallelepiped meet in a point, and bisect each other at that point.
2. Prove that if the diagonals of a parallelepiped are equal then it is a cuboid.
3. If a straight line is perpendicular to two planes, prove that the planes are parallel.   (Assume two planes meet and get a contradiction.)
4. Prove that, if two planes are parallel to a third plane, they are parallel to one another.   (Use the result of No. 3.)
5. Prove that two parallel planes cut any other plane in two parallel straight lines.   What is the exception to this?   (Assume the two lines meet and get a contradiction.)

6. Prove that if two planes are both perpendicular to a third plane, then their line of intersection is also perpendicular to the third plane. (Show that their line of intersection is perpendicular respectively to their lines of intersection with the third plane.)

7. P is a point outside a plane π. N is the foot of the perpendicular from P to the plane. M is the foot of the perpendicular from N to a line RS lying in π. Prove that PM is perpendicular to RS. (By use of Pythagoras' theorem prove that $\triangle$PMR is right-angled at M.)

8. Two congruent right pyramids on square bases have all their edges equal. A regular octahedron is formed by placing them together with their bases in contact. Prove that the opposite edges are parallel.

9. VABCDV' is a regular octahedron. Prove that VB is parallel to the plane V'CD and the plane V'AD. What other edges are parallel to the plane V'AD?

10. Prove that if a line is parallel to each of two intersecting planes, then it is parallel to the line of intersection of those two planes.

11. ABCD is a tetrahedron in which AB = AC = BC = 3 cm; AD = BD = 5 cm; CD = 4 cm. Prove that the line CD is perpendicular to the plane ABC.

12. In a regular tetrahedron, prove that the line joining the mid-points of two opposite edges is perpendicular to these edges. (Use the isosceles triangle property.)

13. Prove that if each edge of a tetrahedron is equal to the opposite edge, then the lines joining the mid points of opposite edges are perpendicular to one another.

14. A tetrahedron ABCD has three right-angled corners at D. The perpendicular from D to BC cuts it at N. The edges AB, AC are bisected at L, M. Prove that LM is perpendicular to the plane ADN.

15. Prove that, if a straight line is parallel to a plane, then the shortest distance between it and any line not parallel to it in the plane is constant.

16. Two planes intersect in a line AB. In each plane a line is drawn through B perpendicular to AB. Prove that the perpendiculars at any two points, one on each line, to their respective planes, intersect.

17. Prove that if each edge of a tetrahedron is perpendicular to its opposite edge, then the lines joining the middle points of pairs of opposite edges are equal.

18. Show that it is possible to draw only one line through a given point to meet two skew lines. (Consider the two planes formed by taking the point and each line, respectively.)

19. Find the locus of the middle points of all straight lines whose ends lie on two skew lines.

20. Prove that if the shortest distance between two skew lines AB, CD is the line joining their mid-points, then AC = BD and BC = AD.

21. Prove that, if two pairs of opposite edges of a tetrahedron be at right angles, the third pair must also be at right angles.

22. Prove that the sum of the squares on the edges of any tetrahedron is four times the sum of the squares on the straight lines joining the middle points of opposite edges.

23. One edge AD of a tetrahedron ABCD is perpendicular to the edge BC. Prove that the sum of the squares on AB, CD is equal to the sum of the squares on AC, BD.

Surfaces of revolution

6.9. So far in this chapter the only surfaces we have considered in three dimensions have been planes. A simple type of three-dimensional surface is that generated when a straight line (called a *generator*), or a plane curve, is rotated about a straight line (called the *axis of rotation*) in its plane; this is called a *surface of revolution*. [The reader should distinguish between this and the related idea of a *solid* of revolution, which is generated when part of a plane is rotated about a line in the plane. It is to be understood that in the remainder of this chapter the words 'cylinder', 'cone', 'sphere' mean the associated surfaces.]

A right circular cylinder is generated when a straight line is rotated about a fixed parallel line.

A right circular cone is generated when a straight line is rotated about a fixed line which cuts it at a constant angle which is not a right angle.

A sphere is generated when a semicircle is rotated about its bounding diameter.

Example 10. *Prove that the intersection of a plane and a sphere is a circle.*

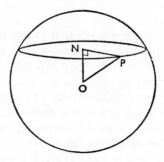

Fig. 6.17

Let P be any point on the intersection of the plane and sphere and let N be the foot of the perpendicular from the centre O on to the plane. Join NP and OP (see Fig. 6.17).

ON is perpendicular to NP so, by Pythagoras' theorem,

$$OP^2 = ON^2 + NP^2.$$

But OP is a radius of the sphere and ON is also fixed in length, therefore PN is fixed in length. Therefore the locus of P is a circle, centre N.

Qu. 20. Prove that the section of a right circular cone by a plane parallel to its base is a circle.

Example 11. *Through four points, not in the same plane, one and only one sphere can be drawn.*

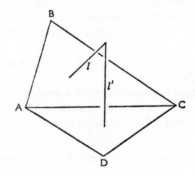

Fig. 6.18

Let A, B, C, D be four points not in the same plane (see Fig. 6.18).

The locus of points equidistant from A, B is the plane which is the perpendicular bisector of AB. Let this be called π_{AB}. Similarly we shall consider the planes π_{BC}, π_{CA}, π_{AD}, π_{DC}.

π_{AB}, π_{BC} meet in a line l which is thus the locus of points equidistant from A, B, C. l lies in π_{AB} and so it is perpendicular to AB, similarly l is perpendicular to BC. Since l is perpendicular to two lines in the plane ABC, it is perpendicular to this plane.

Similarly, π_{AD}, π_{DC} meet in a line l' which is the locus of points equidistant from A, D, C and which is perpendicular to plane ADC.

Now l, l' are coplanar because they both lie in π_{AC} which is the locus of points equidistant from A, C. Therefore l, l' either meet or are parallel. But l, l' are not parallel since they are perpendicular to planes ABC, ADC which are distinct because A, B, C, D are not coplanar. Hence l, l' meet and their point of intersection is the centre of a sphere passing through A, B, C, D. The point is unique since every locus in the proof contains all the points satisfying the appropriate conditions.

Qu. 21. Show that l, l' in the above example meet planes ABC, ADC in the circumcentres of triangles ABC, ADC.

Qu. 22. Show that one and only one sphere can be drawn through the vertices of a right pyramid with a rectangular base.

Example 12. ABCD *is a tetrahedron in which* AB = BD = 2·5 *cm,* AD = 3 *cm, and* AC = BC = CD = 4 *cm. Show that the angle between* CA *and the face* ABD *is* $\cos^{-1} \frac{25}{64}$ *(see Fig. 6.19).*

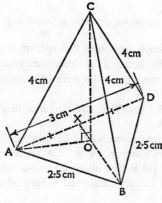

Fig. 6.19

Since C is equidistant from A, B, D, the perpendicular from C to ABD meets ABD in its circumcentre O. Angle CAO is the required angle so we want to calculate R, the radius of the circumcircle of $\triangle$ABD.

$$\frac{BD}{\sin \angle BAD} = 2R,$$

so we must first calculate $\sin \angle BAD$.

Let BO meet AD in X. Since $\triangle$ABD is isosceles, BX is perpendicular to AD. By Pythagoras' theorem in $\triangle$ABX, BX = 2 cm.

$$\therefore \ \sin \angle BAD = \frac{2}{2 \cdot 5}.$$

$$\therefore \ 2R = \frac{2 \cdot 5}{\frac{2}{2 \cdot 5}}.$$

$$\therefore \ R = \frac{2 \cdot 5^2}{2^2},$$

$$= \tfrac{25}{16}.$$

From $\triangle AOC$,

$$\cos \angle OAC = \frac{OA}{AC} = \tfrac{25}{64}.$$

Therefore the angle between CA and the face ABD is $\cos^{-1} \tfrac{25}{64}$.

Qu. 23. In Example 12 calculate the angle between BD and the plane ACD.

Exercise 6d (Miscellaneous)

1. Prove that the common section of two spheres is a circle.

2. Show how to find the centre of the circumscribed sphere of a tetrahedron.

3. Find the radius of the inscribed sphere of a regular tetrahedron of side a cm.

4. Find the radii of the inscribed and circumscribed spheres of a cube of side a cm.

5. Prove that, in general, three spheres have two common points.

6. If the edges of a tetrahedron AB, BC, CA are a cm long and the others are b cm long, find the radius of the circumscribing sphere.

7. Find the radius of the circumscribed sphere of a cone of base radius r cm and height h cm.

8. The angular elevation of a tower at a place A due South of it is $30°$, and at a place B, due West of A and at a distance a from it, the elevation is $15°$. Show that the height of the tower is $a\{(\sqrt{3}-1)/8\}^{\frac{1}{2}}$.

9. ABCD is a tetrahedron in which AB = AC = AD. O is the circumcentre of $\triangle BCD$. Prove that AO is perpendicular to the plane BCD.

10. Prove that the shortest distance between two opposite edges of a regular tetrahedron of side a is equal to $a/\sqrt{2}$.

11. A tetrahedron ABCD has sides AB = BC = 6 cm, AC = 4 cm, AD = BD = CD = 5 cm. Calculate:
    (i) the height of D above the base ABC,
    (ii) the angle between the face CAD and the base ABC.

12. Four equal, rough spheres of radius $2\frac{1}{2}$ cm lie in contact on a horizontal plane with their centres forming a square of side 5 cm. A fifth sphere of radius $3\frac{1}{2}$ cm is placed symmetrically on top of them. Calculate:
    (i) the height of the centre of the top sphere above the plane,
    (ii) the angle which the line joining the centre of the top sphere with the centre of any of the other spheres makes with the horizontal plane.

13. Find the height of the centre of any sloping face of a regular tetrahedron of side a cm above the base.

14. Prove that if the four slant edges of a pyramid on a quadrilateral base are equal, then the vertices of the base must be concyclic.

15. Two spheres of radii r_1 and r_2 have their centres at A and B, respectively. $AB = d$ and $d^2 = r_1^2 + r_2^2$. Prove that the radius, R, of the circle in which the two spheres intersect is given by $Rd = r_1 r_2$.

16. OA, OB, OC are three mutually perpendicular edges of a tetrahedron OABC. $OA = 4$ cm, $OB = 4$ cm, $OC = 3$ cm. Calculate the angle between the planes ABC and OAB.

17. The base ABC of a tetrahedron is an equilateral triangle of side 4 cm. If O is the vertex and $OA = OB = 3$ cm, and $OC = 5$ cm, calculate:
    (i) the angle between OB and ABC,
    (ii) the angle between OAC and ABC.

18. Three spheres of radii 5 cm, 2 cm, and 2 cm, respectively, touch each other and the surface of a table. Prove that the inclination to the table of their common tangent plane is $\tan^{-1}(4/3)$.

19. Points X, Y, Z are marked 4 cm, 3 cm, 1 cm along the concurrent edges of a rectangular box, the distances being measured from a corner A. Calculate the angle between the planes AXY and ZXY.

20. A tetrahedron stands on a horizontal table. Its base ABC is a triangle whose sides AB, AC are each 6 cm long and the angle ABC is 30°. The vertex D is vertically above A and $DA = 8$ cm. Find the inclination of the face DBC to the table and the angle between DBC and DBA.

21. Prove that the lines joining the vertices of a tetrahedron to the centroids of the opposite faces meet.

22. Find the vertical height of a regular tetrahedron of side a cm.

23. Find the locus of a point, the sum of the squares of whose distances from two fixed points is constant.

24. A hemispherical bowl, centre O and radius r, rests with its lowest point C on a horizontal plane. It is tilted until the line CO makes an angle θ with the vertical. Prove that the height of C above the plane is now $2r \sin^2 \frac{1}{2}\theta$.

25. A right circular cone of semi-vertical angle α is inscribed in a sphere of radius R. Prove that the volume of the cone is $(8/3)\pi R^3 \sin^2 \alpha \cos^4 \alpha$.

26. ABC is an equilateral triangle lying in a horizontal plane and of side $a\sqrt{6}$. B′ is a point at a distance $5a$ vertically below B and C′ is a point at a distance a vertically above C. Find the angle C′AB′.

27. A pyramid stands on a horizontal, regular pentagonal base of side a cm. The five triangular faces each slope at an angle a to the horizontal. Calculate the length and the slope of the edges.

28. In a tetrahedron ABCD, AB = BC = 3 cm, AD = CD = 5 cm, AC = BD = 4 cm. Calculate the cosine of the angle between:
    (i) ACB and ACD,
    (ii) ABD and ACD.

29. A tetrahedron stands on a horizontal triangular base whose sides are 5 cm, 5 cm, and 6 cm. The vertex is 3 cm directly above the point dividing the longest side in the ratio 2:1. Calculate:
    (i) the angles of slope of the edges,
    (ii) the angle between the two slant faces.

30. A rectangle ABCD with AB = 8 cm and BC = 6 cm is folded about AC so that the plane ABC is perpendicular to the plane ACD. Calculate the angle BDC in this position.

31. A right prism of height h stands on a triangular base ABC with a right angle at C and the sides BC, CA of length a, b, respectively. It is cut by a plane through AC which is inclined at an angle θ to the base. Show that the ratio of the two volumes into which the plane divides the prism is equal to $(3h - a\tan\theta) : a\tan\theta$ where $a\tan\theta < h$.

32. In a regular tetrahedron ABCD, P is the middle point of AD. Calculate the angle between the planes DBC, PBC.

33. Prove that one and only one sphere can be drawn to contain a given circle and a point not in the plane of the circle.

34. ABCD is a tetrahedron in which AB = AC and DA is perpendicular to the plane ABC. A plane is drawn through BC cutting AD at X, such that the triangle BCX is equilateral. Prove that the angle between planes BCX, BCA is

$$\cos^{-1}\left(\frac{1}{\sqrt{3}}\tan\angle ABC\right).$$

35. ABCD is a tetrahedron in which AD = BC = a, BD = CA = b, CD = AB = c. Find the shortest distance between AD, BC in terms of a, b, c.

36. A ship is observed at the same moment from two places A and B, c km apart, the observed bearings respectively being N a E, N β W and the line AB is in the direction N γ E. Show that the distance of the ship from the line AB is

$$c \sin(\beta + \gamma)\sin(\gamma - a)\operatorname{cosec}(a + \beta) \text{ km.}$$

37. A and B are two points on one bank of a river distance x m apart; P and Q are two points on the opposite bank. PQ is parallel to AB. If $\angle PAQ = \alpha$, $\angle QAB = \beta$, $\angle PBA = \gamma$, prove that

$$PQ = \frac{x \sin \alpha \sin \gamma}{\sin (\alpha + \beta + \gamma) \sin \beta} \text{ m.}$$

38. Three mutually perpendicular lines meet at A. Points X, Y, Z are taken on these lines, and AP is the perpendicular to the plane XYZ. Prove that P is the orthocentre of the $\triangle$XYZ.

CHAPTER 7

SOME INEQUALITIES AND GRAPHS

Some inequalities†

7.1. Anyone who has studied mathematics up to this level will be thoroughly used to manipulating equations, but many readers will not be familiar with inequalities. An inequality is a statement that one number is less than (or greater than) another. Thus the statement, 'The sum of the squares of two numbers is greater than or equal to twice their product', may be written in the form

$$a^2 + b^2 \geqslant 2ab.$$

To prove this

$$\text{L.H.S.} - \text{R.H.S.} = a^2 + b^2 - 2ab = (a-b)^2.$$

But $(a-b)^2$ is a square and so is greater than or equal to zero, and the inequality is proved.

Note that the equality occurs only if $a = b$, therefore we may write

$$a^2 + b^2 > 2ab \quad (a \neq b).$$

Inequalities may be manipulated in much the same way as equations but with certain important reservations (see Qu. 1 to 4). The rules that will be used here are:

   (i) We may add any number (positive or negative) to each side of an inequality.

  (ii) We may multiply each side of an inequality by any *positive* number.

 (iii) If each side of an inequality is multiplied by a *negative* number, the inequality is reversed.

As examples of these, we know that

$$5 < 9.$$

$\therefore$ (i) $5 + 2 < 9 + 2$ and $5 - 13 < 9 - 13$,

      (ii) $5 \times 3 < 9 \times 3$,

but   (iii) $5 \times (-4) > 9 \times (-4)$.

† As elsewhere in Chapters 1 to 21, only real values of the variables are considered.

Note particularly that the last statement is about algebraic (or directed) numbers. If two such numbers are represented by points on an axis going from left to right, the greater is on the right. Thus $-20 > -36$ and $-3 < -1$.

Qu. 1. If $a = x$, then $a^2 = x^2$. Consider the inequality $a < x$ if (i) $a = 3$, $x = 5$, (ii) $a = -7$, $x = 4$. Is $a^2 < x^2$?

Qu. 2. If $b = y$, then $1/b = 1/y$. Consider the inequality $b < y$ if (i) $b = 2$, $y = 4$, (ii) $b = -3$, $y = 1$. Is $1/b < 1/y$?

Qu. 3. If $a = x$, $b = y$, then $a - b = x - y$. Consider the inequalities $a < x$ $b < y$ if $a = 5$, $x = 6$, $b = 4$, $y = 7$. Is $a - b < x - y$?

Qu. 4. If $a < x$, $b < y$, is $ab < xy$? Try $a = -3$, $x = 2$, $b = -4$, $y = 5$.

Example 1. *Find the values of x for which* $\dfrac{3-x}{x+2} > 4$.

$$\frac{3-x}{x+2} > 4.$$

So, provided $x + 2 > 0$,

$$3 - x > 4x + 8.$$
$$\therefore \; -5x > 5.$$
$$\therefore \; x < -1.$$
$$\therefore \; -2 < x < -1.$$

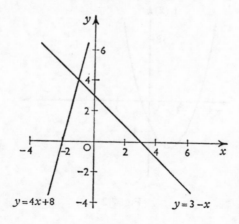

Fig. 7.1

On the other hand, if $x+2<0$,

$$3-x < 4x+8.$$
$$\therefore\ -5x < 5.$$
$$\therefore\ x > -1,$$

which is not compatible with $x+2<0$. Therefore the expression is greater than 4 only when $-2<x<-1$. Figure 7.1 shows the graphs of $3-x$ and $4x+8$. Evidently $3-x>4x+8$ if $x<-1$, but the *ratio* of the values of the functions is negative if $x<-2$.

Example 2. *For what values of x is the function $2x^2+5x-3$, (i) negative, (ii) positive?*

Let $f(x)=2x^2+5x-3=(2x-1)(x+3)$. $f(x)$ vanishes when $x=\frac{1}{2}$ or $x=-3$, so consider the signs of the factors in the following ranges:

	$x < -3$	$-3 < x < \frac{1}{2}$	$\frac{1}{2} < x$
$x+3$	−	+	+
$2x-1$	−	−	+
$f(x)$	+	−	+

Alternatively, sketch the graph of the function $f(x)$. As $x\to\pm\infty$, $f(x)\to+\infty$; $f(\frac{1}{2})=f(-3)=0$. The curve is sketched in Fig. 7.2.

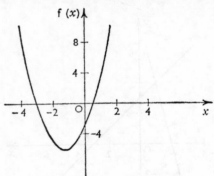

Fig. 7.2

By either method, the function is negative if $-3<x<\frac{1}{2}$ and positive if $x<-3$ or $x>\frac{1}{2}$.

Example 3. *Show that $3x^2+10x+9$ cannot be negative and find its least value.*

Completing the square,

$$3x^2+10x+9 = 3(x^2+\tfrac{10}{3}x+\tfrac{25}{9})+9-\tfrac{25}{3},$$
$$= 3(x+\tfrac{5}{3})^2+\tfrac{2}{3}.$$

Since $(x+\tfrac{5}{3})^2$ is a square, the least value it can take is zero, so that $3x^2+10x+9$ cannot be zero and its least value is $\tfrac{2}{3}$.

A similar method may be applied to functions of more than one variable.

Example 4. *Show that $a^2+b^2+c^2-bc-ca-ab$ cannot be negative. Under what circumstances is it zero?*

$$a^2+b^2+c^2-bc-ca-ab$$
$$= \tfrac{1}{2}(b^2+c^2-2bc + c^2+a^2-2ca + a^2+b^2-2ab),$$
$$= \tfrac{1}{2}\{(b-c)^2+(c-a)^2+(a-b)^2\} \geqslant 0.$$

The equality occurs only when each square is zero, that is when $a=b=c$.

The equation $ax^2+bx+c = 0$

7.2. The method of completing the square provides a proof of an important result which is useful in connection with inequalities and elsewhere. Consider the equation

$$ax^2+bx+c = 0.$$
$$\therefore \quad x^2+\frac{b}{a}\,x = -\frac{c}{a}.$$
$$\therefore \quad x^2+\frac{b}{a}\,x+\frac{b^2}{4a^2} = \frac{b^2}{4a^2}-\frac{c}{a}.$$
$$\therefore \quad \left(x+\frac{b}{2a}\right)^2 = \frac{b^2-4ac}{4a^2}.$$

Hence $x+b/2a$ is real if and only if $b^2-4ac \geqslant 0$. Therefore, the roots of the equation $ax^2+bx+c=0$ are real if and only if

$$b^2-4ac \geqslant 0.$$

The expression b^2-4ac is called the *discriminant* of the equation.

Example 3 may now be solved as follows.

Consider the equation $y=3x^2+10x+9$.

$$\therefore \quad 3x^2+10x+9-y = 0.$$

The roots of this equation are real if and only if

$$10^2 - 4 \times 3 \times (9 - y) \geqslant 0.$$
$$\text{i.e. } 12y - 8 \geqslant 0.$$
$$\text{i.e. } y \geqslant \tfrac{2}{3}.$$

Hence y cannot be negative and the least value of the function is $\tfrac{2}{3}$.

Example 5. *If $(x^2 - x + 1)y = 2x$, within what range of values does y lie?*

Writing $(x^2 - x + 1)y = 2x$ as a quadratic equation in x,

$$x^2 y - x(y + 2) + y = 0.$$

The roots of the equation $ax^2 + bx + c = 0$ are real when $b^2 - 4ac \geqslant 0$. Therefore, for real values of x,

$$\{-(y + 2)\}^2 - 4y^2 \geqslant 0.$$
$$\therefore \quad -3y^2 + 4y + 4 \geqslant 0.$$
$$\therefore \quad (2 + 3y)(2 - y) \geqslant 0.$$

Hence, if x is real, y lies in the range

$$-\tfrac{2}{3} \leqslant y \leqslant 2.$$

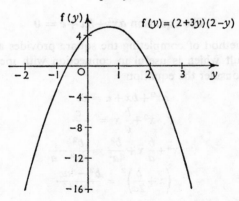

Fig. 7.3

Exercise 7a

For what ranges of values of x do the following inequalities hold?

1. (i) $\dfrac{x + 1}{2 - x} < 1$, (ii) $\dfrac{x + 1}{2 - x} > 1$.

2. (i) $\dfrac{4 - x}{x + 2} < 3$, (ii) $\dfrac{4 - x}{x + 2} > 3$.

3. $(3-x)(x+2) > 0.$

4. $(2x-5)(3x+7) > 0.$

5. $2x^2+x-15 < 0.$

6. $10+x-2x^2 < 0.$

7. $(4x-3)(x+1) > 2.$

8. $(5x-7)(x-3) < 16x.$

9. $\dfrac{2x^2-7x-4}{3x^2-14x+11} > 2.$

10. $\dfrac{(x-1)(x-3)}{(x+1)(x-2)} > 0.$

Prove the following inequalities and find the extreme values of the functions concerned.

11. $x^2-5x+7 > 0.$

12. $4x-x^2-5 < 0.$

13. $2x^2+3x+2 > 0.$

14. $5x-3x^2-3 < 0.$

Find the ranges of values of x and y for which there are no real points on the following loci:

15. $y^2 = x(1-x).$

16. $3x^2+4y^2 = 12.$

17. $y^2 = x(x^2-1).$

18. $y^2(y^2+4) = x^2(4-x^2).$

19. $(x-2)(x-3)y = 2x-5.$

20. $(2x-1)(2x+3)y = 2x+1.$

21. $(x^2+1)y = 3x+4.$

22. $(x^2+x+1)y = 1.$

For what ranges of values of x are the following equations satisfied by real values of θ?

23. $\sin \theta = \dfrac{x-1}{x+1}.$

24. $\cos \theta = \dfrac{x+3}{3-x}.$

25. $\sin \theta = \dfrac{2x-1}{3x+2}.$

26. $\sec \theta = \dfrac{2x+5}{4-x}.$

27. $\sin \theta + \cos \theta = x.$

28. $3 \cos \theta - 4 \sin \theta = 1-x.$

29. Show that $(a+b)^2 \geqslant 4ab.$

30. Verify the identity

$$a^3 + b^3 + c^3 - 3abc = (a+b+c)(a^2+b^2+c^2-bc-ca-ab)$$

and deduce that the arithmetic mean of three unequal positive numbers x, y, z $[\frac{1}{3}(x+y+z)]$ is greater than their geometric mean $[(xyz)^{1/3}]$.

31. Express $5x^2-12xy+9y^2-4x+4$ as the sum of two squares and show that the expression is positive except for one pair of values of x and y.

32. What is the greatest value that can be taken by $-x^2+2xy-5y^2+4y$?

33. Show that $2x^2-4xy+5y^2-2x-4y+6>0$ for all real values of x and y.

34. Show that there is only one real point on the locus

$$x^2+3xy+5y^2-3x+y+5 = 0.$$

Rational functions of two quadratics

7.3. In this section we shall be concerned with rational functions of two quadratics, that is, functions of the form

$$\frac{ax^2+bx+c}{Ax^2+Bx+C},$$

where a, b, c, A, B, C are constants. The method of the last section will be used.

Example 6. *Sketch the curve*

$$y = \frac{(x-1)(x+2)}{(x+1)(x-3)}.$$

First note the following:

  (i) when $y = 0$, $x = 1$ or $x = -2$;

  (ii) when $x = 0$, $y = \frac{2}{3}$;

  (iii) when $x = -1$ or $x = 3$, the denominator of the fraction vanishes so that there is no corresponding value of y. If x differs from 1 or 3 by a small amount, the denominator is small and so y is large. Therefore $y \to \infty$ as $x \to -1$ and $x \to 3$;

  (iv) given any value of x, other than 1 or 3, there exists one and only one value of y. Note that since the equation is a quadratic in x, there are in general two values of x corresponding to each value of y.

  (v) the sign of y may be determined by inspecting the signs of the factors $x+2$, $x+1$, $x-1$, $x-3$;

	$x < -2$	$-2 < x < -1$	$-1 < x < 1$	$1 < x < 3$	$3 < x$
$x+2$	$-$	$+$	$+$	$+$	$+$
$x+1$	$-$	$-$	$+$	$+$	$+$
$x-1$	$-$	$-$	$-$	$+$	$+$
$x-3$	$-$	$-$	$-$	$-$	$+$
y	$+$		$+$	$-$	$+$

  (vi) $y = \dfrac{x^2+x-2}{x^2-2x-3}.$

If x is large, the terms in x and the constants are small compared with x^2 so that $y \simeq x^2/x^2 = 1$. If we substitute $y = 1$ in the equation,

$$x^2 - 2x - 3 = x^2 + x - 2.$$
$$\therefore\ x = -\tfrac{1}{3}.$$

Therefore the graph crosses $y = 1$ at $(-\frac{1}{3}, 1)$.

Our findings are shown in Fig. 7.4; the shading denotes areas where the curve cannot lie (see stage (v)).

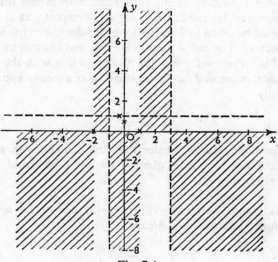

Fig. 7.4

The graph is now sketched as in Fig. 7.5.

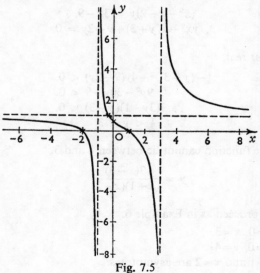

Fig. 7.5

The lines $x = -1$, $x = 3$, $y = 1$, which are represented by broken lines, are called **asymptotes**. Note that the curve approaches them ever more closely, without meeting them, as it recedes from the origin. It is possible, however, for the curve to cut an asymptote, as at $(-\frac{1}{3}, 1)$.

Care should be taken to find from which sides the graph approaches the asymptotes. For $x = -1$ and $x = 3$ this was ensured by examining the sign of y. For $y = 1$ the point of intersection with the graph was found. Another method for the latter is to take a second approximation for y, namely

$$\frac{x^2 + x}{x^2 - 2x}.$$

If $x > 0$, the numerator is greater than the denominator, so that the graph approaches $y = 1$ from above. On the other hand, when x is large and negative, $y < 1$.

Example 7. *Prove that $(3x-9)/(x^2-x-2)$ cannot lie between two certain values. Illustrate graphically.*

Let
$$y = \frac{3x - 9}{x^2 - x - 2}.$$

Regard this equation as a quadratic which gives x in terms of y, then

$$(x^2 - x - 2)y = 3x - 9.$$
$$\therefore \ yx^2 - x(y+3) + 9 - 2y = 0. \tag{1}$$

When x is *not* real,

$$\{-(y+3)\}^2 - 4y(9 - 2y) < 0.$$
$$\therefore \ 9y^2 - 30y + 9 < 0.$$
$$\therefore \ 3(3y - 1)(y - 3) < 0.$$
$$\therefore \ \tfrac{1}{3} < y < 3.$$

Therefore the function cannot lie between $\frac{1}{3}$ and 3.

Now
$$y = \frac{3(x - 3)}{(x + 1)(x - 2)}$$

and we may proceed as in Example 6.

(i) If $y = 0$, $x = 3$.
(ii) If $x = 0$, $y = 4\frac{1}{2}$.
(iii) $x = -1$ and $x = 2$ are asymptotes.

(iv) There is only one value of y for each value of x.

(v) The sign of y is obtained:

	$x < -1$	$-1 < x < 2$	$2 < x < 3$	$3 < x$
$x+1$	−	+	+	+
$x-2$	−	−	+	+
$x-3$	−	−	−	+
y	−	+	−	+

(vi) As $x \to \infty$, $y \to 0$.

(vii) The values of x corresponding to $y=\frac{1}{3}$ and $y=3$ are found from equation (1).

[Note that $y=\frac{1}{3}$ and $y=3$ make the discriminant 'b^2-4ac' $=0$, so that equation (1) has equal roots. The sum of the roots is $(y+3)/y$, therefore $x=\frac{1}{2}(y+3)/y$.]   When $y=\frac{1}{3}$, $x=5$; when $y=3$, $x=1$.

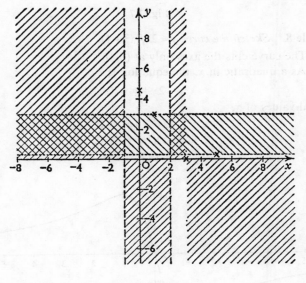

Fig. 7.6

Our findings are shown in Fig. 7.6 and the curve has been sketched in Fig. 7.7, page 132.

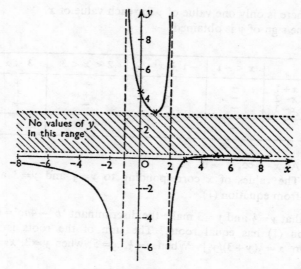

Fig. 7.7

Example 8. *Sketch the curve* $y = 2x/(x^2+1)$.

(i) The curve cuts the axes only at (0, 0).

(ii) As a quadratic in x, the equation is

$$x^2y - 2x + y = 0.$$

For real values of x,

$$(-2)^2 - 4y^2 \geqslant 0.$$
$$\therefore \ 4(1-y)(1+y) \geqslant 0.$$
$$\therefore \ -1 \leqslant y \leqslant +1.$$

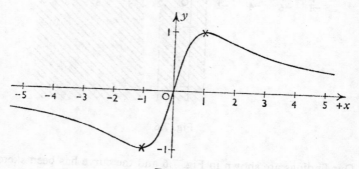

Fig. 7.8

(iii) When $y = -1$, $x = -1$, and when $y = +1$, $x = +1$. Therefore $(-1, -1)$ is a minimum and $(1, 1)$ a maximum.

(iv) As $x \to \infty$, $y \to 0$.

(v) Since $x^2 + 1$ is positive, x and y have the same sign.

The curve has been sketched in Fig. 7.8.

Exercise 7b

Sketch the following curves:

1. $y = \dfrac{x-2}{x+3}$.

2. $y = \dfrac{x}{x-1}$.

3. $y = \dfrac{x}{x^2-1}$.

4. $y = \dfrac{x^2}{x^2-1}$.

5. $y = \dfrac{2x-4}{(x-1)(x-3)}$.

6. $y = \dfrac{(x-1)(x+3)}{(x-2)(x+2)}$.

7. $y = \dfrac{x^2-4x+1}{x^2-4x+4}$.

8. $y = \dfrac{(x-1)^2}{x(x-2)}$.

9. $y = \dfrac{(x-2)^2}{x(x+2)}$.

For each of the following curves, find the range(s) of values within which y *cannot* lie. Illustrate graphically.

10. $y = \dfrac{4}{(x-1)(x-3)}$.

11. $y = \dfrac{3x-6}{x(x+6)}$.

12. $y = \dfrac{1}{x^2+1}$.

13. $y = \dfrac{4x^2-3x}{x^2+1}$.

14. $y = \dfrac{(x-3)(x-1)}{(x-2)^2}$.

15. $y = \dfrac{x^2+1}{x^2-x-2}$.

Find the turning points of the following and sketch the curves

16. $y = \dfrac{x^2-4x}{x^2-4x+3}$.

17. $y = \dfrac{x^2-x-2}{x^2-2x-8}$.

18. $y = \dfrac{x^2-3x}{x^2+5x+4}$.

19. $y = \dfrac{2x^2-9x+4}{x^2-2x+1}$.

Some tests for symmetry

7.4. The remainder of this chapter is devoted to further aids to curve sketching, and the most useful of these is symmetry.

First consider the graph of $y = x^2$ (Fig. 7.9 (i)), which is symmetrical about the y-axis. If the point (h, k) lies on the curve, we have $k = h^2$, and so the point $(-h, k)$ also lies on the curve. In general, if an equation is unaltered by replacing x by $-x$, the curve is symmetrical about the y-axis.

Similarly, if the equation of a curve is unaltered by replacing y by $-y$, there is symmetry about the x-axis.   (Fig. 7.9 (ii).)

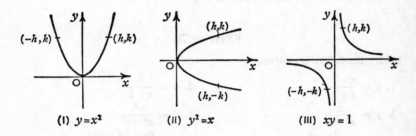

(I) $y=x^2$ (ii) $y^2=x$ (III) $xy=1$

Fig. 7.9

Figure 7.9 (iii) represents the curve $xy=1$, which is symmetrical about a *point*, the origin, and not about an axis, as (i) and (ii).   If (h, k) lies on the locus, so does $(-h, -k)$.   In general, if an equation is unaltered when x and y are replaced by $-x$ and $-y$ respectively, the curve is said to be symmetrical about the origin.

Qu. 5.   Which of the following show symmetry about: (a) the y-axis, (b) the x-axis, (c) the origin?

(i) $4x^2+y^2 = 1$,   (ii) $y^2 = x(x+1)$,
(iii) $x^5+y^5 = 5xy^2$,   (iv) $x^2-3xy+y^2 = 1$,
(v) $y^2 = x^2(x+1)(x-1)$,   (vi) $x^2y-x+y^3 = 0$,
(vii) $y^2 = \cos x$,   (viii) $\tan y = \sin x$.

Qu. 6.   Some equations are unaltered by the following substitutions:

(i) $x = y$, $y = x$;   (ii) $x = -y$, $y = -x$.

About what axes are the corresponding curves symmetrical?

Qu. 7.   Show that a curve which is symmetrical about the x- and y-axes is also symmetrical about the origin.

Example 9.   *Sketch the curve $x^2-y^2=1$.*

(i) The equation shows symmetry about both axes and the origin.
(ii) Since $y^2=x^2-1$, y is not real when x is numerically less than 1.
(iii) When $y=0$, $x=\pm 1$.

(iv) As x increases in magnitude, so does y.

(v) On differentiation,

$$2y \frac{dy}{dx} = 2x.$$

$$\therefore \frac{dy}{dx} = \frac{x}{\sqrt{(x^2 - 1)}}.$$

$$\therefore \text{ as } x \to \pm 1, \quad \frac{dy}{dx} \to \infty,$$

and

$$\text{as } x \to \pm \infty, \quad \frac{dy}{dx} \to \pm 1.$$

(vi) Since $y^2 = x^2 - 1$, when x, y are large, y^2 is nearly equal to x^2. Thus the curve approaches the lines $y = \pm x$.

The curve has been sketched in Fig. 7.10.

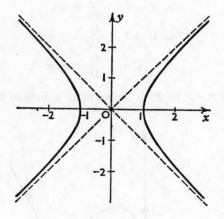

Fig. 7.10

The form $y^2 = f(x)$

7.5. If an equation may be expressed in the form $y^2 = f(x)$ a sketch of the curve may be obtained from a preliminary sketch of the function $f(x)$. It is important to note that $y^2 = f(x)$ shows symmetry about the x-axis.

Example 10. *Sketch the curve $y^2 = x(x-2)^2$.*

First sketch the graph of the function $f(x) \equiv x(x-2)^2$. See the broken line in Fig. 7.11.

Since $y = \pm \sqrt{\{f(x)\}} = \pm (x-2)\sqrt{x}$, we have:

(i) when $x < 0$, y is not real,

(ii) when $x = 0$ or 2, $y = 0$,

(iii) when $0 < f(x) < 1$, $|y| > f(x)$,

(iv) when $f(x) > 1$, $|y| < f(x)$.

$$y^2 = x(x-2)^2.$$

$$\therefore \ 2y \frac{dy}{dx} = (x-2)^2 + x.2(x-2).$$

$$\therefore \ \frac{dy}{dx} = \frac{(x-2)(3x-2)}{\pm 2x^{1/2}(x-2)},$$

$$= \pm \frac{3x-2}{2x^{1/2}}.$$

$$\therefore \ \text{as } x \to 0, \quad \frac{dy}{dx} \to \infty,$$

$$\text{as } x \to 2, \quad \frac{dy}{dx} \to \pm \sqrt{2},$$

and

$$\text{as } x \to \infty, \quad \frac{dy}{dx} \to \pm \infty.$$

The graph of $y^2 = x(x-2)^2$ is shown in an unbroken line in Fig. 7.11.

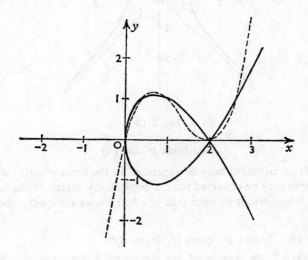

Fig. 7.11

Qu. 8. Find the gradient of

(i) $y^2 = x(x-2)(x-4)$,   (ii) $y^2 = x^2(x+2)$,

at the points where the graphs cut the x-axis.

Sketch the curves by the method of Example 10.

Simple changes of axes

7.6. The equation of a circle, centre C (a, b) and radius r, is (P.M.I., p. 368)

$$(x-a)^2 + (y-b)^2 = r^2,$$

and the equation of an equal circle, centre the origin, is

$$x^2 + y^2 = r^2.$$

Therefore, if new axes CX and CY were taken parallel to Ox and Oy, the equation of the former would become

$$X^2 + Y^2 = r^2.$$

This is equivalent to making the substitutions

$$X = x - a, \qquad Y = y - b,$$

or, as is often more convenient,

$$x = X + a, \qquad y = Y + b.$$

These relationships may easily be verified from a diagram.

Such a change of axes is sometimes helpful in curve sketching. Thus

$$(y-1)^2 = 4(x+2)$$

becomes $\qquad\qquad Y^2 = 4X,$

referred to parallel axes through $(-2, 1)$ and the curve is now easily drawn, as in Fig. 7.12.

Note that the equation

$$y = ax^2 + bx + c$$

may be written

$$y + \frac{b^2}{4a} - c = a\left(x + \frac{b}{2a}\right)^2.$$

Referred to parallel axes through $\left(-\dfrac{b}{2a}, -\dfrac{b^2-4ac}{4a}\right)$ the equation becomes

$$Y = aX^2,$$

which is a parabola (see P.M.I., p. 412).

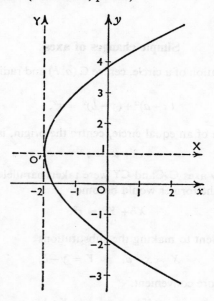

Fig. 7.12

Example 11. *Sketch the function* $1 + 2 \sin (\theta + \tfrac{1}{4}\pi)$ *for values of* θ *from 0 to* 2π.

Write
$$y = 1 + 2 \sin (\theta + \tfrac{1}{4}\pi).$$
$$\therefore \ y - 1 = 2 \sin (\theta + \tfrac{1}{4}\pi).$$

With the substitutions

$$\Theta = \theta + \tfrac{1}{4}\pi, \qquad Y = y - 1, \tag{1}$$

the equation becomes

$$Y = 2 \sin \Theta.$$

The graph of $Y = 2 \sin \Theta$ has been sketched in Fig. 7.13. Writing $\theta = y = 0$ in equations (1), the origin of the θ, y axes is found to be $(\tfrac{1}{4}\pi, -1)$, referred to the Θ, Y axes. The θ, y axes were then drawn to pass through this point.

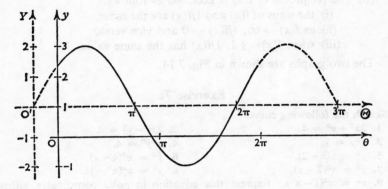

Fig. 7.13

The form $y = 1/f(x)$

7.7.   Example 12.   *Sketch on the same axes the graphs of*

(a) $(x+1)(2x-3)$, (b) $1/\{(x+1)(2x-3)\}$.

(a) The graph of $f(x) \equiv (x+1)(2x-3)$ is a parabola meeting the x-axis at $(-1, 0)$ and $(1\frac{1}{2}, 0)$. As $x \to \pm\infty$, $y \to +\infty$. See the broken line in Fig. 7.14.

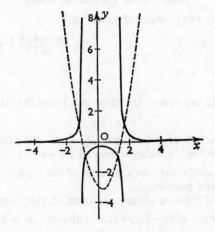

Fig. 7.14

(b) The reciprocal of f(x) is sketched as follows:
   (i) the signs of f(x) and 1/f(x) are the same,
   (ii) as f(x) → ∞, 1/f(x) → 0 and vice versa,
   (iii) when f(x) = ±1, 1/f(x) has the same value.

The two graphs are shown in Fig. 7.14.

Exercise 7c

Sketch the following curves:

1. $2x^2 + y^2 = 4$.
2. $2x^2 - y^2 = 4$.
3. $x^2 y = 1$.
4. $x^2 y^2 = 4$.
5. $y^2 = x(x-2)$.
6. $y^2 = x^3(4-x)$.
7. $y^2 = x^2(2-x)$.
8. $y^2 = x^2(x^2-1)$.
9. $y^2 = x^2(1-x^2)$. Express this equation in polar coordinates, either before or after sketching.
10. $y^2 = x^2(2-x)^3$.
11. $y^2 = (x^2-1)(4-x^2)$.
12. $y = 1/\{(x-2)(5-x)\}$.
13. $y = 1/(x^2-4x+3)$.
14. $xy(x^2-1) = 1$.
15. $x^2 - 1/y^2 = 1$.
16. $y = (x-2)^2 + 1$.
17. $(x-1)(y+2) = 1$.
18. $y = 1 + \frac{1}{2}\cos(x + \frac{1}{6}\pi)$.
19. $y = 3 - 2\sin(2x - \frac{1}{4}\pi)$.
20. $y = 1/(1 + 2\sin x)$, $0 \leqslant x \leqslant 4\pi$.
21. $y = 1/(1 - 2\cos 2x)$, $0 \leqslant x \leqslant 2\pi$.
22. $y = 1/(1 - \sin 2x)$, $0 \leqslant x \leqslant 2\pi$.
23. $y = 1/(1 + \sin^2 x)$, $0 \leqslant x \leqslant 2\pi$.

Exercise 7d (Miscellaneous)

1. Determine for what ranges of values of x

   (i) $x^2 - 4x + 3 > 0$,   (ii) $\dfrac{2x^2 - 4x + 5}{x^2 + 2} > 1$,

   (iii) $\dfrac{x^2 - 4x + 3}{x - 2} > 0$.   L.

2. By completing the square, or otherwise, prove that the inequality

$$x^2 - 2px + q > 0$$

   holds for all values of x if and only if $q > p^2$. Find the range of values of x for which the inequality is broken if $q = p^2 - 1$.   L.

3. Write down conditions that the roots of the equation $ax^2 + bx + c = 0$ may be real and positive.
   Prove that if these conditions are satisfied, the roots of the equation

$$a^2 y^2 + a(3b - 2c)y + (2b - c)(b - c) + ac = 0$$

   are real and positive.   L.

4. Show that, if λ is positive but not greater than 3, the roots of the equation

$$(\lambda-2)x^2-(8-2\lambda)x-(8-3\lambda) = 0$$

are real.

Find the range of values of λ for which one root is real and positive and the other root is real and negative. **L.**

5. Prove that ax^2+bx+c is positive for all real values of x if $a>0$ and $b^2<4ac$.

Find the range of values of k for which $x^2+kx+3+k$ is positive for all real values of x. Deduce the range of values of k for which $k(x^2+kx+3+k)$ is positive for all real values of x. **N.**

6. Show that, if a, b, c are real,

$$a^2+b^2+c^2-bc-ca-ab$$

cannot be negative.

Show also that the roots of the equation

$$3x^2-2x(a+b+c)+bc+ca+ab = 0$$

are real, and find the relation between a, b, c if one root is three times the other. **L.**

7. Find the ranges of values of α in the interval $0 \leqslant \alpha \leqslant 2\pi$ for which the roots of the equation in x

$$x^2 \cos^2 \alpha + ax(\sqrt{3} \cos \alpha + \sin \alpha) + a^2 = 0$$

are real. **C.**

8. Prove that $ax^2+2bx+c$ has the same sign as a except when the equation $ax^2+2bx+c=0$ has real roots and x lies between them.

By using the substitution $x=x'+4$, $y=y'-1$, or otherwise, prove that, for any real values of x and y,

$$x^2+2xy+3y^2-6x-2y \geqslant -11.$$ **N.**

9. Show that the expression $x^2+8xy-5y^2-k(x^2+y^2)$ can be put in the form $a(x+by)^2$ when k has either one or other of two values. Find these values and the values of a and b corresponding to each value of k.

Prove that when the variables x and y are restricted by the relation $x^2+y^2=1$ but are otherwise free, then

$$-7 \leqslant x^2+8xy-5y^2 \leqslant 3.$$ **O.C.**

10. If the roots of the equation $ax^2+2bx+c=0$ are real and unequal, prove that the function

$$(a+c)(ax^2+2bx+c)-2(ac-b^2)(x^2+1)$$

is positive for all real values of x. **N.**

11. Prove, by first squaring, that, if $0 < x < 1$,

$$(1-x)^{\frac{1}{2}} < \frac{4-3x}{4-x},$$

and deduce that

$$(1-x)^{\frac{1}{2}} > \frac{(4-x)(1-x)}{4-3x}.$$

Hence prove that $(1-x)^{\frac{1}{2}}$ differs from $\frac{4-3x}{4-x}$ by less than

$$\frac{x^3}{(4-x)(4-3x)}.$$

By putting $x = \frac{1}{9}$, show that $\sqrt{2}$ differs from $99/70$ by less than $2 . 10^{-4}$.　　　　　　　　　　　　　　　　　　　　　　　O.C.

12. (a) Find what restrictions must be imposed on the values of x and y in order to satisfy both the inequalities

$$x > y, \qquad \frac{x}{x+1} > \frac{y}{y+1}.$$

(b) Show that $b^2 + c^2 \geqslant \frac{1}{2}(b+c)^2$, and hence find the range of values of a for which the simultaneous equations

$$a + b + c = 1,$$
$$a^2 + b^2 + c^2 = 3$$

may be satisfied for real values of b and c.　　　　　　　　　　N.

13. (a) If $pq > 0$, prove that $p/q > 0$.
   Find
   (i) the range of values of x in which

$$\frac{1-4x}{2x-3} > 0,$$

   (ii) the ranges of values of x in which

$$\frac{2x}{x-1} + \frac{x-5}{x-2} > 3.$$

(b) In one diagram sketch the three straight lines

$$x + y - 3 = 0, \qquad y - \tfrac{1}{2}x - 5 = 0, \qquad y - 3x + 10 = 0,$$

and shade the region in which the following three inequalities are all satisfied:

$$x + y - 3 > 0, \qquad y - \tfrac{1}{2}x - 5 > 0, \qquad y - 3x + 10 > 0.$$

In a second diagram shade the region in which none of them is satisfied.

14. Prove that $-\frac{1}{4} \leqslant \dfrac{x}{x^2+4} \leqslant \frac{1}{4}$.

Draw in the same diagram the graphs of

$$y = \frac{x}{x^2+4}, \qquad y = x+1,$$

from $x = -4$ to $x = +4$, and deduce that the equation

$$\frac{x}{x^2+4} = x+1$$

has only one real root in this range.   Find the value of this root as accurately as you can.   o.c.

15. Prove that, if x is real, the function

$$(2x^2 - 5x + 2)/(x-1)$$

can assume all real values.

Sketch the graph of $y = (2x^2 - 5x + 2)/(x-1)$ from $x = -1$ to $x = +3$, omitting the portion given by the values of x very near to $+1$.   L.

16. Find for what value of the constant k the greatest value of the function

$$\frac{x^2 - 2x + 1}{x^2 + k}$$

is 3.

Sketch the graph of the function for this value of k.   L.

17. Prove that, if x is real and a and b are unequal, the expression

$$\frac{x^2 + ax + c}{x^2 + bx + c}$$

can take any value when $c \leqslant 0$.

Sketch the graph of

$$y = \frac{x^2 + 5x - 6}{x^2 - x - 6}.$$

                                                                        c.

18. The curve

$$y = \frac{px+q}{(x+1)(2-x)}$$

has a turning point at $(0, 1)$.   Determine the values of p and q.   Find the other turning point, and draw a sketch of the curve.   N.

19. Find the ranges of values of k for which the equation

$$\frac{3x-1}{x(x+1)} = k$$

has real roots. Hence write down the maximum value and the minimum value of the function

$$y = \frac{3x-1}{x(x+1)},$$

and find the corresponding values of x.

Sketch a graph of the function, showing clearly the behaviour of y as $x \to 0$, as $x \to -1$ and as $x \to \pm \infty$.   N.

20. Show that, if $y = \frac{x^2+1}{x^2-a^2}$, y takes all values twice, except those for which

$$-\frac{1}{a^2} \leqslant y \leqslant 1.$$

Sketch the curve $y = \frac{x^2+1}{x^2-4}$, indicating its asymptotes.   C.

21. (a) Express the function $8 + 2x - x^2$ as a sum or difference of two squares one of which is independent of x. *Deduce* the maximum or minimum value of the function. Sketch the graph of the function.

(b) Find the range, or ranges, of values of the constant k for which the line $y = k$ intersects the graph of the function

$$y = \frac{3x^2 - 10x + 3}{x^2 + 1}$$

in two different (real) points. State the values of k for which the two points of intersection coincide. Deduce the maximum and minimum values of the function.   N.

22. Prove that, if x is real and the constants a, b, c all different, the expression $(x-a)(x-b)/(x-c)$ can assume all real values if c lies between a and b.

Sketch the graphs of the functions

$$\frac{(x-1)(x-3)}{x-2} \quad \text{and} \quad \frac{(x-1)(x-2)}{x-3}. \qquad \text{L.}$$

23. Show that, if $(pz+q)/(z+1)$ is written for x in the expression

$$y \equiv \frac{2x^2 - 2x - 13}{2x^2 + 10x + 17},$$

values of p and q can be found such that the expression when simplified has the form

$$\frac{az^2 + b}{cz^2 + d}.$$

Hence, or otherwise, prove that, for all real values of x, $-1 \leqslant y \leqslant 3$.   C.

24. Prove that the equation

$$\frac{x^2 - x}{x^2 + x - 1} = p,$$

where $p \neq 1$, always has two real roots. If the roots are α and β, show that $\alpha(\alpha - 1)$ and $\beta(\beta - 1)$ have opposite signs unless $p = 0$.

Determine the range of values of p for which α and β are both positive. N.

25. Sketch, preferably on the same sheet of squared paper, the curves represented by the equations

$$\text{(i) } y = \frac{1}{x(x-2)}, \quad \text{(ii) } y^2 = \frac{1}{x(x-2)}. \quad \text{C.}$$

26. Sketch the graph of

$$y = \frac{x}{1 + x}$$

for positive values of x and find the area bounded by that portion of the curve corresponding to values of x from 0 to 2, the line $y = 0$ and the ordinate $x = 2$. Find also the volume of revolution of this region about the line $y = 0$. C.

27. Indicate by a sketch, which need not be exact, the general form of the curve whose equation is $y^2 = x^2(1 - x)$.

Find the equation of the tangent at the point $(\frac{3}{4}, \frac{3}{8})$ and find the coordinates of the point where this tangent meets the curve. C.

28. A curve is given by the equations

$$x = \frac{2t - 1}{1 - t}, \qquad y = \frac{t^2}{1 - t}.$$

Show that
  (i) y cannot lie between 0 and -4;
 (ii) as $t \to 1$, $x - y \to 0$;
(iii) as $t \to \infty$, $x \to -2$.
Sketch the curve, showing how it is described as t varies from $-\infty$ to $+\infty$. L.

29. Prove that the graph of the function

$$x^4 - 2x^3 + 8x - 4$$

(i) crosses the axis of x between $x = 0$ and $x = 1$ and also between $x = -2$ and $x = -1$, (ii) has a minimum point between $x = -1$ and $x = 0$. Find the coordinates of the points of inflexion and sketch the graph. N.

CHAPTER 8

FURTHER EQUATIONS AND FACTORS

Equations reducing to quadratics

8.1. Certain types of equation can be solved by reducing them to a quadratic equation and, as no new principles are involved, this topic is simply illustrated by examples. There is no need to read all the Examples 1 to 7 before attempting Exercise 8a—some readers will prefer to work the corresponding questions before going on to the next example.

Example 1. *Solve the equation*

$$\frac{x^2+4x}{3}+\frac{84}{x^2+4x} = 11.$$

Substitute $y = x^2 + 4x$ in the given equation.

$$\therefore \frac{y}{3}+\frac{84}{y} = 11.$$
$$\therefore y^2-33y+252 = 0.$$
$$\therefore (y-12)(y-21) = 0.$$

(i) If $y=12$,

$$x^2+4x-12 = 0.$$
$$\therefore (x+6)(x-2) = 0.$$
$$\therefore x = -6, 2.$$

(ii) If $y=21$,

$$x^2+4x-21 = 0.$$
$$\therefore (x+7)(x-3) = 0.$$
$$\therefore x = -7, 3.$$

Therefore the roots of the equation are $-7, -6, 2, 3$.

Example 2. *Solve the equation*

$$\sqrt{(5x-25)}-\sqrt{(x-1)} = 2,$$

where the positive values of the square roots are taken.

The method used is to isolate one square root on one side of the equation and then to square both sides.

$$\sqrt{(5x-25)}-\sqrt{(x-1)} = 2.$$
$$\therefore \sqrt{(5x-25)} = 2+\sqrt{(x-1)}.$$

$$\therefore\ 5x-25 = 4+4\sqrt{(x-1)}+x-1.$$
$$\therefore\ 4x-28 = 4\sqrt{(x-1)}.$$
$$\therefore\ x-7 = \sqrt{(x-1)}.$$

Squaring,

$$x^2-14x+49 = x-1.$$
$$\therefore\ x^2-15x+50 = 0.$$
$$\therefore\ (x-5)(x-10) = 0.$$

Now check the solutions $x=5$ and $x=10$.

If $x=5$, L.H.S. $= \sqrt{0}-\sqrt{4} = -2.$

Therefore 5 is not a root of the equation.

If $x=10$, L.H.S. $= \sqrt{25}-\sqrt{9} = 2.$

Therefore the only root of the equation is 10.

Example 2 illustrates the need for checking the roots obtained by squaring both sides of an equation. In Example 3 a similar equation is solved by means of the identity

$$(a-b)(a+b) \equiv a^2-b^2.$$

Note that it is again necessary to check the root obtained.

Example 3. *Solve the equation* $\sqrt{x}+\sqrt{(x-3)}=3.$

In the identity $a-b\equiv(a^2-b^2)\div(a+b)$, take $a=\sqrt{x}$, $b=\sqrt{(x-3)}$; then $a^2-b^2=3$. But $a+b=3$, therefore $a-b=1$. Thus

$$\sqrt{x}+\sqrt{(x-3)} = 3. \tag{1}$$
$$\sqrt{x}-\sqrt{(x-3)} = 1. \tag{2}$$

Adding, $2\sqrt{x} = 4.$
Hence $x = 4.$

Check: L.H.S. $= \sqrt{4}+\sqrt{1} = 3.$

Therefore $x=4$ is a solution of the equation. Now see Qu. 1.

Qu. 1. What value of x is obtained by subtracting the equations (1) and (2) in Example 3? Is this value of x a root of the given equation?

Qu. 2. Solve Example 3 by the method of Example 2.

Qu. 3. Show that, in solving Example 2 by the method of Example 3, the second equation is

$$\sqrt{(5x-25)}+\sqrt{(x-1)} = 2x-12.$$

Complete the solution by this method.

Example 4. *Find the coordinates of the points of intersection of the circles* $x^2+y^2-6x+4y-13 = 0$ *and* $x^2+y^2-10x+10y-15 = 0$.

To find the points of intersection, we solve simultaneously the equations

$$x^2+y^2-6x+4y-13 = 0, \tag{1}$$
$$x^2+y^2-10x+10y-15 = 0.$$

Subtracting, $\qquad\qquad 4x-6y+2 = 0.$

$$\therefore \ 2x = 3y-1 \quad \text{and} \quad 4x^2 = 9y^2-6y+1.$$

$4 \times (1)$: $\qquad\qquad 4x^2+4y^2-24x+16y-52 = 0.$

$$\therefore \ (9y^2-6y+1)+4y^2-12(3y-1)+16y-52 = 0.$$

$$\therefore \ 13y^2-26y-39 = 0.$$
$$\therefore \ 13(y+1)(y-3) = 0.$$

Substituting $\qquad y = -1 \quad \text{and} \quad y = 3 \qquad \text{in } 2x=3y-1,$
we obtain $\qquad x = -2 \quad \text{and} \quad x = 4, \qquad \text{respectively.}$

Therefore the circles meet at $(-2, -1)$, $(4, 3)$.

The next example uses one of the results (P.M.I., p. 178) that if α, β are the roots of the equation

$$ax^2+bx+c = 0,$$

then $\qquad\qquad \alpha+\beta = -b/a, \qquad \alpha\beta = c/a.$

Example 5. *Find where the normal at $(at^2, 2at)$ to the parabola $y^2 = 4ax$ cuts the curve again.*

The gradient at $(at^2, 2at)$ is given by

$$\frac{dy}{dx} = \frac{dy}{dt} \div \frac{dx}{dt} = \frac{2a}{2at} = \frac{1}{t}.$$

Therefore the gradient of the normal is $-t$ and its equation is

$$tx+y-at^3-2at = 0.$$

To solve simultaneously with $y^2 = 4ax$, multiply the former by $4a$:

$$t.4ax+4ay-4a^2t^3-8a^2t = 0.$$
$$\therefore \ ty^2+4ay-4a^2t^3-8a^2t = 0.$$

Now one root of the equation is $y = 2at$ (since the normal meets the curve at $(at^2, 2at)$). But the sum of the roots is $-4a/t$. Therefore the other root is

$$-\frac{4a}{t} - 2at = -2a(t + 2/t).$$

$$x = y^2/4a = a(t + 2/t)^2.$$

Therefore the normal meets the parabola again at

$$\big(a(t + 2/t)^2, \ -2a(t + 2/t)\big).$$

(For another method, see Exercise 9a, No. 7.)

Example 6. *Show that, if the equations*

$$x^2 + ax + 1 = 0 \quad and \quad x^2 + x + b = 0$$

have a common root, then

$$(b - 1)^2 = (a - 1)(1 - ab).$$

If x_1 is a root of both equations,

$$x_1{}^2 + ax_1 + 1 = 0, \qquad (1)$$
$$x_1{}^2 + x_1 + b = 0. \qquad (2)$$

We must now eliminate x_1 from equations (1), (2).

Subtracting, $\qquad x_1(a - 1) + 1 - b = 0.$

$$\therefore \ x_1 = \frac{(b - 1)}{(a - 1)}. \qquad (3) \qquad (a \neq 1)$$

We may now substitute for x_1 in (1) or (2), or proceed as follows.

$a \times (2) - (1): \qquad x_1{}^2(a - 1) + (ab - 1) = 0$

$$\therefore \ x_1{}^2 = \frac{(1 - ab)}{(a - 1)}. \qquad (4) \qquad (a \neq 1)$$

From (3) and (4), $\qquad \dfrac{(b - 1)^2}{(a - 1)^2} = \dfrac{1 - ab}{a - 1}.$

$$\therefore \ (b - 1)^2 = (a - 1)(1 - ab). \qquad (a \neq 1)$$

The next example shows a method which can be used to solve a quartic equation whose coefficients are arranged symmetrically in the form

$$Ax^4 + Bx^3 + Cx^2 + Bx + A = 0.$$

Example 7. *Use the substitution* $y = x + 1/x$ *to solve the equation*

$$2x^4 - 9x^3 + 14x^2 - 9x + 2 = 0.$$

$$y = x + \frac{1}{x}.$$

$$\therefore \ y^2 = x^2 + 2 + \frac{1}{x^2}.$$

Dividing both sides of the given equation by x^2,

$$2x^2 - 9x + 14 - \frac{9}{x} + \frac{2}{x^2} = 0.$$

$$\therefore \ 2\left(x^2 + 2 + \frac{1}{x^2}\right) - 9\left(x + \frac{1}{x}\right) + 10 = 0.$$

$$\therefore \ 2y^2 - 9y + 10 = 0.$$
$$\therefore \ (y - 2)(2y - 5) = 0.$$

(i) If $y = 2$,

$$x + \frac{1}{x} = 2.$$

$$\therefore \ x^2 - 2x + 1 = 0.$$
$$\therefore \ (x - 1)^2 = 0.$$
$$\therefore \ x = 1.$$

(ii) If $y = \frac{5}{2}$,

$$x + \frac{1}{x} = \frac{5}{2}.$$

$$\therefore \ 2x^2 - 5x + 2 = 0.$$
$$\therefore \ (2x - 1)(x - 2) = 0.$$
$$\therefore \ x = \frac{1}{2}, 2.$$

Therefore the roots of the equation are $\frac{1}{2}$, 1, 1, 2.

Qu. 4. Show that the substitution $y = x + 1/x$ transforms the quadratic equation $ay^2 + by + c = 0$ to the form

$$ax^2 + bx + d + b/x + a/x^2 = 0.$$

Exercise 8a

Solve the equations in Nos. 1 to 19.

1. $(x^2 - 2x)^2 + 24 = 11(x^2 - 2x)$.
2. $x^2 + 2x = 34 + 35/(x^2 + 2x)$.
3. $2 - 5e^{-x} + 2e^{-2x} = 0$.
4. $4^x - 5.2^x + 4 = 0$.
5. $x^2 + 9/x^2 = 10$.
6. $x^{\frac{2}{3}} + 16x^{-\frac{2}{3}} = 17$.
7. $\sqrt{(x+1)} + \sqrt{(x-2)} = 3$.
8. $\sqrt{(x-5)} + \sqrt{x} = 5$.
9. $\sqrt{(3x-3)} - \sqrt{x} = 1$.
10. $2\sqrt{(x+4)} - \sqrt{(x-1)} = 4$.
11. $2\sqrt{(x-1)} + \sqrt{(x-4)} = x$.
12. $\sqrt{(x-1)} + 2\sqrt{(x-4)} = 4$.
13. $2\sqrt{(2x-12)} - \sqrt{(2x-3)} = 3$.
14. $xy = 4$, $x - 2y - 2 = 0$.
15. $x^2 + y^2 - 2x - 2y - 23 = 0$, $x - 7y + 31 = 0$.
16. $3x - 4y - 5 = 0$, $x^2 + y^2 + 2x + 4y - 20 = 0$.

17. $x^2 + y^2 - 8x + 6y = 0$, $x^2 + y^2 - 5x + 10y = 0$.
18. $x^2 + y^2 + 8x - 4y + 15 = 0$, $x^2 + y^2 + 6x + 2y - 15 = 0$.
19. $4x^2 + 25y^2 = 100$, $xy = 4$.
20. One root of the equation in x

$$bx^2 - x(ab + 2a + 2b) + 2a(a + b) = 0$$

is a. Use the formulae for (i) the sum, (ii) the product of the roots of a quadratic equation to find the other root of the equation.
21. Repeat No. 20 for the equations:
   (i) $cx^2 - acx + dx + cx - ad - ac = 0$,
   (ii) $ax^2 - bx^2 - a^2x + abx + ax + bx - a^2 - ab = 0$.
22. Find the equation of the normal to $xy = c^2$ at $(ct, c/t)$ and obtain the coordinates of the point where the normal cuts the curve again.
23. A chord of gradient 2 passes through the point $(ap^2, 2ap)$ on the parabola $y^2 = 4ax$. Find the coordinates of the other end of the chord.
24. A line with gradient t cuts the rectangular hyperbola at $(ct, c/t)$. Find the coordinates of the other intersection.
25. For the ellipse $b^2x^2 + a^2y^2 = a^2b^2$, find the coordinates of the other end of the chord through $(a, 0)$ with gradient a/b.
26. Find the condition that the equations

$$x^2 + 2x + a = 0, \qquad x^2 + bx + 3 = 0$$

should have a common root.
27. Show that, if the equations

$$x^2 + 2px + q = 0, \qquad x^2 + 2Px + Q = 0,$$

have a common root, then,

$$(q - Q)^2 + 4(P - p)(Pq - pQ) = 0.$$

28. Solve the equations

$$ay + bx + c = 0, \qquad Ay + Bx + C = 0$$

and deduce the condition that the equations

$$ax^2 + bx + c = 0, \qquad Ax^2 + Bx + C = 0$$

should have a common root.
29. If the equations

$$x^2 + ax + 6 = 0, \qquad x^2 + x + b = 0$$

have a common root, and the difference between the other two roots of the equations is 6, find the possible values of b.
30. Show that, if the equations

$$x^3 + ax^2 + bx + c = 0, \qquad x^2 + ax + 2b = 0$$

have a common root, then $c^2 + abc + 2b^3 = 0$. Show further that if a, b, c are real $|a| > 2\sqrt{2b}$.

Solve the equations in Nos. 31 to 33.

31. $6x^4 - 35x^3 + 62x^2 - 35x + 6 = 0$.

32. $4x^4 + 17x^3 + 8x^2 + 17x + 4 = 0$.

33. $5x^4 - 21x^3 - 16x^2 - 21x + 5 = 0$.

34. Express the equation $Ax^4 + Bx^3 + Cx^2 + Bx + A = 0$ in quadratic form by means of a suitable substitution.

35. Use the substitution $y = x + 2/x$ to solve the equation

$$x^4 - 5x^3 + 10x^2 - 10x + 4 = 0.$$

36. In a right-angled triangle, one of the sides including the right angle is 7 cm longer than the other. If the perimeter is 40 cm, find the lengths of the three sides.

37. Two small pulleys are placed 8 cm apart in a horizontal line and an inextensible string of length 16 cm is placed over the pulleys. Equal masses hang symmetrically at each end of the string and the middle point is pulled down vertically until it is in line with the masses. How far does each mass rise?

38. The lines $y = mx + c$ meets the ellipse $b^2x^2 + a^2y^2 = a^2b^2$ at (x_1, y_1) and (x_2, y_2). Write down a quadratic equation in x whose roots are x_1, x_2, and a similar quadratic in y. Hence, or otherwise, find the coordinates of the mid-point of the chord in terms of a, b, c, m.

A theorem about ratios

8.2. There is a theorem about ratios which on occasions can greatly simplify algebraic working and which sometimes shortens working in trigonometry:

If $\dfrac{a}{b} = \dfrac{c}{d} = \dfrac{e}{f} = \ldots$, then for any numbers $l, m, n \ldots$, not all zero,

$$\frac{a}{b} = \frac{c}{d} = \frac{e}{f} = \ldots = \frac{la + mc + ne + \ldots}{lb + md + nf + \ldots}.$$

To prove this, let $\dfrac{a}{b} = \dfrac{c}{d} = \dfrac{e}{f} = \ldots = k$, say.

$$\therefore \quad a = bk, \qquad c = dk, \qquad e = fk \text{ etc.}$$

$$\therefore \quad \frac{la + mc + ne + \ldots}{lb + md + nf + \ldots} = \frac{lbk + mdk + nfk + \ldots}{lb + md + nf + \ldots},$$

$$= \frac{k(lb + md + nf + \ldots)}{lb + md + nf + \ldots},$$

$$= k.$$

$$\therefore \frac{a}{b} = \frac{c}{d} = \frac{e}{f} = \dots = \frac{la+mc+ne+\dots}{lb+md+nf+\dots}.$$

Qu. 5. With $a=3$, $b=5$, $c=6$, $d=10$, $e=12$, $f=20$, $l=5$, $m=-2$, $n=-1$, or other suitable numbers, verify that the above ratios are equal.

Express the result of the theorem in your own words.

Example 8. *If* $\frac{a}{b}=\frac{c}{d}$, *prove that* $\frac{a-c}{a+c}=\frac{b-d}{b+d}$.

Let
$$\frac{a}{b} = \frac{c}{d} = k.$$

$$\therefore a = bk, \quad c = dk.$$

$$\therefore \frac{a-c}{a+c} = \frac{bk-dk}{bk+dk}.$$

$$\therefore \frac{a-c}{a+c} = \frac{b-d}{b+d}.$$

Qu. 6. With $a=6$, $b=8$, $c=3$, $d=4$, or other suitable numbers, check the results of Example 8.

Qu. 7. If $\frac{a}{b}=\frac{c}{d}$, prove that $\frac{a-b}{a+b} = \frac{c-d}{c+d}$.

★ **Qu. 8.** Prove that, if $\frac{a}{b}=\frac{c}{d}$, then

$$\frac{la+mb}{\lambda a+\mu b} = \frac{lc+md}{\lambda c+\mu d},$$

where l, m, λ, μ are any numbers such that λ, μ are not both zero.

Example 9. *If* $\frac{a}{b}=\frac{c}{d}$, *prove that* $\frac{a^2-b^2}{a^2+b^2}=\frac{c^2-d^2}{c^2+d^2}$.

Let
$$\frac{a}{b} = \frac{c}{d} = k, \qquad \therefore a = bk, \quad c = dk.$$

$$\therefore \frac{a^2-b^2}{a^2+b^2} = \frac{b^2k^2-b^2}{b^2k^2+b^2} = \frac{k^2-1}{k^2+1}.$$

Similarly
$$\frac{c^2-d^2}{c^2+d^2} = \frac{k^2-1}{k^2+1},$$

and the result is proved.

(Nos. 1–14 of Exercise 8b are on the above work.)

Example 10. *If* $\dfrac{\sin(\theta+\phi)}{\sin(\theta-\phi)} = \dfrac{b}{a}$, *prove that* $\tan\theta\cot\phi = \dfrac{b+a}{b-a}$.

$$\frac{\sin(\theta+\phi)}{\sin(\theta-\phi)} = \frac{b}{a}.$$

$$\therefore \frac{b+a}{b-a} = \frac{\sin(\theta+\phi)+\sin(\theta-\phi)}{\sin(\theta+\phi)-\sin(\theta-\phi)},$$

$$= \frac{2\sin\theta\cos\phi}{2\cos\theta\sin\phi},$$

$$\therefore \frac{b+a}{b-a} = \tan\theta\cot\phi.$$

Example 11. *Prove that, in triangle* ABC,

$$\frac{b-c}{b+c} = \tan\frac{B-C}{2}\cot\frac{B+C}{2}.$$

By the sine formula,

$$\frac{b}{\sin B} = \frac{c}{\sin C}.$$

$$\therefore \frac{b-c}{b+c} = \frac{\sin B-\sin C}{\sin B+\sin C},$$

$$= \frac{2\cos\frac{1}{2}(B+C)\sin\frac{1}{2}(B-C)}{2\sin\frac{1}{2}(B+C)\cos\frac{1}{2}(B-C)},$$

$$\therefore \frac{b-c}{b+c} = \cot\frac{1}{2}(B+C)\tan\frac{1}{2}(B-C).$$

(Now work Nos. 15–26 of Exercise 8b.)

Example 12. *Solve the equations*:

$$\frac{2x+z}{3} = \frac{y+3z}{4} = \frac{2x+3y}{5}, \qquad x+y+z = 1.$$

$$\frac{2x+z}{3} = \frac{y+3z}{4} = \frac{2x+3y}{5} = \frac{(2x+z)+(y+3z)+(2x+3y)}{3+4+5},$$

$$= \frac{4x+4y+4z}{12} = \tfrac{4}{12} = \tfrac{1}{3}.$$

$$\therefore \quad 2x+z = 1, \tag{1}$$
$$y+3z = \tfrac{4}{3}, \tag{2}$$
$$2x+3y = \tfrac{5}{3}. \tag{3}$$

$(3)-(1):$
$$3y-z = \tfrac{2}{3}. \tag{4}$$

$(2)+3\times(4):$
$$10y = 3\tfrac{1}{3}.$$
$$\therefore \quad y = \tfrac{1}{3}.$$

Hence $\qquad x = \tfrac{1}{3}, \quad y = \tfrac{1}{3}, \quad z = \tfrac{1}{3}.$

Note the method for solving three linear simultaneous equations: eliminate one variable so that two equations in two variables remain. There now follows another example.

Example 13. *Solve the simultaneous equations.*

$$2x+3y+4z = 8, \qquad 3x-2y-3z = -2, \qquad 5x+4y+2z = 3.$$

$$2x+3y+4z = 8, \tag{1}$$
$$3x-2y-3z = -2, \tag{2}$$
$$5x+4y+2z = 3. \tag{3}$$

y and z may be eliminated equally easily. We eliminate y as follows:

$2\times(2)+(3):$
$$11x-4z = -1. \tag{4}$$

$3\times(3)-4\times(1):$
$$7x-10z = -23. \tag{5}$$

$5\times(4)-2\times(5):$
$$41x = 41.$$

Hence $\qquad x = 1, \quad z = 3, \quad y = -2.$

Qu. 9. Solve Example 13 by eliminating z instead of y.

(Now work Nos. 27–35 of Exercise 8b.)

Exercise 8b

In Nos. 1–14 it is given that $\dfrac{a}{b} = \dfrac{c}{d} = \dfrac{e}{f}$.

Complete the statements in Nos. 1 to 4.

1. $\dfrac{a}{b} = \dfrac{c}{d} = \dfrac{2a-c}{}.$

2. $\dfrac{a}{b} = \dfrac{c}{d} = \dfrac{}{3b-4d}.$

3. $\dfrac{a-c}{b-d} = \dfrac{2a+3c}{}.$

4. $\dfrac{a+c}{b+d} = \dfrac{}{3b-d}.$

Prove the results in Nos. 5 to 14.

5. $\dfrac{a+2c}{b+2d} = \dfrac{3a+c}{3b+d}.$

6. $\dfrac{a-c+e}{b-d+f} = \dfrac{a+c-e}{b+d-}$

7. $\dfrac{a^2+c^2}{b^2+d^2} = \dfrac{c^2+e^2}{d^2+f^2}$.

8. $\dfrac{a^2}{b^2} = \dfrac{c^2-e^2}{d^2-f^2}$.

9. $\dfrac{a+2b}{c+2d} = \dfrac{2a+b}{2c+d}$.

10. $\dfrac{3c+2e}{3d+2f} = \dfrac{c+e}{d+f}$.

11. $\dfrac{a^2+b^2}{a^2-b^2} = \dfrac{c^2+d^2}{c^2-d^2}$.

12. $\dfrac{a^3+c^3+e^3}{b^3+d^3+f^3} = \dfrac{ace}{bdf}$.

13. $\sqrt{\dfrac{a^2+c^2}{b^2+d^2}} = \dfrac{a+c}{b+d}$.

14. $\sqrt{\dfrac{ac}{bd}} = \dfrac{a+c+e}{b+d+f}$.

If $\tan\theta = \dfrac{a}{b}$, prove the results in Nos. 15 to 18:

15. $\dfrac{a}{a+b} = \dfrac{\sin\theta}{\sqrt 2\,\sin(\theta+\frac14\pi)}$.

16. $\dfrac{a+b}{a-b} = -\tan(\theta+\frac14\pi)$.

17. $\dfrac{a^2+b^2}{a^2-b^2} = -\sec 2\theta$.

18. $\left(\dfrac{a+b}{a-b}\right)^2 = \dfrac{1+\sin 2\theta}{1-\sin 2\theta}$.

19. If $\dfrac{\sin 2\theta}{\sin 2\phi} = \dfrac{p}{q}$, prove that

$$\dfrac{p+q}{p-q} = \tan(\theta+\phi)\cot(\theta-\phi).$$

20. If $\dfrac{\sin\theta+\cos\theta}{\sin\theta-\cos\theta} = \dfrac{c}{d}$, prove that:

(i) $\dfrac{c+d}{c-d} = \tan\theta$,

(ii) $\dfrac{c^2+d^2}{c^2-d^2} = \operatorname{cosec} 2\theta$.

21. If $\dfrac{b-c}{b+c} = \dfrac{\tan\theta}{\tan\phi}$, prove that $\dfrac{b}{c} = \dfrac{\sin(\theta+\phi)}{\sin(\phi-\theta)}$.

Use the sine formula to prove that, in triangle ABC,

22. $\sin\frac12(B-C) = \dfrac{b-c}{a}\cos\frac12 A$.

23. $\cos\frac12(B-C) = \dfrac{b+c}{a}\sin\frac12 A$.

24. $\dfrac{a+b-c}{a-b+c} = \tan\frac12 B\cot\frac12 C$.

25. $\dfrac{a+b+c}{a+b-c} = \cot\frac12 A\cot\frac12 B$.

26. $a\cos 2B + 2b\cos A\cos B = c\cos B - b\cos C$.

Solve the equations:

27. $\dfrac{a}{3} = \dfrac{b}{4} = -\dfrac{c}{6}$, $\quad 2a+3b-c = 2$.

28. $-15a = 10b = 6c$, $\quad 4a+3b+c = 6$.

29. $\dfrac{y+z}{5} = \dfrac{z+x}{8} = \dfrac{x+y}{9}$, $\quad 6(x+y+z) = 11$.

30. $\dfrac{y+2z}{9} = \dfrac{z+2x}{10} = \dfrac{x+2y}{14}$, $\quad x+y+z = 11$.

31. $2a+b+3c = 11$,
    $a+2b-2c = 3$,
    $4a+3b+c = 15$.

32. $3p+2q+5r = 7$,
    $2p-4q+9r = 9$,
    $6p-8q+3r = 4$.

33. $2x+3y+4z = -4,$
$4x+2y+3z = -11,$
$3x+4y+2z = -3.$

34. $a-3b+6c = 5,$
$a+6b+2c = 4,$
$2a+b+c = 7.$

35. $d-2e+3f = 4,$
$5d+6e-7f = 8,$
$7d-5e+6f = 4.$

Homogeneous expressions

8.3. An expression, or equation, is said to be **homogeneous** if every term is of the same degree. For instance, x^2z+3x^3, $3x+2y-4z$, $1/x+1/y+1/z$ are homogeneous expressions of degree 3, 1, -1 respectively.

Qu. 10. Which of the following expressions are homogeneous? State the degree of those that are:

(i) $x^3+y^3+z^3$,

(ii) $3yz+zx-2xy$,

(iii) $x^2+y^2+2x+2y$,

(iv) $\dfrac{1}{yz}+\dfrac{1}{zx}+\dfrac{1}{xy}$,

(v) $x^3+y^3z+z^3x$,

(vi) $x^4+y^2z^2+xyz^2$.

Example 14. *Solve, for the ratio $x:y$, the equation $x^2-3xy-40y^2=0$.*

$$x^2-3xy-40y^2 = 0.$$
$$\therefore (x-8y)(x+5y) = 0.$$

$$\therefore x-8y = 0 \quad \text{or} \quad x+5y = 0.$$

$$\therefore x:y = 8 \text{ or } -5.$$

Example 15. *Solve for $x:y:z$ the equations*

$$6x-5y-6z = 0, \qquad 10x+7y-33z = 0.$$

$$6x-5y-6z = 0. \tag{1}$$
$$10x+7y-33z = 0. \tag{2}$$

$5\times(1):$ $\quad 30x-25y-30z = 0.$
$-3\times(2):$ $\quad -30x-21y+99z = 0.$
Adding, $\qquad\qquad -46y+69z = 0.$
$$\therefore -2y+3z = 0. \tag{3}$$

Substituting $3z=2y$ in (1),

$$6x-5y-4y = 0.$$
$$\therefore 6x-9y = 0.$$
$$\therefore 2x-3y = 0. \tag{4}$$

Writing (4), (3) as ratios,

$$x:y = 3:2, \qquad y:z = 3:2.$$

Writing these ratios so that the number corresponding to y is the same in each,

$$x:y = 9:6, \qquad y:z = 6:4.$$

$$\therefore \; x:y:z = 9:6:4.$$

It may be noticed that a number of the equations used in coordinate geometry are homogeneous in x, y, a, b, c, e.g.: the parabola $y^2 = 4ax$, the ellipse $b^2x^2 + a^2y^2 = a^2b^2$, the rectangular hyperbola $xy = c^2$. One advantage of using homogeneous equations is that equations and expressions derived from them are homogeneous, which provides a method of detecting slips. For example the following equations cannot have been derived correctly from equations homogeneous in x, y, a, b.

$$x^2 + y^2 = ax + b; \qquad x - y - ab = 0;$$

$$\frac{a^2x^2 + b^2y^2}{xy} = \frac{x^2 + y^2}{x}; \qquad \frac{x}{a} + y = b.$$

The reader who is used to checking dimensions in applied mathematics and physics will see that the expressions, 'A homogeneous expression of degree n', and 'An expression whose terms all have dimensions $[L^n]$', are equivalent. Thus the terms of the equations $ay^2 = x^3$, $x^2 + y^2 = a^2$, $b^2x^2 - a^2y^2 = a^2b^2$ have dimensions $[L^3]$, $[L^2]$, $[L^4]$, and they are homogeneous expressions in x, y, a, b of degree 3, 2, 4, respectively.

Consider now the equation of the tangent to $y^2 = 4ax$ (homogeneous in x, y, a) at $P(at^2, 2at)$. The coordinates of P are lengths, so that t is a ratio, or it may be said to have dimensions $[L^0]$.

$$\frac{dy}{dx} = \frac{dy}{dt} \div \frac{dx}{dt} = \frac{2a}{2at} = \frac{1}{t}.$$

Check: dy/dx is a ratio, so is $1/t$.

Therefore the tangent at P is

$$x - ty + at^2 = 0.$$

Check: the equation is homogeneous in x, y, a.

Symmetrical and cyclic expressions

8.4. Symmetrical functions of α, β were introduced in P.M.I., p. 181. They are expressions such as $\alpha + \beta + 2\alpha\beta$, $\alpha/\beta + \beta/\alpha$, which are unchanged when α, β are interchanged. This idea can easily be extended to functions of several variables. For instance, $yz + zx + xy$, $x^3 + y^3 + z^3 + a^3$ are unchanged when any two of the variables included in the expressions are interchanged. These, then, are said to be symmetrical in x, y, z and x, y, z, a respectively.

On the other hand, expressions such as

$$(b-c)(c-a)(a-b), \qquad a^2(b-c) + b^2(c-a) + c^2(a-b)$$

are changed if two of a, b, c are interchanged, but are unaltered if a is replaced by b, b by c, and c by a according to the diagram below:

Fig. 8.1

Such expressions are said to be **cyclic** in a, b, c.

When dealing with cyclic expressions in, say, three variables a, b, c, the $\sum$ notation (P.M.I., p. 230) gives a convenient shorthand. Thus

$$\sum bc = bc + ca + ab, \tag{1}$$

$$\sum a^2(b-c) = a^2(b-c) + b^2(c-a) + c^2(a-b). \tag{2}$$

The term after the $\sum$ determines the others which are written down according to the method indicated in Fig. 8.1. Note the order of the terms: in (1) the position of a term is determined by the letter it *lacks*, in (2) the letter which is squared is also the letter lacking in the brackets, and it determines the positions of the terms.

Qu. 11. Write in full the cyclic expressions in the three variables a, b, c given by:

(i) $\sum a^3$,   (ii) $\sum a(b+c)$,   (iii) $\sum \dfrac{1}{a}$,   (iv) $\sum ab^2c^2$.

Some useful identities

8.5. Certain identities will be needed occasionally in this book and for convenience they are grouped together here, which is appropriate because the identities are given in homogeneous forms:

$$(a+b)^3 \equiv a^3+3a^2b+3ab^2+b^3. \tag{1}$$
$$(a-b)^3 \equiv a^3-3a^2b+3ab^2-b^3. \tag{2}$$

$$a^3+b^3 \equiv (a+b)(a^2-ab+b^2). \tag{3}$$
$$a^3-b^3 \equiv (a-b)(a^2+ab+b^2). \tag{4}$$

$$a^3+b^3+c^3-3abc \equiv (a+b+c)(a^2+b^2+c^2-bc-ca-ab). \tag{5}$$

(1) and (2) follow from the binomial theorem and are easily written down with the help of Pascal's triangle (P.M.I., p. 238).

(3) and (4) may be obtained as follows. Consider a^3+b^3 as a function of a. When we substitute $a=-b$ the expression vanishes and so, by the remainder theorem (P.M.I., p. 184) $(a+b)$ is a factor. The other factor is found by inspection or long division.

Similarly, a^3-b^3 vanishes when the substitution $b=a$ is made, therefore $(a-b)$ is a factor.

Qu. 12. Verify the identities (3) and (4).

Identity (5) may be verified by long multiplication of the right-hand side, but it is more instructive to proceed as indicated in the table below.

Term in	Factors from brackets:		Result
	First	Second	
a^3	a	a^2	a^3
a^2b	$\begin{cases} a \\ b \end{cases}$	$\begin{cases} -ab \\ a^2 \end{cases}$	0
abc	$\begin{cases} a \\ b \\ c \end{cases}$	$\begin{cases} -bc \\ -ca \\ -ab \end{cases}$	$-3abc$

Not only is the term in a^2b zero, but since the right-hand side of (5) is symmetrical in a, b, c, the terms in a^2c, b^2c, b^2a, c^2a, c^2b are zero too. Hence the right-hand side is $a^3+b^3+c^3-3abc$.

Qu. 13.   Verify the following identities by the method above:

   (i) $(a+b+c)^2 = \sum a^2 + 2 \sum bc$.

   (ii) $(a+b+c)^3 = \sum a^3 + 3 \sum a^2(b+c) + 6abc$.

   (iii) $(a+b+c)^4 = \sum a^4 + 4 \sum a^3(b+c) + 12 \sum a^2bc + 6 \sum b^2c^2$.

Qu. 14.   Check the identities in Qu. 13 by the substitutions $a=b=c=1$. (This check does not prove they are correct but it is worth doing when you have expanded expressions like these.)

Example 16.   *Factorize*:

$$(i)\ 27a^3b^6 - 8c^3,$$
$$(ii)\ a^3 + b^3 + c^3 + 3ac(a+c).$$

(i) $27a^3b^6 - 8c^3 \equiv (3ab^2)^3 - (2c)^3$,
$$\equiv (3ab^2 - 2c)(9a^2b^4 + 6ab^2c + 4c^2).$$

(ii) $a^3 + b^3 + c^3 + 3ac(a+c) \equiv b^3 + a^3 + 3a^2c + 3ac^2 + c^3$,
$$\equiv b^3 + (a+c)^3,$$
$$\equiv \{b + (a+c)\}\{b^2 - b(a+c) + (a+c)^2\},$$
$$\equiv (a+b+c)(a^2+b^2+c^2-bc+2ca--ab).$$

Exercise 8c

Solve the equations in Nos. 1–6 for the ratio $x : y$.

1. $6x^2 - xy - 12y^2 = 0$.

2. $2x^2 - 7xy - 30y^2 = 0$.

3. $x^3 - 3x^2y + 4y^3 = 0$.

4. $6x^3 + 7x^2y - 7xy^2 - 6y^3 = 0$.

5. $4x^4 - 37x^2y^2 + 9y^4 = 0$.

6. $(x^2 - 2xy)/y^2 - 15y^2/(x^2 - 2xy) = 14$.

Solve for the ratios $x : y : z$:

7. $2x + 3y - z = 0$, $3x - 2y + 4z = 0$.

8. $4x - 5y + 6z = 0$, $2x + 3y - 4z = 0$.

9. $ax + by + cz = 0$, $bx + ay - cz = 0$.

10. $a^2x + ay + z = 0$, $x + ay + a^2z = 0$.

11. $x \cos \theta - y \sin \theta + z = 0$, $x \sin \theta + y \cos \theta - z = 0$.

12. $x - t_1y + t_1^2z = 0$, $x - t_2y + t_2^2z = 0$.

Write in full the cyclic functions in x, y, z given by:

13. $\sum x^4$.

14. $\sum 1/yz$.

15. $\sum x^2(y+z)$.

16. $\sum x^2y$.

17. $\sum xy^2$.

18. $\sum x^3yz$.

Show that:

19. $\sum x(y-z) = 0$.

20. $\sum x^2(y+z) = \sum x(y^2+z^2)$.

21. $\sum x(y+z) = 2 \sum yz$.

22. $\sum x^3(y-z) = - \sum x(y^3-z^3)$.

Factorize:

23. $1 - t^3$.
24. $64x^3 + y^3$.
25. $8 + 27z^3$.
26. $125y^3 - z^6$.
27. $a^3 + 6a^2b + 12ab^2 + 8b^3$.
28. $27u^3 - 27u^2 + 9u - 1$.
29. $(a+b)^3 - (a+c)^3$.
30. $(x-y)^3 + (x+y)^3$.
31. $a^2 + b^2 + c^2 + 2bc + 2ca + 2ab$.
32. $a^2 + b^2 + c^2 - 2bc - 2ca + 2ab$.
33. $a^3 + b^3 + c^3 + 3bc(b+c)$.
34. $a^3 + 8b^3 + 27c^3 - 18abc$.
35. $a^6 - b^6$.

*36. Use the result of No. 35, together with the identity

$$a^6 - b^6 \equiv (a^2 - b^2)(a^4 + a^2b^2 + b^4)$$

to show that

$$a^4 + a^2b^2 + b^4 \equiv (a^2 + ab + b^2)(a^2 - ab + b^2).$$

*37. Find the sum of the geometric progression

$$x^{n-1} + x^{n-2}a + \ldots + xa^{n-2} + a^{n-1}.$$

Hence, or otherwise, show that

$$x^n - a^n \equiv (x-a)(x^{n-1} + x^{n-2}a + \ldots + xa^{n-2} + a^{n-1}).$$

*38. Given the polynomial

$$f(x) \equiv b_n x^n + b_{n-1} x^{n-1} + \ldots + b_1 x + b_0,$$

where b_r is a constant, use the result of No. 37 to show that

$$f(x) - f(a) \equiv (x-a)(c_{n-1} x^{n-1} + \ldots + c_1 x + c_0),$$

where c_r is a constant. (This proves the remainder theorem.)

Roots of cubic equations

8.6. It has been shown (P.M.I., p. 178) that if α, β are the roots of the equation $ax^2 + bx + c = 0$, then $\alpha + \beta = -b/a$, $\alpha\beta = c/a$.

We now consider the cubic equation

$$ax^3 + bx^2 + cx + d = 0 \quad (a \neq 0).$$

Let α, β, γ be the roots of the equation

$$ax^3 + bx^2 + cx + d = 0.$$

Now the equation with roots α, β, γ may be written

$$(x-\alpha)(x-\beta)(x-\gamma) = 0.$$

$$\therefore \quad x^3 - (\alpha+\beta+\gamma)x^2 + (\beta\gamma+\gamma\alpha+\alpha\beta)x - \alpha\beta\gamma = 0.$$

Writing the original equation as

$$x^3 + \frac{b}{a} x^2 + \frac{c}{a} x + \frac{d}{a} = 0$$

and equating coefficients of x^2, x, and the constant terms,

$$\alpha+\beta+\gamma = -b/a,$$
$$\beta\gamma+\gamma\alpha+\alpha\beta = c/a,$$
$$\alpha\beta\gamma = -d/a.$$

Qu. 15. Write down the sum, the sum of the products in pairs, and the product of the roots of the equations:

(i) $3x^3 - 4x^2 + 2x + 5 = 0$,   (ii) $x^3 = 1$,

(iii) $7x^3 + 6x - 5 = 0$,   (iv) $(x+1)^3 = (x+2)^2$,

(v) $x^3 - 5x^2 + 2 = 0$,   (vi) $x^3 + x^2 + x + 1 = 0$.

Qu. 16. Write down the equations whose roots have the following sums, sums of products in pairs, and products respectively:

(i) 6, 11, 6;   (ii) 0, -13, -12,   (iii) 14, 0, -288.

Example 17. *The equation* $3x^3 + 6x^2 - 4x + 7 = 0$ *has roots* α, β, γ. *Find the equations with roots* (i) $1/\alpha$, $1/\beta$, $1/\gamma$, (ii) $\beta+\gamma$, $\gamma+\alpha$, $\alpha+\beta$.

(i) If x is a root of the given equation

$$3x^3 + 6x^2 - 4x + 7 = 0.$$

then $y = 1/x$ is a root of the required equation. Substituting $x = 1/y$, it follows that the required equation is

$$\frac{3}{y^3} + \frac{6}{y^2} - \frac{4}{y} + 7 = 0.$$

i.e. $7y^3 - 4y^2 + 6y + 3 = 0.$

(ii) If x is a root of the given equation, then $y = \alpha + \beta + \gamma - x$ is a root of the required equation.

$$\alpha + \beta + \gamma = -6/3.$$

$$\therefore\ y = -2 - x\ \text{ or }\ x = -y - 2.$$

$$3x^3 + 6x^2 - 4x + 7 = 0.$$

$$\therefore\ 3(-y^3 - 6y^2 - 12y - 8) + 6(y^2 + 4y + 4) - 4(-y - 2) + 7 = 0.$$

$$\therefore\ -3y^3 - 12y^2 - 8y + 15 = 0.$$

Therefore the required equation is

$$3y^3 + 12y^2 + 8y - 15 = 0.$$

Qu. 17. What substitutions would have been required in Example 17 to find the equations with roots:

(i) α^2, β^2, γ^2,   (ii) $\alpha-2$, $\beta-2$, $\gamma-2$.

(iii) $2\alpha+1$, $2\beta+1$, $2\gamma+1$,   (iv) $\dfrac{1}{\beta\gamma}$, $\dfrac{1}{\gamma\alpha}$, $\dfrac{1}{\alpha\beta}$?

Example 18. *Solve the equations*

$$\alpha+\beta+\gamma = 4, \qquad \alpha^2+\beta^2+\gamma^2 = 66, \qquad \alpha^3+\beta^3+\gamma^3 = 280.$$

The method used is to form an equation with roots α, β, γ; in order to find the values of $\sum \beta\gamma$ and $\alpha\beta\gamma$ we use the identities:

$$(\alpha+\beta+\gamma)^2 = \alpha^2+\beta^2+\gamma^2+2(\beta\gamma+\gamma\alpha+\alpha\beta), \tag{1}$$

$$\alpha^3+\beta^3+\gamma^3-3\alpha\beta\gamma = (\alpha+\beta+\gamma)(\alpha^2+\beta^2+\gamma^2-\beta\gamma-\gamma\alpha-\alpha\beta). \tag{2}$$

From (1): $\qquad\qquad 16 = 66+2\sum \beta\gamma.$

$$\therefore \sum \beta\gamma = -25.$$

From (2): $\qquad\qquad 280-3\alpha\beta\gamma = 4(66+25).$

$$\therefore \alpha\beta\gamma = -28.$$

Therefore α, β, γ are the roots of the equation

$$x^3-4x^2-25x+28 = 0.$$

The left-hand side vanishes when $x=1$, therefore $x-1$ is a factor. Hence

$$(x-1)(x^2-3x-28) = 0.$$
$$\therefore (x-1)(x+4)(x-7) = 0.$$
$$\therefore x = -4, 1, 7.$$

Therefore the equations are satisfied by $\alpha=-4, \beta=1, \gamma=7$ and the other five permutations of these numbers.

It is worth noting that the product of the degrees of the three given equations is 6 and that six solutions are obtained.

Repeated roots

8.7. For the quadratic equation $ax^2+bx+c=0$ with roots α, β,

$$\alpha+\beta = -b/a, \qquad \alpha\beta = c/a.$$

If the roots are equal, substitute $\beta=\alpha$, then

$$2\alpha = -\frac{b}{a}, \qquad \alpha^2 = \frac{c}{a}.$$

$$\therefore 4\alpha^2 = \frac{b^2}{a^2} = \frac{4c}{a}.$$

$$\therefore b^2 = 4ac.$$

The cubic equation $ax^3 + bx^2 + cx + d = 0$ may be treated similarly, but it is more instructive to consider the problem graphically. Figure 8.2 shows two cubic curves of the form $y = ax^3 + bx^2 + cx + d$ $(a > 0)$. The y-axis is not shown: the point to emphasize is that, if the equation $y = 0$ has a repeated root, the x-axis is a tangent.

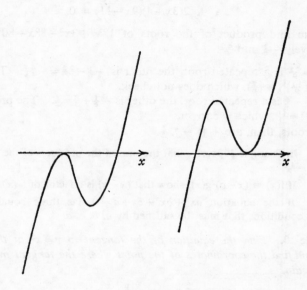

Fig. 8.2

In this case, $dy/dx = 0$ *has a root in common with* $y = 0$.

Hence, if the equation $ax^3 + bx^2 + cx + d = 0$ has a repeated root, the equations

$$ax^3 + bx^2 + cx + d = 0,$$
$$3ax^2 + 2bx + c = 0,$$

have a root in common.

Example 19. *Given that the equation*

$$18x^3 + 3x^2 - 88x - 80 = 0$$

has a repeated root, solve the equation.

If $18x^3 + 3x^2 - 88x - 80 = 0$ has a repeated root, then it is also a root of the equation

$$\frac{\mathrm{d}}{\mathrm{d}x}(18x^3 + 3x^2 - 88x - 80) = 0.$$

$$\therefore \quad 54x^2 + 6x - 88 = 0.$$
$$\therefore \quad 2(3x+4)(9x-11) = 0.$$

The sum and product of the roots of $18x^3 + 3x^2 - 88x - 80 = 0$ are respectively $-\frac{1}{6}$ and $\frac{40}{9}$.

If $x = \frac{11}{9}$ is a repeated root, the other is $-\frac{1}{6} - \frac{22}{9} = -\frac{47}{18}$. The product is $(\frac{11}{9})^2(-\frac{47}{18})$, which does not check.

If $x = -\frac{4}{3}$ is a repeated root, the other is $-\frac{1}{6} + \frac{8}{3} = \frac{5}{2}$. The product is $(-\frac{4}{3})^2(\frac{5}{2}) = \frac{40}{9}$, which is correct.

The roots, then, are $-\frac{4}{3}, -\frac{4}{3}, \frac{5}{2}$.

Qu. 18. In Example 19, verify that the sum of the products of the roots in pairs is $-\frac{44}{9}$.

Qu. 19. If $f(x) \equiv (x-a)^2 g(x)$, show that $(x-a)$ is a factor of $f'(x)$.

Qu. 20. If the equation $ax^3 + bx^2 + cx + d = 0$ has three equal roots, find the conditions that must be satisfied by a, b, c, d.

Example 20. *Find the equation of the tangent to $y = x^3$ at the point (t, t^3) and find the coordinates of the point where the tangent meets the curve again.*

$$y = x^3. \qquad \therefore \quad \frac{\mathrm{d}y}{\mathrm{d}x} = 3x^2.$$

Therefore the tangent at (t, t^3) has gradient $3t^2$. Therefore its equation is

$$3t^2 x - y - 2t^3 = 0.$$

To find the point of intersection with $y = x^3$, solve the equations simultaneously.

$$3t^2 x - x^3 - 2t^3 = 0.$$
$$\therefore \quad x^3 - 3t^2 x + 2t^3 = 0.$$

Now, since the line is a tangent to the curve, two of the roots are t, t. The sum of the roots is zero, so the other root is $-2t$. Therefore the line meets the curve again at $(-2t, -8t^3)$.

Qu. 21. Check that $(-2t, -8t^3)$ lies on the tangent in Example 20. Also check that the product of the roots is correct.

Exercise 8d

1. Write down the sum, the sum of the products in pairs, and the product of the roots of the equations:

   (i) $3x^3 - 4x^2 - x + 2 = 0$,　　　(ii) $4x^3 + 5x - 6 = 0$,

   (iii) $(x+1)^3 + 2(x+1)^2 - 3(x+1) + 4 = 0$,

   (iv) $3(x+1)^3 = 2(x-1)^2$.

2. Write down the equations whose roots are α, β, γ, when $\sum \alpha$, $\sum \beta\gamma$, $\alpha\beta\gamma$ are, respectively:

   (i) 2, 0, -5;　　　(ii) -3, 2, 6;　　　(iii) 0, -1, 5.

3. The equation $2x^3 + 3x^2 - 13x - 7 = 0$ has roots α, β, γ. Find the equations with roots:

   (i) $\alpha + 1$, $\beta + 1$, $\gamma + 1$;　　　(ii) $1/\alpha$, $1/\beta$, $1/\gamma$;

   (iii) $\alpha - 2$, $\beta - 2$, $\gamma - 2$;　　　(iv) $\beta\gamma$, $\gamma\alpha$, $\alpha\beta$.

4. The equation $x^3 + 3x^2 - 3x - 10 = 0$ has roots α, β, γ. Find the equations with roots:

   (i) $-\alpha + \beta + \gamma$, $\alpha - \beta + \gamma$, $\alpha + \beta - \gamma$;　　　(ii) $2\alpha + 1$, $2\beta + 1$, $2\gamma + 1$;

   (iii) $\dfrac{1}{\beta\gamma}$, $\dfrac{1}{\gamma\alpha}$, $\dfrac{1}{\alpha\beta}$.

5. If the equation $x^3 + 3hx + g = 0$ has roots α, β, γ, find the equations with roots:

   (i) α^2, β^2, γ^2;　　　(ii) $1/\alpha^2$, $1/\beta^2$, $1/\gamma^2$;　　　(iii) α^3, β^3, γ^3.

6. Repeat No. 5 for the equation $ax^3 + bx^2 + cx + d = 0$.

7. For the equation $x^3 + 3hx + g = 0$, with roots α, β, γ, find:

   (i) $\sum \alpha^2$,　　　(ii) $\sum \beta^2\gamma^2$,　　　(iii) $\sum \alpha^4$,

   given that $\sum \alpha^4 = (\sum \alpha^2)^2 - 2 \sum \beta^2\gamma^2$.

8. For the equation $x^3 + 3hx + g = 0$, with roots α, β, γ,

   (i) show that $\alpha^3 = -3h\alpha - g$, and use similar expressions for β, γ to deduce that $\sum \alpha^3 = -3h \sum \alpha - 3g$,

   (ii) show that $\alpha^4 = -3h\alpha^2 - g\alpha$ and deduce that $\sum \alpha^4 = -3h\sum \alpha^2 - g \sum \alpha$.

   Find $\sum \alpha^2$, $\sum \alpha^3$, $\sum \alpha^4$ in terms of g and h.

9. For the equation $x^3 + 3ax^2 + 3bx + c = 0$, with roots α, β, γ find:

   (i) $\sum \alpha^2$,　　　(ii) $\sum \alpha^3$,　　　(iii) $\sum \alpha^4$.

10. Find the relation between a, b, c, d if the roots of the equation

    $$ax^3 + bx^2 + cx + d = 0$$

    are in (i) arithmetic, (ii) geometric progression.

    Find the equation whose roots are the reciprocals of the roots of the given equation and deduce the condition that the roots of the given equation are in harmonic progression.

11. Find the relation between a, b, c, d if one root of the equation

    $$ax^3 + bx^2 + cx + d = 0$$

    is equal to the sum of the other two.

12. Solve the equation $54x^3 - 111x^2 + 74x - 16 = 0$, given that the roots are in geometric progression.

13. Solve the equation $64x^3 - 240x^2 + 284x - 105 = 0$, given that the roots are in arithmetic progression.

Solve the equations in Nos. 14 to 16, given that each has a repeated root.

14. $12x^3 - 52x^2 + 35x + 50 = 0$.

15. $18x^3 + 21x^2 - 52x + 20 = 0$.

16. $12x^3 - 20x^2 - 21x + 36 = 0$.

17. Find the condition that the equation $x^3 - 3hx + g = 0$ should have a repeated root. What conditions must be satisfied if all its roots are equal?

18. Sketch the graph of $y^2 = x^3$. Find the point where the tangent at (t^2, t^3) meets the curve again and show that the axes divide the chord in constant ratios.

19. Find the equation of the normal at $(at^2, 2at)$ to the parabola $y^2 = 4ax$. This equation is a cubic in t. Find the condition that it should have two equal roots.

20. Find the equation of the tangent to $x^2y = 1$ at the point $(t, 1/t^2)$. Also find the ratios in which the axes divide the segment of the tangent bounded by the curve.

Solve the following simultaneous equations.

21. $\alpha + \beta + \gamma = -2$, $\alpha^2 + \beta^2 + \gamma^2 = 14$, $\alpha\beta\gamma = 6$.

22. $\alpha + \beta + \gamma = 4$, $\alpha^2 + \beta^2 + \gamma^2 = 38$, $\alpha^3 + \beta^3 + \gamma^3 = 106$.

23. $\alpha + \beta + \gamma = 0$, $\alpha^2 + \beta^2 + \gamma^2 = 42$, $\alpha^3 + \beta^3 + \gamma^3 = -60$.

24. $\alpha + \beta + \gamma = 2$, $\alpha^2 + \beta^2 + \gamma^2 = 14$, $\alpha^3 + \beta^3 + \gamma^3 = 20$.

25. The equation $ax^4 + bx^3 + cx^2 + dx + e = 0$ has roots α, β, γ, δ. Find, in terms of a, b, c, d, e, the sum of the roots, the sums of the products of the roots in pairs and threes, and the product of the roots. Find also, the equations with roots

(i) $1/\alpha$, $1/\beta$, $1/\gamma$, $1/\delta$;   (ii) α^2, β^2, γ^2, δ^2.

26. Prove that $\tan 3x = (3 \tan x - \tan^3 x)/(1 - 3 \tan^2 x)$. Hence, or otherwise, solve the equation $t^3 - 6t^2 - 3t + 2 = 0$ correct to two significant figures.

27. Use the identity $\sin 3\theta = 3 \sin \theta - 4 \sin^3 \theta$ to solve the equation

$$8x^3 - 6x + 1 = 0$$

correct to four significant figures.

28. Use the substitution $x = 2 \sin \theta$ to solve the equation $3x^3 - 9x + 2 = 0$ correct to three significant figures.

29. The equation $x^3 + 3x^2 + 3 = 0$ has a root somewhere in the range $-5 < x < 3$. Find between which two integral values of x in this range the function $f(x) = x^3 + 3x^2 + 3$ changes sign and find the root of the equation to the nearest integer.

*30. Show that the cubic equation $t^3 + 3at^2 + 3bt + c = 0$ can be reduced to the form $x^3 + 3fx + g = 0$ by means of the substitution $t = x - a$. Obtain f and g in terms of a, b, c and explain how any cubic equation may be solved graphically by drawing a straight line to cut the graph $y = x^3$.

31. Find the condition that the equation

$$ax^3 + bx^2 + cx + d = 0$$

should have a repeated root.

Exercise 8e (Miscellaneous)

1. Solve the equations:
   (i) $9x^{2/3} + 4x^{-2/3} = 37$;
   (ii) $x + 2y = 3$, $3x^2 + 4y^2 + 12x = 7$.   o.c.

2. Solve the equations:
   (i) $\dfrac{1}{x} - \dfrac{1}{x+3} = \dfrac{1}{k} - \dfrac{1}{k+3}$,   (ii) $2\sqrt{x} + \sqrt{(2x+1)} = 7$,

   where the *positive* values of the square roots are taken.   o.c.

3. Solve the equations:
   (i) $3^{2(x+1)} - 10 \cdot 3^x + 1 = 0$;
   (ii) $x^2 + 4y^2 = 25$, $xy + 6 = 0$.   o.c.

4. (i) If $(x+1)^2$ is a factor of $2x^4 + 7x^3 + 6x^2 + Ax + B$, find the values of A and B.

   (ii) Prove that, if the equations

   $$x^2 + ax + b = 0 \quad \text{and} \quad cx^2 + 2ax - 3b = 0$$

   have a common root and neither a nor b is zero, then

   $$b = \frac{5a^2(c-2)}{(c+3)^2}.$$
       o.c.

5. (i) Prove that, if two polynomials $P(x)$ and $Q(x)$ have a common linear factor $x - p$, then $x - p$ is a factor of the polynomial

   $$[P(x) - Q(x)].$$

   Hence prove that, if the equations

   $$ax^3 + 4x^2 - 5x - 10 = 0, \quad ax^3 - 9x - 2 = 0$$

   have a common root, then $a = 2$ or 11.

   (ii) Prove that, if $x + \dfrac{1}{x} = y + 1$, then

   $$\frac{(x^2 - x + 1)^2}{x(x-1)^2} = \frac{y^2}{y-1}.$$

   Hence solve the equation

   $$(x^2 - x + 1)^2 - 4x(x-1)^2 = 0.$$
       o.c.

6. (i) Show that the equation $\sqrt{(x^2+2)} - \sqrt{(x^2+2x+5)} = 1$ has no solution if it is assumed that the square roots are positive.

(ii) Show that, if $x^3 + 3px + q = 0$ and $x = y - p/y$, then y^3 satisfies a certain quadratic equation.

By solving the quadratic equation in the case $p = q = 2$, obtain *one* root of the equation

$$x^3 + 6x + 2 = 0,$$

leaving your answer in surd form.   O.C.

7. Find the real roots of the equations

$$\left(x - \frac{1}{y}\right)^2 - \left(y - \frac{1}{x}\right)\left(x - \frac{1}{y}\right) = 9x,$$

$$x - y = 1.$$ L.

8. Solve the equations

$$4\left(\frac{x^2}{y} + y\right)\left(\frac{y^2}{x} + x\right) = 25xy,$$

$$x + y = 6.$$ L.

9. (i) Solve the equations

$$\frac{x}{2} + \frac{2}{y} = 1, \qquad xy - 3x = 4.$$

(ii) Prove that, if a, b, c are real and non-zero, then the roots of the equation

$$x^2(b^2 + c^2) + 2(a^2 + b^2 + c^2)x + a^2 + b^2 = 0$$

are real and distinct.

If the sum and product of the roots are -6 and $2\frac{1}{4}$ respectively, prove that $a^2 = 6b^2 = 3c^2$.   O.C.

10. (i) Given that $\frac{a}{b} = \frac{c}{d}$, prove that each of these ratios equals

$$\frac{ka + lc}{kb + ld},$$

where k, l are any numbers for which $kb + ld \neq 0$.

Solve the simultaneous equations

$$\frac{x}{1} = \frac{x+y}{3} = \frac{x-y+z}{2},$$

$$x^2 + y^2 + z^2 + x + 2y + 4z - 6 = 0.$$

(ii) Find positive integers a, b for which

$$x^4 + 2x^2 + 9 \equiv (x^2 + a)^2 - b^2x^2$$

and hence find the quadratic factors of $x^4 + 2x^2 + 9$.   O.C.

11. Solve the simultaneous equations:
    (i) $x+y = 3$,
    $x^2-xy-y^2-5x+6 = 0$,
    (ii) $x+y+z+2 = 0$,
    $$\frac{x+2y}{-3} = \frac{y+2z}{4} = \frac{z+2x}{5}.$$ o.c.

12. Solve the equations:
    (i) $\sqrt{(2-x)}+\sqrt{(x+3)} = 3$,
    (ii) $\frac{x-y}{4} = \frac{z-y}{3} = \frac{2z-x}{1}$, $x+3y+2z = 4$. o.c.

13. Solve the simultaneous equations:
    $$a_1x+b_1y+c_1z = 0,$$
    $$a_2x+b_2y+c_2z = 0,$$

    for the ratios $x:y:z$.

    Hence, or otherwise,
    (i) solve the equations:
    $$x+4y+2z = 0,$$
    $$2x-y+z = 0,$$
    $$8x+5y+6z = 6,$$

    for x, y, and z; and
    (ii) find the condition that the quadratic equations
    $$a_1x^2+b_1x+c_1 = 0,$$
    $$a_2x^2+b_2x+c_2 = 0,$$

    should have a common root. c.

14. Prove that, if lx^2+mx+n is equal to zero for three distinct values of x, then $l=m=n=0$.

    Hence or otherwise, prove the identity
    $$\frac{a^2(x-b)(x-c)}{(a-b)(a-c)} + \frac{b^2(x-c)(x-a)}{(b-c)(b-a)} + \frac{c^2(x-a)(x-b)}{(c-a)(c-b)} \equiv x^2.$$

    Prove also that
    $$\frac{(x+a)^3}{(a-b)(a-c)} + \frac{(x+b)^3}{(b-c)(b-a)} + \frac{(x+c)^3}{(c-a)(c-b)} \equiv 3x+a+b+c.$$ o.c.

15. Solve the equations:
    (i) $\frac{x}{3}+\frac{3}{y} = \frac{x}{4}-\frac{4}{y} = 1$;
    (ii) $x-y = 1$, $x^2+xy = 6$.

    Prove that
    $$\frac{(2a-b)^3+(a-2b)^3}{a^4+a^2b^2+b^4} = \frac{9(a-b)}{a^2+ab+b^2}.$$ o.c.

16. (i) Solve the equation $x^3 - x^2 - 5x + 2 = 0$.
    (ii) Find the only solution of the equation

$$\sqrt{(4x-2)} + \sqrt{(x+1)} - \sqrt{(7-5x)} = 0. \qquad \text{o.c.}$$

17. (i) Solve $\sqrt{(4x+13)} - \sqrt{(x+1)} = \sqrt{(12-x)}$.
    (ii) One root of the equation $3x^3 + 14x^2 + 2x - 4 = 0$ is rational. Obtain this root and complete the solution of the equation.   c.

18. (i) One of the roots of the equation $21x^3 - 50x^2 - 37x - 6 = 0$ is a positive integer. Find this root and hence solve the equation completely.
    (ii) Three numbers α, β, γ are in arithmetical progression and two of them are roots of the equation $x^2 + ax + b = 0$. Prove that the third is either $-\frac{1}{2}a$ or one of the roots of the equation

$$y^2 + ay + 9b - 2a^2 = 0. \qquad \text{o.c.}$$

19. (i) Show that, if the roots of the equation

$$x^3 - 5x^2 + qx - 8 = 0$$

are in geometric progression, then $q = 10$.
    (ii) If α, β, γ are the roots of the equation

$$x^3 - x^2 + 4x + 7 = 0,$$

find the equation whose roots are $\beta + \gamma$, $\gamma + \alpha$, $\alpha + \beta$.   c.

20. Prove that, if α is a repeated root of the equation $f(x) = 0$, where $f(x)$ is a polynomial, then α is a root of the equation $f'(x) = 0$.

    Given that the equation $4x^4 + x^2 + 3x + 1 = 0$ has a repeated root, find its value.   c.

21. (a) Use the remainder theorem to express

$$x^3 + 2x^2 + x - 18$$

as a product of two factors.
    (b) Find the value of the constant p for which the polynomial

$$x^4 + x^3 + px^2 + 5x - 10$$

has $x + 2$ as a factor.
    (c) Show that if $y = (x-a)^2 V$ where V is a polynomial in x, then dy/dx is a polynomial with $x - a$ as a factor. Hence or otherwise find the values of the constants k and l for which

$$x^4 - 2x^3 + 5x^2 + kx + l$$

has a factor $(x-1)^2$.   N.

22. Sketch the graph of the polynomial

$$x^2(3x - 10)(x - 6)$$

and calculate the coordinates of the turning points. (Graph paper must not be used.)

Determine the range of values of k for which the equation

$$x^2(3x - 10)(x - 6) = k$$

has four real roots.

Show that a real root of the equation

$$x^2(3x - 10)(x - 6) = 330$$

lies between 6·7 and 6·8. Prove that it is nearer to 6·7 than to 6·8. N.

23. Draw the graph of $y = 3^x$ for values of x from -2 to $+2$, plotting at least nine points. [Take 2 cm as unit on the axis of x, and 1 cm as unit on the axis of y.]

Use your graph to find (i) the roots of the equation $3^x = 3x+2$, (ii) the range of values of k for which the equation $3^x = kx$ has no real roots. C.

24. Determine the maximum and minimum values of the function

$$3x^4 - 20x^3 + 36x^2 = k$$

should have four real distinct roots. Prove that when this condition is satisfied the difference between the greatest and the least of the roots is less than $\frac{4}{3}\sqrt{10}$. N.

25. (a) If $x^2 - 3x + 4 = 0$, prove that $x^4 = 3x - 20$.

Hence, or otherwise, if the roots of the quadratic equation $x^2 - 3x + 4 = 0$ are α and β, construct the quadratic equation of which the roots are α^4 and β^4.

(b) Eliminate x from the two equations

$$x^2 + x + a = 0, \qquad x^3 + x + b = 0. \qquad \text{N.}$$

26. Given that

$$c(x - a)^p + b \equiv 2x^3 - 12x^2 + 24x - 13,$$

obtain the values of c, a, p, b.

Hence, or otherwise, find the real root of the equation

$$2x^3 - 12x^2 + 24x - 13 = 0,$$

giving your answer to three decimal places.

Sketch the graph of

$$y = 2x^3 - 12x^2 + 24x - 13. \qquad \text{N.}$$

27. If the roots of the equation $x^2 + bx + c = 0$ are α, β and the roots of the equation $x^2 + \lambda bx + \lambda^2 c = 0$ are γ, δ, prove that

(i) $(\alpha\gamma + \beta\delta)(\alpha\delta + \beta\gamma) = 2\lambda^2 c(b^2 - 2c)$,

(ii) the equation whose roots are $\alpha\gamma + \beta\delta$ and $\alpha\delta + \beta\gamma$ is

$$x^2 - \lambda b^2 x + 2\lambda^2 c(b^2 - 2c) = 0.$$

If b, c, λ are all real, show that the roots of the last equation are all real. N.

28. (a) Find the values of p and q so that

$$x^4 + 3x^3 + 2x^2 + px + q$$

may be the perfect square of a quadratic in x.

(b) Find x, given that it satisfies the two equations

$$8x^3 - 6x^2 - x + 6 = 0,$$
$$4x^3 + 11x^2 - 10x - 12 = 0.$$

(c) If $ab = cd$, verify that

$$2abcd(a^2 + b^2 + c^2 + d^2) = (bc + ad)(ca + bd)(ab + cd)$$ N.

29. The roots of the equation $x^3 = qx + r$ are α, β, γ. Prove that

(i) $\alpha^2 + \beta^2 + \gamma^2 = 2q$,

(ii) $\alpha^3 + \beta^3 + \gamma^3 = 3r$,

(iii) $\alpha^5 = q\alpha^3 + r\alpha^2$,

(iv) $6(\alpha^5 + \beta^5 + \gamma^5) = 5(\alpha^3 + \beta^3 + \gamma^3)(\alpha^2 + \beta^2 + \gamma^2)$. O.C.

CHAPTER 9

COORDINATE GEOMETRY—I

Conic sections

9.1. In this chapter we shall be dealing with three curves, the parabola, ellipse, and hyperbola, which are all known as *conic sections* or *conics*. The Greek mathematicians even before Euclid (third century B.C.) were interested in these curves and examined their properties by pure geometry starting from their definitions by means of sections of a cone. From the point of view of coordinate geometry it is better to start from another definition (which can be shown to be equivalent) that a conic is the locus of a point which moves so that its distance from a fixed point bears a constant ratio to its distance from a fixed line (see Fig. 9.1):

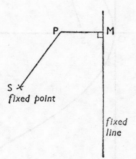

Fig. 9.1

The fixed point S is called the *focus*. The fixed line is called the *directrix*. The constant ratio is called the *eccentricity* and is denoted by e. Thus, if P is a point on the locus, M is the foot of the perpendicular from P to the directrix and if

$$\frac{SP}{PM} = e,$$

then the locus of P is a conic.

175

When: $e = 1$,   the conic is a *parabola*,

        $e < 1$,   the conic is an *ellipse*,

        $e > 1$,   the conic is a *hyperbola*.

We shall first take the parabola, which was briefly mentioned in P.M.I., Chapter 19.

The parabola

9.2. Given the focus S of the parabola and the directrix, we are at liberty to take what axes we find most convenient. First note that the figure formed by the focus and directrix has an axis of symmetry through S perpendicular to the directrix.

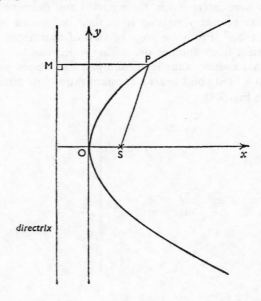

Fig. 9.2

This we take as the x-axis, as shown. If we now plot a few points, using the definition of the locus given in the last section,

$$\frac{SP}{PM} = 1,$$

an indication of the shape of the curve may be obtained (see Fig. 9.2). (The plotting may be done very simply using squared paper and a pair of compasses.) It now seems reasonable to take the y-axis through the point on the axis of symmetry mid-way between the focus and directrix. This point is called the *vertex* of the parabola. Let the distance from the vertex to the focus be a; then

$$\text{the focus S is } (a, 0),$$
and $$\text{the directrix is the line } x = -a.$$

If $P(x, y)$ is any point on the parabola, and M is the foot of the perpendicular from P to the directrix,

$$SP^2 = (x-a)^2 + y^2$$
and $$PM = x + a.$$

But from the definition,

$$\frac{SP}{PM} = 1, \quad \text{so} \quad SP^2 = PM^2.$$

$$\therefore (x-a)^2 + y^2 = (x+a)^2.$$
$$\therefore x^2 - 2ax + a^2 + y^2 = x^2 + 2ax + a^2.$$
$$\therefore y^2 = 4ax,$$

which is the standard equation of a parabola.

Qu. 1. Find the equations of the parabolas:

(i) focus $(-a, 0)$, directrix $x = a$,
(ii) focus $(0, b)$, directrix $y = -b$.

Example 1. *Find, in terms of a, m, the value of c which makes the line $y = mx + c$ a tangent to the parabola $y^2 = 4ax$. Also obtain the coordinates of the point of contact.*

$$y = mx + c.$$

Multiply both sides by $4a$.

$$4ay = m . 4ax + 4ac.$$

Substituting from $y^2 = 4ax$ and collecting terms,

$$my^2 - 4ay + 4ac = 0. \tag{1}$$

The line will be a tangent if this equation has equal roots (see p. 164)

$$\therefore (-4a)^2 = 16mac.$$
$$\therefore c = a/m.$$

When the roots of equation (1) are equal, they will be given by half the sum of the roots,

$$\therefore \quad y = \tfrac{1}{2} \cdot \frac{4a}{m} = \frac{2a}{m}.$$

Now

$$x = \frac{y^2}{4a} = \left(\frac{2a}{m}\right)^2 \cdot \frac{1}{4a} = \frac{a}{m^2}.$$

Therefore the point of contact is

$$\left(\frac{a}{m^2}, \frac{2a}{m}\right).$$

Note that the equation of a general tangent to $y^2 = 4ax$ may be written

$$y = mx + \frac{a}{m}. \quad (m \neq 0.)$$

This last result leads us to a very useful way of representing a point on the parabola. Substituting $t = 1/m$, we see that the tangent

$$y = \frac{x}{t} + at$$

touches the parabola at $(at^2, 2at)$. Since the tangent was a general one, we have shown that *any point on the parabola $y^2 = 4ax$ may be written $(at^2, 2at)$.* $x = at^2$, $y = 2at$ are called the **parametric equations** of the parabola $y^2 = 4ax$.

Qu. 2. Verify, by substitution, that $(at^2, 2at)$ always lies on the parabola $y^2 = 4ax$.

We have found the equation of the tangent at $(at^2, 2at)$, but a more direct method follows in Example 2.

Example 2. *Find the equation of the tangent to $y^2 = 4ax$ at $(at^2, 2at)$.*
To find the gradient at $(at^2, 2at)$,

$$\frac{dy}{dx} = \frac{dy}{dt} \bigg/ \frac{dx}{dt}.$$

But $y = 2at$, $x = at^2$,

$$\therefore \quad \frac{dy}{dx} = \frac{2a}{2at} = \frac{1}{t}.$$

The equation of the tangent is obtained by the method of P.M.I, p. 381, which will be used from now on.

$$x - ty = at^2 - t . 2at,$$

that is
$$x - ty + at^2 = 0.$$

*Qu. 3.   Show that the equation of the normal to $y^2 = 4ax$ at $(at^2, 2at)$ is

$$tx + y - at^3 - 2at = 0.$$

Example 3.   *Show that the equation of the tangent to the parabola $y^2 = 4ax$ at (x_1, y_1) is $yy_1 = 2a(x + x_1)$.*

Differentiating both sides of $y^2 = 4ax$ with respect to x, to find the gradient:

$$2y \frac{dy}{dx} = 4a.$$

Therefore at (x_1, y_1), $\dfrac{dy}{dx} = \dfrac{4a}{2y_1} = \dfrac{2a}{y_1}.$

Therefore the tangent is

$$2ax - y_1 y = 2ax_1 - y_1{}^2.$$
$$\therefore \ y_1 y = 2ax - 2ax_1 + y_1{}^2.$$

Now (x_1, y_1) lies on the parabola, so $y_1{}^2 = 4ax_1$.

$$\therefore \ y_1 y = 2ax - 2ax_1 + 4ax_1.$$
$$\therefore \ yy_1 = 2a(x + x_1).$$

Qu. 4.   Find the equation of the normal to $y^2 = 4ax$ at (x_1, y_1).

Example 4.   *Find the equation of the chord joining the points $(at_1{}^2, 2at_1)$, $(at_2{}^2, 2at_2)$.*

The gradient of the chord is

$$\frac{2at_1 - 2at_2}{at_1{}^2 - at_2{}^2} = \frac{2a(t_1 - t_2)}{a(t_1 - t_2)(t_1 + t_2)},$$
$$= \frac{2}{t_1 + t_2}.$$

Therefore the equation of the chord is

$$2x - (t_1 + t_2)y = 2at_1{}^2 - (t_1 + t_2) . 2at_1,$$
$$= -2at_1 t_2.$$

Therefore the chord is

$$2x - (t_1 + t_2)y + 2at_1 t_2 = 0.$$

*Qu. 5. As $t_2 \to t_1$, the chord approaches the tangent at t_1. Deduce the equation of the tangent from the equation of the chord.

DEFINITIONS: *Any chord of a parabola passing through the focus is called a* focal chord. *The axis of symmetry is usually simply called the* axis *of the parabola. The focal chord perpendicular to the axis is called the* latus rectum.

To find the length of the latus rectum of the parabola $y^2 = 4ax$, substitute $x = a$;

$$y^2 = 4a^2.$$
$$\therefore \ y = \pm 2a.$$

Hence the length of the latus rectum is $4a$.

The reader is advised to work all the questions in Exercise 9a, using the parametric coordinates $(at^2, 2at)$ whenever the opportunity arises. The point $(at^2, 2at)$ is frequently abbreviated to the point t.

Exercise 9a

1. Find the coordinates of the point of intersection of the tangents at the points t_1, t_2 of the parabola $y^2 = 4ax$.
2. Points t_1, t_2 lie on the parabola $y^2 = 4ax$. Find a relation connecting t_1, t_2 if the line joining the points is a focal chord.
3. Prove that the tangents at the ends of a focal chord of a parabola are perpendicular.
4. Find the focus of the parabola $x^2 = 2y$.
5. Find the equation of a parabola whose focus is $(2, 0)$ and directrix $y = -2$.
6. Find the equation of the parabola whose focus is $(-1, 1)$ and directrix $x = y$.
7. Find the gradient of the normal to the parabola $y^2 = 4ax$ at $P(at^2, 2at)$ and the gradient of the chord joining P to $(at_1^2, 2at_1)$. Deduce the coordinates of the point where the normal at P cuts the parabola again.
8. Prove that the foot of the perpendicular from the focus of a parabola on to any tangent lies on the tangent at the vertex.
9. Find the points on the parabola $y^2 = 8x$ where (i) the tangent and (ii) the normal are parallel to the line $2x + y = 1$.
10. The tangents at the end of a focal chord meet each other at P and the tangent at the vertex at Q, R. Show that the centroid of the triangle PQR lies on the line $3x + a = 0$.
11. Find the point of intersection of the normals at the points t_1, t_2 of the parabola $y^2 = 4ax$.
12. Prove that, in general, from any point (h, k) three normals can be drawn to a parabola.

13. If the normals from a point (h, k) meet the parabola $y^2 = 4ax$ at the three point t_1, t_2, t_3, show that $t_1 + t_2 + t_3 = 0$.

14. PQ is a variable chord of a parabola. If the chords joining the vertex A to P and Q are perpendicular, show that PQ meets the axis of the parabola in a fixed point R, and find the length of AR.

15. Find the equations of the tangents to the parabola $y^2 = 4ax$ from the point $(16a, 17a)$.

16. If the tangents at the end of a focal chord of a parabola meet the tangent at the vertex in C, D, prove that CD subtends a right angle at the focus.

Further examples on the parabola

9.3.   Example 5. *Find the focus and directrix of the parabola $y^2 = 2a(x - 4a)$ and give the length of its latus rectum.*

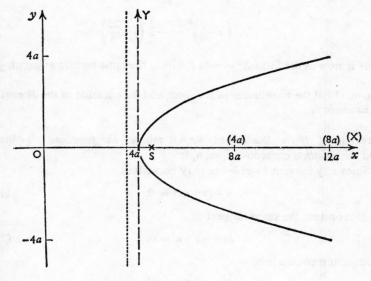

Fig. 9.3

The equation $y^2 = 2a(x - 4a)$ may be written in the form

$$Y^2 = 2aX$$

by the substitutions $y = Y$, $x - 4a = X$. We have thus taken new axes as shown in Fig. 9.3.

The parabola $y^2 = 4bx$ has focus $(b, 0)$ directrix $x = -b$ and latus rectum $4b$. Comparing this with $Y^2 = 2aX$, it follows that the latter has focus $(\frac{1}{2}a, 0)$, directrix $X = -\frac{1}{2}a$ and latus rectum $2a$. Therefore, with the original axes (see Fig. 9.3), the focus is $(9a/2, 0)$, the directrix $x = 7a/2$ and latus rectum $2a$.

Example 6. *Show that the equation* $y = 5x - 2x^2$ *represents a parabola and find the length of its latus rectum.*

We shall try to express the equation in the form $X^2 = -4aY$. The equation may be written as

$$x^2 - \frac{5}{2}x = -\frac{y}{2}.$$

$$\therefore \left(x - \frac{5}{4}\right)^2 = \left(\frac{5}{4}\right)^2 - \frac{y}{2}.$$

$$\therefore \left(x - \frac{5}{4}\right)^2 = -\frac{1}{2}\left(y - \frac{25}{8}\right).$$

This is now in the form $X^2 = -4aY$, giving the latus rectum as length $\frac{1}{2}$.

Qu. 6. Find the coordinates of the focus and the equation of the directrix in Example 6.

Example 7. *If the line* $lx + my + n = 0$ *touches the parabola* $y^2 = 4ax$, *find the equation connecting* l, m, n, a.

Since any tangent to $y^2 = 4ax$ may be written

$$x - ty + at^2 = 0, \tag{1}$$

let it represent the same tangent as

$$lx + my + n = 0. \tag{2}$$

Comparing coefficients,

$$\frac{1}{l} = -\frac{t}{m} = \frac{at^2}{n}.$$

$$\therefore t = -\frac{m}{l} \quad \text{and} \quad t^2 = \frac{n}{al}$$

$$\therefore \frac{m^2}{l^2} = \frac{n}{al}.$$

Therefore the condition is $am^2 = ln$.

The next two examples have been chosen to illustrate the use of symmetrical relationships between t_1, t_2. When symmetry exists, the working is made easier and care should be taken to use symmetrical equations and expressions.

Example 8. *A chord of the parabola $y^2 = 4ax$ subtends a right angle at the vertex. Find the locus of the mid-point of the chord.*

Let the ends of the chord be $P_1(at_1{}^2, 2at_1)$, $P_2(at_2{}^2, 2at_2)$. Then the gradient of the line joining the vertex $O(0, 0)$ to P_1 is

$$\frac{2at_1}{at_1{}^2} = \frac{2}{t_1}.$$

Similarly the gradient of OP_2 is $\dfrac{2}{t_2}$.

P_1P_2 subtends a right angle at O if OP_1, OP_2 are perpendicular,

$$\therefore \; \frac{2}{t_1} \cdot \frac{2}{t_2} = -1,$$

$$\therefore \; t_1 t_2 = -4. \tag{1}$$

The mid-point of P_1P_2 is given by

$$x = \frac{a(t_1{}^2 + t_2{}^2)}{2}, \tag{2}$$

$$y = a(t_1 + t_2). \tag{3}$$

[Note that we have three equations, (1), (2), (3), from which to eliminate the two parameters t_1, t_2. Note, also, that these equations are symmetrical in t_1, t_2. Here, as is often the case, we use the following identity.]

$$(t_1 + t_2)^2 = t_1{}^2 + t_2{}^2 + 2t_1 t_2.$$

Substituting from equations (3), (2), (1):

$$\frac{y^2}{a^2} = \frac{2x}{a} - 8.$$

Therefore the locus is $y^2 = 2a(x - 4a)$.

The next example involves simple algebra but the method often saves time in locus problems.

Example 9. *The tangents to the parabola $y^2 = 4ax$ at the points t_1, t_2 intersect at an angle $\tan^{-1} k$. Find the locus of their point of intersection.*

The tangent at the point t is

$$x - ty + at^2 = 0. \tag{1}$$

The gradients at t_1, t_2 are $\dfrac{1}{t_1}$, $\dfrac{1}{t_2}$ and the tangents at t_1, t_2 intersect at an angle of $\tan^{-1} k$, and so (see P.M.I., p. 384)

$$\therefore \quad \frac{\dfrac{1}{t_1} - \dfrac{1}{t_2}}{1 + \dfrac{1}{t_1} \cdot \dfrac{1}{t_2}} = k.$$

$$\therefore \quad \frac{t_2 - t_1}{t_1 t_2 + 1} = k. \tag{2}$$

If the tangents at t_1, t_2 intersect at (x, y), equation (1) may be regarded as a quadratic in t whose roots are t_1, t_2.

$$\therefore \quad t_1 + t_2 = \frac{y}{a}, \tag{3}$$

$$t_1 t_2 = \frac{x}{a}. \tag{4}$$

We now want to express equation (2) in terms of $t_1 + t_2$, $t_1 t_2$, k. Equation (2) may be written

$$t_2 - t_1 = k(1 + t_1 t_2).$$

$$\therefore \quad t_2 - t_1 = k\left(1 + \frac{x}{a}\right). \tag{5}$$

Now $(t_2 - t_1)^2$ may be expressed in terms of $t_1 + t_2$, $t_1 t_2$ by the identity

$$(t_1 + t_2)^2 - 4t_1 t_2 = (t_2 - t_1)^2.$$

Substituting from (3), (4) and (5),

$$\frac{y^2}{a^2} - 4\frac{x}{a} = k^2 \left(\frac{x}{a} + 1\right)^2.$$

Therefore the locus is

$$y^2 - 4ax = k^2(x + a)^2.$$

Alternatively, equations (3), (4) could have been obtained by solving the equations of the tangents at t_1, t_2 simultaneously.

Example 10. *A variable tangent is drawn to the parabola* $y^2 = 4ax$. *If the perpendicular from the vertex meets the tangent at* P, *find the locus of* P.

Let the variable tangent be

$$x - ty + at^2 = 0. \tag{1}$$

Then the perpendicular from the vertex $(0, 0)$ is

$$tx + y = 0. \tag{2}$$

$P(x, y)$ satisfies equations (1), (2) so that the locus of P may be found by eliminating t from these equations. [Note that it is *not* necessary to solve them to find the coordinates of P in terms of t.]

From (2), $$t = -\frac{y}{x}.$$

Substituting in (1),

$$x + \frac{y^2}{x} + a\frac{y^2}{x^2} = 0.$$

So the locus of P is $$x^3 + xy^2 + ay^2 = 0.$$

Exercise 9b

1. Show that the equation $x^2 + 4x - 8y - 4 = 0$ represents a parabola whose focus is at $(-2, 1)$. Find the equation of the tangent at the vertex.
2. Prove that $x = 3t^2 + 1$ and $y = \frac{1}{2}(3t + 1)$ are the parametric equations of a parabola and find its vertex and the length of the latus rectum.
3. Find the focus of the parabola $y = 2x^2 + 3x - 5$.
4. Prove that the line $y = mx + \frac{3}{4}m + 1/m$ touches the parabola $y^2 = 4x + 3$ whatever the value of m.
5. If $ax + by + c = 0$ touches the parabola $x^2 = 4y$, find an equation connecting a, b, c.
6. A parabola, symmetrical about the axis of y, passes through the points $(1, 3)$ and $(2, 0)$. Find its equation and that of the tangent at $(1, 3)$.
7. Prove that the circles which are drawn on a focal chord of a parabola as diameter touch the directrix.
8. A variable chord of the parabola $y^2 = 4ax$ has a fixed gradient k. Find the locus of the mid-point.
9. A chord of the parabola $y^2 = 4ax$ is drawn to pass through the point $(-a, 0)$. Find the locus of the point of intersection of the tangents at the ends of the chord.
10. The difference of the ordinates of two points on the parabola $y^2 = 4ax$ is constant and equal to k. Find the locus of the point of intersection of the tangents at the two points.

11. Find the locus of the mid-points of focal chords of the parabola $y^2 = 4ax$.

12. The tangent at any point P of the parabola $y^2 = 4ax$ meets the tangent at the vertex at the point Q. S is the focus and SQ meets the line through P parallel to the tangent at the vertex at the point R. Find the locus of R.

13. Show that $y = ax^2 + bx + c$ is the equation of a parabola. Find its focus and directrix.

14. Two tangents to the parabola $y^2 = 4ax$ pass through the point (x_1, y_1). Find the equation of their chord of contact.

15. Find the points of contact on the parabola of the tangents common to the circle $(x - a)^2 + y^2 = 4a^2$ and the parabola $y^2 = 4ax$. [Start by writing down the equation of the tangent at $(at^2, 2at)$.]

16. The normal at the point P of the parabola $y^2 = 4ax$ meets the curve again at Q. The circle on PQ as diameter goes through the vertex. Find the x-coordinate of P.

17. Prove that rays of light parallel to the axis of a parabolic mirror are reflected through the focus.

18. A variable chord of the parabola $y^2 = 4ax$ passes through the point (h, k). Find the locus of the orthocentre of the triangle formed by the chord and the tangents at the two ends.

19. A tangent to the parabola $y^2 = 4ax$ meets the parabola $y^2 = 8ax$ at P, Q. Find the locus of the mid-point of PQ.

20. Find the locus of the mid-point of a variable chord through the point $(a, 2a)$ of the parabola $y^2 = 4ax$.

21. A variable tangent meets the directrix of the parabola $y^2 = 4ax$ at the point R and touches the curve at P. Q is the other end of the focal chord through P. Find the locus of the mid-point of QR.

22. $P_1 P_2$ is a focal chord of the parabola $y^2 = 4ax$. The tangents at P_1, P_2 meet the tangent at the vertex in Q_1, Q_2. Find the locus of the point of intersection of $P_1 Q_2$, $P_2 Q_1$.

23. P, Q, R are three points on the parabola $y^2 = 4ax$. The circle through P, Q, R passes through the vertex. Prove that the normals at P, Q, R are concurrent.

The ellipse

9.4. An ellipse was defined at the beginning of this chapter. Given a fixed point S, the *focus*, and a fixed line, the *directrix*, if P is a point on the locus and M is the foot of the perpendicular from P on to the directrix, then

$$\frac{SP}{PM} = e. \qquad (e < 1).$$

e is called the *eccentricity* of the ellipse.

Qu. 7. On a sheet of squared paper, rule the directrix along one line near the edge, take the focus 2·7 cm in and plot an ellipse with a pair of compasses, taking $e = 4/5$. Measure the width of the ellipse parallel and perpendicular to the directrix.

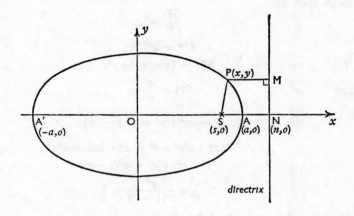

Fig. 9.4

The result of Qu. 7 should be like the ellipse in Fig. 9.4, only larger. It follows from the definition that an ellipse is symmetrical about the line through S perpendicular to the directrix, so we take the x-axis along this axis of symmetry. Let the x-axis cut the ellipse in A′, A, as shown in Fig. 9.4. It appears from our drawing that there may be an axis of symmetry parallel to the directrix, so we shall take the y-axis passing through the mid-point of A′A, parallel to the directrix.

Let A be $(a, 0)$ so that A′ is $(-a, 0)$; let S be $(s, 0)$ and let the x-axis cut the directrix at N$(n, 0)$. We shall now find s, n in terms of a, e.

A′, A lie on the ellipse and so, by the definition of the locus,

$$\frac{SA'}{A'N} = e, \qquad \frac{SA}{AN} = e.$$

Hence
$$a + s = e(n + a),$$
$$a - s = e(n - a).$$

Adding, $2a = 2en$, $\therefore \ n = \dfrac{a}{e}.$

Subtracting, $2s = 2ae$, $\therefore \ s = ae.$

Therefore S is the point $(ae, 0)$ and the equation of the directrix is $x = \dfrac{a}{e}$.

To find the equation of the ellipse, let $P(x, y)$ be any point on the locus, then

$$\frac{SP}{PM} = e.$$

$$\therefore \; SP^2 = e^2 PM^2.$$

But

$$SP^2 = (x - ae)^2 + y^2,$$

and

$$PM = \frac{a}{e} - x.$$

$$\therefore \; (x - ae)^2 + y^2 = e^2 \left(\frac{a}{e} - x\right)^2.$$

$$\therefore \; x^2 - 2aex + a^2 e^2 + y^2 = a^2 - 2aex + e^2 x^2.$$

$$\therefore \; x^2(1 - e^2) + y^2 = a^2(1 - e^2).$$

$$\therefore \; \frac{x^2}{a^2} + \frac{y^2}{a^2(1 - e^2)} = 1.$$

Therefore the equation of the ellipse is

$$\frac{x^2}{a^2} + \frac{y^2}{b^2} = 1,$$

where

$$b^2 = a^2(1 - e^2).$$

Note that we have also found that the focus S is $(ae, 0)$ and the directrix is $x = a/e$; but since the equation of the ellipse is unaltered by replacing x by $-x$, it follows that there is another focus $(-ae, 0)$ and another directrix $x = -a/e$. Hence

the foci are $(ae, 0)$ and $(-ae, 0)$,

the directrices are $x = \dfrac{a}{e}$ and $x = -\dfrac{a}{e}$.

The axes of symmetry meet at the *centre* of the ellipse. Any chord passing through the centre is called a *diameter*.

The diameter through the foci is the *major axis* and the perpendicular diameter is called the *minor axis*.

Qu. 8. Show that the lengths of the axes are $2a$, $2b$.

Qu. 9. Find the length of the semi-axes of the ellipse

$$\frac{x^2}{16} + \frac{y^2}{9} = 1.$$

Qu. 10.   Find the eccentricity of the ellipse

$$\frac{x^2}{25}+\frac{y^2}{16} = \frac{1}{4}.$$

Qu. 11.   Find the foci of the ellipse

$$x^2+4y^2 = 9.$$

Parametric coordinates for an ellipse

9.5.   When dealing with an ellipse

$$\frac{x^2}{a^2}+\frac{y^2}{b^2} = 1,$$

working is generally made easier by using a parameter, but the question arises of what parameter to use.   Now an equation in the form

$$(\quad)^2+(\quad)^2 = 1$$

suggests the identity

$$\cos^2 \theta+\sin^2 \theta = 1.$$

Thus, if we write

$$x = a \cos \theta, \qquad y = b \sin \theta,$$

the equation

$$\frac{x^2}{a^2}+\frac{y^2}{b^2} = 1$$

will always be satisfied.   We therefore take as a general point on the ellipse

$$(a \cos \theta, \, b \sin \theta).$$

θ is called the *eccentric angle* of the point.

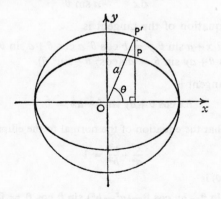

Fig. 9.5

In P.M.I., p. 403, we saw that the parameter θ for a circle in

$$x = a \cos \theta, \qquad y = a \sin \theta$$

could be interpreted in terms of an angle.   This is not so simple for an ellipse but it will now be done.

In Fig. 9.5, P is the point $(a \cos \theta, b \sin \theta)$ on the ellipse $\dfrac{x^2}{a^2} + \dfrac{y^2}{b^2} = 1$ and P′ is a point on the circle $x^2 + y^2 = a^2$ (called the auxiliary circle) such that OP′ makes an angle θ with Ox.   Since P, P′ have the same x-coordinate $a \cos \theta$, PP′ is perpendicular to the major axis of the ellipse. Therefore the eccentric angle θ of any point P may be found as follows: draw the ordinate of P to meet the auxiliary circle at P′, join P′ to the origin, then OP′ makes an angle θ with the positive x–axis.

Qu. 12.   Show how to obtain the y-coordinate of the point $(a \cos \theta, b \sin \theta)$ from a circle of radius b.   Draw two concentric circles and hence plot an ellipse.

Example 11.   *Find the equation of the tangent to the ellipse*

$$\frac{x^2}{a^2} + \frac{y^2}{b^2} = 1 \quad at \quad (a \cos \theta, b \sin \theta).$$

The gradient $$\frac{dy}{dx} = \frac{dy}{d\theta} \Big/ \frac{dx}{d\theta}.$$

$$x = a \cos \theta, \qquad y = b \sin \theta.$$

$$\therefore \frac{dy}{dx} = \frac{b \cos \theta}{-a \sin \theta}.$$

Therefore the equation of the tangent is

$$b \cos \theta \, x + a \sin \theta \, y = b \cos \theta . a \cos \theta + a \sin \theta . b \sin \theta.$$
$$\therefore \ bx \cos \theta + ay \sin \theta = ab(\cos^2 \theta + \sin^2 \theta).$$

Therefore the tangent is

$$bx \cos \theta + ay \sin \theta - ab = 0.$$

*Qu. 13.**   Show that the equation of the normal to the ellipse

$$\frac{x^2}{a^2} + \frac{y^2}{b^2} = 1$$

at $(a \cos \theta, b \sin \theta)$ is

$$ax \sin \theta - by \cos \theta - (a^2 - b^2) \sin \theta \cos \theta = 0.$$

If the general point on the curve is taken to be (x_1, y_1), it is frequently necessary to bring in the extra equation

$$\frac{x_1{}^2}{a^2} + \frac{y_1{}^2}{b^2} = 1.$$

This is why working is usually easier when parameters are used.

*Qu. 14. Show that the equation of the tangent at (x_1, y_1) to the ellipse

$$\frac{x^2}{a^2} + \frac{y^2}{b^2} = 1$$

is

$$\frac{xx_1}{a^2} + \frac{yy_1}{b^2} = 1.$$

Verify that this gives the equation found in Example 11 for the tangent at $(a \cos \theta, b \sin \theta)$.

Example 12. *Find the equation of the chord of the ellipse*

$$\frac{x^2}{a^2} + \frac{y^2}{b^2} = 1$$

joining the points whose eccentric angles are θ, ϕ.

The ends of the chord are $(a \cos \theta, b \sin \theta)$, $(a \cos \phi, b \sin \phi)$, therefore the gradient of the chord is

$$\frac{b \sin \theta - b \sin \phi}{a \cos \theta - a \cos \phi} = \frac{2b \cos \frac{1}{2}(\theta + \phi) \sin \frac{1}{2}(\theta - \phi)}{-2a \sin \frac{1}{2}(\theta + \phi) \sin \frac{1}{2}(\theta - \phi)},$$

$$= -\frac{b \cos \frac{1}{2}(\theta + \phi)}{a \sin \frac{1}{2}(\theta + \phi)}.$$

Therefore the equation of the chord is

$$b \cos \tfrac{1}{2}(\theta + \phi)\, x + a \sin \tfrac{1}{2}(\theta + \phi)\, y$$
$$= b \cos \tfrac{1}{2}(\theta + \phi) . a \cos \theta + a \sin \tfrac{1}{2}(\theta + \phi) . b \sin \theta.$$

$$\begin{aligned}
\text{R.H.S.} &= ab\{\cos \tfrac{1}{2}(\theta + \phi) \cos \theta + \sin \tfrac{1}{2}(\theta + \phi) \sin \theta\}, \\
&= ab \cos \{\tfrac{1}{2}(\theta + \phi) - \theta\}, \\
&= ab \cos \tfrac{1}{2}(\phi - \theta), \\
&= ab \cos \tfrac{1}{2}(\theta - \phi).
\end{aligned}$$

Therefore the equation of the chord is

$$bx \cos \tfrac{1}{2}(\theta + \phi) + ay \sin \tfrac{1}{2}(\theta + \phi) - ab \cos \tfrac{1}{2}(\theta - \phi) = 0.$$

*Qu. 15. Show, by putting $\phi = \theta$, that the equation of the chord approaches the equation of the tangent at θ as $\phi \to \theta$.

Example 13. *A tangent to the ellipse*

$$\frac{x^2}{a^2} + \frac{y^2}{b^2} = 1$$

at the point P *meets the minor axis at* L. *If the normal at* P *meets the major axis at* M, *find the locus of the mid-point of* LM.

Let P be the point $(a \cos \theta, b \sin \theta)$, then the tangent at P has equation

$$bx \cos \theta + ay \sin \theta - ab = 0.$$

This meets the minor axis $x = 0$ at L $\left(0, \dfrac{b}{\sin \theta}\right)$.

The normal at P is

$$ax \sin \theta - by \cos \theta - (a^2 - b^2) \sin \theta \cos \theta = 0.$$

This meets the major axis $y = 0$ at M $\left(\dfrac{a^2 - b^2}{a} \cos \theta, 0\right)$

The mid-point of LM is given by

$$x = \frac{a^2 - b^2}{2a} \cos \theta, \qquad y = \frac{b}{2 \sin \theta}.$$

θ can be eliminated from these equations by means of the identity

$$\cos^2 \theta + \sin^2 \theta = 1.$$

Therefore the locus of the mid-point of LM is

$$\left(\frac{2ax}{a^2 - b^2}\right)^2 + \left(\frac{b}{2y}\right)^2 = 1.$$

Exercise 9c

1. Find the foci and directrices of the ellipse:
   (i) $4x^2 + 9y^2 = 36$, (ii) $x^2 + 16y^2 = 25$.

2. Write down the equation of the tangent to:
   (i) $\dfrac{x^2}{9} + \dfrac{y^2}{4} = 1$ at $(3 \cos \theta, 2 \sin \theta)$,
   (ii) $9x^2 + 16y^2 = 25$ at $(1, 1)$.

3. Find the equation of the normal to:
   (i) $9x^2 + 16y^2 = 25$ at $(1, 1)$,
   (ii) $x^2 + 2y^2 = 9$ at $(1, -2)$.

4. A point moves so that its distance from $(3, 2)$ is half its distance from the line $2x + 3y = 1$. Why is the locus an ellipse? Find the equation of the major axis.

5. P is any point on an ellipse; S, S' are the foci. Prove directly from the focus-directrix definition of the ellipse that $SP + S'P = 2a$, where $2a$ is the length of the major axis.

6. Find the relation between the eccentric angles of the points which are at the ends of a focal chord.

7. Prove that the chord joining points of an ellipse whose eccentric angles are $(\alpha + \beta)$, $(\alpha - \beta)$ is parallel to the tangent at the point whose eccentric angle is α.

8. Find the equation of the tangent to the ellipse $\dfrac{x^2}{a^2} + \dfrac{y^2}{b^2} = 1$ at the end of the latus rectum which lies in the first quadrant.

9. The tangent at P to an ellipse meets a directrix at Q. Prove that lines joining the corresponding focus to P and Q are perpendicular.

10. Find the coordinates of the point of intersection of the tangents to the ellipse $\dfrac{x^2}{a^2} + \dfrac{y^2}{b^2} = 1$ at the points whose eccentric angles are θ, ϕ.

11. P is any point on an ellipse and S, S' are the foci. Prove that the normal at P bisects the angle S'PS.

12. Find the locus of the mid-point of the line joining the focus $(ae, 0)$ to any point on the ellipse

$$\frac{x^2}{a^2} + \frac{y^2}{b^2} = 1.$$

13. The eccentric angles of two points P, Q differ by a constant k. Find the locus of the mid-point of PQ.

14. The normal at the point $(a \cos \theta, b \sin \theta)$ on the ellipse

$$\frac{x^2}{a^2} + \frac{y^2}{b^2} = 1$$

meets the axes at L, M. Find the locus of the mid-point of LM.

15. A variable tangent to the ellipse $\dfrac{x^2}{a^2} + \dfrac{y^2}{b^2} = 1$ meets the axes at R, S. Find the locus of the mid-point of RS.

16. Prove that the tangents to the ellipse $\dfrac{x^2}{a^2} + \dfrac{y^2}{b^2} = 1$ at points whose eccentric angles differ by a right angle meet on a concentric ellipse and find its equation.

17. Prove that perpendicular tangents to the ellipse

$$\frac{x^2}{a^2} + \frac{y^2}{b^2} = 1$$

meet on the circle $x^2 + y^2 = a^2 + b^2$ (called the *director circle*).

18. The tangents to the ellipse

$$\frac{x^2}{a^2} + \frac{y^2}{b^2} = 1$$

at P, Q meet at the point (x_1, y_1). Show that the equation of the chord of contact PQ is

$$\frac{xx_1}{a^2} + \frac{yy_1}{b^2} = 1.$$

[Use the results of No. 10 and Example 12.]

Further examples on the ellipse

9.6. Example 14. *Find the condition that the line* $y = mx + c$ *should touch the ellipse*

$$\frac{x^2}{a^2} + \frac{y^2}{b^2} = 1.$$

The equation of any tangent to the ellipse may be written

$$bx \cos \theta + ay \sin \theta - ab = 0.$$

Let this equation represent the same tangent as the given line which we shall write as

$$mx - y + c = 0.$$

Comparing coefficients,

$$\frac{b \cos \theta}{m} = \frac{a \sin \theta}{-1} = \frac{-ab}{c}.$$

$$\therefore \cos \theta = -\frac{am}{c}, \qquad \sin \theta = \frac{b}{c}.$$

But $\cos^2 \theta + \sin^2 \theta = 1.$

$$\therefore \frac{a^2m^2}{c^2} + \frac{b^2}{c^2} = 1.$$

Therefore $y = mx + c$ touches the ellipse if

$$c^2 = a^2m^2 + b^2.$$

Qu. 16. Work Example 14 by eliminating y between the two equations and writing down the condition that the resulting quadratic in x should have equal roots.

Example 15. *Prove that perpendicular tangents to the ellipse*

$$\frac{x^2}{a^2} + \frac{y^2}{b^2} = 1$$

meet on a circle and find its equation.

From Example 14 we see that the equation of a general tangent to the ellipse may be written

$$y = mx + (a^2m^2 + b^2)^{\frac{1}{2}}.$$
$$\therefore \ (y - mx)^2 = a^2m^2 + b^2.$$
$$\therefore \ m^2(x^2 - a^2) - 2xym + y^2 - b^2 = 0. \tag{1}$$

If (x, y) is a point of intersection of two perpendicular tangents to the ellipse, we may regard equation (1) as a quadratic equation for m, the gradient of the tangents. Since the tangents are perpendicular the product of the roots of the equation is -1,

$$\therefore \ \frac{y^2 - b^2}{x^2 - a^2} = -1.$$
$$\therefore \ y^2 - b^2 = a^2 - x^2.$$

Therefore the equation of the locus is

$$x^2 + y^2 = a^2 + b^2.$$

This is called the *director circle* of the ellipse.

Example 16. *A variable straight line with constant gradient m meets the ellipse*

$$\frac{x^2}{a^2} + \frac{y^2}{b^2} = 1$$

at Q, R. *Find the locus of* P, *the mid-point of* QR.

Let the equation of the line be

$$y = mx + c. \tag{1}$$

To find the coordinates of Q, R, we would solve the equation of the line and the equation of the ellipse

$$b^2x^2 + a^2y^2 = a^2b^2$$

simultaneously:

$$b^2x^2 + a^2(m^2x^2 + 2mxc + c^2) - a^2b^2 = 0.$$
$$\therefore \ x^2(b^2 + a^2m^2) + 2a^2mcx + a^2c^2 - a^2b^2 = 0.$$

The x-coordinates of Q, R, say x_1, x_2, are the roots of this equation. But if P is the point (X, Y),

$$X = \tfrac{1}{2}(x_1 + x_2).$$
$$\therefore \ X = \tfrac{1}{2} \cdot \frac{-2a^2mc}{b^2 + a^2m^2}. \tag{2}$$

Now the coordinates of P satisfy equation (1), so

$$Y = mX + c. \qquad (3)$$

Therefore we may find the locus of P by eliminating c between the equations (2), (3). Substituting

$$c = Y - mX$$

in equation (2) rearranged as

$$X(b^2 + a^2m^2) = -a^2mc,$$

we obtain

$$X(b^2 + a^2m^2) = -a^2m(Y - mX).$$

Therefore the locus of P is

$$b^2x + a^2my = 0,$$

which is a diameter of the ellipse.

Exercise 9d

1. Write down the equations of the tangents to:
   (i) $\dfrac{x^2}{4} + \dfrac{y^2}{9} = 1$ with gradient 2,
   (ii) $x^2 + 3y^2 = 3$ with gradient -1,
   (iii) $4x^2 + 9y^2 = 144$ with gradient $\frac{1}{2}$.

2. *Without* solving the equations completely, find the coordinates of the mid-points of the chords formed by the intersection of:
   (i) $x - y - 1 = 0$ and $\dfrac{x^2}{9} + \dfrac{y^2}{4} = 1$,
   (ii) $10x - 5y + 6 = 0$ and $4x^2 + 5y^2 = 20$,
   (iii) $2x + 3y - 4 = 0$ and $y^2 = 8x$.

3. Prove that the line $x - 2y + 10 = 0$ touches the ellipse $9x^2 + 64y^2 = 576$.

4. Find the equations of the tangents to the ellipse $x^2 + 4y^2 = 4$ which are perpendicular to the line $2x - 3y = 1$.

5. The line $y = x - c$ touches the ellipse $9x^2 + 16y^2 = 144$. Find the value of c and the coordinates of the point of contact.

6. Find the condition for the line $y = mx + c$ to cut the ellipse

$$\frac{x^2}{a^2} + \frac{y^2}{b^2} = 1$$

in two distinct points.

7. The line $y = mx + c$ touches the ellipse

$$\frac{x^2}{a^2} + \frac{y^2}{b^2} = 1.$$

Prove that the foot of the perpendicular from a focus on to this line lies on the auxiliary circle $x^2 + y^2 = a^2$.

8. Find the locus of the foot of the perpendicular from the centre of the ellipse

$$\frac{x^2}{a^2} + \frac{y^2}{b^2} = 1$$

on to any tangent.

9. Find the equation of the normal at the point (x_1, y_1) on the ellipse

$$\frac{x^2}{a^2} + \frac{y^2}{b^2} = 1.$$

10. Find the coordinates of the mid-point of the chord formed by the intersection of:
   (i) $y = mx + c$ and $b^2x^2 + a^2y^2 = a^2b^2$,
   (ii) $lx + my + n = 0$ and $y^2 = 4ax$.

11. Find the equation of the diameter bisecting the chord $3x + 2y = 1$ of the ellipse $4x^2 + 9y^2 = 16$.

12. Find the equation of the line with gradient m passing through the focus $(ae, 0)$ of the ellipse

$$b^2x^2 + a^2y^2 = a^2b^2.$$

If the line meets the ellipse in P, Q, find the coordinates of the mid-point of PQ and show that they satisfy the equation

$$a^2my + b^2x = 0.$$

By substituting the value of m obtained from this equation into the equation of PQ, find the locus of the mid-point of PQ.

13. A variable line passes through the point $(a, 0)$. Find the locus of the mid-point of the chord formed by the intersection of this line and the ellipse

$$b^2x^2 + a^2y^2 = a^2b^2.$$

14. Find the locus of points from which the tangents to the ellipse

$$b^2x^2 + a^2y^2 = a^2b^2$$

are inclined at 45°.

15. Lines of gradient m are drawn to cut the ellipse

$$b^2x^2 + a^2y^2 = a^2b^2.$$

Prove that the mid-points of the chords so formed lie on a straight line through the origin with gradient $-b^2/a^2m$. Deduce the equation of the chord whose mid-point is (h, k).

16. Show that a general tangent to the circle

$$x^2 + y^2 - a^2 = 0$$

may be written

$$y = mx \pm a\sqrt{(1 + m^2)}.$$

A variable tangent to the circle

$$x^2 + y^2 - a^2 = 0$$

meets the ellipse

$$b^2x^2 + a^2y^2 = a^2b^2 \qquad (b > a)*$$

at P, Q.   Find the locus of the mid-point of PQ.

17. A variable tangent to the ellipse

$$b^2x^2 + a^2y^2 = a^2b^2$$

meets the parabola $y^2 = 4ax$ at L, M.   Find the locus of the mid-point of LM.

18. The chord of contact of the point (x_1, y_1) with respect to the ellipse

$$\frac{x^2}{a^2} + \frac{y^2}{b^2} = 1$$

cuts the axes at L, M.   If the locus of the mid-point of LM is the circle $x^2 + y^2 = 1$, find the locus of (x_1, y_1).   [Use the result of Exercise 9c, No. 18.]

The hyperbola

9.7.   In § 9.4, p. 188, certain results were obtained for the ellipse. The working is so similar for the hyperbola that it is left to the reader to obtain the corresponding results.   Starting with the focus–directrix definition with $e > 1$ he should work through the following questions.

Qu. 17.   On a sheet of squared paper, rule the directrix along one line near the middle, take the focus 4 cm out towards the nearer edge and plot part of a hyperbola (there are two branches of it) taking $e = 2$.

Qu. 18.   Show that, with suitable choice of axes, the equation of a hyperbola may be written

$$\frac{x^2}{a^2} - \frac{y^2}{b^2} = 1,$$

where $$b^2 = a^2(e^2 - 1),$$

the foci are $$(ae, 0) \quad \text{and} \quad (-ae, 0),$$

and the directrices $$x = \frac{a}{e} \quad \text{and} \quad x = -\frac{a}{e}.$$

Qu. 19.   Show that any point on the hyperbola

$$\frac{x^2}{a^2} - \frac{y^2}{b^2} = 1$$

may be written

$$(a \sec \theta, b \tan \theta).$$

* The major axis lies along the y-axis.

Qu. 20. Show that at $(a \sec \theta, b \tan \theta)$ on the hyperbola

$$\frac{x^2}{a^2} - \frac{y^2}{b^2} = 1,$$

the equation of the tangent is

$$bx - ay \sin \theta - ab \cos \theta = 0,$$

and the equation of the normal is

$$ax \sin \theta + by - (a^2 + b^2) \tan \theta = 0.$$

Qu. 21. Show that the equation of the tangent at (x_1, y_1) to the hyperbola

$$\frac{x^2}{a^2} - \frac{y^2}{b^2} = 1$$

is

$$\frac{xx_1}{a^2} - \frac{yy_1}{b^2} = 1.$$

Show that the equation of the tangent in Qu. 20 may be deduced from this.

Asymptotes to a hyperbola

9.8. Example 17. *Find c in terms of a, b, m if $y = mx + c$ is a tangent to the hyperbola*

$$\frac{x^2}{a^2} - \frac{y^2}{b^2} = 1.$$

Solving the two equations simultaneously,

$$b^2 x^2 - a^2 y^2 = a^2 b^2.$$
$$\therefore \quad b^2 x^2 - a^2(m^2 x^2 + 2mcx + c^2) - a^2 b^2 = 0.$$
$$\therefore \quad x^2(b^2 - a^2 m^2) - 2a^2 mcx - a^2(b^2 + c^2) = 0. \tag{1}$$

The line is a tangent if and only if this equation has equal roots.

i.e. if and only if $(-2a^2 mc)^2 = -4(b^2 - a^2 m^2)a^2(b^2 + c^2)$.

i.e. $\quad\quad a^2 m^2 c^2 = -(b^2 - a^2 m^2)(b^2 + c^2)$.

i.e. $\quad\quad a^2 m^2 c^2 = -b^4 - b^2 c^2 + a^2 m^2 b^2 + a^2 m^2 c^2$.

i.e. $\quad\quad b^2 c^2 = a^2 m^2 b^2 - b^4$.

Therefore $y = mx + c$ is a tangent to the hyperbola if and only if

$$c^2 = a^2 m^2 - b^2.$$

[Compare the method of Example 14, p. 194.]

In Example 17, the value of x at the point of contact is given by half the sum of the roots of equation (1) since the roots are equal.

$$\therefore \ x = \frac{a^2mc}{b^2 - a^2m^2},$$
$$= \mp \frac{a^2m\sqrt{(a^2m^2 - b^2)}}{a^2m^2 - b^2}.$$

Therefore, at the point of contact,

$$x = \mp \frac{a^2m}{\sqrt{(a^2m^2 - b^2)}}.$$

Hence as

$$m \to \pm\frac{b}{a}, \qquad x \to \infty \ \text{ and, since } c^2 = a^2m^2 - b^2, \ c \to 0,$$

so that

$$y = \pm\frac{b}{a}x$$

may be regarded as the limit of a tangent to the hyperbola as the point of contact tends to infinity.

One way of remembering the equation of the asymptotes

$$\frac{x^2}{a^2} - \frac{y^2}{b^2} = 0$$

is that, when x, y are very large, terms other than those of the highest degree may be neglected in comparison.

The rectangular hyperbola

9.9. There is a special case of the hyperbola which has interesting properties and so receives special attention. A *rectangular hyperbola* is one whose asymptotes are perpendicular. The asymptotes of

$$\frac{x^2}{a^2} - \frac{y^2}{b^2} = 1 \quad \text{are} \quad y = \pm\frac{b}{a}x,$$

and these are perpendicular when

$$-\frac{b}{a} \cdot \frac{b}{a} = -1,$$

that is when $b = a$. Hence

$$x^2 - y^2 = a^2$$

represents a rectangular hyperbola and its asymptotes are

$$x - y = 0, \qquad x + y = 0.$$

A special property of the rectangular hyperbola enables us to write its equation in a very simple way. Let $P(x, y)$ be any point on the rectangular hyperbola

$$x^2 - y^2 = a^2$$

and let X, Y be the perpendicular distances from P on to the asymptotes

$$x - y = 0, \qquad x + y = 0.$$

Then

$$X = \pm \frac{x-y}{\sqrt{2}}, \qquad Y = \pm \frac{x+y}{\sqrt{2}},$$

$$\therefore XY = \pm \frac{x^2 - y^2}{2}.$$

$$\therefore XY = \pm \frac{a^2}{2}.$$

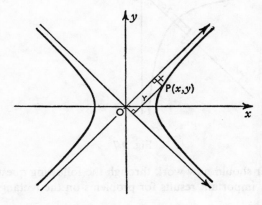

Fig. 9.6

Suppose, now, that we take axes of coordinates along the asymptotes in the directions indicated by the arrows in Fig. 9.6. We have shown that with these new axes, the product of the x- and y-coordinates is constant and equal to $\pm \frac{1}{2} a^2$. But the axes have been chosen so that the product is positive and so we may write the equation of a rectangular hyperbola as

$$xy = c^2$$

when referred to its asymptotes as coordinate axes. The relation $c^2 = \frac{1}{2} a^2$ will be used very soon.

The eccentricity of $x^2 - y^2 = a^2$ is given by $a^2 = a^2(e^2 - 1)$ from which we find that $e = \sqrt{2}$ and hence the foci are $(\pm\sqrt{2}a, 0)$. Now $a = \sqrt{2}c$, so the foci of $xy = c^2$ are on the major axis at a distance $2c$ from the centre (see Fig. 9.7). Therefore the coordinates of the foci of $xy = c^2$ are $(\sqrt{2}c, \sqrt{2}c)$ and $(-\sqrt{2}c, -\sqrt{2}c)$.

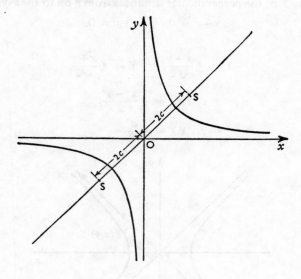

Fig. 9.7

The reader should now work through the following questions which contain very important results for problems on the rectangular hyperbola.

Qu. 22. Show that any point on the rectangular hyperbola $xy = c^2$ may be represented by

$$\left(ct, \frac{c}{t}\right).$$

Qu. 23. Show that the gradient of the hyperbola at $(ct, c/t)$ is

$$-\frac{1}{t^2};$$

that the equation of the tangent is

$$x + t^2 y - 2ct = 0;$$

and that the equation of the normal is

$$t^2x - y - ct^3 + c/t = 0.$$

Qu. 24. Show that the gradient of the chord joining the points $(ct_1, c/t_1)$, $(ct_2, c/t_2)$ on the hyperbola $xy = c^2$ is

$$-\frac{1}{t_1 t_2},$$

and that the equation of the chord is

$$x + t_1 t_2 y - c(t_1 + t_2) = 0.$$

Qu. 25. Verify that the equation of the chord in Qu. 24 becomes the equation of the tangent in Qu. 23 when $t_1 = t_2 = t$.

Further examples on the hyperbola

9.10. The following examples do not illustrate any new principles but rather serve to show that the same methods that were used for problems about the ellipse may also be used in connection with the hyperbola.

Example 18. *A tangent to a hyperbola at* P *meets a directrix at* Q. *If* S *is the corresponding focus, prove that* PQ *subtends a right angle at* S.

Let P be the point $(a \sec \theta, b \tan \theta)$ on the hyperbola

$$\frac{x^2}{a^2} - \frac{y^2}{b^2} = 1.$$

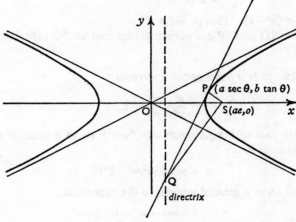

Fig. 9.8

The tangent at P is

$$bx - ay \sin \theta - ab \cos \theta = 0.$$

This meets the directrix $x = a/e$ at a point given by

$$\frac{ba}{e} - ay \sin \theta - ab \cos \theta = 0.$$

$$\therefore \ y \sin \theta = \frac{b(1 - e \cos \theta)}{e}.$$

$$\therefore \ Q \text{ is } \left(\frac{a}{e}, \frac{b(1 - e \cos \theta)}{e \sin \theta} \right).$$

Therefore the gradient of QS

$$m_1 = \frac{\dfrac{b(1 - e \cos \theta)}{e \sin \theta}}{\dfrac{a}{e} - ae} = \frac{b(1 - e \cos \theta)}{a(1 - e^2) \sin \theta}.$$

The gradient of PS

$$m_2 = \frac{b \tan \theta}{a \sec \theta - ae} = \frac{b \sin \theta}{a(1 - e \cos \theta)}.$$

$$m_1 m_2 = \frac{b(1 - e \cos \theta)}{a(1 - e^2) \sin \theta} \cdot \frac{b \sin \theta}{a(1 - e \cos \theta)},$$

$$= \frac{b^2}{a^2(1 - e^2)} = -1,$$

since $b^2 = a^2(e^2 - 1)$. (See p. 198.)

Therefore SQ and SP are perpendicular and so PQ subtends a right angle at S.

Example 19. S *is a focus of the hyperbola*

$$\frac{x^2}{a^2} - \frac{y^2}{b^2} = 1$$

and P *is the foot of the perpendicular from* S *on to a variable tangent. Find the locus of* P.

$$y = mx \pm (a^2 m^2 - b^2)^{\frac{1}{2}} \tag{1}$$

is the equation of a general tangent to the hyperbola.

$$x + my - ae = 0 \tag{2}$$

is the equation of the perpendicular from S to the tangent.

The coordinates of P satisfy equations (1), (2), and so we may find the locus by eliminating m between these equations.

From (1)

$$y^2 - 2mxy + m^2x^2 = a^2m^2 - b^2.$$

From (2)

$$m^2y^2 + 2mxy + x^2 = a^2e^2.$$

Adding,

$$y^2(1+m^2) + x^2(1+m^2) = a^2m^2 - b^2 + a^2e^2.$$
$$\text{R.H.S.} = a^2m^2 + a^2 \quad \text{since} \quad b^2 = a^2(e^2 - 1). \quad \text{(See p. 198.)}$$
$$\therefore \ y^2(1+m^2) + x^2(1+m^2) = a^2(1+m^2).$$

Therefore the locus of P is $x^2 + y^2 = a^2$, which is the *auxiliary circle*.

Example 20. PQ *is a chord of the rectangular hyperbola* $xy = c^2$ *and* R *is its mid-point. If* PQ *has a constant length* k, *find the locus of* R.

Let P be $\left(cp, \dfrac{c}{p}\right)$ and Q be $\left(cq, \dfrac{c}{q}\right)$. Then, if R is (x, y),

$$x = \tfrac{1}{2}c(p+q), \tag{1}$$
$$y = \frac{c(p+q)}{2pq}. \tag{2}$$

Since the length of PQ is given to be k,

$$\text{PQ}^2 = (cp - cq)^2 + \left(\frac{c}{p} - \frac{c}{q}\right)^2 = k^2.$$
$$\therefore \ c^2(p-q)^2 + c^2\frac{(q-p)^2}{p^2q^2} = k^2.$$
$$\therefore \ c^2(p-q)^2(1 + p^2q^2) = k^2p^2q^2. \tag{3}$$

From (1), (2), $\qquad p+q = \dfrac{2x}{c}, \qquad pq = \dfrac{x}{y}.$

Now $\qquad\qquad (p-q)^2 \equiv (p+q)^2 - 4pq,$

so that (3) becomes

$$c^2\{(p+q)^2 - 4pq\}(1 + p^2q^2) = k^2p^2q^2.$$

Substituting for $p+q$ and pq,

$$c^2\left(\frac{4x^2}{c^2}-\frac{4x}{y}\right)\left(1+\frac{x^2}{y^2}\right) = k^2\frac{x^2}{y^2}.$$

$$\therefore \left(4x^2-\frac{4c^2x}{y}\right)(y^2+x^2) = k^2x^2.$$

Therefore the locus of P is

$$4(xy-c^2)(x^2+y^2) = k^2xy.$$

Exercise 9e

1. P is any point on the rectangular hyperbola $xy=c^2$. Show that the line joining P to the centre, and the tangent at P, are equally inclined to the asymptotes.

2. P is any point on the hyperbola

$$\frac{x^2}{a^2}-\frac{y^2}{b^2} = 1$$

and Q is the point (a, b). Find the locus of the point dividing PQ in the ratio $2:1$.

3. Prove that the product of the lengths of the perpendiculars from any point of a hyperbola to its asymptotes is constant.

4. The normal at any point of a hyperbola meets the axes at E, F. Find the locus of the mid-point of EF.

5. Find the coordinates of the point at which the normal at $(ct, c/t)$ meets the rectangular hyperbola $xy=c^2$ again.

6. Any tangent to the rectangular hyperbola $xy=c^2$ meets the asymptotes at L and M. Find the locus of the mid-point of LM.

7. The normal at any point on the hyperbola $xy=c^2$ meets the x-axis at A, and the tangent meets the y-axis at B. Find the locus of the mid-point of AB.

8. Find the equation of the chord of the hyperbola $xy=c^2$ whose mid-point is (x_1, y_1).

9. Show that, in general, four normals can be drawn from any point to the rectangular hyperbola $xy=c^2$.

10. The normal at any point P of the rectangular hyperbola $xy=c^2$ meets the y-axis at A, and the tangent meets the x-axis at B. Find the co-ordinates of the fourth vertex Q of the rectangle APBQ in terms of t, the parameter of P.

11. Find the locus of the foot of the perpendicular from the origin on to a tangent to the rectangular hyperbola $xy=c^2$.

12. Find the condition that the line $lx+my+n=0$ should touch the rectangular hyperbola $xy=c^2$.

13. Prove that the locus of middle points of parallel chords of the rectangular hyperbola $xy = c^2$ is a diameter.

14. Find the locus of the point of intersection of perpendicular tangents to the hyperbola

$$\frac{x^2}{a^2} - \frac{y^2}{b^2} = 1.$$

15. Find the locus of the foot of the perpendicular from the origin to a tangent of the hyperbola

$$\frac{x^2}{a^2} - \frac{y^2}{b^2} = 1.$$

16. PQ is a variable chord of the hyperbola

$$\frac{x^2}{a^2} - \frac{y^2}{b^2} = 1$$

with constant gradient m_1. Show that the locus of the mid-point of PQ is a diameter with gradient m_2 such that $m_1 m_2 = b^2/a^2$.

17. The chord AB of a hyperbola meets the asymptotes at M, N. Prove that AM = BN. [Show that AB and MN have the same mid-point.]

18. Find the equation of the chord joining the points $(a \sec \theta, b \tan \theta)$ $(a \sec \phi, b \tan \phi)$ on the hyperbola

$$\frac{x^2}{a^2} - \frac{y^2}{b^2} = 1.$$

19. The tangent at any point P on the hyperbola

$$\frac{x^2}{a^2} - \frac{y^2}{b^2} = 1$$

meets the asymptotes at Q and Q'. Prove that PQ = PQ'.

20. The normal to the hyperbola

$$\frac{x^2}{a^2} - \frac{y^2}{b^2} = 1$$

at $(a \sec \theta, b \tan \theta)$ meets the asymptotes at G, H. Find the locus of the mid-point of GH.

21. P, Q, R are three points on a rectangular hyperbola such that PQ subtends a right angle at R. Show that PQ is perpendicular to the tangent at R.

22. Any normal to the rectangular hyperbola $xy = c^2$ meets the asymptotes at L and M. Find the locus of N, the fourth vertex of the rectangle MOLN, where O is the centre.

23. P is any point on a rectangular hyperbola whose foci are S and S'. Prove that SP.S'P = OP², where O is the origin.

24. The chord joining the points $(ct_1, c/t_1)$, $(ct_2, c/t_2)$ of the rectangular hyperbola $xy = c^2$ subtends a right angle at the focus. Find the locus of the mid-point of the chord.

Exercise 9f (Miscellaneous)

1. The distance between the foci of an ellipse is 8 and between the directrices is 18. Find its equation in the simplest form.

2. R, S are two fixed points a distance 5 units apart and the point P moves so that $PR + PS$ is constant and equal to 12 units. Find the locus of P.

3. P is any point on a parabola and S is the focus. Prove that the circle on SP as diameter touches the tangent at the vertex.

4. The equation of a parabola is $y^2 = 12x - 12$. Find the equations of the straight lines that pass through the origin and cut the parabola where $x = 4$. Find also the equations of the tangents to the parabola that are parallel to these lines.

5. Show that the line $5y - 4x = 25$ touches the ellipse

$$\frac{x^2}{25} + \frac{y^2}{9} = 1$$

and find the equation of the normal to the ellipse at the point of contact. What is the eccentricity of the ellipse?

6. Find the equations of the tangents to the hyperbola $3x^2 - 4y^2 = 1$ which make equal angles with the axes.

7. The perpendiculars from the foci of the hyperbola $b^2x^2 - a^2y^2 = a^2b^2$ on to any tangent are of length p_1, p_2. Prove that $|p_1p_2| = b^2$.

8. The gradients of the tangents to the parabola $y^2 = 4ax$ and the rectangular hyperbola $xy = c^2$ at the point at which they cut are m_1 and m_2 respectively. Prove that $m_2 = -2m_1$.

9. Find the coordinates of the mid-point of the chord $x + y - 1 = 0$ of the parabola $y^2 = 6x$.

10. Show that the equation $2x^2 + y^2 = 6y$ represents an ellipse with eccentricity $\frac{1}{2}\sqrt{2}$. Find the coordinates of the centre and the length of the minor axis.

11. Show that $x^2 + 4x - 8y - 4 = 0$ represents a parabola whose focus is at $(-2, 1)$. Find the equation of the tangent at the vertex.

12. A man stands on a ladder which rests on smooth horizontal ground against a smooth vertical wall. Prove that his feet will describe part of an ellipse as the ladder falls.

13. Find the eccentricity and focus of the curve $y^2 - 4y + 2x + 2 = 0$ and write down the equations of the tangent and normal at the point $(\frac{1}{2}, 1)$.

14. P $(a \sec \theta, b \tan \theta)$ is any point on the hyperbola

$$\frac{x^2}{a^2} - \frac{y^2}{b^2} = 1$$

and N is the foot of the perpendicular from P to the x-axis. NT is drawn to touch the auxiliary circle at T. Prove that the line joining T to the centre of the circle makes an angle with the x-axis equal to θ.

15. Find the locus of the point of intersection of perpendicular tangents to the ellipse

$$\frac{x^2}{a^2} + \frac{y^2}{b^2} = 1.$$

What are the areas of the greatest and least rectangles circumscribing the ellipse and circumscribed by its director circle?

16. Prove that the tangents at the ends of a focal chord of the rectangular hyperbola $xy = c^2$ meet on a line and find the equation of this line.

17. The parameters of the ends of two perpendicular chords of the rectangular hyperbola $xy = c^2$ are t_1, t_2 and t_3, t_4. Prove that $t_1 t_2 t_3 t_4 = -1$ and deduce that the orthocentre of the triangle formed by three points on a rectangular hyperbola also lies on the hyperbola.

18. The normals at the points P, Q of the parabola $y^2 = 4ax$ meet on the parabola. Find the locus of the mid-point of PQ.

19. Find the locus of mid-points of chords of the ellipse

$$\frac{x^2}{a^2} + \frac{y^2}{b^2} = 1$$

which subtend a right angle at the centre.

20. A tangent to the hyperbola $xy = 4$ meets the ellipse $x^2 + 4y^2 = 1$ at P and Q. Find the locus of the point of intersection of the tangents at P and Q to the ellipse. [See Exercise 9c, No. 18.]

21. Find the locus of the point of intersection of perpendicular normals of the parabola $y^2 = 4ax$.

22. Perpendiculars are drawn from the point $(a, 0)$ to tangents to the ellipse

$$\frac{x^2}{a^2} + \frac{y^2}{b^2} = 1.$$

Find the locus of the feet of the perpendiculars.

23. If a particle is projected under gravity from a point on a level plane with velocity V at an angle of elevation α, the range

$$R = \frac{V^2 \sin 2\alpha}{g}$$

and the greatest height

$$H = \frac{V^2 \sin^2 \alpha}{2g}.$$

With axes through the point of projection, the equation of the parabolic trajectory is

$$y = x \tan \alpha - \frac{gx^2}{2V^2 \cos^2 \alpha}.$$

Show that this equation may be written

$$(x - \tfrac{1}{2}R)^2 = \frac{2V^2 \cos^2 \alpha}{g} (H - y),$$

and determine the coordinates of the focus, and the equation of the directrix.

CHAPTER 10

SERIES FOR e^x AND $\log_e (1+x)$

10.1. The expansion of functions of a variable as series has considerable theoretical and practical importance. There are some problems that are most easily tackled by means of series, for instance estimating the value of the constant e, and there are problems in science and engineering which have no practicable solution except by series. Further, the development of computers in recent years has considerably added to the practical importance of approximate numerical solutions to problems. So far in this book only the function $(1+x)^n$ has been expanded in a series and in this chapter two more functions, e^x and $\log_e (1+x)$ will be considered.

The exponential series

The fundamental property of the function e^x is that

$$\frac{d}{dx} (e^x) = e^x.$$

If two assumptions are made:

  (i) that e^x can be expanded as a series of ascending powers of x and
 (ii) that the nth derivative of such a series is the sum to infinity of the nth derivatives of the individual terms,

it is easy to find the coefficients of the terms in the series.
Suppose that

$$e^x = a_0 + a_1 x + a_2 x^2 + a_3 x^3 + \ldots + a_n x^n + \ldots \tag{1}$$

Differentiating (1) one, two, and three times respectively,

$$e^x = a_1 + 2a_2 x + 3a_3 x^2 + \ldots + na_n x^{n-1} + \ldots \tag{2}$$

$$e^x = 2a_2 + 3.2a_3 x + \ldots + n(n-1)a_n x^{n-2} + \ldots \tag{3}$$

$$e^x = 3.2a_3 + \ldots + n(n-1)(n-2)a_n x^{n-3} + \ldots \tag{4}$$

Differentiating (1) n times,

$$e^x = n!a_n + \ldots \tag{5}$$

Now substituting $x = 0$ in (1), (2), (3), (4), (5),

$$1 = a_0,$$
$$1 = a_1,$$
$$1 = 2a_2,$$
$$1 = 3.2a_3,$$
$$1 = n! a_n.$$

Substituting the values we have just found for a_0, a_1, a_2, a_3, a_n into equation (1),

$$e^x = 1 + x + \frac{x^2}{2!} + \frac{x^3}{3!} + \ldots + \frac{x^n}{n!} + \ldots$$

This series is often denoted by exp x and it is valid for all values of x (see below).

Qu. 1. Write down the first four terms and the general terms in the expansions of: (i) e^{-x}, (ii) e^{x^2}, (iii) e^{3x} in ascending powers of x; (iv) $e^{1/x}$, (v) e^{-1/x^2} in descending powers of x.

Qu. 2. (Another method of proof.) Find the coefficients of the terms in the expansion of e^x by equating coefficients in equations (1) and (2) above.

Alternatively, the expansion of e^x can be obtained by integration. The assumption is made that $x^n \to 0$ as $n \to \infty$ when $|x| < 1$. This has already been assumed in connection with infinite geometrical progressions (P.M.I., p. 248). Most readers, however, will prefer to leave this proof until the second reading, in which case they should proceed to Example 1, p. 214.

Let the variable x lie in the range of values from 0 to c, where c is any positive constant, thus

$$0 < x < c.$$

Now $e^0 = 1$,

$$\therefore \; 1 < e^x < e^c.$$

Integrating from 0 to x,

$$x < e^x - 1 < x \, e^c.$$

Again integrating from 0 to x,

$$\tfrac{1}{2} x^2 < e^x - 1 - x < \tfrac{1}{2} x^2 \, e^c.$$

Integrating a further $n - 2$ times,

$$\frac{x^n}{n!} < e^x - 1 - \frac{x^2}{2!} - \ldots - \frac{x^{n-1}}{(n-1)!} < \frac{x^n}{n!} \, e^c.$$

When $n \to \infty$, $x^n/n! \to 0$ (proved below),

$$\therefore \; e^x - 1 - x - \frac{x^2}{2!} - \ldots - \frac{x^{n-1}}{(n-1)!} \to 0.$$

Therefore the difference between e^x and the series

$$1 + x + \frac{x^2}{2!} + \ldots + \frac{x^r}{r!} + \ldots + \frac{x^{n-1}}{(n-1)!}$$

approaches zero as $n \to \infty$.

$$\therefore \; e^x = 1 + x + \frac{x^2}{2!} + \ldots + \frac{x^r}{r!} + \ldots.$$

To prove the series for negative values of x, take

$$a < x < 0.$$

$$\therefore \; e^a < e^x < 1.$$

Integrating from x to 0,

$$-x \, e^a < 1 - e^x < -x.$$

Again integrating from x to 0,

$$\tfrac{1}{2}x^2 \, e^a < e^x - 1 - x < \tfrac{1}{2}x^2.$$

It is left to the reader to complete the proof.

Note that the expansion of e^x is valid for all values of x. It is clear that exp x must have a finite sum for any value of x, since it has been shown that the sum to infinity is e^x, which is finite.

To show that $x^n/n! \to 0$ as $n \to \infty$, let $u_r = x^r/r!$

$$\therefore \; \frac{u_{r+1}}{u_r} = \frac{x^{r+1}}{(r+1)!} \cdot \frac{r!}{x^r} = \frac{x}{r+1}.$$

Let k be the first integer greater than or equal to $2x$, then if $r > k$,

$$\frac{u_{r+1}}{u_r} < \tfrac{1}{2}.$$

$$\therefore \; u_{k+1} < \tfrac{1}{2}u_k, \qquad u_{k+2} < (\tfrac{1}{2})^2 u_k, \ldots, u_n < (\tfrac{1}{2})^{n-k} u_k,$$

$$\therefore \; u_n < \frac{x^k}{k!} (\tfrac{1}{2})^{n-k}.$$

But $x^k/k!$ is finite and $(\tfrac{1}{2})^{n-k} \to 0$ as $n \to \infty$, therefore $x^n/n! \to 0$ as $n \to \infty$.

Example 1. *Find the value of* e *correct to four places of decimals.*

Substituting $x = 1$ in the series for e^x,

$$e^1 = 1 + 1 + \frac{1}{2!} + \frac{1}{3!} + \ldots + \frac{1}{r!} + \ldots.$$

The working is shown. Each term in the series
is obtained from the previous one by dividing by
$1, 2, 3, \ldots, 9, \ldots$. The working has been taken to
five places of decimals. The value obtained for
e is $2\cdot 7183$, correct to four places of decimals.

$$\begin{array}{l}
1\cdot 00000 \\
1\cdot 00000 \\
0\cdot 50000 \\
0\cdot 16667 \\
0\cdot 04167 \\
0\cdot 00833 \\
0\cdot 00139 \\
0\cdot 00020 \\
0\cdot 00002 \ (5) \\
0\cdot 00000 \\
\hline
2\cdot 71828
\end{array}$$

(Tables give $e = 2\cdot 718\ 28\ldots$.)

It can be shown that e is irrational, and it can also be shown that e is
transcendental, that is, e satisfies no algebraic equation in the form

$$a_0 + a_1 x + \ldots + a_n x^n = 0$$

where the coefficients $a_0, a_1, \ldots, a_n$ are integers.

Example 2. *Find the first four terms in the expansions in ascending
powers of x of* (i) e^{1-x^2}, (ii) e^{x-x^2}, *giving the general term in* (i).

(i) $\quad e^{1-x^2} = e^1 . e^{-x^2}$,

$$= e\left\{1 + (-x^2) + \frac{(-x^2)^2}{2!} + \frac{(-x^2)^3}{3!} + \ldots + \frac{(-x^2)^r}{r!} + \ldots\right\}$$

$$\therefore\ e^{1-x^2} = e\{1 - x^2 + \tfrac{1}{2}x^4 - \tfrac{1}{6}x^6 + \ldots + (-1)^r x^{2r}/r! + \ldots\}.$$

(ii) $\quad e^{x-x^2} = 1 + (x - x^2) + \frac{(x-x^2)^2}{2!} + \frac{(x-x^2)^3}{3!} + \ldots,$

$$= 1 + x - x^2 + \tfrac{1}{2}x^2 - x^3 + \ldots + \tfrac{1}{6}x^3 + \ldots,$$

$$\therefore\ e^{x-x^2} = 1 + x - \tfrac{1}{2}x^2 - \tfrac{5}{6}x^3 + \ldots.$$

Example 3. *Find the sum to infinity of the series*

$$1 + \frac{3x}{1!} + \frac{5x^2}{2!} + \frac{7x^3}{3!} + \ldots.$$

The general term is $\dfrac{1+2n}{n!} x^n$. We aim to find terms in the form $\dfrac{x^r}{r!}$, so the general term is split up as

$$\frac{x^n}{n!} + \frac{2n}{n!} x^n = \frac{x^n}{n!} + 2x \cdot \frac{x^{n-1}}{(n-1)!} \qquad (n \geqslant 1.)$$

Therefore the series may be written:

$$1 + (x + 2x \cdot 1) + \left(\frac{x^2}{2!} + 2x \cdot x\right) + \left(\frac{x^3}{3!} + 2x \cdot \frac{x^2}{2!}\right) + \dots$$

$$= 1 + x + \frac{x^2}{2!} + \frac{x^3}{3!} + \frac{x^4}{4!} + \dots + 2x \cdot 1 + 2x \cdot x + 2x \cdot \frac{x^2}{2!} + 2x \cdot \frac{x^3}{3!} + \dots$$

$$= e^x + 2x\, e^x = (1 + 2x)\, e^x.$$

Exercise 10a

1. Use the expansion exp x to find the values of (i) $e^{0 \cdot 1}$, (ii) $1/e$, (iii) $\sqrt{e}$, giving your answers correct to four places of decimals.

In Nos. 2 to 12, expand the functions of x as far as the fourth non-zero terms and give the general terms.

2. e^{x^3}.   **3.** $\sqrt[3]{(e^x)}$.   **4.** $(1/e^x)^2$.

5. e^{2+x}.   **6.** $1/\sqrt{(e^x)}$.   **7.** $(1+x)e^x$.

8. $(1+2x)\,e^{-2x}$.   **9.** $\dfrac{e^{3x} \cdot e^{2x}}{e^x}$.   **10.** $\dfrac{e^{3x} + e^{2x}}{e^x}$.

11. $\dfrac{e^{2x} + e^x - 2}{e^x - 1}$.   **12.** $(1 + e^x)(1 + e^{2x})$.

13. Find the greatest terms in the expansion of e^x when $x = 10$.

In Nos. 14 to 17 expand the functions in ascending powers of x as far as the term in x^3.

14. $e^{x^2 + 2x}$.   **15.** $e^{x^2 - 3x + 1}$.

16. $\dfrac{e^x}{1+x}$.   **17.** $\dfrac{1 - e^{-x}}{e^x - 1}$.

18. Find the limits of the following functions as x approaches zero:

     (i) $\dfrac{e^x - (1+x)}{e^{2x} - (1+2x)}$.   (ii) $\dfrac{e^{2x} - (1+4x)^{\frac{1}{2}}}{e^{-x} - (1-3x)^{\frac{1}{3}}}$,

     (iii) $\dfrac{e^x + e^{-x} - 2}{e^{x^2} - 1}$.

Find the sums to infinity of the following series:

19. $1 + \dfrac{2x}{1!} + \dfrac{3x^2}{2!} + \dfrac{4x^3}{3!} + \ldots$

20. $1 + \dfrac{3x}{2!} + \dfrac{9x^2}{3!} + \dfrac{27x^3}{4!} + \ldots$

21. $1 + \dfrac{x^2}{2!} + \dfrac{x^2}{4!} + \ldots$ (Start by writing down the series for e^z and e^{-z}.)

22. $x + \dfrac{x^3}{3!} + \dfrac{x^5}{5!} + \ldots$ **23.** $\dfrac{x^2}{2!} + \dfrac{2x^3}{3!} + \dfrac{3x^4}{4!} + \ldots$

24. $\dfrac{1}{2} + \dfrac{1}{2.4} + \dfrac{1}{2.4.6} + \ldots$ **25.** $\dfrac{1}{6} + \dfrac{1}{6.9} + \dfrac{1}{6.9.12} + \ldots$

26. $\dfrac{3}{1!} + \dfrac{4}{2!} + \dfrac{5}{3!} + \ldots$ **27.** $\dfrac{1}{1!} + \dfrac{4}{2!} + \dfrac{7}{3!} + \ldots$

28. $\dfrac{3}{2} + \dfrac{5}{2.4} + \dfrac{7}{2.4.6} + \ldots$

The logarithmic series

10.2.   The geometric series

$$1 - u + u^2 - u^3 + \ldots$$

has a sum to infinity (P.M.I., p. 232) of $1/(1+u)$. So we may write

$$\frac{1}{1+u} = 1 - u + u^2 - u^3 + \ldots.$$

Assuming that the integral of the sum of an infinite series is the sum of the integrals of its terms, integrate between 0 and x:

$$\log_e(1+x) = x - \frac{x^2}{2} + \frac{x^3}{3} - \frac{x^4}{4} + \ldots.$$

The nth term of the geometric series is $(-1)^{n-1}u^{n-1}$ so that the nth term of the logarithmic series is $(-1)^{n-1}x^n/n$. Since the geometric series only has a sum if $|u| < 1$, we should expect that the logarithmic series would be valid when $|x| < 1$ but it can also be shown that the series has a sum when $x = 1$ (see Exercise 10b, No. 37; see also Fig. 18.6, p. 400). Thus

$$\log_e(1+x) = x - \frac{x^2}{2} + \frac{x^3}{3} - \ldots + (-1)^{n-1}\frac{x^n}{n} + \ldots,$$

provided $-1 < x \leqslant 1$.

Note that if x is replaced by $-x$ in this series,

$$\log_e(1-x) = -x - \frac{x^2}{2} - \frac{x^3}{3} - \ldots - \frac{x^n}{n} - \ldots,$$

provided $-1 \leqslant x < 1$.

We can, however, prove the expansion of $\log_e(1+x)$ without making the assumption about integrating an infinite series which was made at the beginning of this section. It is suggested that most readers should omit the following proof on first reading and proceed to Example 4, p. 218.

Consider the sum of n terms of the geometric progression

$$1 - u + u^2 - \ldots + (-1)^{n-1}u^{n-1} = \frac{1 - (-u)^n}{1+u}.$$

$$\therefore \frac{1}{1+u} = 1 - u + u^2 - \ldots + (-1)^{n-1}u^{n-1} + \frac{(-1)^n u^n}{1+u}.$$

Integrating from 0 to x,

$$\log_e(1+x) = x - \frac{x^2}{2} + \frac{x^3}{3} - \ldots + (-1)^{n-1}\frac{x^n}{n} + R_n,$$

where
$$R_n = \int_0^x \frac{(-1)^n u^n}{1+u}\, du = (-1)^n \int_0^x \frac{u^n}{1+u}\, du.$$

We now examine what happens to R_n as $n \to \infty$.

Case (i), $0 < x \leqslant 1$. Consider the function $\dfrac{u^n}{1+u}$, where u lies in the range $0 \leqslant u \leqslant x$. The least value of the denominator is 1, so

$$0 \leqslant \frac{u^n}{1+u} \leqslant u^n.$$

Integrating with respect to u from 0 to x,

$$0 < \int_0^x \frac{u^n}{1+u}\, du < \int_0^x u^n\, du.$$

$$\therefore 0 < |R_n| < \left[\frac{u^{n+1}}{n+1}\right]_0^x = \frac{x^{n+1}}{n+1}.$$

$$\therefore |R_n| < \frac{x^{n+1}}{n+1} \leqslant \frac{1}{n+1}.$$

Hence if $0 < x \leqslant 1$, $R_n \to 0$ as $n \to \infty$.

Case (ii), $-1 < x < 0$. Consider the function $\dfrac{(-1)^n u^n}{1+u}$, where u lies in

the range $x \leqslant u \leqslant 0$. We now have

$$0 \leqslant \frac{(-1)^n u^n}{1+u} \leqslant \frac{(-1)^n u^n}{1+x}.$$

Integrating with respect to u from x to 0,

$$0 < \int_x^0 \frac{(-1)^n u^n}{1+u}\, du < \int_x^0 \frac{(-1)^n u^n}{1+x}\, du.$$

$$\therefore\ 0 < |R_n| < \left[\frac{(-1)^n u^{n+1}}{(1+x)(n+1)}\right]_x^0.$$

$$\therefore |R_n| < \frac{(-1)^{n+1} x^{n+1}}{(1+x)(n+1)} < \frac{1}{(1+x)(n+1)}.$$

Hence if $-1 < x < 0$ (but *not* for $x = -1$), $R_n \to 0$ as $n \to \infty$.

Case (iii), $x = 0$. Both sides of the expansion are zero.

Therefore $\log_e (1+x) = x - \dfrac{x^2}{2} + \dfrac{x^3}{3} - \ldots + (-1)^{n-1}\, \dfrac{x^n}{n} + \ldots,$

provided $-1 < x \leqslant 1$.

Example 4. *Expand as series in ascending powers of x:*

(i) $\log_e (2+x)$,   (ii) $\log_e (2+x)^3$,   (iii) $\log_e (x^2 - 3x + 2)$.

(i) $\log_e (2+x) = \log_e \{2(1+\tfrac{1}{2}x)\},$

$= \log_e 2 + \log_e (1+\tfrac{1}{2}x),$

$= \log_e 2 + \tfrac{1}{2}x - \dfrac{(\tfrac{1}{2}x)^2}{2} + \ldots + (-1)^{n-1} \dfrac{(\tfrac{1}{2}x)^n}{n} + \ldots,$

$$\therefore\ \log_e (2+x) = \log_e 2 + \frac{x}{2} - \frac{x^2}{8} + \ldots + (-1)^{n-1} \frac{x^n}{2^n . n} + \ldots.$$

The expansion is valid if $-1 < \tfrac{1}{2}x \leqslant 1$, i.e. if $-2 < x \leqslant 2$.

(ii) $\log_e (2+x)^3 = 3 \log_e (2+x).$

Therefore, using the result of part (i),

$$\log_e (2+x)^3 = 3 \log_e 2 + \frac{3x}{2} - \frac{3x^2}{8} + \ldots + (-1)^{n-1} \frac{3x^n}{2^n . n} + \ldots.$$

The expansion is again valid if $-2 < x \leqslant 2$.

(iii) $$\log_e (x^2 - 3x + 2) = \log_e \{(1-x)(2-x)\},$$
$$= \log_e (1-x) + \log_e (2-x).$$

$$\log_e (1-x) = -x - \frac{x^2}{2} - \dots - \frac{x^n}{n} - \dots.$$

From (i),

$$\log_e (2-x) = \log_e 2 - \frac{x}{2} - \frac{x^2}{8} - \dots - \frac{x^n}{2^n \cdot n} - \dots.$$

Adding,

$$\log_e (x^2 - 3x + 2) = \log_e 2 - \tfrac{3}{2}x - \tfrac{5}{8}x^2 - \dots - \frac{x^n}{n}\{1 + (\tfrac{1}{2})^n\} - \dots.$$

For the expansions to be valid, x must satisfy both $-1 \leqslant x < 1$ and $-2 \leqslant x < 2$, i.e. $-1 \leqslant x < 1$.

Qu. 3. Expand in ascending powers of x:

(i) $\log_e (1 + \tfrac{1}{4}x)$,   (ii) $\log_e (3 - x)$,   (iii) $\log_e (x^2 - 2x + 1)$.

Give the first three terms and the general term and state the ranges of values of x for which the expansions are valid.

Example 5. *If $|x| > 1$, show that*

$$\log_e (1+x) = \log_e x + \frac{1}{x} - \frac{1}{2x^2} + \dots + \frac{(-1)^{n-1}}{nx^n} + \dots.$$

$$\log_e (1+x) = \log_e \{x(1 + 1/x)\},$$
$$= \log_e x + \log_e (1 + 1/x).$$

$$\therefore \log_e (1+x) = \log_e x + \frac{1}{x} - \frac{1}{2x^2} + \dots + \frac{(-1)^{n-1}}{nx^n} + \dots.$$

Other series have been devised for the calculation of logarithms and one of these will now be obtained.

$$\log_e (1+x) = x - \frac{x^2}{2} + \frac{x^3}{3} - \frac{x^4}{4} + \dots + \frac{x^{2n-1}}{2n-1} - \frac{x^{2n}}{2n} + \dots.$$

$$\log_e (1-x) = -x - \frac{x^2}{2} - \frac{x^3}{3} - \frac{x^4}{4} - \dots - \frac{x^{2n-1}}{2n-1} - \frac{x^{2n}}{2n} - \dots.$$

The expansions are valid if $-1 < x \leqslant 1$, $-1 \leqslant x < 1$, respectively, so for both to be valid, $-1 < x < 1$.

Subtracting,

$$\log_e \frac{1+x}{1-x} = 2\left(x + \frac{x^3}{3} + \dots + \frac{x^{2n-1}}{2n-1} + \dots\right).$$

Dividing by 2 and writing

$$\tfrac{1}{2} \log_e \left(\frac{1+x}{1-x}\right) = \log_e \sqrt{\left(\frac{1+x}{1-x}\right)},$$

$$\log_e \sqrt{\left(\frac{1+x}{1-x}\right)} = x + \frac{x^3}{3} + \ldots + \frac{x^{2n-1}}{2n-1} + \ldots,$$

provided $-1 < x < +1$.

The advantage of this series may be seen by attempting to calculate, say, $\log_e 1 \cdot 5$ by two methods.

(i) Substitute $x = \tfrac{1}{2}$ in

$$\log_e (1+x) = x - \tfrac{1}{2}x^2 + \tfrac{1}{3}x^3 - \ldots + (-1)^{n-1} x^n/n + \ldots.$$

$$\log_3 1 \cdot 5 = \tfrac{1}{2} - \tfrac{1}{8} + \tfrac{1}{24} - \tfrac{1}{64} + \tfrac{1}{160} - \tfrac{1}{384} + \tfrac{1}{896} - \tfrac{1}{2048} + \ldots.$$

(ii) Substitute $x = \tfrac{1}{5}$ in

$$\log_e \left(\frac{1+x}{1-x}\right) = 2\{x + \tfrac{1}{3}x^3 + \tfrac{1}{5}x^5 + \ldots + x^{2n-1}/(2n-1) + \ldots\}.$$
$$\log_e 1 \cdot 5 = 2 \left(\tfrac{1}{5} + \tfrac{1}{375} + \tfrac{1}{15625} + \ldots\right).$$

It is clear that the value correct to four places of decimals can be obtained far more rapidly by the second series.

Working to six places of decimals,

$$\tfrac{1}{5} = 0 \cdot 200000$$
$$\tfrac{1}{375} = \tfrac{1}{125} \div 3 = 0 \cdot 008 \div 3 = 0 \cdot 002667$$
$$\tfrac{1}{15625} = \left(\tfrac{1}{125}\right)^2 \qquad = 0 \cdot 000064$$

$$\overline{\qquad 0 \cdot 202731}$$

$$\therefore \log_e 1 \cdot 5 = 0 \cdot 4055 \quad \text{to 4 places of decimals.}$$

Note that, using $\log_{10} 1 \cdot 5 = \log_{10} e \times \log_e 1 \cdot 5$, $\log_{10} 1 \cdot 5$ can be obtained.

Example 6. *Find the first three terms in the expansion of*

$$\frac{\log_e (1+x)}{\log_e (1-x)}$$

in ascending powers of x.

Let $\qquad \dfrac{\log_e (1+x)}{\log_e (1-x)} = a_0 + a_1 x + a_2 x^2 + \ldots,$

where a_0, a_1, a_2 are constants to be determined.

$$\therefore \log_e (1+x) = \log_e (1-x) (a_0 + a_1 x + a_2 x^2 + \ldots).$$

$$\therefore \ x - \frac{x^2}{2} + \frac{x^3}{3} - \ldots = \left(-x - \frac{x^2}{2} - \frac{x^3}{3} - \ldots \right)(a_0 + a_1 x + a_2 x^2 + \ldots).$$

Equating coefficients of x, x^2, x^3:

$$1 = -a_0,$$
$$-\tfrac{1}{2} = -\tfrac{1}{2}a_0 - a_1,$$
$$\tfrac{1}{3} = -\tfrac{1}{3}a_0 - \tfrac{1}{2}a_1 - a_2,$$

from which we obtain

$$a_0 = -1, \qquad a_1 = 1, \qquad a_2 = -\tfrac{1}{2}.$$

$$\therefore \ \frac{\log_e (1+x)}{\log_e (1-x)} = -1 + x - \tfrac{1}{2}x^2 + \ldots.$$

Qu. 4. Write down the first three terms of the expansion of $\dfrac{\log_{10}(1x+)}{\log_{10}(1-x)}$ in ascending powers of x.

Example 7. *Find the sum to infinity of the series*

$$\frac{1}{1.2} \cdot \frac{1}{3} + \frac{1}{2.3} \cdot \frac{1}{3^2} + \frac{1}{3.4} \cdot \frac{1}{3^3} + \ldots.$$

The general term is

$$\frac{1}{n(n+1)} \cdot \frac{1}{3^n}$$

which may be expressed in partial fractions as

$$\left(\frac{1}{n} - \frac{1}{n+1} \right) \frac{1}{3^n}.$$

Therefore the series may be written

$$1 \cdot \frac{1}{3} + \frac{1}{2} \cdot \frac{1}{3^2} + \frac{1}{3} \cdot \frac{1}{3^3} + \ldots + \frac{1}{n} \cdot \frac{1}{3^n} + \ldots$$
$$- \frac{1}{2} \cdot \frac{1}{3} - \frac{1}{3} \cdot \frac{1}{3^2} - \frac{1}{4} \cdot \frac{1}{3^3} - \ldots - \frac{1}{n+1} \cdot \frac{1}{3^n} - \ldots$$
$$= -\log_e (1 - \tfrac{1}{3}) + S,$$

where

$$S = -\frac{1}{2} \cdot \frac{1}{3} - \frac{1}{3} \cdot \frac{1}{3^2} - \frac{1}{4} \cdot \frac{1}{3^3} - \ldots - \frac{1}{n+1} \cdot \frac{1}{3^n} - \ldots.$$

$$\therefore \quad -\frac{1}{3} + \frac{1}{3} S = -\frac{1}{3} - \frac{1}{2} \cdot \frac{1}{3^2} - \frac{1}{3} \cdot \frac{1}{3^3} - \cdots - \frac{1}{n+1} \cdot \frac{1}{3^{n+1}} - \cdots,$$

$$= \log_e (1 - \tfrac{1}{3}).$$

$$\therefore \quad S = 3 \log_e \tfrac{2}{3} + 1.$$

Therefore the sum of the series is $2 \log_e \tfrac{2}{3} + 1$.

Exercise 10b

1. Expand the following functions in ascending powers of x, giving the first three or four terms, as indicated, and the general term. State the ranges of values of x for which the expansions are valid.

   (i) $\log_e (3 + x)$, (4),   (ii) $\log_e (1 - \tfrac{1}{2}x)$, (4),   (iii) $\log_e (2 - 5x)$, (4),

   (iv) $\log_e (1 - x^2)$, (4),   (v) $\log_e \left(\dfrac{3+x}{3-x}\right)$, (3),   (vi) $\log_e \left(\dfrac{4-3x}{4+3x}\right)$, (3).

Find the first three terms and the general terms in the expansions of the functions in Nos. 2–8. State the necessary restrictions on the values of x.

2. $\log_e \left(\dfrac{2-x}{3-x}\right)$.

3. $\log_e \dfrac{1}{3 - 4x - 4x^2}$.

4. $\log_e \left\{\dfrac{(1+4x)^3}{(1+3x)^4}\right\}$.

5. $\log_e \sqrt{(x^2 + 3x + 2)}$.

6. $\log_e (1 + x + x^2)$ (see p. 226 for a hint).

7. $\log_e \{(1+x)^{1/x}\}$.

8. $\log_e (1 - x + x^2)$.

Expand the following functions in ascending powers of x as far as the terms indicated. State the ranges of values of x for which the expansions are valid.

9. $\dfrac{\log_e (1+x)}{1-x}$, (x^3).

10. $e^x \log_e (1+x)$, (x^3).

11. $\dfrac{x + x^2}{\log_e (1+x)}$, (x^2).

12. $\{\log_e (1-x)\}^2$, (x^4).

13. By substituting $x = \tfrac{1}{3}$ in the expansion of $\log_e \{(1+x)/(1-x)\}$ in ascending powers of x, find the value of $\log_e 2$ correct to four significant figures. [HINT: find the values of $1/3^3$, $1/3^5$, $1/3^7$, $1/3^9$ by successive divisions of $0.333\,333$ by 9. Then the values of the first six terms of the series may be found without difficulty.] Taking $\log_e 1.5 = 0.4055$, estimate the value of $\log_e 3$.

In Nos. 14–16, take $\log_e 2 = 0.693\,147$ and $\log_e 3 = 1.098\,612$.

14. Find $\log_e 10$ correct to four places of decimals by substituting $x = \tfrac{1}{9}$ in the expansion of $\log_e (1+x)$. [HINT: find $\tfrac{1}{9}$ correct to six places of decimals and find $1/9^2$, $1/9^3$, $1/9^4$, $1/9^5$ by successive divisions by 9.] Deduce an approximate value of $\log_{10} e$.

15. Find the value of $\log_e 7$ by substituting $x = \frac{1}{3}$ in the expansion of $\log_e \{(1+x)/(1-x)\}$. Give your answer correct to four places of decimals.

16. Find the value of $\log_{10} 11$ correct to four places of decimals. Use the expansion of $\log_e \{(1+x)/(1-x)\}$ with $x = 0.1$. Take $\log_{10} e = 0.434\ 29$.

17. Find the limits of the following functions as x approaches zero:

(i) $\log_e \{(1-x^2)^{1/x^2}\}$,

(ii) $\dfrac{\log_e (1+x) - x}{\log_e (1-x) + x}$,

(iii) $\dfrac{\log_e \{(1+x)^2\} + x^2 - 2x}{\log_e (1-x^3)}$,

(iv) $\dfrac{\log_e (1-x) + x\sqrt{(1+x)}}{\log_e (1+x^2)}$.

Find the sums to infinity of the following series:

18. $\dfrac{1}{3} - \dfrac{1}{2}\cdot\dfrac{1}{9} + \dfrac{1}{3}\cdot\dfrac{1}{27} - \dfrac{1}{4}\cdot\dfrac{1}{81} + \cdots$.

19. $\dfrac{1}{2} + \dfrac{1}{2}\cdot\dfrac{1}{2^2} + \dfrac{1}{3}\cdot\dfrac{1}{2^3} + \dfrac{1}{4}\cdot\dfrac{1}{2^4} + \cdots$.

20. $\dfrac{1}{4} + \dfrac{1}{3}\cdot\dfrac{1}{4^3} + \dfrac{1}{5}\cdot\dfrac{1}{4^5} + \dfrac{1}{7}\cdot\dfrac{1}{4^7} + \cdots$.

21. $1 - \dfrac{1}{2}\cdot\dfrac{2}{5} + \dfrac{1}{3}\cdot\dfrac{2^2}{5^2} - \dfrac{1}{4}\cdot\dfrac{2^3}{5^3} + \cdots$.

22. $1 + \dfrac{1}{3\cdot2^2} + \dfrac{1}{5\cdot2^4} + \dfrac{1}{7\cdot2^6} + \cdots$.

23. $\dfrac{1}{2}\cdot\dfrac{2}{3} + \dfrac{1}{3}\cdot\dfrac{2^2}{3^2} + \dfrac{1}{4}\cdot\dfrac{2^3}{3^3} + \cdots$.

24. $\dfrac{1}{3} + \dfrac{1}{5}\cdot\dfrac{3^2}{7^2} + \dfrac{1}{7}\cdot\dfrac{3^4}{7^4} + \dfrac{1}{9}\cdot\dfrac{3^6}{7^6} + \cdots$.

25. $\dfrac{2}{1}\cdot\dfrac{1}{2} - \dfrac{3}{2}\cdot\dfrac{1}{2^2} + \dfrac{4}{3}\cdot\dfrac{1}{2^3} - \dfrac{5}{4}\cdot\dfrac{1}{2^4} + \cdots$.

26. $1 - \dfrac{3}{2}\cdot\dfrac{1}{2^2} + \dfrac{5}{3}\cdot\dfrac{1}{2^3} - \dfrac{7}{4}\cdot\dfrac{1}{2^4} + \cdots$.

27. $1 - \dfrac{4}{2}\cdot\dfrac{1}{3^2} + \dfrac{7}{3}\cdot\dfrac{1}{3^3} - \dfrac{10}{4}\cdot\dfrac{1}{3^4} + \cdots$.

28. $\dfrac{3}{1}\cdot\dfrac{1}{5} + \dfrac{4}{2}\cdot\dfrac{1}{5^2} + \dfrac{5}{3}\cdot\dfrac{1}{5^3} + \cdots$.

29. $\dfrac{1}{3\cdot2} + \dfrac{3}{5\cdot2^3} + \dfrac{5}{7\cdot2^5} + \cdots$.

30. $\dfrac{5}{1\cdot4}\left(\dfrac{1}{2}\right) + \dfrac{8}{2\cdot6}\left(\dfrac{1}{2}\right)^2 + \dfrac{11}{3\cdot8}\left(\dfrac{1}{2}\right)^3 + \cdots$.

31. $\dfrac{5}{1\cdot6}\left(\dfrac{1}{3}\right) - \dfrac{8}{2\cdot9}\left(\dfrac{1}{3}\right)^2 + \dfrac{11}{3\cdot12}\left(\dfrac{1}{3}\right)^3 + \cdots$.

32. $\dfrac{1}{1\cdot2}\left(\dfrac{1}{2}\right)^2 + \dfrac{1}{2\cdot3}\left(\dfrac{1}{2}\right)^3 + \dfrac{1}{3\cdot4}\left(\dfrac{1}{2}\right)^4 + \cdots$.

33. $\dfrac{1}{1.2}\left(\dfrac{1}{4}\right) - \dfrac{1}{2.3}\left(\dfrac{1}{4}\right)^2 + \dfrac{1}{3.4}\left(\dfrac{1}{4}\right)^3 + \ldots.$

34. $\dfrac{4}{1.3}\left(\dfrac{1}{5}\right) + \dfrac{6}{2.4}\left(\dfrac{1}{5}\right)^2 + \dfrac{8}{3.5}\left(\dfrac{1}{5}\right)^3 + \ldots.$

*35. Integrate the inequalities

$$\frac{1}{(1+t)^2} < \frac{1}{1+t} < 1 \quad (t > 0)$$

from 0 to u and deduce that

$$\frac{u}{1+u} < \log_e(1+u) < u \quad (u > 0).$$

Sketch the graph of $y = 1/x$ and illustrate the latter inequalities graphically.

Also prove that, if $-1 < u < 0$,

$$\frac{u}{1+u} < \log_e(1+u) < u.$$

36. Sketch the graph of $y = 1/x$ and show that, when n is a positive integer greater than 1,

$$\tfrac{1}{2} + \tfrac{1}{3} + \ldots + \frac{1}{n} < \log_e n < 1 + \tfrac{1}{2} + \ldots + \frac{1}{n-1}.$$

37. Let s_n denote the sum of n terms of the series

$$1 - \tfrac{1}{2} + \tfrac{1}{3} - \tfrac{1}{4} + \tfrac{1}{5} - \tfrac{1}{6} + \ldots.$$

By considering the terms of the series in pairs, show that s_{2n} increases as $n \to \infty$. By considering the terms of the series after 1 in pairs, show that s_{2n+1} is less than 1 and decreases as $n \to \infty$. Show that $|s_{2n}| - |s_{2n+1}| \to 0$ as $n \to \infty$. What can you conclude about s_n as $n \to \infty$?

Exercise 10c (Miscellaneous)

1. By expanding the integrand of

$$\int_0^x \frac{1}{1+x}\,\mathrm{d}x$$

as a series of powers of x and integrating term by term, find the series for $\log_e(1+x)$, assuming your method to be valid provided that $|x| < 1$.

Write down the series for $\log_e(1-x)$, obtain the series for $\log_e \dfrac{1+x}{1-x}$ and deduce a series for $\log_e \dfrac{m}{n}$ in terms of $\dfrac{m-n}{m+n}$.

Hence calculate $\log_e 8$ correct to five places of decimals, given that $\log_e 7 = 1.945\,910$.   **N.**

2. Assuming that $|x| < 1$, write down
   (i) the sum of the infinite geometric series

$$1 + x^2 + x^4 + \ldots,$$

   (ii) the first three terms of the series for $\log_e (1+x)$.
   Obtain the first two terms of the series for

$$\tfrac{1}{2} \log_e \left(\frac{1+x}{1-x}\right).$$

Assuming also that x is positive, show that the sum of the remaining terms of this series is less than

$$\frac{x^5}{5(1-x^2)}. \hspace{4cm} \text{N.}$$

3. Write down the expansions of $\log_e (1+x)$ and $\log_e (1-x)$, stating for what values of x they are valid.
   Prove that, (i) if $-\tfrac{1}{2} < x \leqslant \tfrac{1}{2}$, then

$$\log_e (1 + x - 2x^2) = x - \tfrac{5}{2}x^2 + \tfrac{7}{3}x^3 - \tfrac{17}{4}x^4 \ldots;$$

   (ii) if m/n is positive,

$$\log_e \frac{m}{n} = 2\left\{\left(\frac{m-n}{m+n}\right) + \tfrac{1}{3}\left(\frac{m-n}{m+n}\right)^3 + \tfrac{1}{5}\left(\frac{m-n}{m+n}\right)^5 + \ldots\right\}. \hspace{0.5cm} \text{L.}$$

4. Prove that, when $a > 0$,
   (i) $\log_e (a+x) = \log_e a + \log_e \left(1 + \dfrac{x}{a}\right)$;
   (ii) $a^x = e^{x \log_e a}$.
   Write down the first three terms of the expansions of $\log_e (1+y)$ and e^y.
   Prove that the expansion of

$$a^x - 1 - x \log_e (a+x)$$

as a series of ascending powers of x begins with a term in x^2, and find the coefficient of x^2 in this term.   O.C.

5. (i) Write down the expansion of $\log_e (1+x)$ in ascending powers of x, giving the first three terms and the coefficient of x^m; state the limitations on the value of x.
   Prove that

$$2 \log_e n - \log_e (n+1) - \log_e (n-1) = \frac{1}{n^2} + \frac{1}{2n^4} + \frac{1}{3n^6} + \ldots,$$

stating the necessary restriction on the value of n.
   Given that

$$\log_e 10 = 2 \cdot 302 \, 59 \quad \text{and} \quad \log_e 3 = 1 \cdot 098 \, 61,$$

calculate the value of $\log_e 11$ correct to four places of decimals.

(ii) Find the coefficient of x^n in the expansion of $(1+3x)\,e^{-3x}$ as a series of ascending powers of x.   O.C.

6. Find the sum of the first n terms of a geometric progression of which the first term is a and the common ratio is r.

If p is any odd positive integer and q any even positive integer, and if $x > 0$, prove that

$$1 - x + x^2 - \ldots - x^p < \frac{1}{1+x} < 1 - x + x^2 - \ldots + x^q.$$

Deduce that

$$\frac{x}{1} - \frac{x^2}{2} + \frac{x^3}{3} - \ldots - \frac{x^{p+1}}{p+1} < \log_e (1+x) < \frac{x}{1} - \frac{x^2}{2} + \frac{x^3}{3} - \ldots + \frac{x^{q+1}}{q+1}.$$

By taking $p = 5$, $q = 4$, $x = 0\cdot1$, calculate the value of $\log_e (1\cdot1)$ correct to six places of decimals.   N.

7. (i) Write down the expansions of e^x and $\log_e (1+x)$ in series of ascending powers of x, giving in each case the first three terms and the nth term.

(ii) By considering the factors of $(1 - x^3)$, obtain the coefficients of x^{3n}, x^{3n+1}, x^{3n+2} in the expansion of $\log_e (1 + x + x^2)$.

(iii) Obtain the first two non-vanishing terms in the expansion of

$$(x-1)(1 - e^x) - \log_e (1+x),$$

in ascending powers of x.   O.C.

8. (i) Prove that if $x^p = (xy)^q = (xy^2)^r$ for all values of x and y, then $2pr = q(p+r)$.

(ii) Assuming the expansion for $\log_e (1+x)$ in ascending powers of x, prove that

$$\log_e \sqrt{\left(\frac{1+x}{1-x}\right)} = x + \tfrac{1}{3}x^3 + \tfrac{1}{5}x^5 + \ldots,$$

and, when $0 < \theta < \tfrac{1}{2}\pi$, deduce that

$$\sin\theta + \tfrac{1}{3}\sin^3\theta + \tfrac{1}{5}\sin^5\theta + \ldots = \log_e (\tan\theta + \sec\theta).$$

(iii) Establish the identity

$$n^2 \equiv (n+2)(n+1) - 3(n+2) + 4$$

and hence find the sum of the series

$$\frac{1^2}{3!} + \frac{2^2}{4!} + \frac{3^2}{5!} + \ldots.$$ O.C.

9. (i) Prove the identity

$$1 + 2x + 2x^2 + x^3 \equiv \frac{(1+x)(1-x^3)}{1-x}$$

and hence expand $\log_e (1 + 2x + 2x^2 + x^3)$ in ascending powers of

x as far as the term in x^6, stating the necessary restrictions on the values of x.

(ii) Write down the series for e^z and e^{-z} in ascending powers of x. Prove that, if x^4 and higher powers of x are neglected, then

$$\frac{x}{1-e^{-2x}} - \tfrac{1}{2} = \tfrac{1}{2}x + \tfrac{1}{6}x^2.$$ o.c.

10. Assuming that x is sufficiently small, find the values of p and q, other than zero, for which

$$(1+x)^p - \log_e (1+qx) = 1 + ax^3 + \ldots$$

where the terms omitted contains powers of x higher than the third. Determine the value of the coefficient a.   N.

11. Write down the expansion of $\log_e (1+x)$ in ascending powers of x, giving the general term and stating for what real values of x the expansion is valid.

Determine a and b so that the expansion of

$$\frac{1+ax}{1+bx} \log_e (1+x)$$

may contain no term in x^2 or x^3, and show that with these values

$$\frac{1+bx}{1+ax} = 1 - \frac{x}{2} + \frac{x^2}{3} - \frac{2x^3}{9},$$

neglecting powers of x above the third.   o.c.

12. Give the expansion of $\log_e (1+x)$ in ascending powers of x and state for what range of values of x the expansion is valid.

By taking logarithms or otherwise, verify that, when n is large, an approximate value of $\left(1+\frac{1}{n}\right)^n$ is

$$e\left(1 - \frac{1}{2n} + \frac{11}{24n^2} - \frac{7}{16n^3}\right).$$ o.c.

13. (i) Expand $\log_e \dfrac{2+x}{1-x}$ in ascending powers of x, giving the first four terms and the general term.

(ii) Show that the first non-zero coefficient in the expansion of

$$e^{-z} - \frac{1-x}{(1-x^2)^{1/2}(1-x^3)^{1/3}}$$

in ascending powers of x is that of x^5.   L.

14. Write down the series for $\log_e (1+x)$ in ascending powers of x and state the range of values of x for which it is valid.

Prove that, if $n > 1$,

$$\log_e \frac{n}{n-1} > \frac{1}{n} > \log_e \frac{n+1}{n};$$

and deduce that, if n is a positive integer,

$$1 + \log_e n > 1 + \tfrac{1}{2} + \tfrac{1}{3} + \ldots + \frac{1}{n} > \log_e (n+1).$$ C.

15. Write down the expansion in ascending powers of x of $\log_e \left(\frac{1+x}{1-x} \right)$ where $-1 < x < 1$.

By using partial fractions, obtain the sum of the series

$$\sum_{n=1}^{\infty} \frac{x^{2n}}{(2n-1)(2n+1)}$$

when $0 < x < 1$.

Find the sum of the first N terms of the series when $x = 1$ and deduce that

$$\sum_{n=1}^{\infty} \frac{1}{(2n-1)(2n+1)} = \tfrac{1}{2}.$$ C.

16. State the first four terms in the series expansions of $(1+x)^n$, $\log_e (1+x)$.

Find the sum of the infinite series

$$\frac{1}{2} + \frac{1}{2} \cdot \frac{1}{2^2} + \frac{1}{3} \cdot \frac{1}{2^3} + \ldots + \frac{1}{n} \cdot \frac{1}{2^n} + \ldots.$$ N.

17. (i) If $0 < x < 1$ and $f(x)$ is the sum of the infinite series

$$1 + \frac{x^2}{3} + \frac{x^4}{5} + \frac{x^6}{7} + \ldots$$

show that, for x in this range,

$$f\left(\frac{2x}{1+x^2} \right) = (1+x^2) f(x).$$

(ii) Sum to infinity the series

$$\frac{3^2}{1!} + \frac{4^2}{2!} + \frac{5^2}{3!} + \ldots.$$ L.

18. Sum to infinity each of the following series:

(i) $1 + \dfrac{2x}{1!} + \dfrac{3x^2}{2!} + \dfrac{4x^3}{3!} + \ldots,$

(ii) $\dfrac{x}{1.2} + \dfrac{x^2}{2.3} + \dfrac{x^3}{3.4} + \ldots,$ if $|x| < 1;$

(iii) $1 + \dfrac{5}{3} + \dfrac{5.7}{3.6} + \dfrac{5.7.9}{3.6.9} + \ldots.$ L.

19. (a) Prove, by induction or otherwise, that

$$1^2 + 3^2 + 5^2 + \ldots + (2n-1)^2 = \tfrac{1}{3}n(4n^2 - 1).$$

(b) Find the sum to infinity of the series

$$2 + \frac{3}{1!} + \frac{4}{2!} + \ldots + \frac{n+1}{(n-1)!} + \ldots .$$ **N.**

20. Show that the coefficient of x^n in the expansion of $e^{(e^x)}$ is

$$\frac{1}{n!} \left(\frac{1^n}{1!} + \frac{2^n}{2!} + \frac{3^n}{3!} + \ldots + \frac{r^n}{r!} + \ldots \right).$$

Hence find the sums of the infinite series

$$\frac{1^3}{1!} + \frac{2^3}{2!} + \frac{3^3}{3!} + \ldots ,$$

$$\frac{1^4}{1!} + \frac{2^4}{2!} + \frac{3^4}{3!} + \ldots .$$ **L.**

CHAPTER 11

FURTHER DIFFERENTIATION

11.1. The object of the first three sections of this chapter is to extend the reader's powers of differentiation and to increase the domain in which he can apply his knowledge of the calculus. In the course of this work we shall also discuss how to integrate certain functions.

Logarithmic differentiation

Logarithmic differentiation is a powerful method which can considerably simplify the differentiation of:

(i) products (and quotients) of a number of functions,
(ii) certain exponential functions.

It is best introduced by examples but first it is advisable to revise some of the properties of logarithms and how to differentiate functions of y with respect to x.

Qu. 1.
$$\log_e (a^3 \sqrt{b}/c^2) = \log_e a^3 + \log_e \sqrt{b} - \log_e c^2,$$
$$= 3 \log_e a + \tfrac{1}{2} \log_e b - 2 \log_e c.$$

(See p. 30.) Write in a similar form:

(i) $\log_e (a^2 b)$,
(ii) $\log_e (a^3/b^3)$,
(iii) $\log_e \sqrt{(abc)}$,
(iv) $\log_e (a\sqrt[3]{b}/c^3)$,
(v) $\log_e (1/c^4)$,
(vi) $\log_e (a^b)$.

Qu. 2. $\log_{10} 10\,000 = \log_{10} 10^4 = 4$. Simplify in a similar manner:
(i) $\log_{10} 1000$,
(ii) $\log_{10} (1/100)$,
(iii) $\log_2 (2^4)$,
(iv) $\log_e (e^2)$,
(v) $\log_e (e^{2x})$,
(vi) $\log_e (e^{3x^2})$.

Qu. 3. Differentiate with respect to x:
(i) $\log_e x$,
(ii) $\log_e (1 + 2x)$,
(iii) $\log_e (1 - x)$,
(iv) $\log_e 4x^3$,
(v) $\log_e \sin x$,
(vi) $\log_e \tan x$.

When differentiating functions of y with respect to x we can, if need be, use the formula

$$\frac{\mathrm{d}z}{\mathrm{d}x} = \frac{\mathrm{d}z}{\mathrm{d}y} \cdot \frac{\mathrm{d}y}{\mathrm{d}x}.$$

Thus if
$$z = y^4,$$
$$\frac{dz}{dy} = 4y^3.$$
$$\therefore \frac{dz}{dx} = 4y^3 \cdot \frac{dy}{dx}.$$

Qu. 4. Differentiate with respect to x:

(i) $3y^2$,   (ii) y^3,   (iii) $\cos y$,   (iv) $\log_e y$.

Express, in your own words, a rule which will help you to differentiate any function of y with respect to x. Use this rule to differentiate with respect to x:

(v) $5y^4$,   (vi) $3/y^2$,   (vii) $\sqrt{y}$,   (viii) $\tan y$.

Example 1. *Differentiate* $\dfrac{e^{x^2} \sqrt{(\sin x)}}{(2x+1)^3}$.

Let
$$y = \frac{e^{x^2} \sqrt{(\sin x)}}{(2x+1)^3}.$$

$$\therefore \log_e y = \log_e (e^{x^2}) + \log_e \sqrt{(\sin x)} - \log_e (2x+1)^3,$$
$$= x^2 + \tfrac{1}{2} \log_e \sin x - 3 \log_e (2x+1).$$

Differentiating with respect to x,

$$\frac{1}{y} \frac{dy}{dx} = 2x + \frac{\cos x}{2 \sin x} - \frac{6}{2x+1}.$$

$$\therefore \frac{dy}{dx} = \frac{e^{x^2} \sqrt{(\sin x)}}{(2x+1)^3} \left\{ 2x + \frac{\cos x}{2 \sin x} - \frac{6}{2x+1} \right\}.$$

(There are occasions when this is the most convenient form in which to use the derivative, but here we shall go on to simplify the expression in brackets.)

$$2x + \frac{\cos x}{2 \sin x} - \frac{6}{2x+1} = \frac{4x^2 + 2x - 6}{2x+1} + \frac{\cos x}{2 \sin x}.$$

$$\therefore \frac{dy}{dx} = \frac{e^{x^2}}{2\sqrt{(\sin x)}(2x+1)^4} \{(8x^2 + 4x - 12) \sin x + (2x+1) \cos x\}.$$

Qu. 5. Use the method of Example 1 to differentiate with respect to x:

(i) $\sqrt[3]{\dfrac{x+1}{x-1}}$,   (ii) $\dfrac{\sqrt{(x^2+1)}}{(2x-1)^2}$,   (iii) $\dfrac{x^2 e^x}{(x-1)^3}$.

Example 2. *Differentiate 10^x with respect to x.*

Let
$$y = 10^x.$$
$$\therefore \log_e y = \log_e 10^x,$$
$$= x \log_e 10.$$

Differentiating with respect to x,

$$\frac{1}{y}\frac{dy}{dx} = \log_e 10.$$
$$\therefore \frac{dy}{dx} = 10^x \log_e 10.$$

Example 3. *Differentiate, with respect to x, (i) 2^{x^2}, (ii) x^x.*

(i) Let
$$y = 2^{x^2}.$$
$$\therefore \log_e y = \log_e 2^{x^2},$$
$$= x^2 \log_e 2.$$

Differentiating with respect to x,

$$\frac{1}{y}\frac{dy}{dx} = 2x \log_e 2.$$
$$\therefore \frac{dy}{dx} = 2^{x^2} 2x \log_e 2.$$

(ii) Let
$$y = x^x.$$
$$\therefore \log_e y = \log_e x^x,$$
$$= x \log_e x.$$

Differentiating with respect to x,

$$\frac{1}{y}\frac{dy}{dx} = x \cdot \frac{1}{x} + 1 \cdot \log_e x,$$
$$= 1 + \log_e x.$$
$$\therefore \frac{dy}{dx} = (1 + \log_e x)x^x.$$

Qu. 6. We have shown in Example 2 that the derivative of 10^x is $10^x \log_e 10$. Write down a function whose derivative is 10^x. What is $\int 10^x \, dx$?

Qu. 7. Differentiate with respect to x:

(i) 2^x,   (ii) 3^x,   (iii) $(\frac{1}{2})^x$,   (iv) 10^{5x},   (v) 10^{x^2}.

Qu. 8. From your answers to Qu. 7, write down:

(i) $\int 2^x \, dx$,   (ii) $\int 3^x \, dx$,   (iii) $\int (\frac{1}{2})^x \, dx$,   (iv) $\int 10^{5x} \, dx$.

Integration by trial

11.2. In Qu. 6 we had an example of what may be called "integration by trial". This was a procedure which was discussed in Chapter 1, p. 1, and Chapter 2, p. 35; its stages are shown in the next two examples.

Example 4. *Integrate 2^{-x} with respect to x.*
(*Stage* 1: make a guess. From the last section it is to be expected that the integral involves 2^{-x}.)

Let
$$y = 2^{-x},$$
$$\therefore \log_e y = -x \log_e 2.$$

(*Stage* 2: differentiate.)

$$\therefore \frac{1}{y} \frac{dy}{dx} = -\log_e 2.$$

$$\therefore \frac{dy}{dx} = -2^{-x} \log_e 2.$$

(*Stage* 3: compare with the given functions.) We have an extra constant factor of $-\log_e 2$.
(*Stage* 4: alter the guessed function.)

$$\frac{d}{dx}\left(\frac{2^{-x}}{-\log_e 2} \right) = 2^{-x}.$$

$$\therefore \int 2^{-x} dx = -\frac{2^{-x}}{\log_e 2} + c.$$

Example 5. *Integrate $x \log_e x$ with respect to x.*
(*Stage* 1: make a guess.) When we *differentiate* a product, we differentiate each function in turn and multiply by the other; so, to integrate $x \log_e x$, try integrating one factor and multiply by the other. As we do not know how to integrate $\log_e x$, we had better try $\frac{1}{2}x^2 \log_e x$.

Let
$$y = \frac{1}{2}x^2 \log_e x.$$

(*Stage* 2: differentiate.)

$$\therefore \frac{dy}{dx} = x \log_e x + \frac{1}{2}x^2 \cdot \frac{1}{x}.$$

$$\therefore \frac{d}{dx} (\frac{1}{2}x^2 \log_e x) = x \log_e x + \frac{1}{2}x.$$

(*Stage* 3: compare with the given function.)　We have an extra term of $\frac{1}{2}x$ on the right-hand side.

(*Stage* 4: alter the guessed function.)

$$\frac{d}{dx}\left(\tfrac{1}{2}x^2\log_e x-\tfrac{1}{4}x^2\right)=x\log_e x+\tfrac{1}{2}x-\tfrac{1}{2}x,$$

$$=x\log_e x.$$

$$\therefore \int x\log_e x\,dx=\tfrac{1}{2}x^2\log_e x-\tfrac{1}{4}x^2+c.$$

Qu. 9. Integrate the following functions with respect to x by trial:

(i) $(3x+1)^{\frac{1}{2}}$, 　　　(ii) $\sin x\cos^5 x$, 　　　(iii) $\dfrac{\sin x}{(1+\cos x)^3}$,

(iv) 5^x, 　　　　　(v) 2^{2x}, 　　　　　(vi) $\log_e x$.

Inverse trigonometrical functions

11.3.　The functions $\sin^{-1} x$, $\tan^{-1} x$ have been introduced in Chapter 1.　We now turn to the problem of differentiating such inverse trigonometrical functions.　This will be illustrated by examples but, for some readers, a little revision may be advisable.

Qu. 10.　$y=\sin^{-1} x$ means 'y is the angle (or the number) whose sine is x' so that $\sin y=x$.　Rewrite:

(i) $y=\tan^{-1} x$, 　　　(ii) $\sec^{-1} x=y$, 　　　(iii) $\cos^{-1} p=q$.

Qu. 11.　Differentiate with respect to x:

(i) y^2, 　　　(ii) $\sin y$, 　　　(iii) $\tan y$, 　　　(iv) $\sec y$.

Example 6.　*Differentiate with respect to* x:

$$(i)\ \sin^{-1} x,\qquad (ii)\ \tan^{-1}(x^2+1).$$

(i) Let 　　　　　　　　$y=\sin^{-1} x$.

$$\therefore \sin y=x.$$

Differentiating with respect to x, $\cos y\dfrac{dy}{dx}=1$. 　　$1-\sin^2 y=1-x^2$.

$$\therefore \cos^2 y=1-x^2.$$

(y was our own introduction, so we must get $\dfrac{dy}{dx}$ in terms of x.)

$$\therefore \ \sqrt{(1-x^2)} \frac{dy}{dx} = 1.$$

$$\therefore \ \frac{dy}{dx} = \frac{1}{\sqrt{(1-x^2)}}.$$

(ii) Let $\qquad\qquad y = \tan^{-1}(x^2+1).$

$$\therefore \ \tan y = x^2+1. \qquad\qquad 1+\tan^2 y = 1+(x^2+1)^2.$$

$$\therefore \ \sec^2 y = x^4+2x^2+2.$$

Differentiating with respect to x,

$$\sec^2 y \frac{dy}{dx} = 2x.$$

(We must again express $\frac{dy}{dx}$ in terms of x.)

$$\therefore \ (x^4+2x^2+2)\frac{dy}{dx} = 2x.$$

$$\therefore \ \frac{dy}{dx} = \frac{2x}{x^4+2x^2+2}.$$

Qu. 12. Differentiate with respect to x:

(i) $\cos^{-1} x$,   (ii) $\cot^{-1} x$,   (iii) $\sin^{-1}(2x+1)$.

Exercise 11a

1. Express in the form $p \log_e a + q \log_e b + r \log_e c$:
   (i) $\log_e (a^3b^4)$,   (ii) $\log_e (a/b)$,   (iii) $\log_e \sqrt{(a^3/b)}$,
   (iv) $\log_e (a^2b/\sqrt{c})$,   (v) $\log_e \sqrt{(ab/c)}$,   (vi) $\log_e \{1/\sqrt{(abc)}\}$.

2. Write the following in a form which does not use the logarithm notation:
   (i) $\log_{10} 100\,000$,   (ii) $\log_2 8$,   (iii) $\log_e e^4$,
   (iv) $\log_e \sqrt{e}$,   (v) $\log_e e^{x^3}$,   (vi) $\log_e (1/e^{2x})$.

Differentiate Nos. 3 to 16 with respect to x, using logarithmic differentiation:

3. $\sqrt{\dfrac{(2x+3)^3}{1-2x}}.$

4. $\dfrac{e^{x/2} \sin x}{x^4}.$

5. $\dfrac{1}{\sqrt{(x^2+1)}\sqrt[3]{(x^2-1)}}.$

6. $\dfrac{1}{x\, e^x \cos x}.$

7. $\sqrt{\{(x+1)(x+2)(x+3)\}}.$

8. $\dfrac{(x^2+1)^2(2x+1)^3}{(3x-1)^2}.$

9. $\dfrac{(x-1)^2\, e^{2x}}{(2x-1)^3}.$

10. $x \sin^3 x \cos^2 x.$

11. $7^x.$ 12. $10^{3x}.$ 13. $10^{-x/2}.$

14. $1/10^x.$ 15. $4^{x+1}.$ 16. $3^{x^2}.$

Integrate with respect to x:

17. 5^x. **18.** 8^x. **19.** $(\frac{1}{3})^x$.

20. 3^{2x}. **21.** 7^{-x}. **22.** $1/10^{2x}$.

23. Convince yourself that $e^{\log_e a} = a$. (See p. 30.) Write a^x in the form $e^{x \log_e a}$ and hence find $\dfrac{d}{dx}(a^x)$.

24. Find $\int a^x \, dx$ by writing $a^x = e^{x \log_e a}$.

Differentiate with respect to x:

25. $\tan^{-1} x$. **26.** $\sec^{-1} x$. **27.** $\sin^{-1}(x+1)$.

28. $\cos^{-1}(2x-1)$. **29.** $\tan^{-1}(1/x^2)$. **30.** $\operatorname{cosec}^{-1}(1+x^2)$.

31. $\frac{1}{3}\tan^{-1} 3x$. **32.** $2\cos^{-1} 5x$. **33.** $\frac{1}{2}\sin^{-1}(2x+1)$.

34. Find: (i) $\dfrac{d}{dx}(\sin^{-1} x + \cos^{-1} x)$,

  (ii) $\dfrac{d}{dx}(\tan^{-1} x + \cot^{-1} x)$.

  Explain these answers.

35. Show that

$$\frac{d}{dx}\left\{\tan^{-1}\left(\frac{1}{x}\right)\right\} = \frac{d}{dx}\{\cot^{-1} x\}.$$

36. Find $\dfrac{d}{dx}(\sin^{-1} x)$ and hence write down:

  (i) $\dfrac{d}{dx}(\sin^{-1} 2x)$, (ii) $\dfrac{d}{dx}(\sin^{-1} x^2)$, (iii) $\dfrac{d}{dx}\{\cos^{-1}\sqrt{(1-x^2)}\}$.

Integrate with respect to x by trial:

37. $\sqrt{(4x+3)}$. **38.** $x(2x^2+1)^3$. **39.** $\sec^3 x \tan x$.

40. $\log_e x$. **41.** $x^2 \log_e x$. **42.** $x \sin x$.

43. $\sin^{-1} x$. [Find $\dfrac{d}{dx}(x \sin^{-1} x)$.]

44. $\tan^{-1} x$. **45.** $x \cos 2x$. **46.** $x\,e^x$.

Differentiate with respect to x:

47. x^{-x}. **48.** $x^{\sin x}$. **49.** $x^{\log_e x}$.

50. $(\log_e x)^x$. **51.** $(\cos x)^x$. **52.** $\log_e(x^x)$.

Maxima and minima

11.4. In P.M.I. we considered a number of simple cases of stationary points, and used the sign of dy/dx to establish whether the graph concerned had a maximum, minimum, or point of inflexion. (If in doubt about the definitions of the terms involved in this section, the reader should refer to P.M.I., p. 70.) Some harder examples will now be considered and we shall use a simple test to distinguish between maxima and minima. The test is easy to use, but the reader is warned that it will be important to understand the exact conditions under which it works.

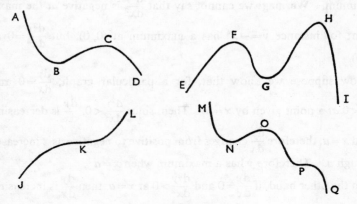

Fig. 11.1

Qu. 13. Copy the curves shown in Fig. 11.1. For each curve, mark the sign of $\frac{dy}{dx}$ above the curve, showing the points where $\frac{dy}{dx} = 0$. Also mark the sign of $\frac{d^2y}{dx^2}$ below the curve, showing the points where $\frac{d^2y}{dx^2} = 0$.

First consider a maximum point such as is shown in Fig. 11.2. As x increases (to the right), the sign of $\frac{dy}{dx}$ (marked in Fig. 11.2) changes

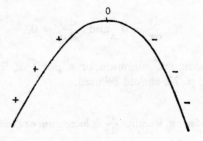

Fig. 11.2

from $+$ to $-$; therefore $\frac{dy}{dx}$ is decreasing in the neighbourhood of the maximum and hence $\frac{d^2y}{dx^2}$ is negative in the neighbourhood of the maximum. **Warning:** we cannot say that $\frac{d^2y}{dx^2}$ is negative *at* the maximum; for instance $y = -x^4$ has a maximum at $(0, 0)$, but $\frac{d^2y}{dx^2} = 0$ at $(0, 0)$.

Now suppose we know that, for a particular graph, $\frac{dy}{dx} = 0$ and $\frac{d^2y}{dx^2} < 0$ *at* a point given by $x = a$. Then, since $\frac{d^2y}{dx^2} < 0$, $\frac{dy}{dx}$ is decreasing when $x = a$, therefore $\frac{dy}{dx}$ changes from positive to negative as x increases through a. Therefore y has a maximum when $x = a$.

On the other hand, if $\frac{dy}{dx} = 0$ and $\frac{d^2y}{dx^2} > 0$ at $x = a$, then $\frac{dy}{dx}$ is increasing when $x = a$. Therefore $\frac{dy}{dx}$ changes from negative to positive as x increases through a. Hence y has a minimum when $x = a$.

In brief, at $x = a$,

$$\text{if} \quad \frac{dy}{dx} = 0 \quad \text{and} \quad \frac{d^2y}{dx^2} < 0, \quad y \text{ has a } \textbf{maximum};$$

$$\text{if} \quad \frac{dy}{dx} = 0 \quad \text{and} \quad \frac{d^2y}{dx^2} > 0, \quad y \text{ has a } \textbf{minimum};$$

$$\text{if} \quad \frac{dy}{dx} = 0 \quad \text{and} \quad \frac{d^2y}{dx^2} = 0,$$

y may have a maximum, minimum or a point of inflexion and the method of P.M.I., p. 72, should be used.

Qu. 14. By considering whether $\frac{dy}{dx}$ is increasing or decreasing, show that, when y has a point of inflexion, $\frac{d^2y}{dx^2} = 0$.

Example 7. *Find the maximum and minimum values of $x^2 e^x$, and sketch the graph of the function.*

Let
$$y = x^2 e^x.$$

$$\therefore \frac{dy}{dx} = 2x.e^x + x^2.e^x,$$

$$= (2x + x^2) e^x.$$

To find when $\frac{dy}{dx} = 0$,

$$(2x + x^2) e^x = 0.$$

But $e^x > 0$ for all values of x

$$\therefore 2x + x^2 = 0.$$
$$\therefore x = 0 \quad \text{or} \quad -2.$$

Differentiating $\frac{dy}{dx}$ as a product,

$$\frac{d^2y}{dx^2} = (2 + 2x).e^x + (2x + x^2).e^x,$$

$$= (2 + 4x + x^2) e^x.$$

(i) If $x = 0$, $\qquad \frac{dy}{dx} = 0$

and
$$\frac{d^2y}{dx^2} = (2 + 4.0 + 0^2) e^0 = 2.$$

Since $\frac{d^2y}{dx^2} > 0$, y has a minimum value which is 0.

(ii) If $x = -2$, $\qquad \frac{dy}{dx} = 0$

and
$$\frac{d^2y}{dx^2} = \{2 + 4(-2) + (-2)^2\} e^{-2}$$

$$= -2 e^{-2}.$$

$e^{-2} > 0$, therefore $\frac{d^2y}{dx^2} < 0$ and so y has a maximum value which is $4 e^{-2} \simeq 0 \cdot 54$.

To sketch the graph, we observe that:

$\qquad$ (a) $\quad$ when $x \to +\infty$, $\quad y \to +\infty$.
$\qquad$ (b) $\quad$ when $x \to -\infty$, $\quad y \to 0$.

The graph is sketched in Fig. 11.3.

The same procedure can be carried out with more difficult examples but also bear in mind the method of P.M.I., p. 72, if the task of obtaining $\dfrac{d^2y}{dx^2}$ looks laborious. A neat device to avoid complicated working is illustrated in Example 8.

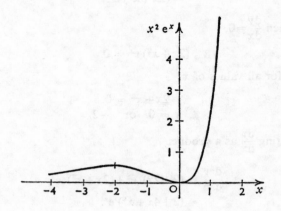

Fig. 11.3

Example 8. *Determine any maxima and minima of the function*

$$\frac{e^x}{x^2-8}.$$

Let
$$y = \frac{e^x}{x^2-8}.$$

Differentiating as a quotient,

$$\frac{dy}{dx} = \frac{(x^2-8)\,.\,e^x - e^x\,.\,2x}{(x^2-8)^2},$$

$$= \frac{(x^2-2x-8)\,e^x}{(x^2-8)^2},$$

which we may write as $(x^2-2x-8)v$, where $v = e^x/(x^2-8)^2$. [Here $x^2-2x-8=0$ gives all the roots of the equation $\dfrac{dy}{dx}=0$, and so we have separated x^2-2x-8 from the rest of the expression.]

To find when $\dfrac{dy}{dx} = 0$,

$$x^2 - 2x - 8 = 0.$$
$$\therefore \ (x+2)(x-4) = 0.$$
$$\therefore \ x = -2 \quad \text{or} \quad 4.$$

Now $\qquad \dfrac{d^2y}{dx^2} = v.(2x-2) + (x^2 - 2x - 8).\dfrac{dv}{dx}$.

NOTE: (a) we do not simplify this expression,

(b) when $\dfrac{dy}{dx} = 0$, $\ x^2 - 2x - 8 = 0$, so we need not concern ourselves with $\dfrac{dv}{dx}$,

(c) $e^z > 0 \quad$ and $\quad (x^2 - 8)^2 > 0$, so $\ v > 0$;

therefore $\dfrac{d^2y}{dx^2}$ has the same sign as $(2x - 2)$.

(i) If $x = -2$, $2x - 2 = -6 < 0$, therefore $\dfrac{d^2y}{dx^2} < 0$ and so y has a maximum value which is $-\frac{1}{4}e^{-2}$.

(ii) If $x = 4$, $2x - 2 = 6 > 0$, therefore $\dfrac{d^2y}{dx^2} > 0$ and so y has a minimum value which is $e^4/8$.

Exercise 11b

Find the nature of the stationary points of:

1. $x(x-3)^2$.

2. $x + \dfrac{4}{x^2}$.

3. $x - 2 + \dfrac{1}{x-3}$.

4. $x - \log_e x$.

5. $\dfrac{x^3 + 2x^2 - 6}{(x-1)^2}$.

6. $(x-2)^2(x+3)^3$.

7. Sketch the graphs of Nos. 1 to 6.

In Nos. 8 to 11, find the maxima and minima of the functions of θ in the range $0 \leqslant \theta \leqslant 2\pi$:

8. $\sin \theta + \frac{1}{2} \sin 2\theta$. Sketch the graph.

9. $(\sin \theta)/(1 + \sin \theta)$. Sketch the graph.

10. $\log_e \cos \theta - \cos \theta$.

11. $\cos \theta - \frac{1}{3} \cos 3\theta$. Sketch the graph.

12. Find the turning point of the function $x\,e^{-z}$ and determine its nature. Show that there is a point of inflexion when $x = 2$ and sketch the curve.

13. Find the turning point of the function $10 \tan^{-1} x - \frac{1}{2}x^2$ and sketch the curve.

14. Show that the function $x \log_e x$ has a minimum at $(1/e, -1/e)$. Given that $x \log_e x \to 0$ as $x \to 0$, sketch the graph of the function.

15. Show that $e^x \cos x$ has turning points at intervals of π in x. Distinguish between maxima and minima and show that these values are in a geometrical progression with common ratio $- e^\pi$.

16. A right circular cylinder is inscribed in a sphere of given radius a. Show that the volume of the cylinder is $\pi h(a^2 - \frac{1}{4}h^2)$, where h is the height of the cylinder. Find the ratio of the height to the radius of the cylinder when its volume is greatest.

17. A right circular cylinder is inscribed in a given sphere. Show that, when the area of the curved surface is greatest, the height of the cylinder is equal to its diameter.

18. A funnel is in the form of a right circular cone. If the funnel is to hold a given quantity of fluid, find the ratio of the height to the radius when the area of the curved surface is a minimum.

19. A right circular cone of vertical angle 2θ is inscribed in a sphere of radius a. Show that the area of the curved surface of the cone is $\pi a^2(\sin 3\theta + \sin \theta)$ and prove that its greatest area is $8\pi a^2/3\sqrt{3}$.

20. An open box has a square horizontal cross-section. If the box is to hold a given amount of material and the internal surface area of the box is to be a minimum, find the ratio of height to the length of the sides.

21. A right circular cone is circumscribed about a sphere of radius a. If h is the distance from the centre of the sphere to the vertex of the cone, show that the volume of the cone is $\frac{1}{3}\pi a^2(a+h)^2/(h-a)$. Find the vertical angle of the cone when the volume is a minimum.

22. Show that the volume of the greatest right circular cone which can be inscribed in a sphere of radius a is $32\pi a^3/81$.

Exercise 11c (Miscellaneous)

Differentiate the functions in Nos. 1 to 7 with respect to x:

1. (i) $\dfrac{x^2}{1-x^2}$, (ii) $\sin^{-1} \dfrac{1}{x}$. (iii) $e^{2x} \cos 3x$.

2. (i) $\dfrac{x+1}{\sqrt{(x-1)}}$, (ii) $\sin^3 x \cos^2 x$, (iii) $x \log_e x - x$.

3. (i) $\dfrac{1}{(2x+1)(3x-2)}$, (ii) $e^{\cos 2x}$, (iii) $\log_e \sin 2x$.

4. (i) $\sqrt{\dfrac{2x+1}{2x-1}}$, (ii) $\tan^{-1}\left(\dfrac{1+x}{1-x}\right)$, (iii) $\dfrac{e^x}{e^{2x} e^{3x}}$.

5. (i) $(x^2+1)^2(x^3+1)^3$, (ii) $\cos^{-1}(\tan x)$, (iii) $\dfrac{\log_e x}{x^2}$.

6. (i) $\dfrac{\sqrt{(3x^2 - 2x)}}{\sqrt[3]{(x^3 + 1)}}$,   (ii) $\dfrac{\sin^4 x}{\cos^5 x}$,   (iii) $\sin^{-1}\dfrac{x}{\sqrt{(1+x^2)}}$.

7. (i) 2^{x^2},   (ii) x^{2x},   (iii) $(\sin x)^{\cos x}$.

8. If $y = e^{2x} \cos 3x$, show that

$$\frac{d^2y}{dx^2} - 4\frac{dy}{dx} + 13y = 0.$$

9. If $y = x\, e^{-x}$, show that

$$\frac{d^2y}{dx^2} + 2\frac{dy}{dx} + y = 0,$$

and that

$$\frac{d^3y}{dx^3} = 3\frac{dy}{dx} + 2y.$$

10. (i) Show that the gradient of the ellipse $b^2x^2 + a^2y^2 = a^2b^2$ at the point $(a \cos \theta, b \sin \theta)$ is $(-b/a) \cot \theta$ and find an expression for $\dfrac{d^2y}{dx^2}$ at that point.   [For the method, see P.M.I., p. 130.]

(ii) The equation of a curve is given parametrically by the equations

$$x = \frac{t^2}{1+t^3}, \qquad y = \frac{t^3}{1+t^3}.$$

Show that

$$\frac{dy}{dx} = \frac{3t}{2 - t^3}$$

and that

$$\frac{d^2y}{dx^2} = 48 \quad \text{at the point } (\tfrac{1}{2}, \tfrac{1}{2}).$$

11. Find $\dfrac{d^2y}{dx^2}$ in terms of the parameter when

(i) $x = t^2 - 4$, $y = t^3 - 4t$,

(ii) $x = \cos^3 t$, $y = \sin^3 t$,

(iii) $x = a(\theta - \sin \theta)$, $y = a(1 - \cos \theta)$.

12. (i) Differentiate $\sin x$ from first principles.

(ii) Prove that

$$\frac{d}{dx}(uv) = v\frac{du}{dx} + u\frac{dv}{dx}.$$

13. (i) Differentiate $\tan x$ from first principles.

(ii) Prove that

$$\frac{d}{dx}\left(\frac{u}{v}\right) = \left(v\frac{du}{dx} - u\frac{dv}{dx}\right)\Big/v^2.$$

14. Differentiate $\sin^{-1} x + x\sqrt{(1-x^2)}$ with respect to x and hence find

$$\int \sqrt{(1-x^2)}\ \mathrm{d}x.$$

15. Differentiate $x \tan^{-1} x - \frac{1}{2} \log_e (1+x^2)$ with respect to x. Hence write down $\int \tan^{-1} x\ \mathrm{d}x$.

16. Find the maximum and minimum values of the function

$$(a+b \sin x)/(b+a \sin x)$$

where $b > a > 0$ in the range $0 \leqslant x \leqslant 2\pi$. Sketch the graph when $a = 4$, $b = 5$.

17. Find the maximum and minimum values of y given by the equation $x^3 + y^3 - 3xy = 0$.

18. Find the length of the longest chord of the ellipse

$$x^2 + 2xy + 6y^2 + 2x + 2y = 0$$

which is parallel to the y-axis.

19. Write down the nth derivatives of:

    (i) e^x,   (ii) x^n,   (iii) $\log_e x$.

20. Show that

    (i) $\dfrac{\mathrm{d}}{\mathrm{d}x} (\sin x) = \sin (x + \frac{1}{2}\pi)$,   (ii) $\dfrac{\mathrm{d}^n}{\mathrm{d}x^n} (\sin x) = \sin (x + n\pi/2)$,

and find similar expressions for

$$\frac{\mathrm{d}}{\mathrm{d}x} (\cos x) \quad \text{and} \quad \frac{\mathrm{d}^n}{\mathrm{d}x^n} (\cos x).$$

21. Prove that

    (i) $\dfrac{\mathrm{d}^n}{\mathrm{d}x^n} (e^x \cos x) = 2^{n/2} e^x \cos (x + n\pi/4)$,

    (ii) $\dfrac{\mathrm{d}^n}{\mathrm{d}x^n} (e^{ax} \sin bx) = (a^2 + b^2)^{n/2} e^{ax} \sin (bx + n\alpha)$,

    where $\alpha = \tan^{-1} (b/a)$.

22. Show that the function $y = \tan x - 8 \sin x$ has two stationary values between $x = 0$ and $x = 2\pi$. Draw a rough graph of the function between these values of x and show that, if the equation

$$\tan x - 8 \sin x = b$$

has four real roots between 0 and 2π, then

$$-3\sqrt{3} < b < 3\sqrt{3}. \qquad \text{o.}$$

23. By first putting the expression into partial fractions, or otherwise, find the first and second derivatives of the function

$$\frac{3x-1}{(4x-1)(x+5)}.$$

Find the coordinates of any maxima, minima and points of inflexion that the function may have, and draw a rough sketch of its graph.   o.

24. Prove that, if $f'(a) = 0$ and $f''(a)$ is negative, then the graph of the function $f(x)$ has a maximum at the point whose abscissa is a.

Prove that the function

$$y \equiv \frac{\sin x \cos x}{1 + 2 \sin x + 2 \cos x}$$

has turning points in the range $0 \leqslant x \leqslant 2\pi$ when $x = \frac{1}{4}\pi$ and $x = \frac{5}{4}\pi$, distinguishing between maximum and minimum values.

Prove that the tangents at the origin and at the point $(\frac{1}{2}\pi, 0)$ meet at a point whose abscissa is $\frac{1}{4}\pi$.   O.C.

25. Find the shortest distance between two points, one of which lies on the parabola $y^2 = 4ax$, and the other on the circle $x^2 + y^2 - 24ay + 128a^2 = 0$.

26. Chords of the hyperbola $xy = c^2$ cut both the branches and pass through the point $(2\sqrt{3}c, 0)$. Find the length of the shortest of these chords.

27. (i) A right circular cylinder is inscribed in a fixed sphere of radius a. Prove that the total area of its surface (including the ends) is $2\pi a^2(\sin 2\theta + \cos^2 \theta)$, where $a \cos \theta$ is the radius of an end. Hence prove that the maximum value of the total area is $\pi a^2(\sqrt{5} + 1)$.

(ii) Prove that the function $x^{-1} e^x$ has a minimum value e at $x = 1$. By considering a rough graph of the function, show that the equation $e^x = kx$ has two real roots if $k > e$, one real root if $k = e$ or $k < 0$, no real roots if $0 \leqslant k < e$.   O.

28. Prove that, if $f(x)$ is a function of x which has a first and second derivative and

$$f'(x_0) = 0, \qquad f''(x_0) < 0,$$

then $f(x)$ has a maximum value at $x = x_0$.

A metal box has a square base of side a with vertical sides of height h. It is closed by a lid which is one-half of a hollow, closed right circular cylinder obtained by bisecting the cylinder by a plane drawn through the axis. The length and diameter of the cylinder are each of length a. Prove that when the box has the greatest possible volume for a given area of metal then $a = 2h$.   O.

29. Prove that the function $f(x)$ has a maximum value at $x = a$ when $f'(a) = 0$ and $f''(a)$ is negative.

A straight tree trunk of circular cross-section is 12 m long and tapers uniformly in radius from 50 cm at one end to 25 cm at the other. Show that the pole of greatest volume with *uniform* cross-section which can be cut from the tree trunk is 8 m long.   O.C.

In Nos. 30 to 32, the following abbreviations will be used:

$$u_n = \frac{\mathrm{d}^n u}{\mathrm{d}x^n}, \qquad v_n = \frac{\mathrm{d}^n v}{\mathrm{d}x^n}.$$

30. Show that

(i) $\dfrac{\mathrm{d}^2(uv)}{\mathrm{d}x^2} = u_2 v + 2u_1 v_1 + uv_2,$

(ii) $\dfrac{\mathrm{d}^3(uv)}{\mathrm{d}x^3} = u_3 v + 3u_2 v_1 + 3u_1 v_2 + uv_3.$

31. If $u = e^{ax}$, $v = e^{bx}$, show that

$$\frac{\mathrm{d}^n(uv)}{\mathrm{d}x^n} = \frac{\mathrm{d}^n(e^{(a+b)x})}{\mathrm{d}x^n}$$

$$= u_n v + \binom{n}{1} u_{n-1} v_1 + \ldots + \binom{n}{r} u_{n-r} v_r + \ldots + uv_n,$$

where $\dbinom{n}{r} = \dfrac{n!}{(n-r)!\, r!}$

*32. Find the term in $u_{n+1-r}\, v_r$ in

$$\frac{\mathrm{d}}{\mathrm{d}x}\left\{ \binom{n}{r-1} u_{n-r+1} v_{r-1} + \binom{n}{r} u_{n-r} v_r \right\}.$$

Use the method of induction to prove Leibnitz's theorem:

$$\frac{\mathrm{d}^n}{\mathrm{d}x^n}(uv) = u_n v + \binom{n}{1} u_{n-1} v_1 + \ldots + \binom{n}{r} u_{n-r} v_r + \ldots + uv_n.$$

33. Use Leibnitz's theorem (see No. 32) to write down:

(i) $\dfrac{\mathrm{d}^3}{\mathrm{d}x^3}(x^3\, e^x),$

(ii) $\dfrac{\mathrm{d}^4}{\mathrm{d}x^4}(x \cos x),$

(iii) $\dfrac{\mathrm{d}^n}{\mathrm{d}x^n}(x^2\, e^{2x}),$

(iv) $\dfrac{\mathrm{d}^n}{\mathrm{d}x^n}\{y(1-x^2)\}.$

CHAPTER 12

FURTHER TRIGONOMETRY

12.1. The first part of this chapter is concerned with the general solution of trigonometrical equations, but it is hoped that the examples and Exercise 12a will, nevertheless, provide useful revision work for those who do not require a knowledge of general solutions.

General solutions

12.2. So far we have solved an equation such as $\cos \theta = \frac{1}{2}$ for a restricted range of values of θ. Thus, between 0° and 360°, $\theta = 60°$, 300° or, using radians, for values of θ between 0 and 2π, $\theta = \frac{1}{3}\pi$, $\frac{5}{3}\pi$. The general solution of an equation is a formula, or formulae, from

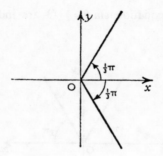

Fig. 12.1

which all solutions of the equation can be written down. Consider a radius, starting at Ox, rotating about O in Fig. 12.1. The two thick lines show all the positions of the radius when it has rotated through an angle which is a solution of the equation $\cos \theta = \frac{1}{2}$. The formula

$$\theta = 2n\pi \pm \tfrac{1}{3}\pi,$$

where n is any positive or negative integer or zero, is the general solution of the equation.

Solutions of the equation $\tan \theta = \sqrt{3}$ are indicated in the same way by the thick lines in Fig. 12.2 from which it may be seen that

$$\theta = n\pi + \tfrac{1}{3}\pi$$

is the general solution.

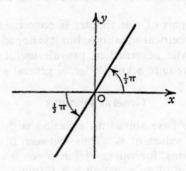

Fig. 12.2

Solutions of the equation $\sin \theta = \tfrac{1}{2}\sqrt{3}$ are indicated by the thick lines in Fig. 12.3.

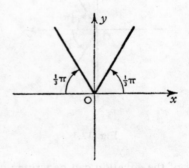

Fig. 12.3

The general solution may be written as

$$\theta = 2n\pi + \tfrac{1}{3}\pi, \qquad 2n\pi + \tfrac{2}{3}\pi,$$

but a neater way is

$$\theta = n\pi + (-1)^n . \tfrac{1}{3}\pi.$$

Qu. 1.   Write down the general solutions of the following equations:

  (i) $\sin \theta = 0$,   (ii) $\cos \theta = -1$,   (iii) $\tan \theta = 1$,
 (iv) $\sin \theta = 1$,   (v) $\cos \theta = 0$,   (vi) $\sin \theta = \frac{1}{2}$,
(vii) $\tan \theta = -1$,   (viii) $\sin \theta = 1/\sqrt{2}$,   (ix) $\cos \theta = -1/\sqrt{2}$.

Example 1.   *Find the general solution of the equation* $\cos 2\theta + \sin \theta = 0$.

$$\cos 2\theta + \sin \theta = 0.$$
$$\therefore\ 1 - 2\sin^2 \theta + \sin \theta = 0.$$
$$\therefore\ 2\sin^2 \theta - \sin \theta - 1 = 0.$$
$$\therefore\ (\sin \theta - 1)(2\sin \theta + 1) = 0.$$

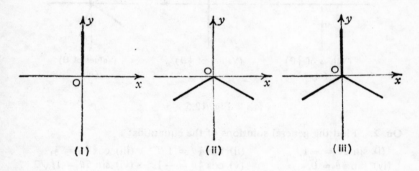

Fig. 12.4

(i) If $\sin \theta = 1$ (see Fig. 12.4 (i)),

$$\theta = (2n + \tfrac{1}{2})\pi = (4n + 1)\pi/2.$$

(ii) If $\sin \theta = -\frac{1}{2}$ (see Fig. 12.4 (ii)),

$$\theta = n\pi - (-1)^n \tfrac{1}{6}\pi.$$

These may be combined (see Fig. 12.4 (iii)) as

$$\theta = \tfrac{1}{2}\pi + \tfrac{2}{3}n\pi.$$

Example 2.   *Find the general solution of the equation* $\sin 3\theta + \sin 2\theta = 0$.

$$\sin 3\theta + \sin 2\theta = 0.$$
$$\therefore\ 2\sin \tfrac{5}{2}\theta \cos \tfrac{1}{2}\theta = 0.$$

(i) If $\sin \frac{5}{2}\theta = 0$ (see Fig. 12.5 (i)),

$$\tfrac{5}{2}\theta = n\pi.$$
$$\therefore\ \theta = 2n\pi/5.$$

(ii) If $\cos \frac{1}{2}\theta = 0$ (see Fig. 12.5 (ii)),

$$\frac{1}{2}\theta = n\pi + \frac{1}{2}\pi.$$
$$\therefore \ \theta = (2n+1)\pi.$$

Therefore the general solution is $\theta = 2n\pi/5$ or $(2n+1)\pi$.

The solutions of the equations are indicated in Fig. 12.5 (iii).

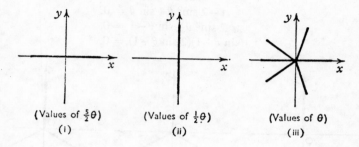

(Values of $\frac{5}{2}\theta$)   (Values of $\frac{1}{2}\theta$)   (Values of θ)

(i)   (ii)   (iii)

Fig. 12.5

Qu. 2. Find the general solutions of the equations:

(i) $\sin 2\theta = -1$,   (ii) $\tan \frac{2}{3}\theta = 1$,   (iii) $\cos \frac{2}{5}\theta = \frac{1}{2}$,

(iv) $\tan \frac{3}{2}\theta = 0$,   (v) $\cos \frac{4}{3}\theta = -1$,   (vi) $\sin 5\theta = 1/\sqrt{2}$.

The equation $a \cos \theta + b \sin \theta = c$

12.3. There are two good methods of solving equations in the form $a \cos \theta + b \sin \theta = c$, where a, b, c are constants. One was given in P.M.I., p. 290, Example 6, and the other was indicated in No. 47 on the same page. To compare the two methods, the same equation will be solved by each; but first we shall prove the formulae

$$\sin \theta = \frac{2t}{1+t^2}, \quad \cos \theta = \frac{1-t^2}{1+t^2},$$

where $t = \tan \frac{1}{2}\theta$.

Multiplying numerators and denominators by $\cos^2 \frac{1}{2}\theta$,

$$\frac{2 \tan \frac{1}{2}\theta}{1 + \tan^2 \frac{1}{2}\theta}, \qquad \frac{1 - \tan^2 \frac{1}{2}\theta}{1 + \tan^2 \frac{1}{2}\theta},$$
$$= \frac{2 \sin \frac{1}{2}\theta \cos \frac{1}{2}\theta}{\cos^2 \frac{1}{2}\theta + \sin^2 \frac{1}{2}\theta}, \qquad = \frac{\cos^2 \frac{1}{2}\theta - \sin^2 \frac{1}{2}\theta}{\cos^2 \frac{1}{2}\theta + \sin^2 \frac{1}{2}\theta},$$
$$= \sin \theta, \qquad\qquad = \cos \theta,$$

the denominators being 1.

A useful check when recalling the formulae from memory is that

$$\frac{\sin \theta}{\cos \theta} = \frac{\dfrac{2t}{1+t^2}}{\dfrac{1-t^2}{1+t^2}} = \frac{2t}{1-t^2},$$

that is,

$$\tan \theta = \frac{2 \tan \frac{1}{2}\theta}{1 - \tan^2 \frac{1}{2}\theta}.$$

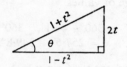

Fig. 12.6

Another aid to memory is the triangle (Fig. 12.6) whose three sides are connected by the identity

$$(1+t^2)^2 = (1-t^2)^2 + (2t)^2.$$

Example 3. *Solve the equation* $4 \cos x - 6 \sin x = 5$, *for values of* x *between* $0°$ *and* $360°$ *correct to* $0\cdot 1°$.

Method (i). $4 \cos x - 6 \sin x = 5.$

Divide both sides by $\sqrt{(4^2+6^2)} = \sqrt{52} = 2\sqrt{13}$.

$$\therefore \frac{2}{\sqrt{13}} \cos x - \frac{3}{\sqrt{13}} \sin x = \frac{5}{2\sqrt{13}}.$$

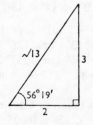

Fig. 12.7

To the nearest minute, $\tan 56° 19' = \frac{3}{2}$ and so (see Fig. 12.7)

$$\cos x \cos 56° 19' - \sin x \sin 56° 19' = 0·6934.$$
$$\therefore \cos (x + 56° 19') = 0·6934.$$
$$\therefore \cos (x + 56° 19') = \cos 46° 06'.$$

Now if
$$0° < x < 360°,$$

$$56·3° < x + 56·3° < 416·3°.$$

So for the required range of values

$$x + 56·3° = 360° - 46·1°, \quad 360° + 46·1°.$$
$$\therefore x = 257·6°, \quad 349·8°.$$

Method (*ii*) $4 \cos x - 6 \sin x = 5.$

Let $t = \tan \frac{1}{2}x,$

then $\cos x = \dfrac{1-t^2}{1+t^2}, \quad \sin x = \dfrac{2t}{1+t^2}.$

Side work
$$-12·00$$
$$+10·39$$

$$\therefore \frac{4-4t^2}{1+t^2} - \frac{12t}{1+t^2} = 5.$$

$2\underline{|-1·61}$
$9\underline{|-0·805}$

$$\therefore 4-4t^2-12t = 5+5t^2.$$
$$\therefore 9t^2+12t+1 = 0.$$

$$-0·0894$$

$$\therefore t = \{-12 \pm \sqrt{(144-36)}\}/18.$$
$$= \{-12 \pm \sqrt{108}\}/18.$$
$$= -0·0894, \quad -1·2439.$$

$$-12·00$$
$$-10·39$$

Now if
$$0° < x < 360°,$$
$$0° < \tfrac{1}{2}x < 180°.$$

$2\underline{|-22·39}$
$9\underline{|-11·195}$

$$-1·2439$$

$$\therefore \tfrac{1}{2}x = 180° - 5° 07' \quad \text{or} \quad 180° - 51° 12',$$
$$= 174° 53' \quad \text{or} \quad 128° 48',$$
$$\therefore x = 349° 46' \quad \text{or} \quad 257° 36'.$$
$$\therefore x = 349·8°, \quad 257·6° \quad \text{correct to } 0·1°.$$

Note that the side work is taken to 4 decimal places. The maximum error in the values of $-12 \pm \sqrt{108}$ is 0·005, so that the maximum error in the values of t is $0·005 \div 18 \simeq 0·0003$. This corresponds to an error of $1'$ and $\frac{3}{8}'$ in the two values of $\frac{1}{2}x$, so that the values of x are correct within $2'$ and $\frac{3}{4}'$. Thus the answer $349 \cdot 8°$ might be $0 \cdot 1°$ less, but the second is correct.

The reader is warned that Method (ii) breaks down in certain circumstances; see Qu. 3.

Qu. 3. Show that the substitution of Example 3, Method (ii) reduces the equation

$$a \cos \theta + b \sin \theta + a = 0 \qquad (1)$$

to

$$bt + a = 0. \qquad (2)$$

Verify that $\theta = \pi$ is a root of equation (1) but not in general of equation (2).

Exercise 12a

Solve the following equations. Solutions are given in radians unless tables have been used, in which case the answers are given in degrees.

1. $\cos 2\theta = \frac{1}{2}$.
2. $\tan \theta = \cos 130°$.
3. $\cos \theta = -\tan 146°$.
4. $\sin 2\theta = \sin \theta$.
5. $\cos 2\theta = \sin \theta$.
6. $\sin 2\theta = \cos \theta$.
7. $\cos \theta = 3 \tan \theta$.
8. $2 \tan \theta = \tan 2\theta$.
9. $\cos \theta = \cos 3\theta$.
10. $\sin 3\theta + \sin \theta = 0$.
11. $\sin \theta + \sin 3\theta + \sin 5\theta = 0$.
12. $\cos \theta + \sin 2\theta - \cos 3\theta = 0$.
13. $\cos 4\theta + 1 = 2 \cos^2 \theta$.
14. $1 - \sin \theta = 2 \cos^2 \theta$.
15. $2 \cos^2 \theta = 1 + \sin \theta$.
16. $2 \tan \theta = 1 - \tan^2 \theta$.
17. $3 \tan^2 \theta = 2 \sin \theta$.
18. $2 \sin^2 \theta + 3 \cos \theta = 3$.
19. $3 \cot^2 \theta + 5 = 7 \operatorname{cosec} \theta$.
20. $\sin 2\theta = \cot \theta$.
21. $2 \operatorname{cosec}^2 \theta = 5(\cot \theta + 13)$.
22. $\cos 2\theta = 5 \sin \theta + 3$.
23. $\cos \theta + \sqrt{3} \sin \theta = 1$.
24. $7 \cos \theta = 5 + \sin \theta$.
25. $7 \sin \theta - 24 \cos \theta = 25$.
26. $3 \cos \theta - 4 \sin \theta = 2$.
27. $\sin^2 \theta - \sin \theta \cos \theta - \cos^2 \theta = 1$.
28. $\sin 2\theta \sin \theta + \cos^2 \theta = 1$.
29. $\sin \theta + \cos \theta = \operatorname{cosec} \theta$.
30. $2 \cos (x - \frac{1}{4}\pi) = \sin (x + \frac{1}{4}\pi)$.
31. $\sin (x + \alpha) = \cos (x - \alpha)$.

Inverse functions

12.4. The notation $\tan^{-1} x$ has been used to denote the angle (or number) whose tangent is x. But consider, for instance, $\tan^{-1} (-\sqrt{3})$. This might be taken to mean $-\frac{1}{3}\pi, \frac{2}{3}\pi, -\frac{4}{3}\pi, \frac{5}{3}\pi, \ldots$ or any of the other values given by the formula $n\pi + \frac{2}{3}\pi$. Thus 'the angle whose tangent is x' has an infinity of values corresponding to every real value of x.

If we want a shorthand for this expression, we write $\text{Tan}^{-1}x$, reserving $\tan^{-1}x$ for the unambiguous function whose range is defined below. The need for this function was pointed out in §1.11, p. 18 when integration by trigonometrical substitution was being discussed.

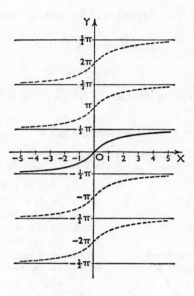

Fig. 12.8

Consider the graph of $\text{Tan}^{-1} x$ which is sketched in Fig. 12.8. The whole range of values of x from $-\infty$ to $+\infty$ is covered by the branch of the curve passing through the origin. This branch is shown with a continuous curve and represents the function $\tan^{-1} x$.
Thus

$$-\tfrac{1}{2}\pi \leqslant \tan^{-1} x \leqslant \tfrac{1}{2}\pi,$$

the equality signs being necessary to assign unambiguous meanings to $\tan^{-1} \pm \infty$.

Similarly, $\text{Sin}^{-1} x$, $\text{Cos}^{-1} x$ are many-valued expressions, while $\sin^{-1} x$, $\cos^{-1} x$ are reserved for functions of x. This time, however, x can only lie in the range $-1 \leqslant x \leqslant +1$. In defining $\sin^{-1} x$, $\cos^{-1} x$ we want values of the functions which will cover this range of values of x and it is natural to take values as near the origin as possible, preferring positive values to negative. Thus it will be seen from Figs. 12.8, 12.9 that

$$-\tfrac{1}{2}\pi \leqslant \sin^{-1} x \leqslant \tfrac{1}{2}\pi$$

$$0 \leqslant \cos^{-1} x \leqslant \pi.$$

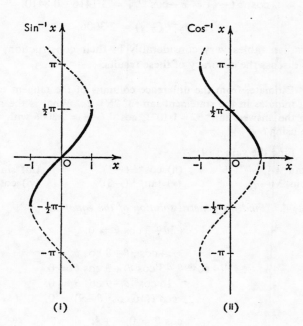

Fig. 12.9

Qu. 4. Write down the values of:

  (i) $\tan^{-1}(-1)$,   (ii) $\sin^{-1}(-\tfrac{1}{2})$,   (iii) $\cos^{-1}(-1)$,
  (iv) $\tan^{-1}(\sqrt{3})$,   (v) $\sin^{-1} 1$,   (vi) $\cos^{-1} 1$.

When finding the values of expressions such as $\tan^{-1} 2$, $\cos^{-1}(-\tfrac{2}{3})$, it is important to remember that they are to be regarded as numbers or angles in radians. The situation is complicated by the fact that the tangent tables are tabulated for angles in degrees. Thus, with four-figure tables,

$$\tan 63° 26' = 2.$$
$$63° 26' = 1·1071 \text{ rad.}$$
$$\therefore \tan^{-1} 2 = 1·1071.$$

Again,
$$\cos 48° 11' = 0·6667.$$
$$48° 11' = 0·8410 \text{ rad.}$$
$$\therefore \cos^{-1} \tfrac{2}{3} = 0·8410.$$

But
$$\cos^{-1}(-\tfrac{2}{3}) = \pi - \cos^{-1} \tfrac{2}{3} = 3·1416 - 0·8410.$$

$$\therefore \cos^{-1}(-\tfrac{2}{3}) = 2·3006.$$

Since radian tables vary considerably in their construction, it is not easy to discuss the accuracy of these results.

Qu. 5. Estimate, from the difference columns of the tangent tables, the error in minutes in the statement $\tan 63° 26' = 2$. What is the maximum error in the answers $\tan^{-1} 2 = 1·1071$, $\cos^{-1}(-\tfrac{2}{3}) = 2·3006$ with the tables you are using?

Qu. 6. Find the values of:

(i) $\tan^{-1} \tfrac{1}{3}$,

(ii) $\cos^{-1}(-\tfrac{1}{4})$,

(iii) $\sin^{-1}(0·01)$,

(iv) $\sin^{-1}(-\tfrac{1}{2})$,

(v) $\tan^{-1}(-3)$,

(vi) $\cos^{-1} \tfrac{3}{5}$.

Example 4. *Find the general solution of the equation*
$$4 \cos 3\theta + 3 \cos \theta = 0.$$

$$4 \cos 3\theta + 3 \cos \theta = 0.$$
$$\therefore 4(4 \cos^3 \theta - 3 \cos \theta) + 3 \cos \theta = 0.$$
$$\therefore 16 \cos^3 \theta - 9 \cos \theta = 0.$$
$$\therefore \cos \theta(16 \cos^2 \theta - 9) = 0.$$

$$\therefore \cos \theta = 0, \quad \pm \tfrac{3}{4}.$$

$$\therefore \theta = n\pi + \tfrac{1}{2}\pi, \quad n\pi \pm \cos^{-1} \tfrac{3}{4}.$$

Example 5. *Find the general solution of the equation* $2 \cos(\theta + \alpha) = \cos \theta$.

$$2 \cos(\theta + \alpha) = \cos \theta.$$

$$\therefore 2 \cos \theta \cos \alpha - 2 \sin \theta \sin \alpha = \cos \theta.$$

$$\therefore \cos \theta(2 \cos \alpha - 1) = 2 \sin \theta \sin \alpha.$$

$$\therefore \tan \theta = \frac{2 \cos \alpha - 1}{2 \sin \alpha}.$$

$$\therefore \theta = n\pi + \tan^{-1}\{(2 \cos \alpha - 1)/(2 \sin \alpha)\}.$$

Now consider the equations:

(i) $\sin \theta = \sin \alpha$,   (ii) $\cos \theta = \cos \alpha$,   (iii) $\tan \theta = \tan \alpha$,

where α is given.

(i) If $\sin \theta = \sin \alpha$,

$$\theta = n\pi + (-1)^n \alpha,$$

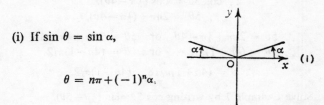

(ii) If $\cos \theta = \cos \alpha$,

$$\theta = 2n\pi \pm \alpha.$$

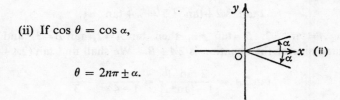

(iii) If $\tan \theta = \tan \alpha$,

$$\theta = n\pi + \alpha.$$

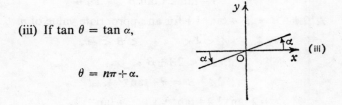

Fig. 12.10

Example 6. *Solve the equation* $\sin 3\theta = \sin \theta$.

$$\sin 3\theta = \sin \theta.$$
$$\therefore \ 3\theta = n\pi + (-1)^n \theta.$$
$$\therefore \ 3\theta = 2m\pi + \theta \quad \text{or} \quad 3\theta = (2m+1)\pi - \theta.$$
$$\therefore \ 2\theta = 2m\pi \quad \quad \text{or} \quad 4\theta = (2m+1)\pi$$
$$\therefore \ \theta = m\pi, \quad (2m+1)\pi/4.$$

Qu. 7. Solve the equation $\sin 3\theta = \sin \theta$ by means of the identity

$$\sin 3\theta = 3 \sin \theta - 4 \sin^3 \theta.$$

Example 7. *Solve the equation* $\cos 5\theta = \sin 4\theta$.

The identity $\sin \phi = \cos (\frac{1}{2}\pi - \phi)$ is used.

$$\cos 5\theta = \sin 4\theta.$$
$$\therefore \cos 5\theta = \cos (\frac{1}{2}\pi - 4\theta).$$
$$\therefore 5\theta = 2n\pi \pm (\frac{1}{2}\pi - 4\theta).$$

$$\therefore 5\theta = 2n\pi + \frac{1}{2}\pi - 4\theta \quad \text{or} \quad 5\theta = 2n\pi - \frac{1}{2}\pi + 4\theta.$$
$$\therefore 9\theta = (4n+1)\pi/2 \quad \text{or} \quad \theta = (4n-1)\pi/2.$$

$$\therefore \theta = (4n+1)\pi/18, \quad (4n-1)\pi/2.$$

Qu. 8. Solve Example 7 by writing $\cos 5\theta = \sin (\frac{1}{2}\pi - 5\theta)$.

Example 8. *Show that*

$$2 \tan^{-1} 2 + \tan^{-1} 3 = \pi + \tan^{-1} \tfrac{1}{3}.$$

Let $A = \tan^{-1} 2$, $B = \tan^{-1} 3$, then $\tan A = 2$, $\tan B = 3$, and the left-hand side of the equation is $2A + B$. We shall find $\tan (2A + B)$.

$$\tan 2A = \frac{2 \tan A}{1 - \tan^2 A} = \frac{4}{1-4} = -\frac{4}{3}.$$

$$\tan (2A + B) = \frac{\tan 2A + \tan B}{1 - \tan 2A \tan B} = \frac{-\frac{4}{3}+3}{1+4} = \frac{1}{3}.$$

$$\therefore 2A + B = n\pi + \tan^{-1} \tfrac{1}{3} \text{ for an appropriate value of } n.$$

Now
$$\tfrac{1}{4}\pi < A < \tfrac{1}{2}\pi, \quad \tfrac{1}{4}\pi < B < \tfrac{1}{2}\pi.$$

$$\therefore \tfrac{3}{4}\pi < 2A + B < \tfrac{3}{2}\pi.$$

$$\therefore 2A + B = \pi + \tan^{-1} \tfrac{1}{3}.$$

$$\therefore 2 \tan^{-1} 2 + \tan^{-1} 3 = \pi + \tan^{-1} \tfrac{1}{3}.$$

Exercise 12b

1. Write down the values of:

    (i) $\sin^{-1} (\frac{1}{2}\sqrt{3})$,   (ii) $\cos^{-1} (\frac{1}{2}\sqrt{2})$,   (iii) $\tan^{-1} (1/\sqrt{3})$,

    (iv) $\cos^{-1} 0$,   (v) $\tan^{-1} (-\sqrt{3})$,   (vi) $\sin^{-1} (-1)$.

2. Use tables to evaluate:

    (i) $\tan^{-1} 2$,   (ii) $\sin^{-1} \frac{3}{5}$,   (iii) $\cos^{-1} \frac{2}{5}$,

    (iv) $\sin^{-1} (-\frac{1}{3})$,   (v) $\tan^{-1} (-\frac{1}{2})$,   (vi) $\cos^{-1} \frac{1}{4}$.

Find the general solutions of the following equations in θ:

3. $3 \sin 2\theta = \sin \theta$.   **4.** $3 \cos 2\theta + 2 \cos \theta = 1$.

5. $9 \sin 3\theta = 2 \sin \theta$.   **6.** $\tan 2\theta = 4 \tan \theta$.

7. $\cos\theta+\sin\theta = \sqrt{2}.$ **8.** $\cos\theta-\sqrt{3}\sin\theta = 1.$

9. $\sin(\theta+\alpha) = 2\sin(\theta-\alpha).$ **10.** $3\cos(\theta-\alpha) = 4\sin(\alpha-\theta).$

11. $1+2\cos 2\theta = \cos 2\alpha+2\cos\theta\cos\alpha.$

12. $4\cos\theta-7\sin\theta+8 = 0.$ **13.** $3\cos\theta = 7(\sin\theta-1).$

14. $\cos 3\theta = \cos\theta.$ **15.** $\sin 3\theta = \sin 2\theta.$

16. $\tan 4\theta = \tan\theta.$ **17.** $\sin 3\theta = \cos 2\theta.$

18. $\cos 4\theta = \sin 3\theta.$ **19.** $\tan 4\theta+\tan 2\theta = 0.$

20. $\sin 2\theta+\cos 3\theta = 0.$

21. Find the general solution of the equation $\cos 3\theta=\cos 2\theta.$ Also express the equation as an equation for $\cos\theta$ and hence show that $\cos\frac{2}{5}\pi=\frac{1}{4}(-1+\sqrt{5}).$ Find a similar expression for $\cos\frac{4}{5}\pi.$

Prove the relations in Nos. 22 to 27.

22. $2\tan^{-1}\frac{1}{2} = \tan^{-1}\frac{4}{3}.$

23. $\tan^{-1}\frac{1}{2}+\tan^{-1}\frac{1}{3} = \frac{1}{4}\pi.$

24. $\frac{1}{4}\pi-\tan^{-1}\frac{2}{3} = \tan^{-1}\frac{1}{5}.$

25. $3\tan^{-1}2 - \pi = \tan^{-1}\dfrac{2}{11}.$

26. $2\tan^{-1}3+2\tan^{-1}2 = 3\pi/2.$

27. $2\tan^{-1}5+\tan^{-1}\dfrac{5}{12} = \pi.$

Find the values of:

28. $\tan^{-1}\frac{1}{2}+\tan^{-1}\frac{1}{5}+\tan^{-1}\frac{1}{8}.$

29. $4\tan^{-1}\frac{1}{5}-\tan^{-1}\frac{1}{239}.$

Simplify:

30. $\tan^{-1}x+\tan^{-1}\left(\dfrac{1-x}{1+x}\right).$

31. $\tan^{-1}\dfrac{x-y}{xy+1}+\tan^{-1}\dfrac{y-z}{yz+1}+\tan^{-1}\dfrac{z-x}{zx+1}.$

32. $\tan^{-1}\sqrt{\dfrac{(w-y)(w-z)}{w(w-x)}}+\tan^{-1}\sqrt{\dfrac{(w-z)(w-x)}{w(w-y)}}+\tan^{-1}\sqrt{\dfrac{(w-x)(w-y)}{w(w-z)}},$
where $2w=x+y+z.$

Properties of triangles

12.5. The last section of this chapter contains a number of properties of triangles and quadrilaterals, some of which can be proved by geometry, others by trigonometry, or a combination of the two. The reader should be able to prove most of these properties for himself and hints have been provided to assist him with some of the harder ones. In the exercise which follows, a number of results have been asterisked:

these are important enough to be regarded as theorems, but it is suggested that other results should also be used when they help to prove properties found later in the exercise.

Exercise 12c

*1. The perpendicular bisectors of the sides of a triangle meet at a point. This point is called the *circumcentre* and is denoted by O. [The locus of a point equidistant from two given points is the perpendicular bisector of the line joining the two given points.]

*2. The medians of a triangle meet at a point. This point is called the *centroid* and is denoted by G. [If BB', CC' are medians, prove triangles BGC, B'G'C' similar. Show that G trisects the median.]

*3. In any triangle, the sum of the squares on two sides is equal to twice the square on half the third side together with twice the square on the median which bisects the third side. This is known as Apollonius' theorem. [Use the cosine formula in both the triangles which have the median as one side.]

4. In any parallelogram, the sum of the squares on its sides is equal to the sum of the squares on its diagonals.

*5. The altitudes of a triangle meet in a point. This point is called the *orthocentre* and is denoted by H. [Let altitudes BE, CF meet at H. Let AH meet BC at D. Prove angle ADC is a right angle by means of concyclic points.]

In what follows, the feet of the altitudes through A, B, C will be denoted by D, E, F. Triangle DEF is called the *pedal* triangle.

6. The angles of triangle DEF are $\pi - 2A$, $\pi - 2B$, $\pi - 2C$.

7. The circumcircle of the pedal triangle bisects the sides of triangle ABC.

8. $EF = a \cos A = R \sin 2A$. Use the sine formula to deduce the circumradius of the pedal triangle.

9. $AH = 2R \cos A$.

10. The distance between the orthocentre and a vertex is twice the perpendicular distance from the circumcentre to the opposite side.

11. If P and A' are the mid-points of AH and BC respectively, OH and A'P bisect each other.

*12. If A', B', C' are the mid-points of BC, CA, AB, and P, Q, R the mid-points of AH, BH, CH, there is a circle with centre at N, the mid-point of OH, with A'P, B'Q, C'R as diameters and passing through D, E, F. What is the radius of this circle, the *nine-point circle*, in terms of R?

13. Show that G lies on ONH and that OG : GN : NH = 2 : 1 : 3.

14. The radii of the circumcircles of ABH and ABC are equal.

15. AH and AO are equally inclined to the bisector of angle A.

16. $AD = bc/2R$.

17. $HD = 2R \cos B \cos C$.

18. AB is a fixed chord of a fixed circle. C is any point on the minor arc AB. Find the locus of the orthocentre of triangle ABC.

19. If AH produced meets the circumcircle in P, prove that $HD = DP$. [Prove triangles HDC, PDC congruent.]

20. $AD \cos A + BE \cos B + CF \cos C = (a^2 + b^2 + c^2)/4R$.

21. $EF + FD + DE = 4R \sin A \sin B \sin C$.

*22. In any triangle, the internal and external bisectors of one angle divide the opposite side in the ratio of the other two sides. [If the internal bisector of angle A meets BC in P, $BP : BC = \triangle ABP : \triangle APC$.]

*23. The internal bisectors of the angles of a triangle meet in a point. This point is called the *incentre* and is denoted by I. [The locus of a point equidistant from two given lines is the pair of lines which bisect the angles between the given lines.]

*24. In triangle ABC, the internal bisector of angle A meets the external bisectors of angles B and C in a point called an *e(x)centre* and denoted by I_1. I_2 and I_3 are defined similarly.

*25. Circles centres I, I_1, I_2, I_3 with radii r, r_1, r_2, r_3 can be drawn to touch the sides (produced if necessary) of triangle ABC. The first is called the *inscribed circle* or *incircle* and the others are called *escribed circles* or *excircles*.

26. The tangents from A to the incircle are of length $s - a$. [Let the tangents from A, B, C be of lengths x, y, z.]

27. The tangents from A to the escribed circle with centre I_1 are of length s.

28. The tangents from B to the escribed circle, centre I_1, are of lengths $s - c$.

*29. $r = 4R \sin \frac{1}{2}A \sin \frac{1}{2}B \sin \frac{1}{2}C$.
[Let the incircle touch BC at X. Then from triangles BIX, CIX, $a = r \cot \frac{1}{2}B + r \cot \frac{1}{2}C$. Substitute $a = 2R \sin A$, solve for r and simplify.]

*30. $r_1 = 4R \sin \frac{1}{2}A \cos \frac{1}{2}B \cos \frac{1}{2}C$.
[Let the escribed circle, centre I_1, touch BC at X_1. From triangles BI_1X_1, CI_1X_1, $a = r_1 \tan \frac{1}{2}B + r_1 \tan \frac{1}{2}C$.]

31. Find the angles of triangle I_1, I_2, I_3.

32. $I_2I_3 = 4R \cos \frac{1}{2}A = a \operatorname{cosec} \frac{1}{2}A$.

33. Why is I_1A perpendicular to I_2I_3? Use No. 8 to deduce the circumradius of triangle $I_1I_2I_3$.

34. The radius of the inscribed circle of triangle DEF is $2R \cos A \cos B \cos C$.

35. $\dfrac{1}{AD} + \dfrac{1}{BE} + \dfrac{1}{CF} = \dfrac{1}{r}$.
[See P.M.I., p. 321, and p. 322, No. 10.]

36. $r_1(s-a)=\Delta$ where Δ is the area of triangle ABC.
   [ΔABC$=\Delta I_1$CA$+\Delta I_1$AB$-\Delta I_1$BC.]   Find an expression for r_1 in terms of a, b, c, s.

37. $\dfrac{1}{r_1}+\dfrac{1}{r_2}+\dfrac{1}{r_3} = \dfrac{1}{r}$.

38. $\Delta I_1 I_2 I_3 = 8R^2 \cos \frac{1}{2}A \cos \frac{1}{2}B \cos \frac{1}{2}C = 8R^2 s\Delta/abc$.

*39. If ABCD is a cyclic quadrilateral, then AB.CD$+$AD.BC$=$AC.BD. (Ptolemy's theorem.)   [Take E on BD such that angles DAE and BAC are equal.   Prove triangles DAE and BAC similar; also triangles BAE, CAD.]

40. Use Ptolemy's theorem to obtain the formula for $\sin (A+B)$.

41. If a point P is taken on the minor arc AD of the circumcircle of the rectangle ABCD which has AB$=$2BC, prove that
$$2(PC^2 - PB^2) = PB.PD - PA.PC.$$

42. The diagonal BD of a cyclic quadrilateral ABCD bisects the angle ABC. Prove that angle ABC$=2 \cos^{-1} \dfrac{AB+BC}{2BD}$.

43. If P is a point on the circumcircle of a square ABCD, P lying between C and D, prove that PB$-$PD$=\sqrt{2}$PC.

Exercise 12d (Miscellaneous)

1. Find all the angles between $0°$ and $360°$ which satisfy the equations:
   (i) $\sin \theta = \cos 127°$,   (ii) $3 \cot^2 \theta = 2 \cos \theta$,
   (iii) $3 \sin \theta - 4 \cos \theta = 2$.   o.c.

2. Find all the values of x between $0°$ and $180°$ inclusive for which:
   (i) $\sin 3x = \sin x$,   (ii) $2 \cos^2 x - \sin^2 x = 1$,
   (iii) $\sin 2x + \cos x = 0$.

3. Find all the values of x between $0°$ and $360°$ inclusive which satisfy the equations:
   (i) $\sin x + \cos x = \sin 18° + \cos 18°$;
   (ii) $\sin 2x = \cos x \sin 3x$.   L.

4. Find the general solutions of the equations:
   (i) $\cos x + \cos 3x = \cos 2x$,   (ii) $4 \cos x - 3 \sin x = 2$.   c.

5. Find all the values of x (in radian measure) which satisfy the following equations:
   (i) $\sin x + \cos 4x = 0$,   (ii) $\cos 2x - 5 \cos x + 3 = 0$,
   (iii) $\tan 2x + \tan 4x = 0$.   o.c.

6. Give the general solutions of the following equations:
   (i) $2 \sin 3\theta - 7 \cos 2\theta + \sin \theta + 1 = 0$,
   (ii) $\cos \theta - \sin 2\theta + \cos 3\theta - \sin 4\theta = 0$.   c.

7. (i) Express $\cos 5\theta$ in terms of $\cos \theta$.   Hence prove that
$$\cos 18° = \tfrac{1}{4}\sqrt{(10+2\sqrt{5})},$$
   and evaluate $\cos 54°$, leaving your answer in surd form.

(ii) Solve the equation

$$\sin 4x - \sin 3x + \sin 2x = 0$$

completely, expressing your answers in radian measure.　　o.c.

8. Write down and solve the quadratic equation in x whose roots p and q are given by the relations

$$p+q = \tfrac{1}{2}, \qquad pq = -\tfrac{1}{2}.$$

Use your result to find the simultaneous values of θ and ϕ which satisfy the equations

$$\cos\theta + \cos\phi = \tfrac{1}{2}, \qquad \cos\theta\cos\phi = -\tfrac{1}{2},$$

and lie in the range 0 to π inclusive.　　N.

9. (i) Find all the solutions of the equation

$$\sin 60° + \sin(60° + x) + \sin(60° + 2x) = 0$$

which lie between 0° and 360°.

(ii) Prove that $(\sin 3\theta)/(1 + 2\cos 2\theta) = \sin\theta$ and hence show that $\sin 15° = (\sqrt{3} - 1)/2\sqrt{2}$.　　L.

10. Prove the formula

$$\tan 3\theta = \frac{3\tan\theta - \tan^3\theta}{1 - 3\tan^2\theta}.$$

Find the general solution of the equation $\tan 3\theta + \tan 2\theta = 0$ and show that $\tan^2 \tfrac{1}{5}\pi$, $\tan^2 \tfrac{2}{5}\pi$ are the roots of the quadratic equation

$$x^2 - 10x + 5 = 0.$$　　o.c.

11. (i) Express $\tan 4\theta$ in terms of $\tan\theta$, and hence solve the equation

$$t^4 + 4t^3 - 6t^2 - 4t + 1 = 0.$$

(ii) Prove by induction, or otherwise, that

$$\sum_{k=1}^{n} \frac{\sin k\theta}{\cos^k\theta} = \cot\theta - \frac{\cos(n+1)\theta}{\sin\theta\cos^n\theta}.$$　　o.c.

12. Find (in radian measure) all the values of θ which satisfy the equation $\cos 3\theta = \sin 2\theta$.

Prove that one of the values of $\sin\theta$ is $\sin(\pi/10)$, and find the remaining values of $\sin\theta$.

Prove that $\sin(\pi/10) = \tfrac{1}{4}(\sqrt{5} - 1)$, and find the corresponding expression for $\sin(3\pi/10)$.　　o.c.

13. (i) Find the solutions of the equation

$$\cos(30° + x) = 2\sin(40° + x)$$

which lie between 0° and 360°.

(ii) Prove the identity

$$\cos 4\theta\sin\theta + \sin 2\theta\cos\theta = \sin 3\theta\cos 2\theta.$$　　L.

14. (i) Prove the identity

$$\cos \theta + \cos 3\theta + \cos 5\theta + \cos 7\theta = 4 \cos \theta \cos 2\theta \cos 4\theta.$$

(ii) Find the values of x, between $0°$ and $360°$, such that

$$4 \cos x + 5 \sin x = 6.$$ L.

15. Prove that

$$\tan 2\theta = \frac{2 \tan \theta}{1 - \tan^2 \theta}, \qquad \tan 4\theta = \frac{4 \tan \theta (1 - \tan^2 \theta)}{1 - 6 \tan^2 \theta + \tan^4 \theta}.$$

Give rough sketches (in the same diagram) of the graphs of $4 \tan \theta$ and of $\tan 4\theta$ for the range $0 \leqslant \theta \leqslant \frac{1}{2}\pi$, and calculate the values of $\tan \theta$ at the points where they intersect.   O.C.

16. If $\cos \theta + \cos \psi = a$, $\sin \theta + \sin \psi = b$, prove that

$$\cos (\theta - \psi) = \frac{1}{2}(a^2 + b^2 - 2), \qquad \tan \frac{1}{2}(\theta + \psi) = b/a.$$

Hence solve the simultaneous equations

$$\cos \theta + \cos \psi = -1, \qquad \sin \theta + \sin \psi = 1\cdot5$$

for values of θ and ψ between $0°$ and $180°$.   L.

17. (a) Prove that

$$8 \cos \theta \cos 2\theta \cos 3\theta - 1 = (\sin 7\theta)/\sin \theta,$$

and find all the angles between $0°$ and $180°$ for which

$$\cos \theta \cos 2\theta \cos 3\theta = \frac{1}{4}.$$

(b) If α and β are two values of θ satisfying the equation

$$a \cos 2\theta + b \sin 2\theta = c$$

and such that $\tan \alpha$ and $\tan \beta$ are unequal, prove that

(i) $\tan \alpha + \tan \beta = \dfrac{2b}{a+c}$,

(ii) $\tan \alpha \tan \beta = \dfrac{c-a}{c+a}$,

(iii) $c \tan (\alpha - \beta) = \pm \sqrt{(a^2 + b^2 - c^2)}$.   N.

18. (i) Solve the equation $3 \cos \theta + 4 \sin \theta + 2 = 0$ completely, giving your answers to the nearest tenth of a degree.

(ii) Prove that, if α and β are two roots of the equation

$$a^2 \cos \theta + a \sin \theta + 1 = 0$$

and if α and β do not differ by a multiple of 2π, then

$$(\tan \tfrac{1}{2}\alpha + \tan \tfrac{1}{2}\beta)^2 + 1 - \tan^2 \tfrac{1}{2}\alpha \tan^2 \tfrac{1}{2}\beta = 0,$$

and

$$\cos (\alpha + \beta) = 1 + \cos \alpha + \cos \beta.$$ O.C.

19. Prove that, if $\tan \theta \tan (\theta + \alpha) = k$, then

$$(k+1) \cos (2\theta + \alpha) = (1-k) \cos \alpha.$$

Solve the equation

$$\tan \theta \tan (\theta + \pi/3) = 2.$$

Discuss the equation

$$\tan \theta \tan (\theta + \alpha) + 1 = 0. \qquad \text{c}$$

20. (i) Prove that $5 + 6 \cos A + 2 \cos 2A > 0$.
   (ii) Eliminate A from the equations

$$\cos A + \sin A = a, \qquad \cos 2A = ab, \qquad (a \neq 0). \qquad \text{L.}$$

21. Given that $\sin \alpha \neq 0$ and that

$$x \cos 2\alpha + y \cos 3\alpha = c \sin \beta,$$
$$x \sin 2\alpha + y \sin 3\alpha = c \cos \beta,$$

express x and y in terms of c, α, β.
   Prove that, when $x = y$, $5\alpha + 2\beta$ is an odd multiple of π.   O.C.

22. Express the function $6 \cos^2 \theta + 8 \sin \theta \cos \theta$ in terms of $\cos 2\theta$ and $\sin 2\theta$. Deduce an expression for the function in the form $A + 5 \cos (2\theta - \alpha)$, where A and α are constants.

Hence write down the greatest and least values of the function and find, correct to the nearest minute, one value of θ corresponding to each.   N.

23. Prove that, if $t = \tan \frac{1}{2}\theta$, then

$$\sin \theta = \frac{2t}{1 + t^2} \quad \text{and} \quad \cos \theta = \frac{1 - t^2}{1 + t^2}.$$

By expressing $\dfrac{3 + \cos \theta}{\sin \theta}$ in terms of t, or otherwise, show that this expression cannot have any value between $-2\sqrt{2}$ and $+2\sqrt{2}$.   C.

24. (i) Find all the values of θ which satisfy the equation

$$\cos (3\theta + \alpha) \cos (3\theta - \alpha) + \cos (5\theta + \alpha) \cos (5\theta - \alpha) = \cos 2\alpha.$$

   (ii) Prove the identity

$$\tan (A_1 + A_2 + A_3) = \frac{\tan A_1 + \tan A_2 + \tan A_3 - \tan A_1 \tan A_2 \tan A_3}{1 - \tan A_2 \tan A_3 - \tan A_3 \tan A_1 - \tan A_1 \tan A_2}.$$

Hence show that

$$\tan^{-1} \sqrt{\left(\frac{a(a+b+c)}{bc}\right)} + \tan^{-1} \sqrt{\left(\frac{b(a+b+c)}{ca}\right)} + \tan^{-1} \sqrt{\left(\frac{c(a+b+c)}{ab}\right)}$$
$$= n\pi,$$

where n is an integer, positive or negative, and the positive values of the square roots are taken.   O.C.

25. If A, B, C are the angles of a triangle and the products

$$\cos 2A \cos 2B \cos 2C, \qquad \sin 2A \sin 2B \sin 2C$$

have given values p, q, respectively, prove that $p - q \cot 2A = \cos^2 2A$ and deduce that $\tan 2A$, $\tan 2B$, $\tan 2C$ are the roots of the equation $(pt - q)(t^2 + 1) = t$. Show that, if $p = \frac{1}{2}$ and $q = 0$, then the angles of the triangle are in the ratio $1 : 3 : 4$. L.

26. (a) Prove that

$$\sin 3A = 3 \sin A - 4 \sin^3 A.$$

Deduce that

$$\sin^3 A + \sin^3 (120° + A) + \sin^3 (240° + A) = -\tfrac{3}{4} \sin 3A.$$

(b) The lines CA, CB are perpendicular and subtend angles θ, ϕ, respectively, at a point P which lies in the plane containing the lines and within the angle ACB. If CA $= a$ and CB $= b$, prove that

$$\tan \angle \text{ACP} = \frac{b \cot \phi - a}{a \cot \theta - b}. \qquad \text{N.}$$

27. Prove that, in any triangle ABC,

$$\frac{b + c}{a} = \frac{\cos \frac{1}{2}(B - C)}{\cos \frac{1}{2}(B + C)}.$$

If $b + c = \lambda a$, prove that

$$\cot \tfrac{1}{2}B \cot \tfrac{1}{2}C = \frac{\lambda + 1}{\lambda - 1}.$$

By expressing $\cot \frac{1}{2}A$ in terms of $\cot \frac{1}{2}B$, $\cos \frac{1}{2}C$, or otherwise, deduce that

$$\cot \tfrac{1}{2}B + \cot \tfrac{1}{2}C = \frac{2}{\lambda - 1} \cot \tfrac{1}{2}A. \qquad \text{L.}$$

28. Assuming that, in any triangle ABC,

$$\frac{\sin A}{a} = \frac{\sin B}{b} = \frac{\sin C}{c},$$

prove that

$$\frac{a + b - c}{a + b + c} = \tan \tfrac{1}{2}A \tan \tfrac{1}{2}B.$$

Calculate the value of c for the triangle in which

$$a + b = 18 \cdot 5 \text{ cm}, \qquad A = 72° \; 14', \qquad B = 45° \; 42'. \qquad \text{O.C.}$$

29. Prove that, in any triangle ABC,

(i) $a = r \dfrac{\cos \frac{1}{2}A}{\sin \frac{1}{2}B \sin \frac{1}{2}C}$,

(ii) $\triangle = r^2 \cot \frac{1}{2}A \cot \frac{1}{2}B \cot \frac{1}{2}C$, where r is the radius of the incircle and $\triangle$ is the area of the triangle.

In a triangle $r = 10$ cm, $A = 80°$, $B = 60°$, $C = 40°$. Calculate to three significant figures the side a and the area of the triangle.   O.C.

30. (i) With the usual notation for a triangle, prove that the radius of the inscribed circle is $\triangle/s$.

    (ii) An acute-angled triangle ABC is inscribed in a circle of radius R. The interior bisector of the angle A meets the circle again at E. Prove that the radius of the circle inscribed in triangle BEC is $R \sin A \tan \frac{1}{4}A$.   L.

31. The radii of the inscribed and circumscribed circles of a triangle ABC are r and R, respectively. Prove that $r = 4R \sin \frac{1}{2}A \sin \frac{1}{2}B \sin \frac{1}{2}C$.

The incentre of the triangle is at I. Prove that, if the angle A is 60°, then the radius of the circle circumscribed about the triangle IBC is R, and the radius of the circle inscribed in the triangle is

$$2R\sqrt{3} \sin \tfrac{1}{4}B \sin \tfrac{1}{4}C. \quad\quad\quad \text{O.C.}$$

32. In the acute-angled triangle ABC the perpendicular from A to BC cuts the incircle at P and Q; the centre of the incircle is the point I and the radius is r. Prove that

$$AI = r \operatorname{cosec} \tfrac{1}{2}A;$$

that the perpendicular distance of I from PQ is the positive value of

$$r \operatorname{cosec} \tfrac{1}{2}A \sin \tfrac{1}{2}(C-B),$$

and that the length of PQ is

$$2r \operatorname{cosec} \tfrac{1}{2}A \sqrt{(\cos B \cos C)}. \quad\quad\quad \text{O.C.}$$

33. (i) If $A+B+C=180°$, prove that

$$\sin B + \sin C - \sin A = 4 \cos \tfrac{1}{2}A \sin \tfrac{1}{2}B \sin \tfrac{1}{2}C.$$

    (ii) If $\cos(\theta+\alpha)=p$, $\sin(\theta+\beta)=q$, express $\cos \theta$ and $\sin \theta$ in terms of α, β, p, q and deduce that

$$p^2+q^2+2pq \sin(\alpha-\beta) = \cos^2(\alpha-\beta). \quad\quad\quad \text{L.}$$

34. (a) Prove the identities

$$\cot A - \tan A = 2 \cot 2A,$$

$$\cot A - \tan A - 2 \tan 2A = 4 \cot 4A.$$

    (b) If ABC is a triangle, prove that

$$1 + \cos 2A - \cos 2B - \cos 2C = 4 \sin B \sin C \cos A.$$

    (c) In the triangle ABC the mid-point of BC is D. Prove that

$$4 AD^2 = b^2 + c^2 + 2bc \cos A. \quad\quad\quad \text{N.}$$

35. If G is the centroid of the triangle ABC and Δ its area, prove that

$$9AG^2 = 2b^2 + 2c^2 - a^2$$

and

$$\tan \angle BGC = 12\Delta/(b^2 + c^2 - 5a^2). \qquad \text{L.}$$

36. (a) The lengths of the sides of a trapezium ABCD are given by AB$=8$, BC$=6$, CD$=3$, DA$=5$, the parallel sides being AB and DC. Calculate the angles A and B.

(b) In any quadrilateral ABCD of sides AB$=a$, BC$=b$, CD$=c$, DA$=d$, prove that the angles A and B are connected by the equation

$$2ad \cos A + 2ab \cos B - 2bd \cos (A+B) = a^2 + b^2 - c^2 + d^2. \qquad \text{N.}$$

37. If $\alpha + \beta + \gamma = 90°$, prove that

$$1 - \sin^2 \alpha - \sin^2 \beta - \sin^2 \gamma - 2 \sin \alpha \sin \beta \sin \gamma = 0.$$

A convex quadrilateral ABCD is inscribed in a circle of which DA is a diameter. If $a=$AB, $b=$BC, $c=$CD, $d=$DA, prove that

$$d^3 - (a^2 + b^2 + c^2)d - 2abc = 0. \qquad \text{N.}$$

CHAPTER 13

FURTHER INTEGRATION

Integration by parts

13.1. We have learnt the importance of recognizing such integrals as

$$\int x\, e^{x^2}\, dx = \tfrac{1}{2}e^{x^2}+c, \quad \text{and} \quad \int 2x \cos (x^2+2)\, dx = \sin (x^2+2)+c.$$

When, however, the integrand is the product of two functions of x but is not susceptible to this treatment, e.g. $\int x\, e^x\, dx$, $\int x \cos x\, dx$, we may often successfully apply a technique known as *integration by parts*; this is based upon the idea of differentiating the product of two functions of x.

If u and v are two functions of x,

$$\frac{d}{dx}(uv) = v\frac{du}{dx}+u\frac{dv}{dx}.$$

Integrating each side with respect to x,

$$uv = \int v\frac{du}{dx}\, dx + \int u\frac{dv}{dx}\, dx.$$

$$\therefore \int u\frac{dv}{dx}\, dx = uv - \int v\frac{du}{dx}\, dx.$$

Example 1. *Find* $\int x \cos x\, dx$.

$$\int u\frac{dv}{dx}\, dx = uv - \int v\frac{du}{dx}\, dx.$$

Let $\quad u = x.$

Let $\quad \dfrac{dv}{dx} = \cos x,$

$$\int x \cos x\, dx = x \sin x - \int \sin x\,.\,1\, dx,$$

$$\therefore\ v = \sin x.$$

$$= x \sin x + \cos x + c.$$

269

This method can of course only be attempted if the factor chosen as $\frac{dv}{dx}$ can be integrated; Example 1 illustrates the fact that its successful application usually depends upon the *correct choice of u*, since it is this which determines whether $\int v \frac{du}{dx} dx$ is easier to tackle than the original integral.

Qu. 1. Check the answer to Example 1 by differentiation.

Qu. 2. Attempt Example 1 taking $\cos x$ as u.

Qu. 3. Find the following integrals:

(i) $\int x \sin x \, dx$,   (ii) $\int x \cos 2x \, dx$,

(iii) $\int x \log_e x \, dx$,   (iv) $\int x \, e^x \, dx$.

Qu. 4. Find $\frac{d}{dx} (e^{x^2})$, and deduce $\int x^3 \, e^{x^2} \, dx$.

The integral $\int \tan^{-1} x \, dx$ does not at first sight appear to be susceptible to the method under discussion. However, this is one of a small group of integrals which may be found by taking $\frac{dv}{dx}$ as 1.

Example 2.   *Find* $\int \tan^{-1} x \, dx$.

$$\int u \frac{dv}{dx} dx = uv - \int v \frac{du}{dx} dx.$$

$$\int \tan^{-1} x . 1 \, dx = \tan^{-1} x . x - \int x . \frac{1}{1+x^2} dx,$$

$$= x \tan^{-1} x - \tfrac{1}{2} \log_e (1+x^2) + c.$$

Let $u = \tan^{-1} x$.

Let $\frac{dv}{dx} = 1$,

$\therefore \ v = x$.

Qu. 5. Find $\int \log_e x \, dx$.

Qu. 6. (i) Find $\frac{d}{dx} (\sin^{-1} x)$,   (ii) find $\int \sin^{-1} x \, dx$.

To some integrals it is necessary to apply the method of integration by parts more than once, as is illustrated in the next example.

Example 3. *Find* $\int x^2 \sin x \, dx$.

$$\int x^2 \sin x \, dx = x^2(-\cos x) - \int -\cos x \cdot 2x \, dx, \quad \text{Let } u = x^2.$$

$$= -x^2 \cos x + \int 2x \cos x \, dx. \qquad \text{Let } \frac{dv}{dx} = \sin x,$$

$$\therefore \quad v = -\cos x.$$

$$\int 2x \cos x \, dx = 2x \sin x - \int \sin x \cdot 2 \, dx, \qquad \text{Let } u = 2x.$$

$$= 2x \sin x + 2 \cos x + c. \qquad \text{Let } \frac{dv}{dx} = \cos x,$$

$$\therefore \int x^2 \sin x \, dx = -x^2 \cos x + 2x \sin x + 2 \cos x + c, \qquad v = \sin x.$$

$$= 2x \sin x + (2 - x^2) \cos x + c.$$

Qu. 7. Check the answer to Example 3 by differentiation.

Qu. 8. Find the following integrals:

(i) $\int x^2 \cos x \, dx$,   (ii) $\int x^2 \, e^x \, dx$.

Exercise 13a

1. Find the following integrals, and check by differentiation:

   (i) $\int 2x \sin x \, dx$,   (ii) $\int \frac{1}{2}x \, e^x \, dx$,   (iii) $\int x \sin 2x \, dx$,

  (iv) $\int x^2 \log_e x \, dx$,   (v) $\int x \cos (x+2) \, dx$,

  (vi) $\int x(1+x)^7 \, dx$,   (vii) $\int x \, e^{2x} \, dx$,   (viii) $\int x \, e^{x^2} \, dx$,

  (ix) $\int \frac{\log_e x}{x^2} \, dx$,   (x) $\int x \sec^2 x \, dx$,

  (xi) $\int x^n \log_e x \, dx$,   (xii) $\int x \, 3^x \, dx$.

2. Find the following integrals, and check by differentiation:

   (i) $\int \log_e 2x \, dx$,   (ii) $\int \sin^{-1} 3x \, dx$,   (iii) $\int \log_e y^2 \, dy$,

  (iv) $\int \tan^{-1} \frac{\theta}{2} \, d\theta$,   (v) $\int \cos^{-1} t \, dt$,   (vi) $\int e^{\sqrt{x}} \, dx$.

3. Find the following integrals (see Qu. 4 on p. 270):

   (i) $\int x^5 \, e^{x^3} \, dx$,   (ii) $\int x \, e^{-x^2} \, dx$,   (iii) $\int x^3 \, e^{-x^2} \, dx$,

  (iv) $\int x^3 \cos x^2 \, dx$,   (v) $\int x^3 \sec^2 (x^2) \, dx$.

4. Find the following integrals:

    (i) $\int x^2 \cos 3x \, dx$,

    (ii) $\int x^3 e^x \, dx$,

    (iii) $\int x^2 \sin x \cos x \, dx$,

    (iv) $\int x^2 e^{-x} \, dx$,

    (v) $\int (x \cos x)^2 \, dx$,

    (vi) $\int x (\log_e x)^2 \, dx$.

5. Find the following integrals:

    (i) $\int x \sin x \cos x \, dx$,   (ii) $\int \dfrac{x}{e^x} \, dx$,   (iii) $\int x(1+2x)^5 \, dx$,

    (iv) $\int \dfrac{\log_e y}{y} \, dy$,   (v) $\int u \tan^{-1} u \, du$, (vi) $\int x e^{-x^2} \, dx$,

    (vii) $\int x^3 e^{-x} \, dx$,   (viii) $\int x(1-x^2)^6 \, dx$, (ix) $\int t \sin^2 t \, dt$,

    (x) $\int v e^{3v} \, dv$,   (xi) $\int \cot^{-1} y \, dy$, (xii) $\int \theta \operatorname{cosec}^2 \theta \, d\theta$,

   (xiii) $\int \dfrac{\log_e x}{x^3} \, dx$.

6. (i) Find $\int x \tan^2 x \, dx$.

    (ii) Show that $\int x \sin^{-1} x \, dx = \tfrac{1}{4}(2x^2-1) \sin^{-1} x + \tfrac{1}{4} x \sqrt{(1-x^2)} + c$.

7. Evaluate:

    (i) $\int_0^{\frac{\pi}{2}} x \cos x \, dx$,   (ii) $\int_0^1 x^2 e^x \, dx$,   (iii) $\int_1^{e^2} \log_e x \, dx$,

    (iv) $\int_0^1 \sin^{-1} y \, dy$,   (v) $\int_0^\pi t \sin^2 t \, dt$,   (vi) $\int_1^{10} x \log_{10} x \, dx$.

Involving inverse trigonometrical functions

13.2. In § 11.3 we dealt with the differentiation of inverse trigonometrical functions. The frequency with which we meet inverse sine
and inverse tangent functions in integration is just one good reason why
we should be adept at differentiating these functions on sight.

If $y = \sin^{-1} u$, where u is a function of x,

$$\frac{dy}{dx} = \frac{dy}{du} \cdot \frac{du}{dx} = \frac{1}{\sqrt{(1-u^2)}} \cdot \frac{du}{dx}.$$

Thus $\dfrac{d}{dx}\left\{3 \sin^{-1} \dfrac{x}{2}\right\} = 3 \dfrac{1}{\sqrt{\left(1 - \dfrac{x^2}{4}\right)}} \cdot \dfrac{1}{2} = \dfrac{3}{\sqrt{(4-x^2)}}$,

and $\dfrac{d}{dx}\{2 \tan^{-1} 5x\} = 2 \cdot \dfrac{1}{1+25x^2} \cdot 5 = \dfrac{10}{1+25x^2}$.

Qu. 9. Write down, and simplify where necessary, the derivatives of the following functions:

(i) $\sin^{-1} 3x$,   (ii) $\tan^{-1} 2x$,   (iii) $\sin^{-1} \frac{x}{3}$,

(iv) $\cos^{-1} 2x$,   (v) $\frac{1}{2} \tan^{-1} 3x$,   (vi) $3 \tan^{-1} \frac{x}{2}$,

(vii) $\frac{1}{2} \sin^{-1} (x-1)$,   (viii) $2 \tan^{-1} \left(\frac{x+1}{2}\right)$.

The reader should now be able to write down certain integrals, hitherto obtained by the change of variable $x = k \sin u$ or $x = k \tan u$.

For example,
$$\int \frac{2}{3+4x^2}\, dx,$$

written as
$$\int \frac{\frac{2}{3}}{1+\frac{4}{3}x^2}\, dx$$

is seen to be of the form $k \tan^{-1} \frac{2x}{\sqrt{3}} + c$.

Now
$$\frac{d}{dx}\left(k \tan^{-1} \frac{2x}{\sqrt{3}}\right) = \frac{1}{1+\frac{4}{3}x^2}\cdot\frac{2k}{\sqrt{3}}.$$

Comparing this with the integrand we find $k = \frac{1}{\sqrt{3}}$,

$$\therefore \int \frac{2}{3+4x^2}\, dx = \frac{1}{\sqrt{3}} \tan^{-1} \left(\frac{2x}{\sqrt{3}}\right) + c.$$

Qu. 10. Find the following integrals:

(i) $\int \frac{2}{4+x^2}\, dx$,   (ii) $\int \frac{3}{1+4x^2}\, dx$,   (iii) $\int \frac{4}{\sqrt{(9-x^2)}}\, dx$,

(iv) $\int \frac{1}{\sqrt{(1-9x^2)}}\, dx$,   (v) $\int \frac{1}{2+25x^2}\, dx$,   (vi) $\int \frac{2}{\sqrt{(3-4x^2)}}\, dx$,

(vii) $\int \frac{1}{3-2x+x^2}\, dx$,   (viii) $\int \frac{5}{\sqrt{\{9-(x+2)^2\}}}\, dx$.

The change of variable $t = \tan \frac{x}{2}$

13.3. Of the trigonometrical ratios, two have not yet been integrated in this book, $\sec x$ and $\csc x$.

Now
$$\operatorname{cosec} x = \frac{1}{2 \sin \frac{x}{2} \cos \frac{x}{2}} = \frac{\sec^2 \frac{x}{2}}{2 \tan \frac{x}{2}}$$

$\left(\text{dividing numerator and denominator by } \cos^2 \frac{x}{2}\right)$.

Thus
$$\int \operatorname{cosec} x \, dx = \int \frac{\frac{1}{2} \sec^2 \frac{x}{2}}{\tan \frac{x}{2}} \, dx,$$

$$\therefore \int \operatorname{cosec} x \, dx = \log_e \tan \frac{x}{2} + c. \dagger$$

Furthermore
$$\sin \left(x + \frac{\pi}{2}\right) = \sin x \cos \frac{\pi}{2} + \cos x \sin \frac{\pi}{2} = \cos x,$$

$$\therefore \operatorname{cosec} \left(x + \frac{\pi}{2}\right) = \sec x.$$

Thus
$$\int \sec x \, dx = \int \operatorname{cosec} \left(x + \frac{\pi}{2}\right) \, dx,$$

$$= \log_e \tan \left(\frac{x}{2} + \frac{\pi}{4}\right) + c.$$

The above working suggests a change of variable which is of considerable importance. If we write $\tan \frac{x}{2}$ as t, since

$$\tan 2A = \frac{2 \tan A}{1 - \tan^2 A}, \qquad \tan x = \frac{2t}{1 - t^2}.$$

It is also possible (see § 12.3) to express $\sin x$ and $\cos x$ in terms of t.

$$\sin x = \frac{2t}{1 + t^2} \quad \text{and} \quad \cos x = \frac{1 - t^2}{1 + t^2}.$$

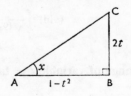

Fig. 13.1

† See footnote to p. 275.

Fig. 13.1 provides a useful mnemonic for these identities. Starting with the fact that $\tan x = \dfrac{2t}{1-t^2}$, one can at once deduce that $AC = 1 + t^2$.

When we make the change of variable $t = \tan \dfrac{x}{2}$,

$$\frac{\mathrm{d}t}{\mathrm{d}x} = \frac{1}{2} \sec^2 \frac{x}{2},$$

$$\frac{\mathrm{d}x}{\mathrm{d}t} = \frac{2}{\sec^2 \dfrac{x}{2}} = \frac{2}{1 + \tan^2 \dfrac{x}{2}},$$

$$\therefore \quad \frac{\mathrm{d}x}{\mathrm{d}t} = \frac{2}{1+t^2}.$$

Qu. 11. Find $\int \operatorname{cosec} x \, \mathrm{d}x$ using the change of variable $t = \tan \dfrac{x}{2}$.

Qu. 12. Find $\int \dfrac{\sin \theta}{1 + \cos \theta} \, \mathrm{d}\theta$ (i) by expressing the integrand in terms of ratios of $\dfrac{\theta}{2}$, (ii) by the change of variable $t = \tan \dfrac{\theta}{2}$.

Qu. 13. Use the change of variable $t = \tan \dfrac{x}{2}$ to show that

$$\int \sec x \, \mathrm{d}x = \log_e \frac{1+t}{1-t} + c.$$

Compare this form of the integrand with that obtained earlier and deduce that

$$\int \sec x \, \mathrm{d}x = \log_e (\sec x + \tan x) + c. \dagger$$

This change of variable is best thought of in more general terms as '$t = \tan$ (half angle)'. For example, when applied to $\int \operatorname{cosec} 4x \, \mathrm{d}x$ it is $t = \tan 2x$; then $1 = 2 \sec^2 2x \dfrac{\mathrm{d}x}{\mathrm{d}t}$, giving $\dfrac{\mathrm{d}x}{\mathrm{d}t} = \dfrac{1}{2(1+t^2)}$. Care must be taken to establish the correct numerical factor in the expression for $\dfrac{\mathrm{d}x}{\mathrm{d}t}$.

† If $\tan \dfrac{x}{2}$ is negative $\int \operatorname{cosec} x \, \mathrm{d}x = \log_e \left(-\tan \dfrac{x}{2} \right) + c$. If $\sec x + \tan x$ is negative $\int \sec x \, \mathrm{d}x = \log_e (-\sec x - \tan x) + c$ (see p. 40).

Example 4. *Find* $\int \dfrac{1}{3+5\cos\frac{1}{2}x}\,dx.$

$$\int \frac{1}{3+5\cos\frac{1}{2}x}\frac{dx}{dt}\,dt = \int \frac{1}{3+5.\dfrac{1-t^2}{1+t^2}}.\frac{4}{1+t^2}\,dt,$$

$$= \int \frac{4}{3(1+t^2)+5(1-t^2)}\,dt,$$

$$= \int \frac{2}{4-t^2}\,dt,$$

$$= \int \left\{\frac{1}{2(2+t)}+\frac{1}{2(2-t)}\right\}\,dt,$$

$$= \tfrac{1}{2}\log_e(2+t)-\tfrac{1}{2}\log_e(2-t)+c,$$

$$= \log_e \frac{k\sqrt{(2+\tan\frac{1}{4}x)}}{\sqrt{(2-\tan\frac{1}{4}x)}}.$$

Let $t = \tan\dfrac{x}{4}$.

$$1 = \tfrac{1}{4}\sec^2\frac{x}{4}.\frac{dx}{dt}$$

$$\frac{dx}{dt} = \frac{4}{1+t^2}.$$

Qu. 14. Find $\dfrac{dx}{dt}$ in terms of t if:

(i) $t = \tan x,$　　　　(ii) $t = \tan 4x,$　　　　(iii) $t = \tan\frac{3}{2}x.$

Qu. 15. Find: (i) $\int \operatorname{cosec} 2x\,dx,$　　(ii) $\int \dfrac{1}{1+\sin 3\theta}\,d\theta,$

(iii) $\int \dfrac{1}{\sqrt{(x^2-1)}}\,dx$　(use $x = \sec u$).

The change of variable $t = \tan x$

13.4. An integrand containing $\sin x$ and $\cos x$, particularly even powers of these, may often be expressed as a function of $\tan x$ and $\sec x$. In such a case the change of variable $t=\tan x$ is worth trying.

Example 5. *Find* $\int \dfrac{1}{1+\sin^2 x}\,dx.$

[In this case we divide the numerator and denominator by $\cos^2 x$.]

$$\int \frac{1}{1+\sin^2 x}\,dx = \int \frac{\sec^2 x}{\sec^2 x+\tan^2 x}\,dx,$$

$$= \int \frac{\sec^2 x}{1+2\tan^2 x}\frac{dx}{dt}\,dt,$$

$$= \int \frac{1+t^2}{1+2t^2}.\frac{1}{1+t^2}\,dt,$$

Let $t = \tan x.$

$$\frac{dx}{dt} = \frac{1}{1+t^2}.$$

$$= \int \frac{1}{1+2t^2}\, dt,$$

$$= \frac{1}{\sqrt{2}} \tan^{-1}(\sqrt{2}\, t) + c,$$

$$= \frac{1}{\sqrt{2}} \tan^{-1}(\sqrt{2} \tan x) + c.$$

Qu. 16.   Find (i) $\int \dfrac{1}{1+\cos^2 x}\, dx$,   (ii) $\int \dfrac{2 \tan x}{\cos 2x}\, dx$.

Splitting the numerator

13.5.   When a fractional integrand with a quadratic denominator cannot be written in simple partial fractions, it may often be usefully expressed as two fractions by splitting the numerator.   To take a simple example, such as the reader has already met in Exercises 1d and 1f,

$$\int \frac{1+x}{1+x^2}\, dx = \int \left(\frac{1}{1+x^2} + \frac{x}{1+x^2} \right) dx,$$
$$= \tan^{-1} x + \log_e \sqrt{(1+x^2)} + c.$$

The key to a more general application of this method is to express the numerator in two parts, one of which is *a multiple of the derivative of the denominator*.

Example 6.   *Find* $\int \dfrac{5x+7}{x^2+4x+8}\, dx$.

Since
$$\frac{d}{dx}(x^2+4x+8) = 2x+4,$$

let $5x+7 \equiv A(2x+4) + B$;   whence $A = \frac{5}{2}$, $B = -3$.

$$\therefore \int \frac{5x+7}{x^2+4x+8}\, dx = \int \left\{ \frac{\frac{5}{2}(2x+4)}{x^2+4x+8} - \frac{3}{x^2+4x+8} \right\} dx,$$

$$= \tfrac{5}{2} \log_e (x^2+4x+8) - 3 \int \frac{1}{(x+2)^2+4}\, dx,$$

$$= \tfrac{5}{2} \log_e (x^2+4x+8) - \tfrac{3}{2} \tan^{-1} \left(\frac{x+2}{2} \right) + c.$$

This method is also appropriate for integrands of the form

$$\frac{a \cos x + b \sin x}{\alpha \cos x + \beta \sin x},$$

since the numerator may be expressed in the form

$$A\ (derivative\ of\ denominator) + B\ (denominator).$$

Example 7. *Find* $\int \dfrac{2 \cos x + 3 \sin x}{\cos x + \sin x}\, dx$.

Let $2 \cos x + 3 \sin x \equiv A(-\sin x + \cos x) + B(\cos x + \sin x)$;

whence $A = -\frac{1}{2}, \qquad B = \frac{5}{2}.$

$\therefore \int \dfrac{2 \cos x + 3 \sin x}{\cos x + \sin x}\, dx = \int \left\{ \dfrac{-\frac{1}{2}(-\sin x + \cos x)}{\cos x + \sin x} + \dfrac{\frac{5}{2}(\cos x + \sin x)}{\cos x + \sin x} \right\} dx,$

$\qquad\qquad = -\frac{1}{2} \log_e (\cos x + \sin x) + \frac{5}{2} x + c.$

Qu. 17. Find: (i) $\int \dfrac{2x+3}{x^2 + 2x + 10}\, dx,$ (ii) $\int \dfrac{1 - 2x}{\sqrt{\{9 - (x+2)^2\}}}\, dx,$

(iii) $\int \dfrac{\sin x}{\cos x + \sin x}\, dx,$ (iv) $\int \dfrac{2 \cos x + 9 \sin x}{3 \cos x + \sin x}\, dx.$

'Integrals and asymptotes'

13.6. There are two types of integrals to be discussed under this heading, and we shall consider them in terms of the area under a curve.

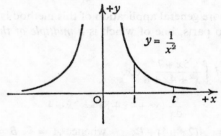

Fig. 13.2

Figure 13.2 shows part of the curve $y = \dfrac{1}{x^2}$, to which the x-axis is an asymptote.   The area under this curve from $x = 1$ to $x = t$ $(t > 1)$ is

$$\int_1^t \frac{1}{x^2}\, dx = \left[-\frac{1}{x} \right]_1^t = 1 - \frac{1}{t};$$

as $t \to \infty$, this area $\to 1$.   Thus although the area 'enclosed' by $y = \dfrac{1}{x^2}$, $x = 1$ and the x-axis is not in fact a finite enclosed area, we see that it can be evaluated as the limiting value of the area $\int_1^t \dfrac{1}{x^2}\, dx$ as $t \to \infty$.

For brevity it is permissible to write

$$\int_1^\infty \frac{1}{x^2}\, dx = \left[-\frac{1}{x}\right]_1^\infty = 1.$$

We are faced with a similar situation when we consider 'the area under the curve $y = \dfrac{1}{\sqrt{(1-x^2)}}$ from $x = 0$ to $x = 1$' (Fig. 13.3), since $x = 1$ is an asymptote to the curve.

The area under this curve from $x = 0$ to $x = t$ $(0 < t < 1)$ is

$$\int_0^t \frac{1}{\sqrt{(1-x^2)}}\, dx = \left[\sin^{-1} x\right]_0^t = \sin^{-1} t;$$

as $t \to 1$, this area $\to \dfrac{\pi}{2}$.

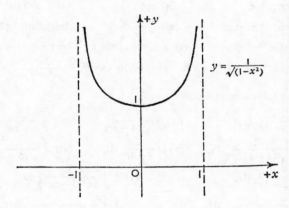

Fig. 13.3

Thus, although the integrand $\dfrac{1}{\sqrt{(1-x^2)}}$ is meaningless when $x = 1$, the limiting process is implied when we write

$$\int_0^1 \frac{1}{\sqrt{(1-x^2)}}\, dx = \left[\sin^{-1} x\right]_0^1 = \frac{\pi}{2}.$$

Qu. 18.   Evaluate:

(i) $\displaystyle\int_1^\infty \frac{1}{x^2}\, dx$, using the change of variable $x = \dfrac{1}{u}$,

(ii) $\displaystyle\int_0^1 \frac{1}{\sqrt{(1-x^2)}}\, dx$, using the change of variable $x = \sin u$.

Qu. 19. Evaluate the following integrals where possible, otherwise show that they are meaningless. Illustrate with a sketch.

(i) $\int_0^1 \frac{1}{x^2}\,dx$,   (ii) $\int_0^\infty \frac{1}{1+x^2}\,dx$,   (iii) $\int_1^\infty \frac{1}{x}\,dx$,

(iv) $\int_0^3 \frac{1}{(x-1)^2}\,dx$,   (v) $\int_0^\infty e^{-x}\,dx$,   (vi) $\int_1^2 \frac{1}{\sqrt{(4-x^2)}}\,dx$.

Exercise 13b

1. Differentiate the following with respect to x:

    (i) $\sin^{-1} 2x$,   (ii) $\tan^{-1}(3x+1)$,   (iii) $\frac{1}{3}\cos^{-1} 2x$,

    (iv) $2\sin^{-1}\left(\frac{x-1}{3}\right)$,   (v) $\frac{1}{2}\tan^{-1}\frac{x}{2}$,   (vi) $\frac{2}{3}\sin^{-1}\frac{3x}{2}$,

    (vii) $\cot^{-1} x$,   (viii) $\sec^{-1} x$,   (ix) $x^2\tan^{-1} x^2$,

    (x) $\cot^{-1} x + \tan^{-1} x$,   (xi) $\frac{\sin^{-1} x}{x}$,   (xii) $\tan^{-1}(2\tan x)$,

    (xiii) $(\sin^{-1} 3x)^2$,   (xiv) $\sin^{-1}(\frac{1}{2}\sin x)$,   (xv) $\cos^{-1} x + \sin^{-1} x$,

    (xvi) $\sin^{-1} x + x\sqrt{(1-x^2)}$,   (xvii) $\tan^{-1}\frac{x}{2} + \frac{2x}{x^2+4}$.

2. Find the following integrals:

    (i) $\int \frac{1}{9+x^2}\,dx$,   (ii) $\int \frac{3}{\sqrt{(4-y^2)}}\,dy$,   (iii) $\int \frac{2}{1+9u^2}\,du$,

    (iv) $\int \frac{2}{\sqrt{(1-16x^2)}}\,dx$,   (v) $\int \frac{2}{3+4t^2}\,dt$,   (vi) $\int \frac{1}{\sqrt{(5-4x^2)}}\,dx$,

    (vii) $\int \frac{1}{2+3y^2}\,dy$,   (viii) $\int \frac{1}{3\sqrt{(3-6x^2)}}\,dx$,

    (ix) $\int \frac{1}{2y^2-8y+17}\,dy$,   (x) $\int \frac{2}{\sqrt{(1+6x-3x^2)}}\,dx$.

3. Find the following integrals:

    (i) $\int \csc\frac{x}{2}\,dx$,   (ii) $\int \sec 2\theta\,d\theta$,   (iii) $\int \csc 3x\,dx$,

    (iv) $\int \sec 4\phi\,d\phi$,   (v) $\int \sec x \csc x\,dx$,

    (vi) $\int \frac{1}{1+\cos y}\,dy$,   (vii) $\int \frac{1}{1+\sin 2x}\,dx$,   (viii) $\int \frac{\sin\theta}{1-\cos\theta}\,d\theta$,

    (ix) $\int \frac{1}{4+5\cos x}\,dx$,   (x) $\int \frac{1}{5+3\cos\frac{\theta}{2}}\,d\theta$,

    (xi) $\int \frac{1}{\sqrt{(x^2-9)}}\,dx$ (use $x = 3\sec u$).

4. Use the change of variable $\tan x = t$ to find the following integrals:

(i) $\int \dfrac{1}{1 + 2 \sin^2 x}\, dx$,

(ii) $\int \dfrac{1}{\cos 2x - 3 \sin^2 x}\, dx$,

(iii) $\int \dfrac{\sin^2 x}{1 + \cos^2 x}\, dx$,

(iv) $\int \dfrac{1}{1 - 10 \sin^2 x}\, dx$.

5. Find the following integrals:

(i) $\int \dfrac{x+5}{x^2 + 3}\, dx$,

(ii) $\int \dfrac{y+4}{y^2 + 6y + 9}\, dy$,

(iii) $\int \dfrac{3u + 8}{u^2 + 2u + 5}\, du$,

(iv) $\int \dfrac{3 - 7x}{\sqrt{(4x - x^2)}}\, dx$,

(v) $\int \dfrac{\cos \theta}{\cos \theta + \sin \theta}\, d\theta$,

(vi) $\int \dfrac{3 \cos x - 2 \sin x}{\cos x + \sin x}\, dx$,

(vii) $\int \dfrac{\cos x + 8 \sin x}{2 \cos x + 3 \sin x}\, dx$.

6. Evaluate:

(i) $\displaystyle\int_3^\infty \dfrac{1}{(x-2)^2}\, dx$, using the change of variable $x - 2 = \dfrac{1}{u}$,

(ii) $\displaystyle\int_0^{2/3} \dfrac{1}{\sqrt{(4 - 9x^2)}}\, dx$, using the change of variable $x = \tfrac{2}{3} \sin u$.

7. Evaluate the following integrals where possible, otherwise show that they are meaningless. Illustrate with a sketch.

(i) $\displaystyle\int_1^2 \dfrac{1}{x-1}\, dx$,

(ii) $\displaystyle\int_2^3 \dfrac{1}{\sqrt{(x-2)}}\, dx$,

(iii) $\displaystyle\int_0^3 \dfrac{1}{(x-3)^2}\, dx$,

(iv) $\displaystyle\int_1^4 \dfrac{1}{(x-2)^2}\, dx$,

(v) $\displaystyle\int_3^\infty \dfrac{1}{(x-1)^2}\, dx$,

(vi) $\displaystyle\int_{-\infty}^0 e^x\, dx$,

(vii) $\displaystyle\int_0^{1/2} \log_e x\, dx$,

(viii) $\displaystyle\int_{-\infty}^0 x\, e^x\, dx$,

(ix) $\displaystyle\int_1^{3/2} \dfrac{1}{\sqrt{(9 - 4x^2)}}\, dx$,

(x) $\displaystyle\int_0^{\pi/2} \sec x\, dx$,

(xi) $\displaystyle\int_{\pi/4}^{\pi/2} \tan x\, dx$,

(xii) $\displaystyle\int_0^\infty \dfrac{1}{4 + 25x^2}\, dx$.

8. The area enclosed by the x-axis, $x = 1$, $x = t$, and the curve $y = \dfrac{1}{x}$ is rotated through 2π radians about the x-axis. What may be said about the volume of the solid so generated (i) as $t \to \infty$, (ii) as $t \to 0$?

*9. Find the area of the ellipse given by the parametric equations

$$x = 5 \cos \theta, \qquad y = 3 \sin \theta.$$

(Use the fact that $\int y\, dx = \int y \dfrac{dx}{d\theta}\, d\theta$.)

10. Find the area of the segment cut off by $x = 8$ from the parabola given by the parametric equations $x = 2t^2$, $y = 4t$.

11. If

$$S \equiv \int_0^{\frac{\pi}{2}} \frac{\sin \theta}{\cos \theta + \sin \theta}\, \mathrm{d}\theta, \quad \text{and} \quad C \equiv \int_0^{\frac{\pi}{2}} \frac{\cos \theta}{\cos \theta + \sin \theta}\, \mathrm{d}\theta,$$

prove that $S = C = \dfrac{\pi}{4}$.

Further integration by parts

13.7. The purpose of this section is to consolidate the method of integration by parts, and to introduce an interesting development in its application to certain integrals in which the original integral appears again. This gives us the opportunity to consider two integrals of great importance in physics,

$$\int e^{az} \cos bx \, \mathrm{d}x \quad \text{and} \quad \int e^{az} \sin bx \, \mathrm{d}x.$$

Example 8. *Find* $\displaystyle\int e^{az} \cos bx \, \mathrm{d}x$.

$$\int u \frac{\mathrm{d}v}{\mathrm{d}x}\, \mathrm{d}x = uv - \int v \frac{\mathrm{d}u}{\mathrm{d}x}\, \mathrm{d}x. \qquad \text{Let} \quad u = \cos bx.$$

Let $\dfrac{\mathrm{d}v}{\mathrm{d}x} = e^{az},$

$$I = \int e^{az} \cos bx \, \mathrm{d}x, \qquad\qquad\qquad \therefore \; v = \frac{1}{a}\, e^{az}.$$

$$= \frac{1}{a}\, e^{az} \cos bx - \int \frac{1}{a}\, e^{az} (-b \sin bx)\, \mathrm{d}x,$$

$$= \frac{1}{a}\, e^{az} \cos bx + \frac{b}{a} \int e^{az} \sin bx \, \mathrm{d}x. \qquad\qquad (1)$$

But

$$\int e^{az} \sin bx \, \mathrm{d}x = \frac{1}{a}\, e^{az} \sin bx - \int \frac{1}{a}\, e^{az}\, b \cos bx \, \mathrm{d}x, \qquad \text{Let} \quad u = \sin bx.$$

Let $\dfrac{\mathrm{d}v}{\mathrm{d}x} = e^{az},$

$$= \frac{1}{a}\, e^{az} \sin bx - \frac{b}{a}\, I. \qquad\qquad\qquad\qquad \therefore \; v = \frac{1}{a}\, e^{az}.$$

Substituting in (1),

$$I = \frac{1}{a}\, e^{az} \cos bx + \frac{b}{a^2}\, e^{az} \sin bx - \frac{b^2}{a^2}\, I,$$

$$\therefore \; a^2 I = a\, e^{az} \cos bx + b\, e^{az} \sin bx - b^2 I,$$

$$\therefore \; I(a^2 + b^2) = e^{az}(a \cos bx + b \sin bx) + k,$$

$$\therefore \; I = \int e^{az} \cos bx \, \mathrm{d}x = \frac{e^{az}}{a^2 + b^2}(a \cos bx + b \sin bx) + c.$$

Qu. 20. Find $\int e^{2x} \sin 3x \, dx$.

Qu. 21. Find $\int e^x \cos 2x \, dx$,

(i) taking e^x as u throughout,

(ii) taking $\cos 2x$ as u in the first step, and $\sin 2x$ as u in the second.

(iii) Can we usefully take $\cos 2x$ as u in the first step, and e^x as u in the second?

Exercise 13c

1. Use the method of Example 8 to find the following integrals:

(i) $\int e^{3x} \cos 2x \, dx$,

(ii) $\int e^{4x} \sin 3x \, dx$,

(iii) $\int e^{-t} \cos \dfrac{t}{2} \, dt$,

(iv) $\int e^x \sin (2x+1) \, dx$,

(v) $\int e^{2\theta} \cos^2 \theta \, d\theta$,

(vi) $\int e^{x/2} \sin x \cos x \, dx$.

2. Find $\int \sec^3 x \, dx$ by first proving it equal to $\frac{1}{2} \sec x \tan x + \frac{1}{2} \int \sec x \, dx$.

3. Find the following integrals:

(i) $\int x^3 \log_e x \, dx$,

(ii) $\int \tan^{-1} 2y \, dy$,

(iii) $\int \dfrac{x}{e^{x^2}} \, dx$,

(iv) $\int x \sin 3x \, dx$,

(v) $\int x^2 \sin 2x \, dx$,

(vi) $\int e^{3x} \sin 2x \, dx$.

(vii) $\int \frac{1}{2} u^3 \, e^{u^2} \, du$,

(viii) $\int x(2x-1)^5 \, dx$,

(ix) $\int x \log_e \sqrt{(x-1)} \, dx$,

(x) $\int \log_e (3x) \, dx$,

(xi) $\int x^2 \, e^{2x} \, dx$,

(xii) $\int e^{-y} \cos \dfrac{y}{2} \, dy$,

(xiii) $\int x^{-3} \log_e x \, dx$,

(xiv) $\int \sin^{-1} \dfrac{t}{3} \, dt$,

(xv) $\int \log_e x^3 \, dx$,

(xvi) $\int y^2 \cos^2 y \, dy$,

(xvii) $\int x \cos x^2 \, dx$,

(xviii) $\int x \log_e (x^2) \, dx$,

(xix) $\int \theta^3 \sin (\theta^2) \, d\theta$,

(xx) $\int x^3 \cos 2x \, dx$,

(xxi) $\int e^{ax} \sin^2 bx \, dx$,

(xxii) $\int e^x \sin^3 x \, dx$,

(xxiii) $\int e^{\frac{1}{2} \sqrt{y}} \, dy$,

(xxiv) $\int x \, 10^x \, dx$.

4. If $C = \int e^{ax} \cos bx \, dx$, and $S = \int e^{ax} \sin bx \, dx$, prove that

$$aC - bS = e^{ax} \cos bx, \quad \text{and} \quad aS + bC = e^{ax} \sin bx.$$

Hence find C and S.

5. Prove that $\int_0^\infty e^{-2x} \sin 3x \, dx = \dfrac{3}{13}$.

6. Find the area enclosed by the x-axis and the curve $y = x(2-x)^5$.

7. A uniform lamina is enclosed by the axes and the curve $y = \cos x$ from $x = 0$ to $x = \dfrac{\pi}{2}$.　Find the coordinates of its centre of gravity.

8. The area under $y = \cos x$ from $x = 0$ to $x = \dfrac{\pi}{2}$ is rotated through four right angles about the x-axis.　Find the centre of gravity of the uniform solid so generated.

9. Prove that $\int \cos^4 x \, dx = \tfrac{1}{4} \sin x \cos^3 x + \tfrac{3}{4} \int \cos^2 x \, dx$.

10. Find the area bounded by the x-axis and that part of the curve $y = e^{3x} \sin x$ from $x = 0$ to $x = \pi$.

Reduction formulae

13.8.　The normal method of finding $\int \cos^2 x \, dx$ is to use the fact that $\cos^2 x = \tfrac{1}{2}(1 + \cos 2x)$; $\int \cos^4 x \, dx$ may be tackled in the same way by expressing the integrand in terms of $\cos 2x$ and $\cos 4x$, but for the integrals of higher even powers of $\cos x$ the working becomes tedious.

It is instructive to find $\int \cos^2 x \, dx$ using integration by parts.　Once again we find that in the process the original integral reappears; this special aspect of integration by parts is found to have a most powerful application, not only in finding the integrals of high powers of $\cos x$ and $\sin x$, but also in establishing general formulae for dealing with integrands of high power.

$$I \equiv \int \cos^2 x \, dx = \cos x \sin x - \int \sin x \, (-\sin x) \, dx,$$

Let $u = \cos x$.
Let $\dfrac{dv}{dx} = \cos x$,

$$= \cos x \sin x + \int (1 - \cos^2 x) \, dx,$$

$\therefore v = \sin x$.

$$\therefore I = \cos x \sin x + x - I,$$

$$\therefore 2I = \cos x \sin x + x + k,$$

$$\therefore I = \tfrac{1}{2} \cos x \sin x + \tfrac{1}{2} x + c.$$

Qu. 22.　Find $\int \cos^4 x \, dx$ by finding it first in terms of $\int \cos^2 x \, dx$ using integration by parts, and then using the above result.

Qu. 23. Show that

$$\int \cos^6 x \, dx = \tfrac{1}{6} \cos^5 x \sin x + \tfrac{5}{24} \cos^3 x \sin x + \tfrac{5}{16} \cos x \sin x + \frac{5.3.1}{6.4.2} x + c.$$

Qu. 24. Find $\int \cos^3 x \, dx$ (i) by finding it first in terms of $\int \cos x \, dx$ using integration by parts, (ii) by another method, giving it as a function of $\sin x$ only.

Now the real value, in terms of economy, of the results we are beginning to establish is apparent when we come to consider definite integrals between certain limits. For example, using the result obtained in Qu. 23,

$$\int_0^{\frac{\pi}{2}} \cos^6 x \, dx = \left[f(x) \right]_0^{\frac{\pi}{2}} + \left[\frac{5.3.1}{6.4.2} x \right]_0^{\frac{\pi}{2}},$$

where each term of $f(x)$ contains both $\cos x$ and $\sin x$, and therefore vanishes at each limit.

Hence

$$\int_0^{\frac{\pi}{2}} \cos^6 x \, dx = \frac{5.3.1}{6.4.2} \frac{\pi}{2} = \frac{5\pi}{32}.$$

Qu. 25. Show that

$$\int \cos^5 x \, dx = \tfrac{1}{5} \cos^4 x \sin x + \tfrac{4}{15} \cos^2 x \sin x + \frac{4.2}{5.3} \sin x + c,$$

and evaluate $\int_0^{\frac{\pi}{2}} \cos^5 x \, dx$.

A pattern is emerging in the last but one term in Qu. 23 and in Qu. 25. We shall now consider the general treatment of this form.

Suppose

$$I_n = \int_0^{\frac{\pi}{2}} \cos^n x \, dx, \quad (n \geqslant 2).$$

Using integration by parts,

$$I_n = \left[\cos^{n-1} x \sin x \right]_0^{\frac{\pi}{2}}$$

Let $u = \cos^{n-1} x$.

$$- \int_0^{\frac{\pi}{2}} \sin x \, (n-1) \cos^{n-2} x \, (-\sin x) \, dx$$

Let $\dfrac{dv}{dx} = \cos x$,

$$\therefore v = \sin x.$$

$$= 0 + (n-1) \int_0^{\frac{\pi}{2}} (1 - \cos^2 x) \cos^{n-2} x \, dx,$$

$$= (n-1) \int_0^{\frac{\pi}{2}} \cos^{n-2} x \, dx - (n-1) \int_0^{\frac{\pi}{2}} \cos^n x \, dx.$$

$$\therefore \; I_n = (n-1)I_{n-2} - (n-1)I_n,$$

$$\therefore \; nI_n = (n-1)I_{n-2},$$

$$\therefore \; I_n = \frac{n-1}{n} I_{n-2}, \; (n \geqslant 2). \tag{1}$$

Since this relationship reduces by 2 the power of cos x in the integrand it is called a **reduction formula**.

Replacing n by $(n-2)$ in (1),

$$I_{n-2} = \frac{n-3}{n-2} I_{n-4}, \quad (n \geqslant 4).$$

Similarly

$$I_{n-4} = \frac{n-5}{n-4} I_{n-6}, \quad (n \geqslant 6).$$

Thus $\quad I_n = \dfrac{n-1}{n} I_{n-2} = \dfrac{(n-1)(n-3)}{n(n-2)} I_{n-4},$

$$= \frac{(n-1)(n-3)(n-5)}{n(n-2)(n-4)} I_{n-6}, \quad (n \geqslant 6).$$

If n is *odd*, e.g. $n = 7$, we obtain a multiple of I_1, which is

$$\int_0^{\frac{\pi}{2}} \cos x \, \mathrm{d}x = 1.$$

If n is *even*, e.g. $n = 6$, we obtain a multiple of I_0, which is

$$\int_0^{\frac{\pi}{2}} 1 \, \mathrm{d}x = \frac{\pi}{2}.$$

Thus $\quad \displaystyle\int_0^{\frac{\pi}{2}} \cos^7 x \, \mathrm{d}x \equiv I_7 = \frac{6.4.2}{7.5.3} \cdot 1 = \frac{16}{35},$

and $\quad \displaystyle\int_0^{\frac{\pi}{2}} \cos^6 x \, \mathrm{d}x \equiv I_6 = \frac{5.3.1}{6.4.2} \frac{\pi}{2} = \frac{5\pi}{32}.$

Qu. 26. Evaluate:

(i) $\displaystyle\int_0^{\frac{\pi}{2}} \cos^8 x \, \mathrm{d}x,$ (ii) $\displaystyle\int_0^{\frac{\pi}{2}} \cos^9 x \, \mathrm{d}x,$ (iii) $\displaystyle\int_0^{\frac{\pi}{2}} \cos^{10} x \, \mathrm{d}x.$

Qu. 27. If $I_n = \displaystyle\int_0^{\frac{\pi}{2}} \sin^n x \, \mathrm{d}x,$ use integration by parts to show that

$$I_n = \frac{n-1}{n} I_{n-2}.$$

Qu. 28. Use the change of variable $x = \frac{\pi}{2} - y$ to show that

$$\int_0^{\frac{\pi}{2}} \sin^n x \, \mathrm{d}x = \int_0^{\frac{\pi}{2}} \cos^n x \, \mathrm{d}x.$$

We can now state a reduction formula which the reader should memorize.

$$\textbf{If} \quad I_n = \int_0^{\frac{\pi}{2}} \cos^n x \, \mathrm{d}x \quad \textbf{or} \quad \int_0^{\frac{\pi}{2}} \sin^n x \, \mathrm{d}x,$$

$$I_n = \frac{n-1}{n} I_{n-2}.$$

Hence, when n is *odd*,

$$I_n = \frac{(n-1)(n-3)\ldots 4.2}{n(n-2)\ldots 5.3},$$

and when n is *even*,

$$I_n = \frac{(n-1)(n-3)\ldots 3.1}{n(n-2)\ldots 4.2} \frac{\pi}{2}.$$

A thorough treatment of reduction formulae is beyond the scope of this book, but as an introduction to this topic the above ideas are developed more fully in Exercise 13d; of particular interest is No. 6, from which the basic formula quoted above may be deduced as a special case.

Exercise 13d

1. Use integration by parts to show that

$$\int \sin^2 x \, \mathrm{d}x = -\tfrac{1}{2} \cos x \sin x + \tfrac{1}{2}x + c.$$

Assuming this result, find $\int \sin^4 x \, \mathrm{d}x$ by the same method, and evaluate

$$\int_0^{\frac{\pi}{2}} \sin^4 x \, \mathrm{d}x.$$

2. Use integration by parts to show that

$$\int \sin^3 \theta \, \mathrm{d}\theta = -\tfrac{1}{3} \cos \theta \sin^2 \theta - \tfrac{2}{3} \cos \theta + c.$$

Assuming this result, find $\int \sin^5 \theta \, \mathrm{d}\theta$ by the same method, and evaluate

$$\int_0^{\frac{\pi}{2}} \sin^5 \theta \, \mathrm{d}\theta.$$

3. Evaluate the following:

(i) $\int_0^{\frac{\pi}{2}} \sin^3 x \, dx$,　　　(ii) $\int_0^{\frac{\pi}{2}} \sin^6 x \, dx$,　　　(iii) $\int_0^{\frac{\pi}{2}} \sin^9 x \, dx$,

(iv) $\int_0^{\frac{\pi}{2}} \cos^4 x \, dx$,　　　(v) $\int_0^{\frac{\pi}{2}} \sin^{10} x \, dx$,　　　(vi) $\int_0^{\frac{\pi}{2}} \sin^7 x \, dx$.

4. Use the change of variable $u = 2\theta$ to prove that

$$\int_0^{\pi} \sin^8 \frac{u}{2} \, du = \frac{35\pi}{128}.$$

Evaluate the following:

(i) $\int_0^{\frac{\pi}{4}} \cos^7 2y \, dy$,　　　(ii) $\int_0^{\frac{3\pi}{2}} \sin^5 \frac{t}{3} \, dt$,　　　(iii) $\int_0^{\frac{\pi}{6}} \cos^6 3x \, dx$.

*5. Demonstrate graphically that:

(a) $\int_0^{\pi} \cos^3 \theta \, d\theta = 0$,　　　　　　(b) $\int_0^{\pi} \sin^3 \theta \, d\theta = 2 \int_0^{\frac{\pi}{2}} \sin^3 \theta \, d\theta$,

(c) $\int_0^{\pi} \cos^4 \theta \, d\theta = 2 \int_0^{\frac{\pi}{2}} \cos^4 \theta \, d\theta$.

Evaluate the following:

(i) $\int_0^{\pi} \sin^7 \theta \, d\theta$,　　　(ii) $\int_{-\frac{\pi}{2}}^{\frac{\pi}{2}} \cos^4 \theta \, d\theta$,　　　(iii) $\int_{-\frac{\pi}{2}}^{\frac{\pi}{2}} \sin^6 \theta \, d\theta$,

(iv) $\int_{-\frac{\pi}{2}}^{\frac{\pi}{2}} \sin^7 \theta \, d\theta$,　　　(v) $\int_0^{\pi} \cos^5 \theta \, d\theta$,　　　(vi) $\int_{-\frac{\pi}{2}}^{\frac{\pi}{2}} \cos^9 \theta \, d\theta$,

(vii) $\int_0^{\pi} \sin^{10} \theta \, d\theta$,　　　(viii) $\int_0^{\pi} \cos^8 \theta \, d\theta$.

*6. (i) Writing

$$\int_0^{\frac{\pi}{2}} \cos^m \theta \sin^n \theta \, d\theta \quad \text{as} \quad \int_0^{\frac{\pi}{2}} c^m s^n \, d\theta, \quad \text{or} \quad I_{m,\,n},$$

use integration by parts (taking u as c^{m-1}) to prove that

$$I_{m,\,n} = \frac{m-1}{m+n} I_{m-2,\,n}, \quad (m \geqslant 2),$$

and write down $I_{m-2,\,n}$, and hence $I_{m,\,n}$, in terms of $I_{m-4,\,n}$, $(m \geqslant 4)$.

(ii) Use the change of variable $x = \frac{\pi}{2} - y$ to prove that $I_{m,\,n} = I_{n,\,m}$, and reduce $I_{m-4,\,n}$ to the form $k I_{m-4,\,n-6}$, $(n \geqslant 6)$.

(iii) Show that

$$\int_0^{\frac{\pi}{2}} \cos^5 \theta \sin^6 \theta \, d\theta = \tfrac{8}{99} \int_0^{\frac{\pi}{2}} \cos \theta \sin^6 \theta \, d\theta,$$

and proceed to evaluate this

(a) by reduction to the form $k I_{1,\,0}$,

(b) by writing the latter integral as a function of $\sin \theta$.

(iv) Evaluate the following:

(a) $\displaystyle\int_0^{\frac{\pi}{2}} \cos^8 \theta \sin^5 \theta \, d\theta$, 　　　　(b) $\displaystyle\int_0^{\frac{\pi}{2}} \cos^6 \theta \sin^8 \theta \, d\theta$,

(c) $\displaystyle\int_0^{\frac{\pi}{2}} \cos^7 \theta \sin^6 \theta \, d\theta$, 　　　　(d) $\displaystyle\int_0^{\frac{\pi}{2}} \cos^5 \theta \sin^7 \theta \, d\theta$.

7. Use a suitable change of variable and the method of No. 6 to evaluate the following:

(i) $\displaystyle\int_0^1 x^5(1-x^2)^6 \, dx$, 　　　　(ii) $\displaystyle\int_0^1 x^4(1-x^2)^{\frac{7}{2}} \, dx$.

8. If $I_n = \displaystyle\int_0^{\infty} x^n \, e^{-ax} \, dx$, (i) obtain a reduction formula for I_n in terms of I_{n-1}, and (ii) evaluate $\displaystyle\int_0^{\infty} x^9 \, e^{-2x} \, dx$.

9. If $I_n = \displaystyle\int_0^{\frac{\pi}{4}} \tan^n \theta \, d\theta$, obtain a reduction formula relating I_n and I_{n-2}, and use it to evaluate the following correct to two significant figures:

(i) $\displaystyle\int_0^{\frac{\pi}{4}} \tan^7 \theta \, d\theta$, 　　　　(ii) $\displaystyle\int_0^{\frac{\pi}{4}} \tan^8 \theta \, d\theta$.

10. (i) If $I_n = \displaystyle\int \sec^n x \, dx$, prove that

$$I_n = \frac{1}{n-1} \tan x \sec^{n-2} x + \frac{n-2}{n-1} I_{n-2} \quad (n \geqslant 2),$$

and use this reduction formula to write down an expression relating I_n to I_{n-6} $(n \geqslant 6)$.

(ii) Use the result obtained in (i) to find $\displaystyle\int_0^{\frac{\pi}{6}} \sec^8 x \, dx$, and check your answer by expressing the integral as a function of $\tan x$ only.

(iii) Prove that

$$\int_0^{\frac{\pi}{3}} \sec^7 x \, dx = \frac{61}{8}\sqrt{3} + \frac{5}{16} \log_e (2 + \sqrt{3}).$$

Exercise 13e (Revision)

No list of 'standard integrals' is given in this chapter, in the belief that the recognition of form is more important than the learning of formulae (see Nos. 13 to 17).

In this exercise,

Nos. 1 to 7 summarize the main methods dealt with in chapters 1, 2, 3, 13.

Nos. 8 to 12 gather together the integrals of some trigonometrical functions and inverse functions, to enable the reader to take stock of his power of handling these integrals.

Nos. **13** to **17** are designed to develop discrimination in choice of method. These questions test the essential skill, recognition of form, and the more experienced reader may confine his attention to these questions, together with some of the less obvious integrals in Nos. 8 to 12.

Find the integrals in Nos. 1 to 6:

1.   (i) $\int x\sqrt{(x^2+1)}\,dx$,   (ii) $\int \dfrac{x^2+1}{\sqrt{(x^3+3x-4)}}\,dx$,

     (iii) $\int \cos^5 u\,du$,   (iv) $\int \sec^6 \theta\,d\theta$,   (v) $\int \sec x \tan^5 x\,dx$,

     (vi) $\int x \sin x^2\,dx$,   (vii) $\int \dfrac{\sec^2 \sqrt{x}}{\sqrt{x}}\,dx$,

     (viii) $\int x(2x^2+3)^{-1}\,dx$,   (ix) $\int \dfrac{x}{e^{x^2}}\,dx$,

     (x) $\int y\,10^{y^2}\,dy$,   (xi) $\int \tan \dfrac{\theta}{2}\,d\theta$,

     (xii) $\int \dfrac{\log_e x}{x}\,dx$,   (xiii) $\int \dfrac{\sec x(\sec x+\tan x)}{\sec x+\tan x}\,dx$.

2. *Change of variable.*

     (i) $\int x\sqrt{(2x-3)}\,dx$,   (ii) $\int 2x(3x-1)^7\,dx$,

     (iii) $\int \dfrac{y(y-8)}{(y-4)^2}\,dy$,   (iv) $\int \dfrac{1}{\sqrt{(4-5y^2)}}\,dy$,

     (v) $\int \dfrac{1}{3+9u^2}\,du$,   (vi) $\int \dfrac{1}{u^2-6u+17}\,du$,

     (vii) $\int \dfrac{1}{\sqrt{(7+4x-2x^2)}}\,dx$,   (viii) $\int \sqrt{(4-y^2)}\,dy$,

     (ix) $\int \dfrac{1}{x\sqrt{(9x^2-1)}}\,dx$,   (x) $\int \dfrac{1}{5+4 \cos \theta}\,d\theta$,

     (xi) $\int \operatorname{cosec} 3u\,du$,   (xii) $\int \dfrac{1}{1+3 \sin^2 x}\,dx$,

     (xiii) $\int \dfrac{1}{\sqrt{(x^2-16)}}\,dx$.

3. *Involving exponential and logarithmic functions.*

     (i) $\int e^{3x}\,dx$,   (ii) $\int 10^y\,dy$,   (iii) $\int \dfrac{x^2}{e^{x^3}}\,dx$,

     (iv) $\int \dfrac{1}{3x}\,dx$,   (v) $\int \dfrac{1}{3x+4}\,dx$,

     (vi) $\int \dfrac{1}{3-2x}\,dx$ $(x > \tfrac{3}{2})$,   (vii) $\int \dfrac{1}{3x+9}\,dx$,

     (viii) $\int \dfrac{1}{1-x^2}\,dx$,   (ix) $\int \log_e x\,dx$,

     (x) $\int e^{\sqrt{x}}\,dx$ $\left(\text{write as} \int x^{1/2}x^{-1/2}e^{x^{1/2}}\,dx\right)$,   (xi) $\int \log_e (x+3)\,dx$.

4. Partial fractions.

(i) $\int \dfrac{2}{9-x^2}\,dx,$

(ii) $\int \dfrac{1}{y(y-3)}\,dy,$

(iii) $\int \dfrac{1}{x^3-x^2}\,dx,$

(iv) $\int \dfrac{x}{(4-x)^2}\,dx,$

(v) $\int \dfrac{2-x^2}{(x+1)^3}\,dx,$

(vi) $\int \dfrac{(x-2)^2}{x^3+1}\,dx.$

5. Integration by parts.

(i) $\int x \cos \dfrac{x}{2}\,dx,$

(ii) $\int \dfrac{x}{2}\,e^x\,dx,$

(iii) $\int y \operatorname{cosec}^2 y\,dy,$

(iv) $\int 2y(1-3y)^6\,dy,$

(v) $\int x\,3^x\,dx,$

(vi) $\int x \log_e 2x\,dx,$

(vii) $\int \log_e t\,dt,$

(viii) $\int \tan^{-1} 3x\,dx,$

(ix) $\int 4^x\,dx,$

(x) $\int x^3 \sin x\,dx,$

(xi) $\int t^3 \sin t^2\,dt,$

(xii) $\int \theta \cos^2 \dfrac{\theta}{2}\,d\theta,$

(xiii) $\int y \tan^{-1} 2y\,dy,$

(xiv) $\int x \tan^2 2x\,dx,$

(xv) $\int e^{3x} \cos 5x\,dx,$

(xvi) $\int \sec^3 2\theta\,d\theta,$

(xvii) Prove $\int \cos^4 \dfrac{x}{2}\,dx = \tfrac{1}{2} \sin \dfrac{x}{2} \cos^3 \dfrac{x}{2} + \tfrac{3}{4} \int \cos^2 \dfrac{x}{2}\,dx.$

6. Splitting the numerator.

(i) $\int \dfrac{2x-1}{4x^2+3}\,dx,$

(ii) $\int \dfrac{1-4y}{\sqrt{(1+2y-y^2)}}\,dy,$

(iii) $\int \dfrac{\cos \theta}{2 \cos \theta - \sin \theta}\,d\theta,$

(iv) $\int \dfrac{\cos x - 2 \sin x}{3 \cos x + 4 \sin x}\,dx.$

7. Evaluate the following:

(i) $\int_{1/2}^{2/3} \dfrac{1}{\sqrt{(4-9x^2)}}\,dx,$

(ii) $\int_{1}^{\sqrt{2}} \dfrac{1}{8+y^2}\,dy,$

(iii) $\int_{5}^{\infty} \dfrac{1}{(x-3)^2}\,dx,$

(iv) $\int_{0}^{\pi/2} \cos^{11} x\,dx,$

(v) $\int_{0}^{\pi/2} \sin^{12} \theta\,d\theta,$

(vi) $\int_{0}^{\pi/8} \cos^6 4y\,dy,$

(vii) $\int_{-\pi/2}^{\pi/2} \sin^8 u\,du,$

(viii) $\int_{0}^{\pi} \cos^7 x\,dx,$

(ix) $\int_{0}^{\pi/2} \cos^9 \theta \sin^{10} \theta\,d\theta,$

(x) $\int_{-1}^{+1} \dfrac{1}{2x-3}\,dx.$

8. Find the following integrals:

(i) $\int \sin 5x\,dx,$

(ii) $\int \cos \dfrac{x}{3}\,dx,$

(iii) $\int \tan 5x\,dx,$

(iv) $\int \cot \tfrac{1}{2}x\,dx,$

(v) $\int \operatorname{cosec} x\,dx \quad \left(\text{use } \tan \dfrac{x}{2} = t\right),$

(vi) $\int \sec x\,dx \quad \left(\text{use } \tan \dfrac{x}{2} = t, \text{ or see No. 1 (xiii)}\right).$

9. Find the following integrals:

(i) $\int \sec^2 \frac{x}{3} \, dx$,   (ii) $\int \csc^2 4x \, dx$,   (iii) $\int \sin^2 x \, dx$,

(iv) $\int \cos^2 x \, dx$,   (v) $\int \tan^2 x \, dx$,   (vi) $\int \cot^2 x \, dx$.

10. Find the following integrals:

(i) $\int \sin^3 x \, dx$,   †(ii) $\int \cos^3 x \, dx$,

(iii) $\int \tan^3 x \, dx$ (use Pythagoras' theorem),

†(iv) $\int \cot^3 x \, dx$ (use Pythagoras' theorem),

(v) $\int \sec^3 x \, dx$ (by reduction),   †(vi) $\int \csc^3 x \, dx$ (by reduction).

11. Find the following integrals ((i) and (ii) by expressing the integrands in terms of cos $2x$, cos $4x$, or by reduction, the remainder by using Pythagoras' theorem):

(i) $\int \sin^4 x \, dx$,   (ii) $\int \cos^4 x \, dx$,   (iii) $\int \tan^4 x \, dx$,

(iv) $\int \csc^4 x \, dx$,   (v) $\int \sec^4 x \, dx$,   (vi) $\int \cot^4 x \, dx$.

12. Find the following integrals using integration by parts (in (v) and (vi) continue by using the change of variable $x = \sec u$):

(i) $\int \sin^{-1} x \, dx$,   (ii) $\int \cos^{-1} x \, dx$,   (iii) $\int \tan^{-1} x \, dx$,

(iv) $\int \cot^{-1} x \, dx$,   (v) $\int \sec^{-1} x \, dx$,   (vi) $\int \csc^{-1} x \, dx$.

Find the integrals in Nos. 13 to 17:

13.   (i) $\int \dfrac{1}{3+4x^2} \, dx$,   (ii) $\int \dfrac{x}{\sqrt{(5+8x^2)}} \, dx$,

(iii) $\int \dfrac{1}{\sqrt{(1+x^2)}} \, dx$,   (iv) $\int \dfrac{x}{2+3x^2} \, dx$,

(v) $\int x\sqrt{(3+x^2)} \, dx$,   (vi) $\int \dfrac{x+1}{3+2x^2} \, dx$,

(vii) $\int \dfrac{x-2}{x^2-4x+7} \, dx$,   (viii) $\int \sqrt{(2+x^2)} \, dx$,

(ix) $\int \dfrac{3x-11}{x^2-4x+5} \, dx$,   (x) $\int x\sqrt{(2+3x)} \, dx$,

(xi) $\int \dfrac{x}{(3+x)^2} \, dx$.

† The change of variable $y = \frac{1}{2}\pi - x$ may be used.

14. (i) $\int \dfrac{1}{\sqrt{(4-5x^2)}}\,dx,$ (ii) $\int \dfrac{x}{\sqrt{(1-3x)}}\,dx,$

(iii) $\int \dfrac{2}{9-x^2}\,dx,$ (iv) $\int \dfrac{3}{(16-x)^2}\,dx,$

(v) $\int x\sqrt{(6-x^2)}\,dx,$ (vi) $\int \dfrac{3x}{4-x^2}\,dx,$

(vii) $\int \sqrt{(4-x^2)}\,dx,$ (viii) $\int \dfrac{x}{\sqrt{(7-2x^2)}}\,dx,$

(ix) $\int \dfrac{x-2}{\sqrt{(3-4x^2)}}\,dx,$ (x) $\int \dfrac{1}{\sqrt{(x^2-9)}}\,dx,$

(xi) $\int \dfrac{x+1}{\sqrt{(13-6x-3x^2)}}\,dx.$

15. (i) $\int \cos x°\,dx,$ (ii) $\int x \sin 2x \cos 2x\,dx,$

(iii) $\int \sec \dfrac{\theta}{2} \operatorname{cosec} \dfrac{\theta}{2}\,d\theta,$ (iv) $\int \cos^6 x \sin^5 x\,dx,$

(v) $\int y \sec^2 y\,dy,$ (vi) $\int x \sin x\,dx,$

(vii) $\int x \sin x^2\,dx,$ (viii) $\int u^2 \cos u\,du,$

(ix) $\int \sin^2 y \cos^2 y\,dy,$ (x) $\int \sin 5x \cos 2x\,dx,$

(xi) $\int \theta \cos^2 \theta\,d\theta,$ (xii) $\int \tan^4 \dfrac{x}{3}\,dx,$

(xiii) $\int \sin^5 \dfrac{x}{2}\,dx,$ (xiv) $\int y \tan^2 y\,dy,$

(xv) $\int \tan \theta \sec \theta \sqrt{\sec \theta}\,d\theta.$

16. (i) $\int \dfrac{1}{1+\cos \theta}\,d\theta,$ (ii) $\int \dfrac{1}{1-5\sin^2 \theta}\,d\theta,$

(iii) $\int \dfrac{1}{1+\sin x}\,dx,$ (iv) $\int \dfrac{2\cos \theta - 4\sin \theta}{\cos \theta + 3\sin \theta}\,d\theta,$

(v) $\int \dfrac{1}{1-\cos \frac{1}{2}x}\,dx,$ (vi) $\int \dfrac{4}{\cos^2 x + 9\sin^2 x}\,dx,$

(vii) $\int \dfrac{1}{\cos^2 2y - \sin^2 2y}\,dy,$ (viii) $\int \dfrac{1+\sin x}{\cos^2 x}\,dx,$

(ix) $\int \dfrac{1}{\sin \theta + 2\cos \theta + 1}\,d\theta,$ (x) $\int \dfrac{1}{1+\tan x}\,dx.$

17. (i) $\int x^3\, e^{-z}\,dx,$ (ii) $\int \log_e (x+2)\,dx,$

(iii) $\int \dfrac{e^{\sqrt{y}}}{\sqrt{y}}\,dy,$ (iv) $\int \dfrac{1}{t \sqrt{\log_e t}}\,dt,$

(v) $\int x \tan^{-1} 3x \, dx,$

(vi) $\int \dfrac{\sin^{-1} x}{\sqrt{(1-x^2)}} \, dx,$

(vii) $\int 4^x \, dx,$

(viii) $\int x \, 10^x \, dx,$

(ix) $\int x^3 \log_e (2x) \, dx,$

(x) $\int x^3 \, e^{x^2} \, dx,$

(xi) $\int \dfrac{1}{(1+4x^2) \tan^{-1} 2x} \, dx,$

(xii) $\int t^2 \, 10^{t^3} \, dt,$

(xiii) $\int \dfrac{\log_e (y+2)}{y+2} \, dy,$

(xiv) $\int \sin^{-1} 2\theta \, d\theta,$

(xv) $\int e^{2x} \sin 3x \, dx,$

(xvi) $\int \log_e (4-x^2) \, dx,$

(xvii) $\int e^y \, y \sin y \, dy,$

(xviii) $\int \dfrac{1}{x \log_e x} \, dx.$

Exercise 13f (Miscellaneous)

1. Integrate the following functions with respect to x:

(i) $\left(x^2 - \dfrac{2}{x}\right)^2,$ (ii) $\sin 3x \cos 5x,$ (iii) $\dfrac{2}{x(1+x^2)}.$

Prove by means of the substitution $t = \tan x$ that

$$\int_0^{\frac{\pi}{4}} \frac{dx}{1 + \sin 2x} = \tfrac{1}{2},$$

and find the value of

$$\int_0^{\frac{\pi}{4}} \frac{dx}{(1 + \sin 2x)^2}. \qquad \text{O.C.}$$

2. Integrate the following functions with respect to x:

(i) $x(1+x^2)^{3/2},$ (ii) $\dfrac{3+2x}{1-4x^2},$ (iii) $x^2 \log_e x.$

Evaluate

$$\int_0^{\frac{\pi}{2}} x \cos^2 3x \, dx. \qquad \text{O.C.}$$

3. Express $\dfrac{1}{1-x^4}$ in partial fractions and hence show that

$$\int_0^{\frac{1}{2}} \frac{dx}{1-x^4} = \tfrac{1}{4} \log_e 3 + \tfrac{1}{2} \tan^{-1} (\tfrac{1}{2}). \qquad \text{L.}$$

4. Integrate the following functions with respect to x:

(i) $\dfrac{1}{x^2 - 3x + 2},$ (ii) $x \tan^{-1} x,$ (iii) $\dfrac{x^2}{x^2 + 2x + 2}.$

By means of the substitution $x = 1 - \dfrac{1}{u^4}$, show that

$$\int_1^{2^{\frac{1}{4}}} \frac{du}{u(2u^4 - 1)^{\frac{1}{2}}} = \frac{\pi}{24}.$$ O.C.

5. Integrate the following functions with respect to x:

   (i) $\dfrac{x+6}{x^2 + 6x + 8}$,   (ii) $x \log_e x$,   (iii) $\sin^{-1} x$.

   Evaluate

   $$\int_{\frac{1}{2}}^1 \frac{(1 + x^{\frac{3}{2}})}{x\sqrt{\{x(1-x)\}}} \, dx.$$

   by means of the substitution $x = \cos^2 \phi$.   O.C.

6. Use the substitution $u = +\sqrt{(1 + x^2)}$ to evaluate

   $$\int_{\frac{4}{3}}^{\frac{12}{5}} \frac{dx}{x(1 + x^2)^{\frac{1}{2}}}.$$ O.C.

7. By making the substitution $x = \pi - y$, or otherwise, prove that

   $$\int_0^\pi x \sin^3 x \, dx = \frac{2\pi}{3}.$$ O.C.

8. Integrate with respect to x:

   (i) $\cos x \operatorname{cosec}^3 x$,   (ii) $\dfrac{x+3}{\sqrt{(7 - 6x - x^2)}}$.

   By making the substitution $x = a \cos^2 \theta + b \sin^2 \theta$, prove that

   $$\int_a^b \frac{x \, dx}{\sqrt{\{(x-a)(b-x)\}}} = \tfrac{1}{2}\pi(a + b).$$ O.C.

9. Integrate with respect to x:

   (i) $\dfrac{1}{\sqrt{(5 - 4x - x^2)}}$,   (ii) $x^3 e^{-x^2}$.

   If $S = \displaystyle\int \frac{\sin x}{a \sin x + b \cos x} \, dx$ and $C = \displaystyle\int \frac{\cos x}{a \sin x + b \cos x} \, dx$,

   find $aS + bC$ and $aC - bS$.   Hence, or otherwise, prove that

   $$\int_0^{\frac{\pi}{2}} \frac{\sin x}{3 \sin x + 4 \cos x} \, dx = \frac{3\pi}{50} + \frac{4}{25} \log_e \left(\tfrac{4}{3}\right).$$ O.C.

10. Evaluate

    $$\int_1^2 \frac{dx}{x^2 \sqrt{(x-1)}}$$

    by means of the substitution $x = \sec^2 \phi$.   O.C.

11. By means of the substitution $x^2 = \dfrac{1}{u}$, evaluate the integral

$$\int_1^2 \frac{dx}{x^2\sqrt{(5x^2-1)}}.$$

O.C.

12. Integrate with respect to x:

    (i) $\dfrac{x^2}{x^2-4}$,   (ii) $x^3 e^{-x^2/2}$,   (iii) $\dfrac{1}{\sqrt{(5+4x-x^2)}}$.

By means of the substitution $u = \tan x$, or otherwise, evaluate the integral

$$\int_0^{\frac{\pi}{2}} \frac{dx}{2+\cos^2 x}.$$

O.C.

13. Prove that, if $C = \displaystyle\int e^{ax} \cos bx \, dx$ and $S = \displaystyle\int e^{ax} \sin bx \, dx$,

$$aC - bS = e^{ax} \cos bx,$$
$$aS + bC = e^{ax} \sin bx.$$

Evaluate

$$\int_0^{\frac{\pi}{2}} e^{2x} \sin 3x \, dx.$$

O.C.

14. Integrate with respect to x:

    (i) $\displaystyle\int (2x-3)^{-\frac{1}{2}} \, dx$,   (ii) $\displaystyle\int x \sin x \, dx$,   (iii) $\displaystyle\int x\sqrt{(x-4)} \, dx$.

By means of the substitution $x = 3\cos^2\theta + 6\sin^2\theta$, or otherwise, evaluate

$$\int_3^6 \frac{dx}{\sqrt{\{(x-3)(6-x)\}}}.$$

O.C.

15. Give a geometrical interpretation of the integral

$$I = \int_a^b f(x) \, dx, \quad (b > a).$$

Without attempting to evaluate them, determine whether the following integrals are positive, negative, or zero:

    (i) $\displaystyle\int_0^1 x^3(1-x^2)^2 \, dx$,   (ii) $\displaystyle\int_0^{\pi} \sin^3 x \cos^3 x \, dx$,

    (iii) $\displaystyle\int_{\frac{1}{2}}^1 e^{-x} \log_e x \, dx$.

O.C.

16. (i) Let

$$I(Z) = \int_1^Z \frac{(x-1)^p(2-x)^p}{x^{p+1}} \, dx, \quad (p > 0).$$

By writing $x = \dfrac{2}{y}$, prove that $2I(\sqrt{2}) = I(2)$.

(ii) Without attempting to evaluate them, determine whether the following integrals are positive, negative, or zero:

$$\int_0^1 x^3(1-x)^3 \, dx, \qquad \int_0^\pi \sin^2 x \cos^3 x \, dx, \qquad \int_0^\pi e^{-z} \sin x \, dx. \qquad \text{O.C.}$$

17. By means of the substitution $y = a - x$ or otherwise, prove that

$$\int_0^a f(x) \, dx = \int_0^a f(a-x) \, dx.$$

Hence prove that

$$\int_0^\pi \frac{x \sin x}{1 + \cos^2 x} \, dx = \int_0^\pi \frac{(\pi - x) \sin x}{1 + \cos^2 x} \, dx = \frac{\pi^2}{4}. \qquad \text{O.C.}$$

18. (i) Evaluate the integral $\int_0^1 \tan^{-1} x \, dx$.

(ii) If

$$I_n = \int_0^1 \frac{dx}{(1+x^2)^n},$$

show that

$$2n I_{n+1} = 2^{-n} + (2n-1) I_n.$$

Deduce the value of

$$\int_0^1 \frac{dx}{(1+x^2)^3}. \qquad \text{N.}$$

19. Prove that, if

$$u_n = \int_0^{\frac{\pi}{2}} \frac{\sin 2n\theta}{\sin \theta} \, d\theta,$$

where n is a positive integer,

$$u_n - u_{n-1} = \frac{2(-1)^{n-1}}{2n-1}.$$

Hence prove that

$$u_n = 2\left\{ 1 - \tfrac{1}{3} + \tfrac{1}{5} - \ldots + \frac{(-1)^{n-1}}{2n-1} \right\}. \qquad \text{O.C.}$$

20. (i) Prove that, if $I_n = \int \sec^{2n} \theta \, d\theta$, and $n > 1$,

$$(2n-1) I_n = 2(n-1) I_{n-1} + \sec^{2n-2} \theta \tan \theta.$$

(ii) Using the result of (i) prove that

$$\int_0^{\frac{\pi}{4}} \sec^{10} \theta \, d\theta = \frac{1328}{315}. \qquad \text{O.C.}$$

21. If $u_n = \int_0^{\frac{\pi}{2}} x^n \sin x \, dx$, prove that, for $n \geqslant 2$,

$$u_n + n(n-1)u_{n-2} = n\left(\frac{\pi}{2}\right)^{n-1}.$$

Calculate u_1 and deduce that $u_3 = \frac{3}{4}\pi^2 - 6$.   O.C.

22. If $I = \int_0^{\frac{\pi}{2}} \log_e \sin x \, dx$, show by means of the substitution $x = \frac{1}{2}\pi - y$

that $I = \int_0^{\frac{\pi}{2}} \log_e \cos y \, dy$, and deduce that

$$2I = \int_0^{\frac{\pi}{2}} \log_e \sin 2x \, dx + \frac{1}{2}\pi \log_e (\frac{1}{2}).$$

Use the further substitution $2x = Z$ to prove that

$$I = \frac{1}{2}\pi \log_e (\frac{1}{2}).$$ O.C.

CHAPTER 14

PROJECTION

Projection on to a straight line

14.1. In the first part of this chapter we shall only be concerned with the projection of points, lines, and curves on to a *coplanar* line.

DEFINITION: *The projection of a point on to a line is the foot of the perpendicular from that point to the line.*

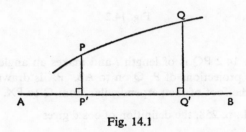

Fig. 14.1

In Fig. 14.1 P′, Q′ are the projections of P, Q on to the line AB. Points on the curve joining P, Q project into points on the line joining P′, Q′. Since the curve joining P, Q is the locus of these points, it follows that the *curve joining* P, Q *projects into the line joining* P′, Q′.

Qu. 1. In fig. 14.1, if *l* is the length of the straight line joining P, Q and PQ makes an acute angle θ with AB, what is the length of the projection of PQ on to AB?

Directed lengths

14.2. The idea of projection must now be extended to include directed lengths. $\overrightarrow{PQ}$ means 'the length measured from P to Q'; when emphasis is not needed, the arrow may be omitted. (See P.M.I., p. 389, for a note about the signs of directed lengths.) It follows that if P′ and Q′ are the projections of P and Q on to a line AB, the projection of PQ is P′Q′ and the projection of QP is Q′P′.

Qu. 2. Show that the sum of the projections of the sides of a triangle taken in cyclic order on to any line lying in its plane is zero.

If PQ *is of length l and makes an angle θ with* AB, *its projection on to* AB, P'Q', *is of length l cos θ, for all values of θ.*

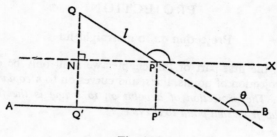

Fig. 14.2

Proof: In Fig. 14.2 PQ is of length l and makes an angle θ with AB. P', Q' are the projections of P, Q on to AB. PX is drawn parallel to AB and N is the foot of the perpendicular from Q to PX.

From P.M.I., p. 258, the definition of $\cos \theta$ gives

$$\cos \theta = \frac{\overrightarrow{PN}}{l}.$$

$$\therefore \ \overrightarrow{PN} = l \cos \theta.$$

But

$$\overrightarrow{P'Q'} = \overrightarrow{PN}.$$

$$\therefore \ \overrightarrow{P'Q'} = l \cos \theta.$$

This result holds for all values of θ.

Qu. 3. If in Fig. 14.2 AC is drawn at an angle $\frac{1}{2}\pi$ with AB (in the same sense as θ), show that the projection of $\overrightarrow{PQ}$ on to AC is equal to $l \sin \theta$, for all values of θ.

Example 1. *A vertical post* BD *stands on a plane inclined at an angle* θ *to the horizontal. The pole is supported by two ropes attached to its top* D *and pegged down to points* C, A *on either side of the pole at equal distances from its base up and down, respectively, the line of greatest slope through the base of the pole.* C *lies below the horizontal through* D *and* AD *and* CD *make angles* α *and* β *respectively with the horizontal. Prove that*

$$\tan \alpha - \tan \beta = 2 \tan \theta.$$

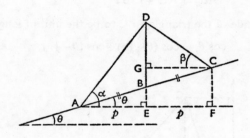

Fig. 14.3

Let E, F be the feet of the perpendiculars drawn from B and C to the horizontal line through A in the plane DAC. Let G be the foot of the perpendicular from C to BD (see Fig. 14.3).

By the mid-point theorem in $\triangle$ACF,

$$AE = EF = p, \quad \text{say.}$$

Project the sides of $\triangle$ADC on to ED:

$$ED + DG + GE = 0.$$

$$\therefore ED - GD - EG = 0.$$

$$ED = p \tan \alpha, \qquad GD = p \tan \beta, \qquad EG = FC = 2p \tan \theta.$$

$$\therefore p \tan \alpha - p \tan \beta - 2p \tan \theta = 0.$$
$$\therefore \tan \alpha - \tan \beta = 2 \tan \theta.$$

Qu. 4. Show that if, in Example 1, C lies above the horizontal through D, then $\tan \alpha + \tan \beta = 2 \tan \theta.$

Example 2.   *By projecting the sides of an equilateral triangle on to a line inclined at an angle θ to one of them, show that*

$$\cos \theta = \cos (\theta + \tfrac{1}{3}\pi) + \cos (\theta - \tfrac{1}{3}\pi),$$

and $\sin \theta + \sin (\theta + \tfrac{2}{3}\pi) + \sin (\theta + \tfrac{4}{3}\pi) = 0.$

In Fig. 14.4, triangle ABC is equilateral and AB, AC, CB are inclined at angles θ, $\theta + \tfrac{1}{3}\pi$, $\theta - \tfrac{1}{3}\pi$ to AX.   Projecting the sides of the triangle on to AX,

projection of AB = projection of AC + projection of CB,

i.e.   AB′ = AC′ + C′B′.

Taking the side of the triangle ABC to be the unit of length,

$$\cos \theta = \cos (\theta + \tfrac{1}{3}\pi) + \cos (\theta - \tfrac{1}{3}\pi).$$

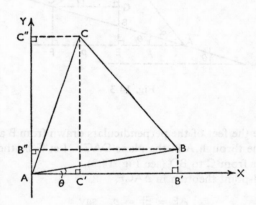

Fig. 14.4

Now if AB, BC, CA are projected on to a line AY, perpendicular to AX, as shown in Fig. 14.4, the sum of their projections AB″, B″C″, C″A will be zero.   But AB, BC, CA are inclined at angles of θ, $\theta + \tfrac{2}{3}\pi$, $\theta + \tfrac{4}{3}\pi$ to AX,

∴ $\sin \theta + \sin (\theta + \tfrac{2}{3}\pi) + \sin (\theta + \tfrac{4}{3}\pi) = 0.$

Qu. 5.   By considering the projections of AB, BC, AC on to AY in Fig. 14.4, prove that

$$\sin \theta + \sin (\theta + \tfrac{2}{3}\pi) = \sin (\theta + \tfrac{1}{3}\pi).$$

Example 3. *Prove the identity*

$$\cos(\theta+\phi) = \cos\theta\cos\phi - \sin\theta\sin\phi$$

by projection, where ϕ is an acute angle.

Suppose a line OP of unit length makes an angle ϕ with another line OC. Rotate the fixed angle POC though an angle θ from the fixed direction OX. Now draw PN perpendicular to OC (see Fig. 14.5).

OP, ON, NP are inclined at angles $\theta+\phi$, θ, $\frac{1}{2}\pi+\theta$ with OX. Projecting the sides of $\triangle$ONP on to OX,

       projection of OP = projection of ON + projection of NP.

$$\therefore \text{ OP}\cos(\theta+\phi) = \text{ON}\cos\theta + \text{NP}\cos(\tfrac{1}{2}\pi+\theta).$$

In $\triangle$ONP, OP$=1$.

$$\therefore \text{ ON} = \cos\phi, \qquad \text{NP} = \sin\phi.$$
$$\therefore \cos(\theta+\phi) = \cos\phi\cos\theta + \sin\phi\cos(\tfrac{1}{2}\pi+\theta).$$
$$\therefore \cos(\theta+\phi) = \cos\theta\cos\phi - \sin\theta\sin\phi.$$

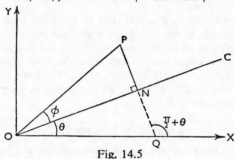

Fig. 14.5

Qu. 6. Show that this proof is valid for all angles θ when ϕ is acute.

Qu. 7. Check the proof against a diagram showing θ, ϕ both obtuse.

Qu. 8. By projecting on to OY, prove the corresponding formula for $\sin(\theta+\phi)$.

Exercise 14a

1. Prove that the area of triangle ABC is $\frac{1}{2}bc\sin A$.
2. A man walks from A to D by way of B and C. From A to B is 170 m at 059°, from B to C is 240 m at 126°, and from C to D is 190 m at 340°. Find the distances of D to the North and East of A, and deduce the distance and bearing (to the nearest degree) of D from A.
3. Show that the sum of the projections of the sides of a closed polygon on to any line is zero, when the sides are taken in order round the perimeter of the polygon.

4. Three forces 3, 5, 4 N are in the directions 050°, 120°, 215°. Find the sums of the resolved parts of these forces in the directions North and East. Hence deduce the magnitude and direction of their resultant.

5. By projecting the sides of an equilateral triangle, prove that:
   (i) $\cos \theta + \cos (\theta + \frac{2}{3}\pi) + \cos (\theta + \frac{4}{3}\pi) = 0$,
   (ii) $\cos (\theta + \frac{1}{6}\pi) = \cos (\frac{1}{6}\pi - \theta) + \sin \theta$.

6. Prove $\sin (A - B) = \sin A \cos B - \cos A \sin B$, by projection. What conditions, if any, must you impose upon A and B?

7. In Example 1, if the ropes are pegged down at unequal distances from B, where AB $= l$ and BC $= m$, prove that

$$l \tan \alpha - m \tan \beta = (l+m) \tan \theta.$$

8. AC is the diameter of a circle centre O and of unit radius, and B is a point on its circumference such that $\angle$BAC $= \theta$. By projecting the sides of $\triangle$AOB on to AC, prove that $\cos 2\theta = 2 \cos^2 \theta - 1$.

   Similarly, by projecting on to a line perpendicular to AC, prove that $\sin 2\theta = 2 \sin \theta \cos \theta$.

9. A rectangle of sides a, b ($a > b$) rests in a vertical plane on one corner with a longer side making an angle α with the ground. Find the height above the ground of its centre and also the horizontal distance of the mid-point of the upper shorter edge from the point of support.

10. ABC is an isosceles triangle in which AB $=$ AC. D is a point on AC such that AD $=$ DB $=$ BC. Mark in all the angles in your figure, and by projecting on to AB and AC, prove that

$$2 \cos 36° = 1 + 2 \cos 72°.$$

11. By projecting the sides of a regular pentagon on to a line which makes an angle θ with one of its sides, prove that

$$\cos \theta + \cos (\theta + \frac{2}{5}\pi) + \cos (\theta + \frac{4}{5}\pi) + \cos (\theta + \frac{6}{5}\pi) + \cos (\theta + \frac{8}{5}\pi) = 0.$$

12. A cylinder of height h and radius r rests in contact with a horizontal table to which its axis is inclined at an angle α, and the centre of the top is at the same height above the table as when the cylinder is upright. Prove that $\alpha = \frac{1}{2}\pi - 2 \tan^{-1} (r/h)$.

13. A man stands at a point A on a straight road whose inclination to the horizontal is i. He measures the angle of depression α, from his eye to a point B on the road. He then walks to B and measures the angle of elevation β, from his eye to A. Prove that

$$2 \tan i = \tan \alpha + \tan \beta.$$

14. A circle radius r rolls on a horizontal straight line. Show that, when the circle has turned through an angle θ, the original point of contact is $r(1 - \cos \theta)$ above, and $r(\theta - \sin \theta)$ forward of its original position.

15. A cuboid of side $2b$ rests symmetrically on a rough, fixed cylinder of radius a. The cuboid rolls without slipping so that the radius to the point of contact turns through an angle θ. Find the new height of the centre of the cuboid above the centre of the cylinder.

16. From O to P is 1 unit in a direction making θ with the x-axis. From P to Q is 1 unit in a direction making 3θ (in the same sense) with the x-axis. By projection of the triangle OPQ, show that:
  (i) $2 \cos \theta \cos 2\theta = \cos \theta + \cos 3\theta$,
  (ii) $2 \cos \theta \sin 2\theta = \sin \theta + \sin 3\theta$.

17. A particle describes a circle centre O in a vertical plane with constant speed v m/s. Show that the motions of its projections on to horizontal and vertical diameters respectively are both simple harmonic and of period $\dfrac{2\pi a}{v}$ s where a m is the radius of the circle.

18. A particle is projected from a point O on level ground with velocity v at elevation α. After time t it is at a point P. Neglecting air resistance, find expressions for the projections of OP on to horizontal and vertical axes through O. If the particle hits the ground at Q, show that there are two positions on the trajectory such that OP is at right angles to PQ if $\tan \alpha > 2$.

19. A particle is describing a circle centre O and of radius a m in a vertical plane with constant angular velocity ω rad/s. At any time t s it is at a point P and OP makes an angle θ with the horizontal. Find the projections of OP on to horizontal and vertical diameters and hence, by differentiation, the horizontal and vertical components of the acceleration of P. Show that the resultant acceleration of the particle is of magnitude $a\omega^2$ and is directed towards O.

Orthogonal projection

14.3. We shall now consider a special type of projection from one plane to another.

DEFINITION: *The orthogonal projection of a point* P *lying in the plane* π *on to the plane* π' *is defined to be the foot of the perpendicular from* P *on to* π'.

Considering a curve lying in the plane π as a locus of points, its projection on to π' is another curve which is the locus of the projections of those points.

The orthogonal projection of a straight line is another straight line.

Proof: In Fig. 14.6, π, π' are two planes which meet at an angle θ. Let PQR be a straight line in π, and let P$'$, Q$'$, R$'$ be the projections of P, Q, R on to π'.

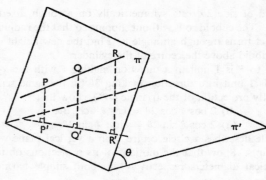

Fig. 14.6

The plane containing PR and PP' contains all the lines parallel to PP' through points in PR. But PP' is perpendicular to π', therefore the plane PRP' contains QQ' and RR'. But the plane PRP' meets π' in a straight line. Hence P'Q'R' is a straight line.

The line in which π and π' meet is called the *common axis* of the two planes, and the angle between π and π' is called the *angle of projection*.

Qu. 9. Show that the line PR meets its projection P'R' on the common axis. [Consider the projection of the point in which PR meets the common axis.]

Example 4. *Find the projected length of a line of length l which is perpendicular to the common axis, when the angle of projection is θ.*

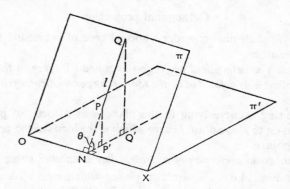

Fig. 14.7

In Fig. 14.7, PQ is of length l and QP produced meets the common axis OX in N, where $\angle$ QNO $= 90°$. The projection of PQ is P'Q', and Q'P' produced also meets OX in N, where $\angle$ Q'NX $= 90°$.

It follows that the inclination of PQ to P'Q' is θ, the angle of projection. The orthogonal projection of PQ on to π' is the same as its projection on to P'Q'.

$$\therefore \; P'Q' = PQ \cos \theta.$$

Qu. 10. If, in Example 4, PQ is parallel to the common axis, what is its projected length?

Example 5. *A line* PQ *of length l, lies in the plane* π *and is projected on to the plane* π', *the angle of projection being* θ. *If* PQ *is inclined at an angle* α *to the common axis, find l', the projected length of* PQ.

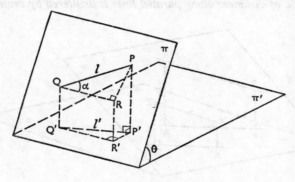

Fig. 14.8

Construct PR and QR in π, perpendicular and parallel to the common axis, respectively. Let P', Q', R' be the projections of P, Q, R on to π'. (See Fig. 14.8.)

From $\triangle$PQR, $PR = l \sin \alpha$.
                                          $QR = l \cos \alpha$.

$\therefore \; P'R' = l \sin \alpha \cos \theta$ (PR is $\perp$ to common axis).
$Q'R' = l \cos \alpha$ (QR is $\parallel$ to common axis).

By Pythagoras' theorem in $\triangle$P'Q'R',

$$l'^2 = l^2 \sin^2 \alpha \cos^2 \theta + l^2 \cos^2 \alpha.$$
$$\therefore \; l' = l\sqrt{(\sin^2 \alpha \cos^2 \theta + \cos^2 \alpha)}.$$

Note that if ϕ is the inclination of PQ to P'Q', then

$$\cos \phi = \sqrt{(\sin^2 \alpha \cos^2 \theta + \cos^2 \alpha)}.$$

Qu. 11. Find the projected length of a line of length l perpendicular to PQ and lying in π.

It has been proved that a line projects into another line and clearly a curve projects into another curve. Various geometrical properties which hold in the plane π are carried through under orthogonal projection and will hold in the projected plane π'.

Qu. 12. Show that parallel lines project into parallel lines, by obtaining a contradiction to their projections meeting in a point in π'.

The ratio of distances along parallel lines is unaltered by projection.

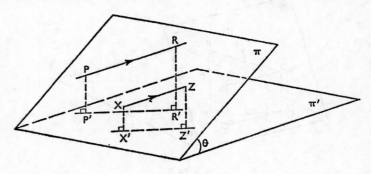

Fig. 14.9

Proof: In Fig. 14.9, PR, XZ are parallel lines in the plane π. Let α be the inclination of PR and XZ to the common axis. Let θ be the angle of projection.

If ϕ is the angle between PR and its projection P'R' on to π', it is also the angle between XZ and its projection X'Z', because (see Example 5) in each case,

$$\cos \phi = \sqrt{(\sin^2 \alpha \cos^2 \theta + \cos^2 \alpha)}.$$

$$\therefore \ P'R' = PR \cos \phi, \qquad X'Z' = XZ \cos \phi.$$

$$\therefore \ \frac{P'R'}{X'Z'} = \frac{PR}{XZ}.$$

This could not, of course, be true if XZ were not parallel to PR, because then ϕ would not be the same in each case. It follows that equal lengths do not project into equal lengths unless the original lengths are inclined at equal angles to the intersection of π and π'.

Qu. 13. If, in Fig. 14.9, Q is a point on PR, prove that PQ : QR is unaltered by projection.

Qu. 14. Prove that the mid-point of a line projects into the mid-point of the projected line.

An area is reduced in the ratio $\cos \theta : 1$ *by projection, where* θ *is the angle of projection.*

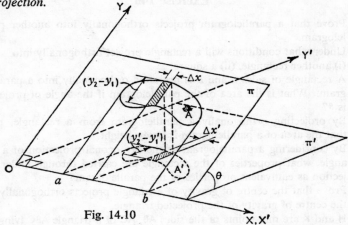

Fig. 14.10

Proof: In Fig. 14.10, A is an area lying in π and A' is the projected area in π'. Take the common axis as the x-axis for each plane, and OY and OY' perpendicular to the common axis as y-axes, one lying in each plane.

Take an element of area in π parallel to the y-axis and of width $\varDelta x$, and let the y-coordinates of its extremities be y_2 and y_1.

$$\text{Element of area in } \pi = (y_2 - y_1)\, \varDelta x.$$

$$\therefore \ \mathsf{A} = \int_a^b (y_2 - y_1)\, \mathrm{d}x.$$

The extremities of the projected element of area in π' has y-coordinates y'_2 and y'_1. As the element is perpendicular to the common axis

$$y'_1 = y_1 \cos \theta, \qquad y'_2 = y_2 \cos \theta.$$

Since $\Delta x'$, Δx are measured parallel to the common axis, $\Delta x' = \Delta x$.

$$\therefore \text{ element of area in } \pi' = (y'_2 - y'_1)\,\Delta x',$$
$$= (y_2 - y_1)\cos\theta.\Delta x.$$
$$\therefore A' = \int_a^b (y_2 - y_1)\cos\theta\,dx,$$
$$= A\cos\theta.$$

Qu. 15. With the notation above, show that $A' = A\cos\theta$ by taking elements of area parallel to the common axis.

Exercise 14b

1. Prove that a parallelogram projects orthogonally into another parallelogram.

2. Under what conditions will a rectangle project orthogonally into (i) another rectangle, (ii) a square.

3. A rectangle of sides a and b is projected orthogonally into a parallelogram. What is the area of the parallelogram if the angle of projection is θ?

4. By projecting orthogonally a parallelogram from a rectangle, prove that the area of a parallelogram is base × height.

5. By considering a parallelogram as the orthogonal projection of a rectangle, what properties of the rectangle are carried through under projection as equivalent properties of the parallelogram?

6. Prove that the centre of gravity of a triangle projects orthogonally into the centre of gravity of the projected triangle.

7. H and K are mid-points of the sides AB, AC of a triangle ABC lying in a plane π. Denoting by dashes the orthogonal projections of points on to a plane π', prove that H'K' is parallel to B'C'.

8. A line is inclined at an angle of 30° to the common axis and the angle of projection is 45°. Calculate the acute angle which the orthogonal projection of this line makes with the common axis.

9. A regular pentagon with one side parallel to the common axis is projected orthogonally into another pentagon. If the angle of projection is 35°, find the angles of the new pentagon.

10. Two triangles ABC and PQR lie in a plane with AB ∥ PQ, BC ∥ QR, AC ∥ PR. They are projected orthogonally on to another plane. Are the projected triangles similar?

    Under what condition will similar triangles project orthogonally into similar triangles?

11. Under what conditions will a square project orthogonally into a rhombus?

12. Find the angle of projection if a rectangle of sides a, b projects into a square of side b.

13. How could you place an isosceles triangle on a sloping plane so that its projection on to a horizontal plane was also an isosceles triangle?

14. The distance between two points on a plane inclined to the horizontal at θ is halved by projection on to the horizontal plane. Prove that the angle between the given line and the common edge of the planes is $\cot^{-1}\sqrt{\{(1-4\cos^2\theta)/3\}}$.

15. If the angle of greatest slope of a plane is θ and a slant path makes an angle α with a line of greatest slope, prove that the angle of slope of the slant path is $\sin^{-1}(\sin\theta\cos\alpha)$ and that the difference in the bearings of the path and of the line of greatest slope is $\tan^{-1}(\sec\theta\tan\alpha)$.

16. Two lines are inclined at angles α, β to the common axis of projection. If the angle of projection is θ, prove that if the projection of these lines make angles α', β' respectively with the common axis, then

$$\tan\alpha' = \tan\alpha\cos\theta,$$
$$\tan\beta' = \tan\beta\cos\theta.$$

If $\beta > \alpha$, prove that

$$\tan(\beta'-\alpha') = \frac{\cos\theta(\tan\beta-\tan\alpha)}{1+\cos^2\theta\tan\alpha\tan\beta}.$$

Prove that it is possible to choose θ such that $\beta'-\alpha'=90°$ only if

$$1+\tan\alpha\tan\beta < 0.$$

17. With the notation of No. 16 find a similar expression for $\tan(\beta-\alpha)$ in terms of α', β', θ. Prove that an angle $(\beta'-\alpha')$ can be orthogonally projected from a right angle if and only if $-1 < \tan\alpha'\tan\beta' < 0$.

18. Two lines lying in a plane π' and inclined at 60° and 160° to the common axis are projected orthogonally from two lines lying in a plane π. Calculate the angle of projection θ such that the two lines lying in π should be at right angles.

19. Show that by a suitable choice of common axis and angle of projection, any angle less than 180° can be orthogonally projected from a right angle.

20. Show that by a suitable choice of common axis and angle of projection, (i) a rectangle, (ii) a rhombus, can always be found which can be projected orthogonally into a given parallelogram.

21. OA, OB are lines in a horizontal plane making an angle α with each other. The point P is vertically above A and Q is vertically above B. Prove that

$$PQ^2 = OP^2 + OQ^2 - 2AP.BQ - 2OA.OB\cos\alpha.$$

Orthogonal projection of the circle and ellipse

14.4. *A circle projects into an ellipse.*

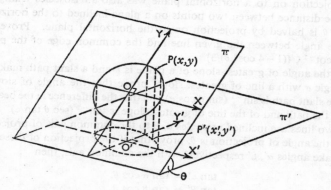

Fig. 14.11

Proof: Consider a circle, centre O, radius a, which lies in π and is projected at an angle θ on to π'. Take axes OX, OY in π, parallel and perpendicular to the common axis. Take O'X', O'Y' the projection of OX, OY on to π' as coordinate axes in π'. Let $P(x, y)$ be a point on the circle in π, and let $P'(x', y')$ be its projection on to π' (see Fig. 14.11).

Then $\qquad\qquad x' = x \quad \text{and} \quad y' = y \cos \theta.$ $\qquad$ (i)

The equation of the circle in π is $x^2 + y^2 = a^2$. On projection, this becomes (using (i))

$$x'^2 + \frac{y'^2}{\cos^2 \theta} = a^2.$$

$$\therefore \frac{x'^2}{a^2} + \frac{y'^2}{a^2 \cos^2 \theta} = 1.$$

$$\therefore \frac{x'^2}{a^2} + \frac{y'^2}{b^2} = 1,$$

where $b = a \cos \theta$.

This is the equation of the locus of P' in π', as P moves round the circle in π. It is the equation of an ellipse. In referring to the equations of the two loci, it is usual to drop the dashes and say that the circle $x^2 + y^2 = a^2$ projects into the ellipse

$$\frac{x^2}{a^2} + \frac{y^2}{b^2} = 1.$$

Qu. 16. Show that by a suitable choice of angle of projection, the ellipse

$$\frac{x^2}{a^2} + \frac{y^2}{b^2} = 1 \quad (a > b),$$

can be projected into the circle $x^2 + y^2 = b^2$.

It follows that for many theorems which can be proved about the circle there are corresponding theorems about the ellipse. Angle properties do not, in general, survive projection; so that angle theorems about the circle will have to be expressed differently and ratios of lengths along parallel lines must be considered when finding the corresponding theorem about the ellipse. We have seen that these ratios are unaltered by projection.

Example 6. *Deduce, by projection, a property of an ellipse corresponding to the fact that the angle in a semi-circle is a right angle.*

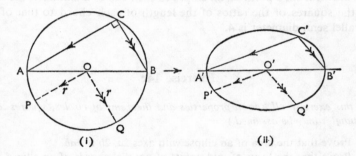

Fig. 14.12

The circle and diameter AB, and the point C on the circle, project into the ellipse and diameter A′B′ and the point C′ on the ellipse.

Since angle properties do not project, we must use the right angle to get a connection between lengths.

$$AC^2 + BC^2 = AB^2. \tag{1}$$

Only ratios of lengths along parallel lines remain unaltered in projection, so we construct radii OP, OQ parallel to CA and CB in Fig. 14.12 (i). These project into O′P′, O′Q′ parallel to C′A′, C′B′ in Fig. 14.12 (ii).

Equation (1) is now modified to read:

$$\frac{AC^2}{OP^2} + \frac{CB^2}{OQ^2} = \frac{AB^2}{r^2},$$

where $OP = OQ = r$.

$$\therefore \frac{AC^2}{OP^2} + \frac{CB^2}{OQ^2} = \frac{4r^2}{r^2}.$$

$$\therefore \frac{AC^2}{OP^2} + \frac{CB^2}{OQ^2} = 4.$$

Under projection, this becomes

$$\frac{A'C'^2}{O'P'^2} + \frac{C'B'^2}{O'Q'^2} = 4$$

because

$$\frac{AC}{OP} = \frac{A'C'}{O'P'}, \qquad \frac{CB}{OQ} = \frac{C'B'}{O'Q'}.$$

The corresponding property of an ellipse is therefore, 'When chords are drawn from a point on an ellipse to each end of a diameter, the sum of the squares of the ratios of the length of each chord to that of its parallel semi-diameter is 4.'

Exercise 14c

(*In this exercise all known properties and theorems of circles, squares and rectangles may be assumed.*)

1. Prove that the area of an ellipse with axes $2a$, $2b$ is πab.
2. Prove that the locus of mid-points of parallel chords of an ellipse is a diameter.
3. A rectangle of sides a, b is projected into a square. Find in its simplest form the equation of the ellipse which is the projection of the circumcircle of the rectangle given that $a > b$.
4. Prove that the tangents at the ends of a diameter of an ellipse are parallel.
5. A chord of an ellipse passes through a fixed point. Find the locus of its mid-point.
6. An ellipse touches the sides of a parallelogram internally. Prove that if the ellipse is projected orthogonally into a circle, then the parallelogram is projected into a rhombus.
7. If a line cuts two concentric ellipses at A, B and C, D, respectively, prove that $AC = BD$.

8. A circle touches all four sides of a square internally. Under what conditions will this project into an ellipse touching all four sides of: (i) a parallelogram, (ii) a rectangle.

9. In an ellipse a diameter PR bisects chords parallel to another diameter QT. Prove that QT bisects chords parallel to PR. (PR and QT are called conjugate diameters.)

10. Two tangents TP, TQ are drawn from an external point T to an ellipse centre C. CT meets the ellipse at R and PQ at S. Prove that $CS . CT = CR^2$.

11. AB and CD are conjugate diameters (see No. 9) of an ellipse

$$\frac{x^2}{a^2} + \frac{y^2}{b^2} = 1.$$

Prove that the area of the figure formed by drawing tangents at A, B, C, D is $4ab$.

12. Prove that the eccentric angles of the points at the ends of a pair of conjugate diameters of an ellipse differ by 90° (see No. 9).

13. Two similar ellipses (the ratio of major to minor axes is the same) lie in the same plane with their major axes parallel. The larger one is touched internally by the smaller one at a point A. Lines are drawn through the point A to cut the ellipses at P, Q and R, S, respectively. Prove that PR is parallel to QS.

14. Two similar ellipses with their major axes parallel and lying in the same plane intersect at P, Q. A is a point on one ellipse and AP, AQ are produced to cut the other ellipse in R, S. Prove that the tangent to the ellipse at A is parallel to RS.

15. Prove that in general under orthogonal projection an ellipse projects into another ellipse.

16. P is a point at one end of the major axis of an ellipse. Show that the locus of mid-points of chords through P is a similar ellipse which touches the original ellipse at P.

17. An ellipse has major and minor axes, a, b, respectively. Prove that the largest quadrilateral which can be inscribed in it is a rectangle, and find its area.

18. Prove that, if a point P lies inside an ellipse but is not the centre, there is one and only one chord QR of the ellipse which is bisected at P.

O.C.

19. A hemisphere and a right cylinder stand on the same circular base, whose radius is a. (They are on the same side and not on opposite sides of the base.) A plane π is inclined at an angle θ to the base and cuts it in a line distant $2a$ from its centre. The plane cuts the hemisphere in a circle whose area is A_1 and cuts the cylinder in an ellipse whose area is A_2. Prove that $A_1 = A_2 \cos 3\theta$. O.C.

20. An ellipse S, of eccentricity $2/\sqrt{5}$, is projected orthogonally on to a plane which is inclined to the plane of S at an angle of $60°$ and which cuts it in a line parallel to the minor axis. Prove (i) that the projected figure is an ellipse S' and (ii) that its eccentricity is $1/\sqrt{5}$.   o.c.

21. Prove that infinitely many triangles can be inscribed in an ellipse so that the tangent at each vertex is parallel to the opposite side. Prove also that the triangles are equal in area, and that the ratio of this area to the area of the ellipse is $3\sqrt{3}:4\pi$.   o.c.

22. Prove that an ellipse can be drawn to touch each side of a given parallelogram at its mid-point. Find the ratio of the area of this ellipse to the area of the parallelogram.   o.c.

23. The centre of an ellipse is C, the major axis is AA', and P is any point on the line of the minor axis BB'. The line AP meets the ellipse again at Q, and RR' is the diameter of the ellipse parallel to AP. Prove that $AP \cdot AQ = 2CR^2$.

     Given further that P lies on CB produced so that $CP = \sqrt{3}\,CB$, prove that APR'R is a parallelogram.   o.c.

CHAPTER 15

MENSURATION AND MOMENTS OF INERTIA

The circle

15.1. It is assumed that in elementary work the reader will have used formulae for the circumference and area of a circle, and for the surface areas and volumes of a right circular cylinder, a right circular cone, and a sphere. In the first part of this chapter this work will be revised and extended.

Qu. 1. A regular, n-sided polygon of perimeter P_i is inscribed in a circle of radius r. A similar polygon of perimeter P_c is circumscribed about the circle. Show that:

   (i) $P_i = 2rn \sin (180°/n)$,   (ii) $P_c = 2rn \tan (180°/n)$,

   (iii) $P_i/P_c \rightarrow 1$ as $n \rightarrow \infty$.

We thus see that P_i, P_c both approach the same limit as $n \rightarrow \infty$. So we may define the length of the circumference of a circle as this limit. If the limit of $n \sin (180°/n)$ and $n \tan (180°/n)$ is called π, it follows that P_i, $P_c \rightarrow 2\pi r$ as $n \rightarrow \infty$.

Qu. 2. If the areas of the polygons in Qu. 1 are A_i, A_c, respectively, show that $A_i/A_c \rightarrow 1$ as $n \rightarrow \infty$. Hence the area of a circle may be defined as the limit of A_i, A_c as $n \rightarrow \infty$.

Show that $A_c = \frac{1}{2}r.P_c$ and deduce the formula for the area of the circle.

Right circular cone

15.2. The formulae for the curved surface area and for the volume of a right circular cone are not difficult to obtain and are left as exercises for the reader in the following questions.

Qu. 3. A right circular cone has base radius r and slant height l. Consider the curved surface to be slit along a generator and spread out in a plane as a sector of a circle. Hence show that the area is πrl.

Qu. 4. A right circular cone has base radius *r* and height *h*. By considering the volume of the solid of revolution generated when one of the right-angled triangles of Fig. 15.1 is rotated through four right angles about the *x*-axis, show that the volume of the cone is $\frac{1}{3}\pi r^2 h$.

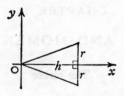

Fig. 15.1

Frustum of a right circular cone

15.3. Consider a right circular cone of base radius *R*, height *X*, vertex O, from which another right circular cone of base radius *r*, height *x* has been removed. The part remaining is called a *frustum* of a cone.

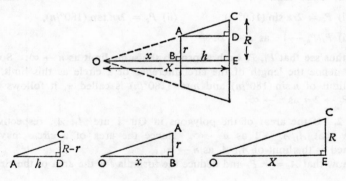

Fig. 15.2

Figure 15.2 shows a section of the cone through its axis OBE. AB, CE are radii of the ends of the frustum and D is the foot of the perpendicular from A to CE.

Volume of frustum = volume of original cone − volume removed
$$= \tfrac{1}{3}\pi R^2 X - \tfrac{1}{3}\pi r^2 x.$$

Now triangles OAB, OCE, ACD are similar so that

$$X = h.\frac{R}{R-r}, \quad x = h.\frac{r}{R-r}.$$

$$\therefore \text{ volume of frustum} = \tfrac{1}{3}\pi\left\{R^2.\frac{hR}{R-r} - r^2.\frac{hr}{R-r}\right\},$$

$$= \tfrac{1}{3}\pi h\left(\frac{R^3-r^3}{R-r}\right),$$

$$= \tfrac{1}{3}\pi h(R^2+Rr+r^2), \tag{1}$$

using the identity $a^3 - b^3 = (a-b)(a^2+ab+b^2)$ (p. 160).

Another form of (1) may be obtained by writing $A = \pi R^2$, $A' = \pi r^2$ so that A, A' are the areas of the ends of the frustum.   (1) may be written

$$\tfrac{1}{3}h(\pi R^2 + \pi Rr + \pi r^2) = \tfrac{1}{3}h(A + \sqrt{(AA')} + A'). \tag{2}$$

Being familiar with the factor $\tfrac{1}{3}h$ in the formula $\tfrac{1}{3}\pi r^2 h$, the reader will not be surprised to see it again here.   The formula may be thought of as

height × 'average area' of cross-section,

where the 'average area' of the cross-section is $\tfrac{1}{3}\{A + \sqrt{(AA')} + A'\}$. The advantage of (2) over (1) is that the result also holds for the frustum of a pyramid: the proof follows similar lines and is left (Exercise 15a, No. 13) to the reader.

Note that $\sqrt{(AA')}$ is the geometric mean of A, A' (see P.M.I., p. 224).

The curved surface area of the frustum of a right circular cone may be deduced from the formula πrl (see Qu. 3).   The method is so similar to that for the volume (and the working is easier), that it is left to the reader to deduce the formula.

Qu. 5.   A frustum of a right circular cone has slant height l and the radii of the ends of the frustum are r, R.   Show that the area of the curved surface is

$$\pi(R+r)l.$$

Qu. 6.   Show that the area of the curved surface of a frustum of a right circular cone is given by

slant height × average circumference of the ends.

Qu. 7.   Check the formulae $\tfrac{1}{3}\pi h(R^2 + Rr + r^2)$, $\pi(R+r)l$ by substituting (i) $r = 0$, (ii) $R = r$.

Surface area of a sphere

15.4. The area of a curved surface is much more difficult to define than the length of a curve and we shall not attempt to justify the following method.

Consider the circle $x^2 + y^2 = a^2$ and draw the radius making θ with the x-axis (Fig. 15.3). Take an element of arc $a\varDelta\theta$ and rotate it

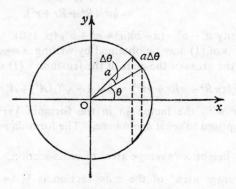

Fig. 15.3

through four right angles about the x-axis. This generates part of the surface of the sphere. We shall approximate to this element by regarding it as the surface of a frustum of a cone so that its area is approximately (see Qu. 5, p. 319)

$$\pi\{y + (y + \varDelta y)\}a\varDelta\theta.$$

Hence the area of the sphere

$$A = \int_0^\pi 2\pi y a\, d\theta.$$

But $y = a \sin \theta$

$$\therefore A = \int_0^\pi 2\pi a^2 \sin \theta\, d\theta,$$

$$= \left[-2\pi a^2 \cos \theta \right]_0^\pi.$$

$$\therefore A = 4\pi a^2.$$

A more general result, the area on the sphere between two parallel planes, can be obtained by integrating between α and β (see Fig. 15.4).

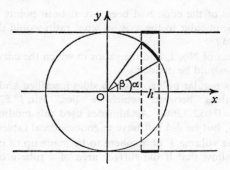

Fig. 15.4

$$\text{Area cut off on sphere} = \Big[-2\pi a^2 \cos \theta \Big]_\alpha^\beta,$$
$$= 2\pi a^2(\cos \alpha - \cos \beta),$$
$$= 2\pi a(a \cos \alpha - a \cos \beta),$$
$$= 2\pi ah,$$

where h is the distance between the two planes. But if a cylinder with its axis perpendicular to the planes is drawn circumscribing the sphere, the area cut off on the cylinder is $2\pi ah$, which is equal to the area cut off on the sphere. This remarkable result was discovered by Archimedes (287–212 B.C.), who did not have the advantage of the calculus as we know it.

The area cut off on a sphere by two parallel planes is called a *zone*.

Qu. 8. Express in your own words the theorem just proved.

Qu. 9. Find the volume of a sphere of radius a by considering the solid generated when that part of the circle $x^2 + y^2 = a^2$ in the first two quadrants is rotated through four right angles about the x-axis.

Exercise 15a

1. A sphere of radius a is divided by a plane whose least distance from the centre is d. Find the ratio in which the plane divides the surface area of the sphere when d is (i) $\frac{1}{3}a$, (ii) $\frac{1}{2}a$, (iii) $\frac{2}{3}a$.

2. A right circular cone is cut by a plane perpendicular to its axis and passing through its mid-point. Show that the volume of the cone is divided in the ratio 1:7. (Continued on p. 322.)

If the axis of the cone had been cut at both points of trisection by planes perpendicular to the axis, in what ratio would the volume have been divided?

3. With the data of No. 2, find the ratios in which the curved surface area of the cone would be divided.

4. A circle has regular polygons of 96 sides inscribed and circumscribed. Deduce values between which π lies. [$\sin 1° 52\frac{1}{2}' = 0.032\,7191$, $\tan 1° 52\frac{1}{2}' = 0.032\,7366$. Archimedes used this method to show that $3\frac{10}{71} < \pi < 3\frac{1}{7}$, but he did not have trigonometrical tables to refer to!]

5. Consider the volume V of a sphere to be made up of concentric shells and hence show that if the surface area of a sphere of radius r is A,

$$\frac{\mathrm{d}V}{\mathrm{d}r} = A.$$

6. Deduce the area of a circle from the formula for the circumference by considering the area to be made up of thin concentric rings of thickness Δr.

★7. The area enclosed by a curve lying in a plane is A. O is any point at distance h from the plane. Show that, if P describes the curve, the volume of the cone generated by OP is $\frac{1}{3}Ah$. [Remember that areas of similar figures are proportional to the squares of corresponding lengths.]

8. Consider the volume of a sphere to be made up of cones (not necessarily circular) with bases on the surface of the sphere and vertices at the centre. Hence show that, for a sphere,

volume = $\frac{1}{3}$ radius × surface area.

9. A sphere of radius r is circumscribed by a right circular cone whose base also touches the sphere. If the volume of the cone is V and the total surface area of the cone is S, show that $V = \frac{1}{3}rS$.

10. A sector of a circle, radius a, contains an angle 2α at the centre. If this sector is rotated through two right angles about its axis of symmetry, what is the volume of the solid generated?

11. Find the volume of a right circular cone of base radius r, height h, by taking an element of volume in the form of a circular lamina perpendicular to the axis of the cone. [Take the *vertex* of the cone as origin and use the theorem that areas of similar figures are proportional to the squares of corresponding lengths.]

12. The base of a right pyramid is a regular polygon of perimeter P. A frustum is formed from the pyramid in such a way that the perimeter of the top is p. Show that, if the shortest distance between corresponding sides of the top and bottom of the frustum is q, the total area of the slant faces is $\frac{1}{2}q(P+p)$.

13. The two ends of the frustum of a pyramid are of area S, S' and the distance between their planes is d. Show that the volume of the frustum is $\frac{1}{3}d\{S + \sqrt{(SS')} + S'\}$. [See hint of No. 11.]

★14. Find the equation of a circle of radius r referred to a tangent as x-axis and the perpendicular diameter as positive y-axis. Hence show that the volume of a cap of height h cut off from a sphere (by a plane at distance $r - h$ from the centre) is

$$\tfrac{1}{3}\pi h^2(3r - h).$$

Check that the formula gives the volume of a sphere and a hemisphere when suitable values of h are substituted.

15. If a sphere is divided as in No. 1, find the ratios in which the volume is divided.

16. The axes of two cylindrical shells of radius a intersect at right angles in a horizontal plane. It is required to find the volume V of the space which lies inside both cylinders. Consider a horizontal plane at height h above the axes.
    (i) Describe the intersection of the plane with one cylinder.
    (ii) Describe the intersection of the plane with the boundary of the space common to both cylinders.
    (iii) Find the area enclosed by the figure in (ii) and hence
    (iv) find V.

17. Find the position of the centroid of (i) a uniform solid pyramid, (ii) a uniform solid cone, both of height h.

18. Find the position of the centroid of a uniform shell in the form of an open right circular cone of height h.

19. Find the position of the centroid of an open uniform hemispherical shell of radius a.

    If the hemisphere is closed by a uniform circular face of the same material, where is the new centroid?

20. A cap of a uniform solid sphere of radius r is of height h. Find the distance of the centroid from the plane surface of the cap.

    Check your result by substituting $h = 0$ and $h = 2r$. What is the result for a hemisphere?

21. Find the position of the centroid of a frustum of a uniform right circular conical shell bounded by circles of radius r, R at a distance h apart. [Give answer as distance from centre of circle of radius R.]

22. A frustum of a uniform solid right circular cone has ends with radii r, R and height h. Show that the centroid is at a distance

$$\tfrac{1}{4}h(R^2 + 2rR + 3r^2)/(R^2 + Rr + r^2)$$

from the face of radius R.

What results do you obtain by substituting (i) $r = 0$, (ii) $r = R$?

Moments of inertia

15.5. Moments of inertia arise from the study of the dynamics of a rigid body.   The commonest approach is to consider the kinetic energy of a rigid body which is free to rotate about a fixed axis.   Figure 15.5

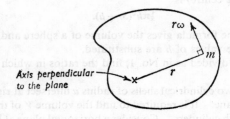

Fig. 15.5

shows a section of such a rigid body in a plane perpendicular to the axis.   Consider a particle of mass m at a distance r from the axis. When the body is rotating with an angular velocity ω, the particle has a velocity $r\omega$, therefore it has a kinetic energy of $\frac{1}{2}m(r\omega)^2$.   Hence the total kinetic energy of the body is

$$\sum \tfrac{1}{2}mr^2\omega^2,$$

where $\sum$ denotes a summation over all the particles of the body.   Now ω is the same for all particles of the body, so we may write the kinetic energy as

$$\tfrac{1}{2}(\sum mr^2)\omega^2.$$

Thus we have obtained an expression in the same form as that for the kinetic energy of a particle

$$\tfrac{1}{2}Mv^2.$$

The velocity v corresponds to the angular velocity ω, and the mass, or *inertia*, M corresponds to the expression $\sum mr^2$ which is called the *moment of inertia* of the body about the axis and is denoted by I.

For theoretical work, the calculus provides a method of carrying out the summation involved in finding moments of inertia.   A number of the results so obtained are conveniently summarized in Routh's rule, which may be found in mechanics text-books.   On the other hand, in practical situations it is often more convenient to find the moment of inertia of a body by experiment.   We shall, since this book is concerned with pure mathematics, stress applications of the calculus.

Moments of inertia from first principles

15.6.   First consider a simple case which does *not* require calculus.

Example 1.   *Find the moment of inertia of a uniform, thin, circular ring, radius a, mass M, about an axis through its centre and perpendicular to the plane of the ring.*

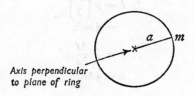

Axis perpendicular
to plane of ring

Fig. 15.6

Consider a particle of mass m at any point on the ring (Fig. 15.6).

$$I = \sum mr^2.$$

r, the distance of the mass from the axis, is equal to a, the radius of the ring.   Therefore

$$I = \sum ma^2.$$

But a is the same for all particles on the ring.

$$\therefore \ I = (\sum m)a^2.$$
$$\therefore \ I = Ma^2.$$

Example 2.   *Find the moment of inertia of a uniform rod, mass M, length 2a, about an axis perpendicular to the rod and passing through it mid-point.*

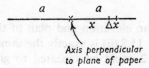

Axis perpendicular
to plane of paper

Fig. 15.7

Consider an element of the rod, length Δx, at a distance x from the mid-point (Fig. 15.7).

$$I = \sum mr^2.$$

Let the mass per unit length of the rod be ρ, then the mass of the element is $\rho\,\Delta x$. The distance of the element from the axis is x; therefore the moment of inertia of the element is $\rho\,\Delta x . x^2$.

$$\therefore \text{ total moment of inertia } I = \int_{-a}^{a} \rho x^2\,\mathrm{d}x.$$

$$\therefore I = \rho \left[\frac{x^3}{3}\right]_{-a}^{a},$$

$$= \tfrac{2}{3}\rho a^3.$$

But $\rho = \dfrac{M}{2a}$, $\therefore I = \dfrac{2}{3} \cdot \dfrac{M}{2a} \cdot a^3$.

$$\therefore I = \tfrac{1}{3}Ma^2.$$

Qu. 10. Find, by taking different limits of the integral, the moment of inertia about a parallel axis through one end of the rod in Example 2.

Now consider the moment of inertia of a uniform rectangular lamina, mass M, sides $2a$, $2b$, about the axis of symmetry parallel to the sides of length $2b$.

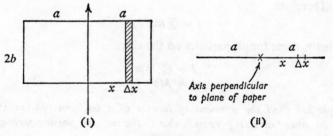

Fig. 15.8

Figure 15.8 shows an elevation and plan of the lamina. You may have noticed that Fig. 15.8 (ii) is exactly the same as Fig. 15.7. So the working of Example 2 could be repeated to give precisely the same answer. This is because the particles of the element are at the *same distance from the axis*.

Qu. 11.   What is the moment of inertia of a uniform cylindrical shell of mass M, radius a, height h (open at both ends), about its axis?

Example 3.   *Find the moment of inertia of a uniform circular disc of mass M, radius a, about an axis through its centre perpendicular to its plane.*

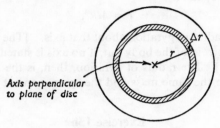

Axis perpendicular
to plane of disc

Fig. 15.9

The formula $\sum mr^2$ is easiest to use if we can find an element such that all the particles in it are at the same distance from the axis. We therefore take an element in the form of a concentric ring of radius r, thickness Δr.

Since the ring is thin, we may write (correct to the term in Δr)

$$\text{area of ring} \simeq \text{circumference} \times \text{thickness}$$
$$= 2\pi r \, . \, \Delta r.$$

[To justify this: $\text{area} = \pi(r+\Delta r)^2 - \pi r^2,$
$$= \pi\{2r\,\Delta r + (\Delta r)^2\},$$
$$\simeq 2\pi r \, \Delta r.]$$

If the mass per unit area of the disc is σ,

$$\text{mass of the element} \simeq \sigma \, . \, 2\pi r \, \Delta r.$$

$\therefore$ moment of inertia of element $\simeq 2\pi\sigma r \, \Delta r \, . \, r^2.$

$\therefore$ total moment of inertia $= \displaystyle\int_0^a 2\pi\sigma r^3 \, \mathrm{d}r,$

$$= 2\pi\sigma \left[\frac{r^4}{4}\right]_0^a,$$
$$= \tfrac{1}{2}\pi a^4 \sigma.$$

But $\sigma = \dfrac{M}{\pi a^2},$ $\therefore I = \tfrac{1}{2}\pi a^4 \, . \, \dfrac{M}{\pi a^2}.$

$\therefore I = \tfrac{1}{2}Ma^2.$

Qu. 12. What is the moment of inertia of a uniform solid circular cylinder of mass M, radius a, height h about its axis?

It is sometimes convenient to introduce another concept, that of *radius of gyration*. If the moment of inertia of a body, mass M, about a certain axis is I, and if

$$I = Mk^2,$$

k is called the radius of gyration about that axis. [The axis need not be through the centroid of the body but, if no axis is stated, it is sometimes assumed to be.] The radius of gyration, then, is the radius of a ring or cylinder with the same mass and the same moment of inertia as the body.

Exercise 15b

1. A light equilateral triangle of side $2a$ has unit masses at its vertices. Find the moment of inertia of the system about (i) an axis of symmetry, (ii) an axis perpendicular to the plane of the triangle and passing through its centre.
2. A light square, side $2a$, has unit masses at its vertices. Find the moment of inertia of the system about (i) an axis joining the mid-points of a pair of opposite sides, (ii) a diagonal, (iii) an axis perpendicular to the plane of the square and passing through its centre.

Find, by integration, the moments of inertia of the bodies in Nos. 3–9 about the axes indicated.

3. A uniform rod of mass M, length a about a perpendicular axis through one end.
4. A uniform lamina of mass M, bounded by concentric circles of radii a, b about an axis perpendicular to the plane of the lamina and passing through the centre of the circles.
   [Show that the mass per unit area $\sigma = M/\{\pi|a^2 - b^2|\}$.]
5. A uniform isosceles triangular lamina of mass M, height h, base $2a$, about an axis parallel to the base and passing through the vertex of the lamina.
6. A uniform rod of mass M, length $2a$, about a perpendicular axis through a point on the rod at a distance h from its centre.
7. A uniform rod of mass M, length $2a$, about a perpendicular axis which does not meet the rod, given that the shortest distance between the centroid of the rod and the axis is h.
8. The lamina of No. 5 about its axis of symmetry.
9. The lamina of No. 5 about its base.

10. A rod has mass M, length $2a$, and its density varies as the distance from one end. Find the moment of inertia of the rod about axes perpendicular to its length (i) through the vertex, (ii) through its centroid.

11. A circular flywheel is made in such a way that its density varies as the distance from its axis. If the radius of the flywheel is a, find its radius of gyration.

12. Find the radius of gyration about its axis of a thin uniform right circular conical shell of base radius a, height h, slant height l.

13. Deduce the moment of inertia in No. 4 from the moment of inertia of a uniform circular disc. [Note that the mass of the annulus is M: find the masses of discs of radii a and b in terms of M.]

14. Find the moment of inertia of a uniform spherical shell of mass M, radius a, about a diameter.

15. The moments of inertia of a lamina about two perpendicular axes lying in its plane are I_x, I_y. If I_z is the moment of inertia of the lamina about an axis perpendicular to its plane and passing through the point of intersection of the first two axes, show that $I_z = I_x + I_y$.

Perpendicular axes theorem

15.7. No. 15 of Exercise 15b has anticipated this theorem. It is, however, desirable to draw attention to the conditions of the theorem and to illustrate its use.

The theorem is sometimes called the *lamina theorem* in order to draw attention to the fact that the theorem only applies to bodies which lie in a plane. (This is perhaps not at all obvious if Fig. 15.5 (p. 324) and Fig. 15.10 are compared.) The theorem may be stated as follows:

If OX, OY *are perpendicular axes in the plane of a lamina and if* I_x, I_y *are the moments of inertia of the lamina about* OX, OY, *then the*

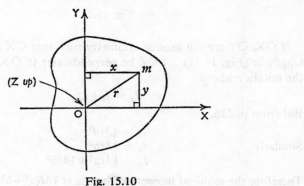

Fig. 15.10

moment of inertia I_z about an axis OZ perpendicular to OX and OY is given by

$$I_z = I_x + I_y.$$

Consider an element, mass m, at (x, y). Note that the distance of the element from OX is y so that

$$I_x = \sum my^2. \tag{1}$$

Similarly

$$I_y = \sum mx^2. \tag{2}$$

But, if r is the distance of the element from O, we have

$$I_z = \sum mr^2, \tag{3}$$

and

$$r^2 = y^2 + x^2. \tag{4}$$

From (3) and (4),

$$I_z = \sum m(y^2 + x^2),$$
$$= \sum my^2 + \sum mx^2.$$

So, from (1) and (2)

$$I_z = I_x + I_y.$$

Example 4. *A rectangle of mass M has sides $2a$, $2b$. Find the moment of inertia about an axis through the centre of the rectangle perpendicular to its plane.*

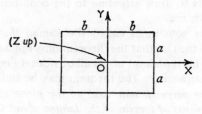

Fig. 15.11

If OX, OY are the axes of symmetry such that OX cuts the sides of length $2a$ (Fig. 15.11), let OZ be perpendicular to OX, OY, then with the notation above

$$I_z = I_x + I_y.$$

But from p. 326,

$$I_x = \tfrac{1}{3}Ma^2.$$

Similarly,

$$I_y = \tfrac{1}{3}Mb^2.$$
$$\therefore \ I_z = \tfrac{1}{3}Ma^2 + \tfrac{1}{3}Mb^2.$$

Therefore the required moment of inertia is $\tfrac{1}{3}M(a^2 + b^2)$.

NOTE: Physics books generally give the moment of inertia of a rectangular bar magnet as $M(l^2 + w^2)/12$ where l is its length and w its width.

Example 5. *Find the moment of inertia of a square, mass M, side 2a, about a diagonal.*

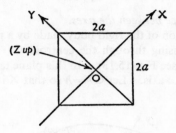

Fig. 15.12

Take axes OX, OY along the diagonals of the square and OZ perpendicular to its plane. Using the same notation for the moments of inertia as before, we may deduce I_z from Example 4 by substituting $b = a$. Thus

$$I_z = \tfrac{2}{3}Ma^2.$$

Now, from symmetry, $I_x = I_y$. But by the perpendicular axes theorem,

$$I_x + I_y = I_z.$$
$$\therefore \ 2I_x = \tfrac{2}{3}Ma^2.$$
$$\therefore \ I_x = \tfrac{1}{3}Ma^2.$$

Therefore the moment of inertia of the square about a diagonal is $\tfrac{1}{3}Ma^2$.

Qu. 13. Write down the moment of inertia of a uniform ring of mass M, radius a, about an axis perpendicular to its plane and passing through its centre. Deduce the moment of inertia of the ring about a diameter.

Parallel axes theorem

15.8. This theorem, together with the perpendicular axes theorem, enables a great range of moments of inertia to be found. The theorem, unlike the previous one, applies to three dimensional bodies as well as laminas. There is, however, an important restriction that one axis must pass through the centroid of the body. The theorem may be stated as follows:

If the moment of inertia of a rigid body of mass M about an axis through its centroid is I_G and the moment of inertia about a parallel axis is I, then

$$I = I_G + Mh^2$$

where h is the distance between the axes.

Consider the section of the rigid body made by a plane perpendicular to the axes and passing through the centroid G. Let the other axis cut this plane in X (see Fig. 15.13). In this plane take axes with origin at G, GX being the *x*-axis. Let $GX = h$ so that X is $(h, 0)$.

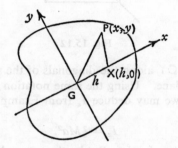

Fig. 15.13

Consider as an element of mass a thin prism through P (x, y) with its axis parallel to the given axes. Let the mass of this prism be m.

$$I = \sum mr^2.$$

With axis at G: $I_G = \sum m GP^2.$

With axis at X: $I_X = \sum m XP^2.$

By the distance formula,

$$GP^2 = x^2 + y^2 \quad \text{and} \quad XP^2 = (x-h)^2 + y^2,$$

$$\therefore I_G = \sum m(x^2 + y^2),$$

and $$I_X = \sum m\{(x-h)^2 + y^2\}.$$

$$\therefore I_X = \sum m\{x^2 + y^2 + h^2 - 2xh\},$$
$$= \sum m(x^2 + y^2) + \sum mh^2 - 2h \sum mx.$$

Now the x-coordinate of the centroid is given by $\bar{x} = (\sum mx)/(\sum m)$ (P.M.I., p. 156). But since G was taken to be the origin, $\bar{x} = 0$; therefore $\sum mx = 0$.

Further,
$$\sum mh^2 = (\sum m)h^2 = Mh^2,$$
$$\therefore\ I_x = I_G + Mh^2.$$

NOTE: (i) As an aid to memory, observe that the term Mh^2 is equivalent to adding a mass equal to that of the body at a distance h from the axis.

(ii) The result may be expressed a little more simply making use of the radius of gyration.   If k_x, k_G are the radii of gyration about axes through X, G respectively,
$$k_x{}^2 = k_G{}^2 + h^2.$$

SUMMARY:

(i) Moments of inertia were defined by the formula $I = \sum mr^2$ (p. 324).

(ii) The following three formulae are frequently required:

a uniform ring or hollow cylinder Ma^2 (pp. 325, 327);

a uniform disc or solid cylinder $\frac{1}{2}Ma^2$ (pp. 327, 328);

a uniform rod or rectangle, length $2a$ $\frac{1}{3}Ma^2$ (pp. 325, 326).

If in doubt about the axes for which the formulae hold, turn to the pages referred to.

(iii) The perpendicular axes theorem (p. 329) and the parallel axes theorem (p. 332), together with these results, enable a great range of moments of inertia to be found.

Example 6.   *Find the moment of inertia of a uniform circular ring of mass M, radius a, about a tangent.*

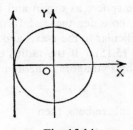

Fig. 15.14

The moment of inertia about an axis through the centre O of the ring and perpendicular to its plane is Ma^2 (Example 1, p. 325). Let I_x, I_y be the moments of inertia about two perpendicular diameters of the ring; then, by the perpendicular axes theorem,

$$I_x + I_y = Ma^2.$$

From symmetry, $I_x = I_y$,
∴ $I_y = \tfrac{1}{2}Ma^2$.

By the parallel axes theorem, the moment of inertia about a tangent parallel to OY is

$$\tfrac{1}{2}Ma^2 + Ma^2.$$

Therefore the moment of inertia about a tangent is $3Ma^2/2$.

Example 7. *Find the moment of inertia of a uniform solid sphere of mass M, radius r, about a diameter.*

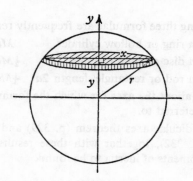

Fig. 15.15

Take the centre of the sphere as origin and let the diameter be the y-axis. As an element, consider the disc formed by sections of the sphere by planes perpendicular to the y-axis and passing through points given by y, $y + \Delta y$ (Fig. 15.15). If the radius of the circle formed by the former section is x, by Pythagoras' theorem,

$$x^2 = r^2 - y^2.$$

Let the density of the sphere be ρ, then

$$\text{mass of element} \simeq \pi x^2 \, \Delta y . \rho.$$

Using the formula $\frac{1}{2}Ma^2$ (p. 327),

moment of inertia of element $\simeq \frac{1}{2}(\pi x^2 \, \Delta y \, \rho)x^2$.

$$\therefore \; I_y = \int_{-r}^{r} \frac{1}{2}(\pi \rho x^4) \, dy, \qquad\qquad (r^2 - y^2)^2$$

$$= \frac{1}{2}\pi\rho \int_{-r}^{r} (r^2 - y^2)^2 \, dy, \qquad = r^4 - 2r^2y^2 + y^4.$$

$$= \frac{1}{2}\pi\rho \left[r^4 y - \frac{2}{3}r^2 y^3 + \frac{1}{5}y^5 \right]_{-r}^{r},$$

$$= \frac{1}{2}\pi\rho . 2 \left[r^4 y - \frac{2}{3}r^2 y^3 + \frac{1}{5}y^5 \right]_{0}^{r},$$

$$= \pi\rho(r^5 - \frac{2}{3}r^5 + \frac{1}{5}r^5), \qquad\qquad 1 - \frac{2}{3} + \frac{1}{5}$$

$$= \frac{8}{15}\pi\rho r^5. \qquad\qquad\qquad = \frac{15 - 10 + 3}{15}$$

But $\qquad\qquad \rho = \dfrac{M}{\frac{4}{3}\pi r^3}. \qquad\qquad\qquad = \frac{8}{15}.$

$$\therefore \; I_y = \frac{8}{15}\,\pi \cdot \frac{3M}{4\pi r^3} \cdot r^5 = \frac{2}{5}\,Mr^2.$$

Therefore the moment of inertia of the sphere about a diameter is $\frac{2}{5}Mr^2$.

Exercise 15c

It is intended that calculus should not be used in Nos. 1–13.

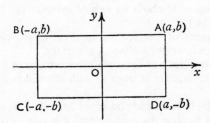

Fig. 15.16

1. Masses m are placed at the vertices of the rectangle in Fig. 15.16. Find, from first principles, the moments of inertia of the system about the two axes and about a perpendicular axis through the origin. Verify that the perpendicular axes theorem holds in this case.

2. With the data of No. 1, write down the moment of inertia of the system about the y-axis. Find, from first principles, the moment of inertia of the system about $x = -a$ and verify that the parallel axes theorem holds in this case.

3. If Fig. 15.16 represents four uniform rods of mass m per unit length, find the moment of inertia of the system about (i) AD, (ii) an axis through the origin perpendicular to the coordinate axes.

4. If Fig. 15.16 represents a uniform lamina of mass M, find the moment of inertia about a perpendicular axis through B (i) by using the perpendicular axes theorem applied at B, (ii) by first finding the moment of inertia about an axis perpendicular to Ox, Oy and passing through O.

5. Find the moment of inertia of a uniform circular disc radius a, mass M, about a diameter.

6. Find the moment of inertia of a uniform ring of radius a, mass M, about a chord which subtends an angle of $120°$ at the centre of the ring.

7. A square is formed from four uniform rods each of mass M and length $2a$. What is the moment of inertia of the system (i) about a perpendicular axis through the centre, (ii) about an axis through the centre, parallel to one pair of sides.

8. A uniform square lamina has mass $2M$ and sides $2a$. Find its moment of inertia about a diagonal. Deduce the moment of inertia, about the hypotenuse, of a uniform right-angled isosceles lamina of mass M and sides containing the right angle of length $2a$.

9. Find the moment of inertia of a uniform equilateral triangular wire of mass M, side $2a$, about (i) an axis of symmetry, (ii) an axis through the centre, parallel to one side.

10. Find the moment of inertia of a uniform closed hemispherical shell of radius r and mass M about its axis of symmetry. [Use the result of Ex. 15b, No. 14.]

11. Find the moment of inertia of a uniform closed cylindrical shell of mass M, radius r, height h about a generator.

12. Deduce from the result of Example 7 the moment of inertia about a diameter of a hollow sphere of mass M with internal radius b and external radius a.

13. A uniform cube has mass M and edges $2a$. Find its moment of inertia about an axis joining the mid-points of a pair of opposite sides of one of its faces (i) when the cube is solid, (ii) when it is made out of six square laminas.

Calculus is to be used in Nos. 14–19.

14. Find the moment of inertia of a uniform solid right circular cone of mass M, base radius a and height h about its axis. [Take the vertex as origin and see Example 7.]

15. Show that the solid ellipsoid of revolution obtained by rotating the ellipse $b^2x^2 + a^2y^2 = a^2b^2$ about the x-axis has a moment of inertia about this axis of $\frac{2}{5}Mb^2$, where M is the mass of the body.

★16. A uniform right circular cylindrical shell (with open ends) has mass M, radius a, height h. Axes are taken with x-axis along the axis of the cylinder and y-axis along a diameter of one end. An element in the form of a ring (thickness $\varDelta x$) is taken with its centre at $(x, 0)$. Using σ as the mass per unit area of the shell, find the moment of inertia of the element (i) about its own diameter parallel to the y-axis, (ii) about the y-axis (using the parallel axes theorem). Hence find the moment of inertia of the cylinder about the y-axis. Deduce the moment of inertia about a parallel axis through the centroid.

17. Repeat No. 16 for a solid cylinder, replacing the element in that question by a disc.

18. Use the method of No. 16 to find the moment of inertia of a uniform right circular conical shell of mass M, base radius r, height h about an axis through the vertex parallel to the base and deduce the moments of inertia of the cone about (i) a parallel axis through the centroid, (ii) a diameter of the base.

19. A solid right square pyramid has mass M, height h and base with sides $2a$. Find the moment of inertia of this body about (i) the bisector of one pair of opposite sides of the base, (ii) a diagonal of the base.

In the following questions, no hints are given as to the method to be employed.

20. Find the moment of inertia of a uniform circular disc of mass M, radius r, about a tangent.

21. A uniform solid prism of mass M, height h, has a cross-section in the form of an equilateral triangle of side a. Find its moment of inertia about an edge parallel to its axis.

22. Prove that the radius of gyration of a uniform spherical shell of radius a is $\sqrt{6}a/3$.

23. Prove that the moment of inertia of a uniform solid sphere of mass M, radius a about a diameter is $2Ma^2/5$.

Show that the moment of inertia of a uniform solid hemisphere of mass M, radius a is the same about all axes through the centre of its plane face, and find it.

24. Two uniform triangular laminas of equal masses lie between two parallel lines with their vertices on one of the lines and their bases on the other. Prove that their moments of inertia about any axis parallel to their bases are equal.

25. A uniform lamina is bounded by the curve

$$y = \frac{1}{\sigma\sqrt{(2\pi)}}\, e^{-x^2/(2\sigma^2)}$$

and the x-axis.

Given that

$$\int_{-\infty}^{\infty} e^{-x^2/(2\sigma^2)}\, dx = \sigma\sqrt{(2\pi)},$$

find the radius of gyration of the lamina about the y-axis.

26. The area bounded by the ellipse $b^2x^2 + a^2y^2 = a^2b^2$ is rotated through two right angles about the x-axis. Show that the radius of gyration about this axis of the uniform solid of revolution is $\sqrt{10}b/5$.

27. The area bounded by the parabola $y^2 = 4ax$ and $x = a$ is rotated through two right angles about the x-axis. Find:
  (i) the position of the centroid of this solid,
  (ii) the radius of gyration about the x-axis.

28. Find the moment of inertia of a uniform solid right square pyramid of mass M, height h, base edge $2a$ about its axis.

29. Find the moment of inertia of a uniform right circular cone, height h, base radius a, mass M, about a diameter of the base.

30. Find the moment of inertia about a generator of a closed uniform right circular cylindrical shell of mass M, height h, base radius a.

31. If Fig. 15.16, p. 335, represents a uniform lamina of mass M, find its moment of inertia about an oblique axis through O perpendicular to Ox and making an angle θ with Oy.

32. Find the moment of inertia of a uniform circular ring of mass M, radius a about an axis through its centre inclined at an angle $\frac{1}{2}\pi - \alpha$ to its plane.

CHAPTER 16

COORDINATE GEOMETRY—II

Pairs of straight lines

16.1. The sections of a right circular cone have, historically, been of considerable interest to mathematicians. Such a cone was defined on p. 115 and it should be emphasized that it is to be thought of as being generated by a line of infinite length thus producing what might loosely be called a 'double cone.' Most of the conic sections have been discussed in Chapter 9; another conic section is produced by the intersection of a plane passing through the vertex of a cone: either a point, or a pair of straight lines (which may be coincident).

Consider the lines

$$3x + 2y - 4 = 0 \qquad (1)$$
$$2x - y + 1 = 0. \qquad (2)$$

These equations may be combined by multiplying together the left-hand sides and the right-hand sides:

$$(3x + 2y - 4)(2x - y + 1) = 0. \qquad (3)$$

Then any point which satisfies equation (1) or equation (2) will satisfy equation (3), and conversely. On multiplying out the brackets in (3),

$$6x^2 + xy - 2y^2 - 5x + 6y - 4 = 0$$

we obtain a quadratic equation in x and y as we have already done for other conics.

Example 1. *Find the equations of the lines which form the line-pair*

$$4x^2 - 7xy - 15y^2 + 2x + 11y - 2 = 0.$$

We begin by factorizing the terms of second degree.

$$4x^2 - 7xy - 15y^2 \equiv (x - 3y)(4x + 5y).$$

The constant terms in the equations of the lines may then be found by inspection [or by factorizing, say, the terms in x, keeping the x and $4x$ obtained above.

$$4x^2 + 2x - 2 \equiv (x+1)(4x-2)].$$

Thus

$$4x^2 - 7xy - 15y^2 + 2x + 11y - 2 \equiv (x-3y+1)(4x+5y-2).$$

This is checked by inspection. Therefore the lines forming the line-pair are

$$x - 3y + 1 = 0, \qquad 4x + 5y - 2 = 0.$$

Line-pairs with vertex at the origin

16.2. In general, two lines may be taken as

$$lx + my + n = 0, \qquad Lx + My + N = 0.$$

On multiplying the left-hand sides and the right-hand sides of the equations, we obtain

$$(lx + my + n)(Lx + My + N) = 0.$$

This equation is cumbersome to deal with, especially if the brackets are expanded, and it is more convenient for most purposes to take two lines parallel to

$$lx + my + n = 0, \qquad Lx + My + N = 0,$$

but passing through the origin:

$$(lx + my)(Lx + My) = 0,$$

i.e.

$$lLx^2 + (lM + mL)xy + mMy^2 = 0,$$

or

$$ax^2 + 2hxy + by^2 = 0$$

where

$$\frac{a}{lL} = \frac{2h}{lM + mL} = \frac{b}{mM}.$$

★**Qu. 1.** Write down the condition that the lines $lx + my = 0$, $Lx + My = 0$ should be perpendicular. Deduce the condition that the line-pair

$$ax^2 + 2hxy + by^2 = 0$$

should be perpendicular.

Qu. 2. The point $(0, 0)$ always satisfies the equation $ax^2 + 2hxy + by^2 = 0$. Find the condition that the equation is satisfied by other real points and show that, when this condition is satisfied, the equation always represents two straight lines (possibly coincident).

Example 2. *Find the acute angle between the lines of the pair*

$$ax^2 + 2hxy + by^2 = 0.$$

Let the equations of the lines be

$$y - m_1 x = 0, \qquad y - m_2 x = 0,$$

so that the angle θ between them is (by the method of P.M.I., p. 384) given by

$$\tan \theta = \frac{m_1 - m_2}{1 + m_1 m_2}.$$

The equation of the line-pair is

$$(y - m_1 x)(y - m_2 x) = 0,$$

i.e.

$$m_1 m_2 x^2 - (m_1 + m_2)xy + y^2 = 0. \tag{1}$$

But the line-pair is also represented by the equation

$$ax^2 + 2hxy + by^2 = 0. \tag{2}$$

Comparing coefficients in equations (1) and (2),

$$\frac{m_1 m_2}{a} = \frac{-(m_1 + m_2)}{2h} = \frac{1}{b},$$

so that

$$m_1 + m_2 = -\frac{2h}{b}, \qquad m_1 m_2 = \frac{a}{b}. \tag{3}$$

$$\tan^2 \theta = \frac{(m_1 - m_2)^2}{(1 + m_1 m_2)^2} = \frac{(m_1 + m_2)^2 - 4m_1 m_2}{(1 + m_1 m_2)^2},$$

$$= \frac{\dfrac{4h^2}{b^2} - \dfrac{4a}{b}}{\left(1 + \dfrac{a}{b}\right)^2},$$

$$= \frac{4(h^2 - ab)}{(a + b)^2}.$$

Therefore the acute angle between the lines is

$$\tan^{-1} \left| \frac{2\sqrt{(h^2 - ab)}}{a + b} \right|.$$

This formula is hardly worth memorizing, but the fact that the lines are perpendicular if $a+b=0$ will be of use in Exercise 16a.

Qu. 3. Check that the above condition is satisfied by the asymptotes of a rectangular hyperbola when the coordinate axes are: (i) the axes of symmetry, (ii) the asymptotes.

Qu. 4. What are the conditions that the pair of lines $ax^2+2hxy+by^2=0$ should be (i) coincident, (ii) real?

Qu. 5. What is the acute angle between the pair of lines

$$ax^2+2hxy+by^2+2gx+2fy+c = 0?$$

Example 3. *Find the equations of the bisectors of the angles between the lines* $ax^2+2hxy+by^2=0$.

Let the equations of the given lines be

$$y-m_1x = 0, \qquad y-m_2x = 0.$$

A point (x, y) on the bisectors of the angles between these lines is equidistant from them so that (see P.M.I., p. 398).

$$\frac{y-m_1x}{\sqrt{(1+m_1{}^2)}} = \pm\frac{y-m_2x}{\sqrt{(1+m_2{}^2)}}.$$

Squaring and clearing fractions,

$$(y-m_1x)^2(1+m_2{}^2) = (y-m_2x)^2(1+m_1{}^2).$$

Expanding and collecting like terms,

$$x^2(m_1{}^2 - m_2{}^2)+y^2(m_2{}^2 - m_1{}^2)-2xy(m_2{}^2m_1+m_1-m_1{}^2m_2-m_2) = 0.$$

$$\therefore\ (x^2-y^2)(m_1-m_2)(m_1+m_2)-2xy(m_1-m_2)(1-m_1m_2) = 0.$$

Dividing by m_1-m_2 and substituting for m_1+m_2 and m_1m_2 from (3) of Example 2,

$$(x^2-y^2)\left(-\frac{2h}{b}\right)-2xy\left(1-\frac{a}{b}\right) = 0.$$

Therefore the locus is

$$h(x^2-y^2)+(b-a)xy = 0.$$

Exercise 16a

1. Find the equation of the line-pair formed by the lines:
   (i) $2x-y = 0$, $x+2y = 0$,
   (ii) $x+y+7 = 0$, $2x-3y+4 = 0$,
   (iii) $x+3y-1 = 0$, $x+3y+4 = 0$.

2. Obtain the equations of the lines which form the line-pair:
   - (i) $3x^2 + 19xy - 14y^2 = 0$,
   - (ii) $x^2 - y^2 - 5x + y + 6 = 0$,
   - (iii) $8x^2 - 2xy - 15y^2 - 8x + 23y - 6 = 0$.
3. Find the angle between the lines:
   - (i) $3x^2 + 2xy - 6y^2 = 0$,
   - (ii) $x^2 + 7xy + y^2 = 0$,
   - (iii) $x^2 - 6xy - 2y^2 - 2x - 16y - 10 = 0$.
4. Obtain the equations of the bisectors of the angles between the line-pairs of Nos. 3 (i) and (ii). What is the condition that the lines of the pair $ax^2 + 2hxy + by^2 = 0$ should be perpendicular? Check that this condition is satisfied by the angle bisectors.

*5. The equations of the circle $x^2 + y^2 - a^2 = 0$ and the straight line $lx + my + n = 0$ may be written

$$x^2 + y^2 = a^2, \qquad n^2 = (lx + my)^2.$$

Multiplying the left-hand sides and the right-hand sides together,

$$n^2(x^2 + y^2) = a^2(lx + my)^2.$$

Show that this locus passes through the points of intersection of the line and the circle. What locus does this equation represent?

*6. In the last question, the equations of the circle and the straight line were combined to obtain a *homogeneous* quadratic equation. Repeat the process to find the equations of the line-pairs joining the origin to the points of intersection of the line $lx + my + n = 0$ with:
   - (i) the ellipse $\dfrac{x^2}{a^2} + \dfrac{y^2}{b^2} = 1$,
   - (ii) the hyperbola $\dfrac{x^2}{a^2} - \dfrac{y^2}{b^2} = 1$,
   - (iii) the parabola $y^2 = 4ax$.

7. Write down the equation of the line-pair joining the origin to the points of intersection of the line $lx + my + n = 0$ and the ellipse

$$b^2x^2 + a^2y^2 = a^2b^2.$$

Find (i) the distance of the chord from the centre of the ellipse, (ii) the condition that the lines of the pair should be perpendicular. Hence show that any chord which subtends a right angle at the centre of the ellipse is at a distance $ab/\sqrt{(a^2 + b^2)}$ from the centre.

8. A chord of the parabola $y^2 = 4ax$ subtends a right angle at the vertex. Find the locus of the foot of the perpendicular from the vertex on to the chord.

9. A chord of the hyperbola $b^2x^2 - a^2y^2 = a^2b^2$ subtends a right angle at the centre. Find the locus of the foot of the perpendicular from the

centre on to the chord. Under what circumstances is this locus real and other than a single point?

10. $P(a \cos \theta, b \sin \theta)$ is any point on the ellipse $b^2x^2 + a^2y^2 = a^2b^2$. Find the line-pairs through the origin
   (i) parallel to the tangent and normal at P,
   (ii) parallel to the bisectors of the line pair joining P to the foci.
   What property of the ellipse can be deduced?

*11. (x_1, y_1) is a point outside the circle $x^2 + y^2 - a^2 = 0$. (X, Y) is a variable point. What does the point

$$P\left(\frac{x_1 + \lambda X}{1 + \lambda}, \frac{y_1 + \lambda Y}{1 + \lambda}\right)$$

represent?

Show that the condition that P should lie on the circle is

$$\lambda^2(X^2 + Y^2 - a^2) + 2\lambda(Xx_1 + Yy_1 - a^2) + (x_1^2 + y_1^2 - a^2) = 0.$$

Obtain the condition that this equation in λ should have equal roots and hence show that any point on the tangents from (x_1, y_1) to the circle $x^2 + y^2 - a^2 = 0$ satisfies the equation

$$(xx_1 + yy_1 - a^2)^2 = (x^2 + y^2 - a^2)(x_1^2 + y_1^2 - a^2).$$

12. Use the method of No. 11 to find the equation of the tangents from (x_1, y_1) to:
   (i) the parabola $y^2 = 4ax$,
   (ii) the ellipse $\dfrac{x^2}{a^2} + \dfrac{y^2}{b^2} = 1$,
   (iii) the hyperbola $\dfrac{x^2}{a^2} - \dfrac{y^2}{b^2} = 1$.

13. Use the results of the last question to find the locus of the point of intersection of perpendicular tangents to each of the three loci. [For the parabola this is called the *directrix*; for the ellipse and hyperbola the locus is called the *director circle*.]

Length of tangents to a circle

16.3. In P.M.I., p. 374, a method was given for finding the length of the tangents from a point to a circle. We shall now use this to obtain a formula for the length of the tangents from (x_1, y_1) to the circle

$$x^2 + y^2 + 2gx + 2fy + c = 0.$$

The centre of the circle is $(-g, -f)$ and its radius $r = \sqrt{(g^2 + f^2 - c)}$. The distance d from the centre to (x_1, y_1) is given by

$$d^2 = (x_1 + g)^2 + (y_1 + f)^2.$$

By Pythagoras' theorem, the length of the tangents t is given by

$$t^2 = d^2 - r^2.$$
$$\therefore \ t^2 = (x_1+g)^2 + (y_1+f)^2 - (g^2+f^2-c).$$
$$\therefore \ t^2 = x_1{}^2 + y_1{}^2 + 2gx_1 + 2fy_1 + c.$$

Therefore the length of the tangents is

$$\sqrt{(x_1{}^2 + y_1{}^2 + 2gx_1 + 2fy_1 + c)}.$$

Qu. 6. Write down, in your own words, a rule for finding the length of the tangents to a circle from an external point.

Qu. 7. Write down the length of the tangents from:

  (i) $(0, 0)$ to $x^2 + y^2 - 6x + 4y + 2 = 0$,
  (ii) $(-3, 1)$ to $x^2 + y^2 - 8x = 0$,
  (iii) (a, b) to $x^2 + y^2 - ax - by + c = 0$.

Qu. 8. If the lengths of the tangents from a point P to the circles

$$x^2 + y^2 - 6x = 0, \qquad x^2 + y^2 - 8x + 6y - 2 = 0$$

are equal, find the locus of P. At what angle does the locus cut the line of centres?

The radical axis

16.4. Given two circles,

$$x^2 + y^2 + 2gx + 2fy + c = 0,$$
$$x^2 + y^2 + 2Gx + 2Fy + C = 0,$$

what is the locus of points from which the lengths of the tangents to the two circles are equal?

Equating the squares of the lengths of the tangents from any point (x, y) on the locus,

$$x^2 + y^2 + 2gx + 2fy + c = x^2 + y^2 + 2Gx + 2Fy + C,$$

$$\therefore \ 2(g-G)x + 2(f-F)y + c - C = 0.$$

This is the equation of a straight line called the *radical axis* of the two circles.

Qu. 9. Find the equation of the radical axis of the circles:

  (i) $x^2 + y^2 - 7x + 6y = 0$, $\quad x^2 + y^2 + 7x - 6y = 0$.
  (ii) $x^2 + y^2 + 8x + 4y - 3 = 0$, $\quad 2x^2 + 2y^2 - 1 = 0$.
  (iii) $x^2 + y^2 + 2ky + c = 0$, $\quad x^2 + y^2 + 2Ky + c = 0$.

*Qu. 10. Prove that, if two circles intersect, their radical axis is also their common chord.

*Qu. 11. Prove that the radical axis of two circles is perpendicular to the line of centres.

Coaxal circles

16.5. The two circles in §16.4 were quite general; if we choose different axes of coordinates, the equations can be simplified and working made easier. We can make use of the symmetry of the figure by taking the x-axis along the line of centres. The radical axis is perpendicular to the line of centres (see Qu. 11) and so we take the y-axis along the radical axis.

Since the centres of the circles lie on the x-axis, their equations may be written

$$x^2+y^2+2gx+c = 0,$$
$$x^2+y^2+2Gx+C = 0.$$

Their radical axis is

$$2(g-G)x+c-C = 0.$$

But this is the y-axis, so that $c=C$. The circles may therefore be written

$$x^2+y^2+2gx+c = 0,$$
$$x^2+y^2+2Gx+c = 0.$$

These equations are now identical except for the coefficients of x.

Now consider another circle of the same form

$$x^2+y^2+2\lambda x+c = 0.$$

The radical axis of this and

$$x^2+y^2+2gx+c = 0$$

is

$$2(g-\lambda)x = 0, \quad \text{i.e. } x = 0.$$

Therefore every pair of circles of the system

$$x^2+y^2+2\lambda x+c = 0,$$

where c is a constant but λ varies from circle to circle (and is therefore called a parameter), has the y-axis as radical axis. Such a system of

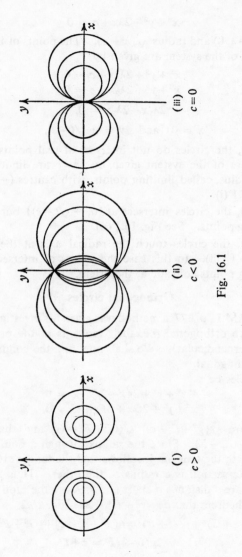

Fig. 16.1

circles is called *coaxal* or *coaxial* because they have the same radical *axis*.

The circle $$x^2+y^2+2\lambda x+c = 0$$

has centre $(-\lambda, 0)$ and radius $\sqrt{(\lambda^2-c)}$. The points of intersection of two members of the system are given by

$$x^2+y^2+2\lambda_1 x+c = 0,$$
$$x^2+y^2+2\lambda_2 x+c = 0.$$

Subtracting, $$2\lambda_1 x-2\lambda_2 x = 0.$$

Hence $$x = 0 \quad \text{and} \quad y = \pm\sqrt{(-c)}.$$

(i) If $c>0$, the circles do not intersect at real points but the two members of the system given by $\lambda^2=c$ are limiting circles of zero radius, called limiting points, with centres $(\pm\sqrt{c}, 0)$. See Fig. 16.1 (i).

(ii) If $c<0$, the circles intersect at $\left(0, \pm\sqrt{(-c)}\right)$ but there are no limiting points. See Fig. 16.1 (ii).

(iii) If $c=0$, the circles touch the radical axis at the origin. See Fig. 16.1 (iii). In this case the points of intersection and the limiting points coincide at the origin.

Orthogonal circles

16.6. In P.M.I., p. 377, a method was given for investigating whether two circles are orthogonal (i.e. their tangents at the points of intersection are perpendicular). We shall now find the condition that two circles are orthogonal.

Let the circles be

$$x^2+y^2+2gx+2fy+c = 0,$$
$$x^2+y^2+2Gx+2Fy+C = 0.$$

Their radii are $\sqrt{(g^2+f^2-c)}$, $\sqrt{(G^2+F^2-C)}$ and their centres are $(-g, -f)$, $(-G, -F)$. Since the radius through a point of contact is perpendicular to the tangent, it follows that the tangent to one circle at a point of intersection is a radius of the other. Thus, if the centres of the circles are a distance d apart and r, R are the radii, it follows by Pythagoras' theorem that $d^2=r^2+R^2$. See Fig. 16.2.

$$\therefore \ (-g+G)^2+(-f+F)^2 = (g^2+f^2-c)+(G^2+F^2-C).$$

$$\therefore \ 2gG+2fF = c+C$$

is the condition that the circles are orthogonal.

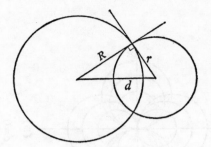

Fig. 16.2

Qu. 12. Which of the following pairs of circles are orthogonal?

(i) $x^2+y^2-1 = 0$, $x^2+y^2-6x+8y+1 = 0$,

(ii) $x^2+y^2+\lambda x+c = 0$, $x^2+y^2+\mu y-c = 0$,

(iii) $x^2+y^2+6x-8y+5 = 0$, $x^2+y^2+8x+6y+5 = 0$.

Orthogonal systems of coaxal circles

16.7. Consider the coaxal system of circles

$$x^2+y^2+2\lambda x+c = 0$$

(c constant, λ a parameter) and another coaxal system, with centres on the radical axis of the first system,

$$x^2+y^2+2\mu y+k = 0.$$

These circles will be orthogonal if (substituting in the condition above)

$$2\lambda.0+2.0.\mu = c+k.$$

So taking $k = -c$, every circle of the system

$$x^2+y^2+2\mu y-c = 0$$

is orthogonal to every circle of the system

$$x^2+y^2+2\lambda x+c = 0.$$

Further, as may be seen from §16.5, the limiting points of one system are the points of intersection of the other system. See Fig. 16.3.

If an electric current flows in opposite directions along two long parallel wires, a cross section of the electromagnetic field in a plane

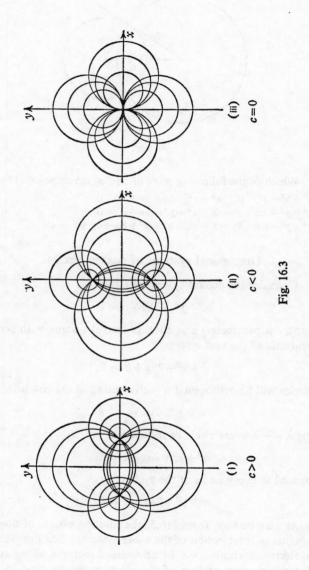

Fig. 16.3

perpendicular to the wires is illustrated in Fig. 16.3 (i). The magnetic lines of force are the circles of the non-intersecting coaxal system, with the position of the wires given by the limiting points; and the electric lines of force are the circles of the intersecting system.

Exercise 16b

1. Show that the circle $x^2 + y^2 - 3x - 7y + 8 = 0$ is orthogonal to both the circles

$$2x^2 + 2y^2 - 10x - 8y + 27 = 0,$$
$$3x^2 + 3y^2 - 12x - 10y + 29 = 0,$$

and that its centre lies on the radical axis of these circles.

2. Find the equation of the circle which passes through the points $(1, 1)$, $(1, -1)$ and is orthogonal to the circle $x^2 + y^2 - 4 = 0$.

3. Find the equation of a circle which is orthogonal to the circles

$$x^2 + y^2 - 8x + 5 = 0,$$
$$x^2 + y^2 + 6x + 5 = 0,$$

and which passes through the point $(3, 4)$.

4. The equations of two circles are $C = 0$, $C' = 0$. Show that any circle with the equation $C + \lambda C' = 0$ is a member of the coaxal system defined by $C = 0$, $C' = 0$. Hence find the member of the coaxal system defined by the limiting points $(1, 2)$, $(-1, 3)$ which passes through the origin.

5. One circle of a coaxal system is $x^2 + y^2 - 6x + 8 = 0$ and $(0, 0)$ is a limiting point. Find the equations of: (i) the radical axis, (ii) the circle of the system which passes through the point $(1, 1)$.

6. A system of coaxal circles has radical axis $2x + 3y - 1 = 0$. If one member of the system is $x^2 + y^2 - 5 = 0$, show that the circle

$$x^2 + y^2 - 5 + k(2x + 3y - 1) = 0$$

belongs to the coaxal system and find the equation of the circle of the system which passes through the origin.

7. In general, three circles taken in pairs have three radical axes. Show that the three radical axes are concurrent or parallel.

8. Prove that, if three circles touch each other, their common tangents are concurrent.

9. L_1, L_2 are the limiting points of a non-intersecting coaxal system and P is any point on the radical axis. Prove that the length of the tangents from P to any circle of the system is equal to PL_1 and PL_2.

10. A variable circle is orthogonal to two fixed circles. Prove that the variable circle passes through two fixed points.

11. P is a variable point and the ratio of the lengths of the tangents from P to two fixed circles is constant. Show that the locus of P is (part of) a circle.

12. Tangents are drawn to two circles from any point on the radical axis. Show that the four points of contact are concyclic and that the chords of contact meet on the radical axis.

13. Obtain the differential equation (see p. 360) satisfied by the circles of the coaxal system $x^2 + y^2 + 2\lambda x + c = 0$ where λ is a parameter. If y'_1 is the gradient of this system at (x, y) and y'_2 is the gradient of the orthogonal system $x^2 + y^2 + 2\mu y - c = 0$, verify that $y'_1 y'_2 = -1$.

14. Two circles of radii a and b have centres at a distance c apart $(c > a + b)$. Find the position of their radical axis and of the limiting points of the coaxal system defined by them.

★15. [A result to be used in Nos. 16, 17.] The quadratic equations $ax^2 + bx + c = 0$, $Ax^2 + Bx + C = 0$ have roots x_1, x_2 and x_3, x_4. Show that the condition that the points $(x_1, 0)$, $(x_2, 0)$ should divide the line joining $(x_3, 0)$ and $(x_4, 0)$ internally and externally in the same ratio is $2(cA + Ca) = bB$

16. [See No. 15.] Show that the equations of two orthogonal circles may be written in the form

$$x^2 + y^2 + 2\lambda_1 x + c = 0,$$
$$x^2 + y^2 + 2\lambda_2 x + c = 0,$$

where $\lambda_1 \lambda_2 = c$.

The line of centres of two orthogonal circles meets one of the circles in A, B and the other in C, D. Show that C, D divide A, B internally and externally in the same ratio.

17. [See No. 15.] If the axis of symmetry of a non-intersecting system of coaxal circles cuts any circle in A, B and if L_1, L_2 are the limiting points, show that L_1, L_2 divide A, B internally and externally in the same ratio.

18. L_1, L_2 are the limiting points of a system of coaxal circles. If any circle of the system has centre O and radius r, show that

$$OL_1 . OL_2 = r^2.$$

Exercise 16c (Miscellaneous)

★1. Show that the point

$$\left(\frac{\lambda_1 x_1 + \lambda_2 x_2}{\lambda_1 + \lambda_2}, \ \frac{\lambda_1 y_1 + \lambda_2 y_2}{\lambda_1 + \lambda_2} \right)$$

divides the line segment joining (x_1, y_1), (x_2, y_2) in the ratio $\lambda_2 : \lambda_1$.

*2. Prove that the acute angle between the lines $y = m_1x + c_1$, $y = m_2x + c_2$ is

$$\tan^{-1}\left|\frac{m_1 - m_2}{1 + m_1m_2}\right|.$$

Deduce the corresponding expression for the acute angle between the lines $a_1x + b_1y + c_1 = 0$, $a_2x + b_2y + c_2 = 0$.

*3. Show that the equation

$$(a_1x + b_1y + c_1) + \lambda(a_2x + b_2y + c_2) = 0$$

represents a straight line through the point of intersection of the lines

$$a_1x + b_1y + c_1 = 0, \quad a_2x + b_2y + c_2 = 0.$$

Hence find the equation of the line joining the origin to the point of intersection of these lines

Find by a similar method, the equation of that circle through the point of intersection of

$$x^2 + y^2 + 2g_1x + 2f_1y + c_1 = 0, \quad x^2 + y^2 + 2g_2x + 2f_2y + c_2 = 0$$

which also passes through the origin.

4. Find the equation of the tangent at any point (h, k) on the curve

$$y = x^4 - x + 3.$$

Find the two real values of h for which the tangent passes through the origin. Give the coordinates of the points of contact P and Q.

Prove that the normals to the curve at P and Q meet at the point $(-\frac{7}{2}, \frac{9}{2})$.

O.C.

5. The line $y = mx$ meets the curve $y = x^3 - 4x^2 + 3x$ at the origin O and again at A and B, A lying between O and B. The abscissae of A and B are x_1 and x_2 respectively: prove that

$$x_1 + x_2 = 4.$$

Show that the tangents to the curve at A and B meet at the point whose abscissa is $\frac{1}{2}(m + 5)$.

If $AB/OA = k$, and $x_2 > x_1$, prove that

$$m = (3k^2 - 4k - 4)/(k + 2)^2.$$

O.C.

6. Prove that the perpendicular distance of the point (h, k) from the line $ax + by + c = 0$ is the numerical value of

$$(ah + bk + c)/\sqrt{(a^2 + b^2)}.$$

Find the equations of the two straight lines which bisect the angles between the lines $ax + by + c = 0$ and $a_1x + b_1y + c_1 = 0$.

The lines $3x-4y+3=0$, $3x+4y-15=0$, and $y=0$ are the sides of a triangle. Find the coordinates of the centre of the circle inscribed in the triangle and write down the equation of this circle. o.c.

7. A curve is given by the parametric equations

$$x = t^3-t^2+2, \quad y = t^2-1.$$

Prove that the equation of the tangent at the point t_1 is

$$(3t_1-2)y = 2x+(t_1-2)(t_1+1)^2,$$

and show that this tangent meets the curve again at the point $t=-\tfrac{1}{2}t_1$.
o.c.

8. Show that the equation

$$x^2+y^2+2kx-4ky = -k^2$$

represents a circle for all real values of k, and find the radius and the coordinates of the centre in terms of k.

Prove that, for all real values of k, the circles touch the x-axis and find the equation of another straight line through the origin which touches all the circles of the family.

Prove that there are two circles of this family which pass through the point $(-1, 2)$ and, if their radii are r_1 and r_2 $(r_1 > r_2)$, show that

$$\frac{r_1}{r_2} = 9+4\sqrt{5}.$$ o.c.

9. Prove that the equation $x^2+y^2+2gx+2fy+c=0$ represents a circle and state the radius and the coordinates of the centre.

Show that the common chord of

$$x^2+y^2-6x+2y = 0 \quad \text{and} \quad x^2+y^2 = 20$$

is a diameter of the first circle, and find the angle at which the circles cut. o.

10. OX, OY are the axes of coordinates and A and B have coordinates $(1, 0)$, $(2, 0)$ respectively. P is any point on OY, and Q is the point of intersection of PB and the perpendicular from O to PA. Prove that, as P moves on OY, Q describes a curve whose equation is

$$x^2+2y^2 = 2x.$$ o.

11. The tangent at P $(3t^2, 2t^3)$ on the curve $4x^3=27y^2$ meets the x-axis at Q. If the normal at P meets the y-axis at R, find the locus of the mid-point of QR.

12. The tangent at a point P on the rectangular hyperbola $x^2-y^2=a^2$ forms with the axes a triangle of area A; the normal at P forms with the asymptotes a triangle of area B. Prove that $A^2B=a^6$. o.

13. Prove that the line $y = mx + c$ is a tangent to the ellipse

$$\frac{x^2}{a^2} + \frac{y^2}{b^2} = 1$$

provided that $c^2 = a^2 m^2 + b^2$.

Find the equations of the tangents from $(-4, 3)$ to the ellipse

$$\frac{x^2}{25} + \frac{y^2}{16} = 1$$

and prove that the tangent of the angle between them is 9/8.   o.c.

14. P is any point on the curve $y^2 = 4ax$, and O is the origin; Q is the foot of the perpendicular from P to the y-axis, R is the foot of the perpendicular from Q to OP, and QR produced meets the x-axis at K. Prove that K is a fixed point, and find its coordinates.

   Prove also that the locus of R is a circle.   L.

15. Prove that the centroid of the triangle whose vertices are (x_1, y_1), (x_2, y_2), (x_3, y_3) is the point

$$\left(\frac{x_1 + x_2 + x_3}{3}, \frac{y_1 + y_2 + y_3}{3} \right).$$

   P is the point $(at^2, 2at)$ on the curve $y^2 = 4ax$. The tangent and normal at P meet the x-axis at T and G, respectively. Show that as t varies, the centroid of the triangle PGT describes the curve

$$9y^2 = 12ax - 8a^2.$$ L.

16. Find the equation of the tangent at the point $(2t^3, 3t^2)$ on the curve $4y^3 = 27x^2$.

   Prove that the locus of the intersection of perpendicular tangents to the curve is $y = x^2 + 1$.   L.

*17. Prove that the equation of the tangent at the origin to the curve

$$ax + by + px^2 + qxy + ry^2 = 0$$

is $ax + by = 0$.

   Deduce the equation of the tangent at the origin to:

   (i) $x^2 + y^2 + 2gx + 2fy = 0$,   (ii) $y^2 - 4ax = 0$,
   (iii) $y = x^2 - 4x$,   (iv) $ay = x(a - x)$.

*18. Prove that the simultaneous equations

$$ax^2 + 2hxy + by^2 + px^3 + qx^2y + rxy^2 + sy^3 = 0 \tag{1}$$

$$y = mx \tag{2}$$

have a repeated solution $x = 0$, $y = 0$ if m satisfies the equation

$$bm^2 + 2hm + a = 0.$$ (Continued.)

If the roots of this equation are m_1, m_2, show that $y = m_1 x$, $y = m_2 x$ are tangents at the origin to the curve whose equation is (1), and that the equation of the line-pair

$$(y - m_1 x)(y - m_2 x) = 0$$

may be written

$$ax^2 + 2hxy + by^2 = 0.$$

Hence write down the equations of the tangents at the origin to:
  (i) $y^2 - x^3 = 0$, (ii) $x^3 + y^3 - 3xy = 0$,
  (iii) $4y^2 = x^2(9 + 4x)$.

19. Find the equation of the normal at the point P(cp, c/p) on the rectangular hyperbola $xy = c^2$, and prove that it cuts the hyperbola again in the point Q whose parameter q is $-1/p^3$.

   If R is the opposite end of the diameter of the hyperbola through P, prove that PR is perpendicular to RQ.　　　　　　　　　　　　N.

20. Show that two circles can be found which go through the two points $(a, 0)$ and $(-a, 0)$ and touch the line

$$lx + my + n = 0.$$

Prove that these two circles cut at right angles if

$$a^2(2l^2 + m^2) = n^2.$$
　　　　　　　　　　　　　　　　　　　　　　　　　　　　　O.

21. P($a \cos \theta$, $b \sin \theta$) is a point on the ellipse

$$\frac{x^2}{a^2} + \frac{y^2}{b^2} = 1.$$

Lines are drawn through P parallel to the axes to meet the ellipse again in Q and R.   QR meets the normal at P to the ellipse in X.   Prove that the coordinates of X are

$$\left(a \, \frac{a^2 - b^2}{a^2 + b^2} \cos \theta, \quad -b \, \frac{a^2 - b^2}{a^2 + b^2} \sin \theta \right).$$

Hence prove that, as P varies on the given ellipse, the locus of X is a second ellipse. Prove also that the two ellipses have the same eccentricity.　　　　　　　　　　　　　　　　　　　　O.

22. Prove that the line $y = mx + a/m$ is a tangent to the parabola $y^2 = 4ax$.

   The tangent at P to this parabola meets the parabola $y^2 = 4bx$ (where $b > a > 0$) in the points Q and R, the one nearer the common vertex of the parabolas being Q.   Prove that the ratio QP/PR is equal to

$$\{b^{1/2} - (b - a)^{1/2}\} / \{b^{1/2} + (b - a)^{1/2}\}.$$
　　　　　　　　　　　　　　　　　　　　　　　　　　　　　O.

23. P is a point on the circle $x^2+y^2-2ax = 0$ and O is the origin. Q is the point on the chord OP such that the mid-point of PQ lies on the diameter $x - a = 0$. Find the equation of the *cissoid* traced out by Q as P moves on the circle. Sketch the curve.

24. P is a point on the line $x - a = 0$, and O is the origin. Q is a point on OP such that $|PQ| = b$. Find the equation of the *conchoid* traced out by Q as P moves on $x - a = 0$. With a ruler, plot the curves given by (i) $a = 1$, $b = 1\frac{1}{2}$, (ii) $a = 1$, $b = \frac{1}{2}$, taking 2 cm as the unit.

25. Prove that the equation

$$x^2 + 2hxy - y^2 = 0$$

represents a pair of perpendicular lines through the origin O.

If A, B are the points in which these lines are cut by the line

$$x \cos \alpha + y \sin \alpha = p,$$

show that the x-coordinate of the mid-point of AB is

$$p\,\frac{(\cos \alpha + h \sin \alpha)}{\cos 2\alpha + h \sin 2\alpha},$$

and find the y-coordinate. Prove that the radius of the circumcircle of the triangle OAB is

$$\frac{p\sqrt{(1+h^2)}}{\cos 2\alpha + h \sin 2\alpha}. \qquad \text{N.}$$

26. Prove that the line $y = mx + a/m$ is a tangent to the parabola $y^2 = 4ax$ for all values of m.

Two tangents to the parabola $y^2 = 4ax$ meet at an angle of $60°$. Prove that the locus of their point of intersection is the curve

$$y^2 - 4ax = 3(x+a)^2. \qquad \text{O.C.}$$

27. Prove that the equation of the chord joining the points $(ct_1, c/t_1)$, $(ct_2, c/t_2)$ on the rectangular hyperbola $xy = c^2$ is

$$x + t_1 t_2 y = c(t_1 + t_2).$$

The vertices of a triangle ABC lie on the rectangular hyperbola $xy = c^2$ and the sides AB; AC touch the parabola $y^2 = 4ax$. Prove that BC also touches the parabola.

Deduce that, if E is the point of intersection of the two conics and if the tangent at E to the parabola meets the hyperbola again in F, then the tangent at F to the hyperbola will touch the parabola. N.

28. Prove that the equations of the tangent and normal at the point $P(ct, c/t)$ on the rectangular hyperbola $xy = c^2$ are, respectively

$$x + t^2 y = 2ct, \qquad t^3 x - ty = c(t^4 - 1).$$

The normal at P meets the hyperbola again at $R(cT, c/T)$; prove that

$$t^3 T = -1.$$

The tangent at P meets the y-axis at Q. Find the area of the triangle PQR in terms of t and prove that the area is a minimum when $t = \pm 1$.   o.c.

29. Prove that the equation of the tangent PT at the point $P(ct, c/t)$ on the rectangular hyperbola $xy = c^2$ is

$$x + t^2 y = 2ct.$$

The perpendicular to PT from the origin O meets PT at Q and the normal at P meets the hyperbola again in R. Prove (i) that, as P varies, the locus of Q is

$$(x^2 + y^2)^2 = 4c^2 xy$$

and (ii) that

$$c^2 . PR = OP^3.$$ o.c.

30. The line L_1 is given by the equation $3x - y - 2 = 0$. Find the equation of the line L_2 which passes through the point $(1, 1)$ and is perpendicular to L_1.

Verify that, for all values of t_1 and t_2, the point P with coordinates $(1 + t_1, 1 + 3t_1)$ lies on L_1 and that the point Q with coordinates $(1 + 3t_2, 1 - t_2)$ lies on L_2.

The points P on L_1 and Q on L_2 vary so that PQ is always 3 units in length. R is the point on PQ such that $PR = 2RQ$. Prove that the locus of R is

$$(3x - y - 2)^2 + 4(x + 3y - 4)^2 = 40.$$ o.c.

31. Find the equation of the pair of lines joining the origin to the points of intersection of the line $x + ky = 1$ and the circle $x^2 + y^2 + 4x - 4y - 1 = 0$.

Determine the values of k for which the equation represents perpendicular lines.

Find also the values of k for which the equation represents coincident lines, and deduce the equations of the tangents to the circle from the point $(1, 0)$.   N.

32. Write down the equation of the radical axis of the circles

$$x^2 + y^2 + 6x + 4 = 0, \qquad x^2 + y^2 - 8x + 4 = 0,$$

and also the equation of the system of circles coaxal with these circles. Show that the system has real limiting points, and state their coordinates.

Find the equations of

(i) the members of the system which touch the line $2x + y = 1$,

(ii) the circle which is orthogonal to all members of the system and passes through the point (3, 1). N.

33. Explain the terms 'limiting points' and 'common points' used in connection with coaxal systems of circles.

One circle of a coaxal system is

$$x^2 + y^2 - 4x - 8y + 10 = 0$$

and the radical axis is

$$2x + 4y - 5 = 0.$$

Show that the equation of any circle of the system may be written in the form

$$x^2 + y^2 + k(2x + 4y - 5) = 0.$$

Find the coordinates of the limiting points of the system.

Verify that the circle

$$x^2 + y^2 - x - 2y = 0$$

is orthogonal to each circle of the system. N.

34. The focus of a parabola is S, and the tangent to the parabola at P meets the directrix in R. Prove that SP and SR are perpendicular.

The line PSQ is a focal chord of the parabola. The line drawn through Q perpendicular to PQ meets the tangent at P in N. Prove that QN is bisected by the directrix. N.

CHAPTER 17

DIFFERENTIAL EQUATIONS

The general problem

17.1. An equation containing any *differential coefficients* such as $\dfrac{dy}{dx}$, $\dfrac{d^2y}{dx^2}$, is called a *differential equation*; a solution of such an equation is an equation relating x and y and containing no differential coefficients.

Given the differential equation $\dfrac{dy}{dx} = 3$, we obtain the *general solution* $y = 3x + c$, which is the equation of all straight lines of gradient 3. If the data also includes the fact that $y = 5$ when $x = 1$, we can determine that $c = 2$, and we obtain the *particular solution* $y = 3x + 2$.

Thus, in simple graphical terms,†

(i) a **differential equation** defines some property common to a family of curves,

(ii) the **general solution**, involving one or more arbitrary constants, is the equation of *any* member of the family,

(iii) a **particular solution** is the equation of *one* member of the family.

Qu. 1. Find the general solution of $\dfrac{d^2y}{dx^2} = 0$. What is the particular solution given by $\dfrac{dy}{dx} = 3$, and $y = -2$ when $x = 1$?

Qu. 2. Find the general solution of $\dfrac{dy}{dx} = 2x$. Illustrate with a sketch.

Qu. 3. For any circle centre the origin $\dfrac{dy}{dx} = -\dfrac{x}{y}$. Solve this equation by writing it as $y\dfrac{dy}{dx} = -x$. $\left(\text{What is } \dfrac{d}{dx}(y^2)?\right)$

---

† The reader may refer to P.M.I., p. 91, Example 1(i), for a simple example in kinematics.

Qu. 4. Find the general solution of $\frac{d^2s}{dt^2} = a$, where a is a constant. What does this become, given the initial conditions $s = 0$ and $\frac{ds}{dt} = u$ when $t = 0$?

DEFINITION: *The **order** of a differential equation is determined by the highest differential coefficient present.*

Thus the equations in Qu. 1 and 4 are of the second order, whereas those in Qu. 2 and 3 are of the first order.

Since each step of integration introduces one arbitrary constant, it is in general true that the order of a differential equation gives us the number of arbitrary constants in the general solution.

This suggests that from an equation involving x, y, and n arbitrary constants there may be formed (by differentiating n times and eliminating the constants) a differential equation of order n.

Qu. 5. Form differential equations by differentiating and eliminating the constants A, B from the following:

(i) $y = Ax + B$,   (ii) $y = Ax$,   (iii) $r = A \cos \theta$,

(iv) $xy = A$,   (v) $y = A e^x$,   (vi) $y = e^{Bx}$,

(vii) $y = A e^{Bx}$,   (viii) $y = A \log_e x$,   (ix) $x = \tan (Ay)$.

Qu. 6. Confirm the given general solution of each of the following differential equations:

(i) $\dfrac{d^2y}{dx^2} - \dfrac{dy}{dx} - 2y = 0$, $y = A e^{2x} + B e^{-x}$,

(ii) $\dfrac{d^2x}{dt^2} - 4 \dfrac{dx}{dt} + 4x = 0$, $x = e^{2t} (A + Bt)$.

We must now classify some of the simpler forms of differential equations.

First order—separating the variables

17.2. The solutions of $\frac{dy}{dx} = f(x)$ and $\frac{dy}{dx} = f(y)$ $\left(\text{which may be written}\right.$ $\frac{dx}{dy} = \frac{1}{f(y)}\left.\right)$ depend upon the integrals $\int f(x)\, dx$ and $\int \frac{1}{f(y)}\, dy$. There are other differential equations equally susceptible to direct integration once they have been written in a suitable form.

Consider
$$\frac{dy}{dx} = xy. \tag{1}$$

We write this as
$$\frac{1}{y}\frac{dy}{dx} = x,$$

then integrating each side with respect to x,
$$\int \frac{1}{y}\frac{dy}{dx}\,dx = \int x\,dx.$$

$$\left[\text{But from § 1.5 we know that } \int f(y)\,dy = \int f(y)\frac{dy}{dx}\,dx. \right]$$

$$\therefore \int \frac{1}{y}\,dy = \int x\,dx. \tag{2}$$

$$\therefore \log_e y + c = \tfrac{1}{2}x^2,$$

$$\therefore \log_e (ky) = \tfrac{1}{2}x^2, \quad \text{or} \quad y = A\,e^{x^2/2}.$$

Note how the arbitrary constant of integration appears in different forms; we have written c as $\log_e k$, and A as $\dfrac{1}{k}$.

Now let us look back at (1) and (2) in the above working. The symbols dx, dy have as yet no meaning for us in isolation; they have been used only in composite symbols such as $\dfrac{dy}{dx}$, $\dfrac{d}{dx}f(x)$, $\int f(x)\,dx$. However, in the present circumstances it is convenient to think of dx as an 'x-factor', and dy as a 'y-factor', and to proceed direct from (1) to (2) by 'separating the variables' and adding the integral sign. The intervening lines provide the justification for this.

Example 1. *Solve* $x^2\dfrac{dy}{dx} = y(y-1)$.

$$x^2\frac{dy}{dx} = y(y-1).$$

Separating the variables,
$$\int \frac{1}{y(y-1)}\,dy = \int \frac{1}{x^2}\,dx,$$

$$\therefore \int \left\{ \frac{1}{y-1} - \frac{1}{y} \right\} dy = \int \frac{1}{x^2}\,dx,$$

$$\therefore \log_e \frac{k(y-1)}{y} = -\frac{1}{x},$$

or
$$k(y-1) = y\,e^{-1/x}.$$

Qu. 7. Solve the following differential equations, and check solutions† by differentiation and elimination of arbitrary constants:

(i) $\dfrac{dy}{dx} = \dfrac{x}{y}$, (ii) $\dfrac{dy}{dx} = \dfrac{y}{x}$, (iii) $\dfrac{dx}{dy} = xy$,

(iv) $x\dfrac{dy}{dx} = \tan y$, (v) $e^{-x}\dfrac{dy}{dx} = y^2 - 1$, (vi) $\sqrt{(x^2+1)}\dfrac{dy}{dx} = \dfrac{x}{y}$.

Qu. 8. $v\dfrac{dv}{ds} = a$, where a is a constant. Solve this equation given that $v = u$ when $s = 0$.

Exercise 17a

1. By differentiating and eliminating the constants A and B from the following equations, form differential equations, and illustrate the geometrical significance of each:

(i) $3x - 2y + A = 0$, (ii) $Ax + 2y + 1 = 0$, (iii) $Ax + By = 0$,
(iv) $x^2 + y^2 = A$, (v) $y = Ax^{-1}$, (vi) $y = A(x-4)$.

2. By differentiating and eliminating the constants A and B from the following equations, form differential equations:

(i) $y = A\cos(3t + B)$, (ii) $y = A + Be^{3t}$, (iii) $y = Ae^{3x} + Be^{-3x}$,

(iv) $y = Ae^{3x} + Be^{-2x}$ (first multiply each side by e^{2x}),

(v) $y = e^{4x}(A + Bx)$ $\left(\text{first show that } \dfrac{dy}{dx} = 4y + Be^{4x}\right)$.

3. Obtain the equation of the straight line of gradient $\tfrac{3}{10}$, which passes through $(5, -2)$, by finding a particular solution of the differential equation $\dfrac{dy}{dx} = \tfrac{3}{10}$.

4. A family of parabolas has the differential equation $\dfrac{dy}{dx} = 2x - 3$. Find the equation of that member of the family which passes through $(4, 5)$.

5. Find the general solution of the differential equation $6t\dfrac{dt}{ds} + 1 = 0$, and the particular solution given by the condition $s = 0$ when $t = -2$.

6. Find the particular solutions of the differential equation

$$\operatorname{cosec} x \, \dfrac{dy}{dx} = e^x \operatorname{cosec} x + 3x$$

given by the conditions (i) $y = 0$ when $x = 0$, (ii) $y = 3$ when $x = \dfrac{\pi}{2}$.

† The reader is advised always to check solutions, even though this is not specifically asked for in the rest of this chapter.

7. Find the general solutions of the following differential equations:

(i) $\dfrac{dy}{dx} = y,$ (ii) $\dfrac{1}{x}\dfrac{dy}{dx} = \sqrt{(x-1)},$

(iii) $(x+2)\dfrac{dy}{dx} = y,$ (iv) $\dfrac{dy}{dx} = \sec^2 y,$ (v) $\dfrac{dv}{du} = v(v-1),$

(vi) $\log_e x \dfrac{dx}{dy} = 1,$ (vii) $\dfrac{dy}{dx} = \tan y,$

(viii) $\tan^{-1} y \dfrac{dy}{dx} = 1,$ (ix) $y \dfrac{dy}{dx} = x-1,$

(x) $(x^2-1)\dfrac{dy}{dx} = y,$ (xi) $\dfrac{d\theta}{dr} = \sin \theta,$ (xii) $x^2 \dfrac{dy}{dx} = y+3,$

(xiii) $x \dfrac{dy}{dx} = y+xy,$ (xiv) $\dfrac{d\phi}{d\theta} = \tan \phi \tan \theta,$

(xv) $\theta \dfrac{d\theta}{dr} = \cos^2 \theta,$ (xvi) $\dfrac{y}{x}\dfrac{dy}{dx} = \log_e x,$

(xvii) $2 \sin \theta \dfrac{d\theta}{dr} = \cos \theta - \sin \theta,$ (xviii) $x \dfrac{dy}{dx} - 3 = 2\left(y + \dfrac{dy}{dx}\right),$

(xix) $e^t \dfrac{dx}{dt} = \sin t,$ (xx) $e^x \dfrac{dy}{dx} + y^2 + 4 = 0.$

8. Find the particular solutions of the following differential equations which satisfy the given conditions:

(i) $(1 + \cos 2\theta)\dfrac{dy}{d\theta} = 2,$ $y = 1$ when $\theta = \dfrac{\pi}{4}$

(ii) $\dfrac{dy}{dx} = x(y-2),$ $y = 5$ when $x = 0,$

(iii) $(1+x^2)\dfrac{dy}{dx} = 1+y^2,$ $y = 3$ when $x = 2$

(iv) $\dfrac{dy}{dx} = \sqrt{(1-y^2)},$ $y = 0$ when $x = \dfrac{\pi}{6}.$

First order exact equations

17.3. The equation

$$2xy \dfrac{dy}{dx} + y^2 = e^{2x}$$

is not one in which the variables may be separated. However, the L.H.S. is $\dfrac{d}{dx}(xy^2)$ and the equation may be solved by integrating each side with respect to x; it is called an **exact equation** and the solution is

$$xy^2 = \tfrac{1}{2} e^{2x} + A.$$

Qu. 9. Solve the following exact equations:

(i) $x^2 \dfrac{dy}{dx} + 2xy = 1$,

(ii) $\dfrac{t^2}{x} \dfrac{dx}{dt} + 2t \log_e x = 3 \cos t$,

(iii) $x^2 \cos u \dfrac{du}{dx} + 2x \sin u = \dfrac{1}{x}$,

(iv) $e^y + x e^y \dfrac{dy}{dx} = 2$.

Integrating factors

17.4. There are some differential equations which are not *exact* as they stand, but which may be made so by multiplying each side by an **integrating factor**.

Example 2. *Solve* $xy \dfrac{dy}{dx} + y^2 = 3x$.

$\left[\text{We cannot separate the variables.}\right.$ Can we find a function whose derivative is the L.H.S. as in § 17.3? No. Then can we find a function whose derivative is $f(x) \times$ L.H.S.?

$$\frac{d}{dx}(xy^2) = y^2 + 2xy \frac{dy}{dx}; \quad \text{this is no good.}$$

$$\frac{d}{dx}(x^2 y^2) = 2xy^2 + 2x^2 y \cdot \frac{dy}{dx} = 2x\left(y^2 + xy \frac{dy}{dx}\right) = 2x \times \text{L.H.S.}$$

The required integrating factor is $2x$.$\Big]$

$$xy \frac{dy}{dx} + y^2 = 3x.$$

Multiplying each side by $2x$,

$$2x^2 y \frac{dy}{dx} + 2xy^2 = 6x^2,$$

$$\therefore \quad x^2 y^2 = 2x^3 + A.$$

Qu. 10. Find the integrating factors required to make the following differential equations into exact equations, and solve them:

(i) $x \dfrac{dy}{dx} + 2y = e^{x^2}$,

(ii) $x e^y \dfrac{dy}{dx} + 2e^y = x$,

(iii) $2x^2 y \dfrac{dy}{dx} + xy^2 = 1$,

(iv) $r \sec^2 \theta + 2 \tan \theta \dfrac{dr}{d\theta} = 2r^{-1}$.

First order linear equations

17.5. A differential equation is *linear in y* if it is of the form

$$\frac{d^n y}{dx^n} + P_1 \frac{d^{n-1}y}{dx^{n-1}} + P_2 \frac{d^{n-2}y}{dx^{n-2}} + \ldots + P_{n-1}\frac{dy}{dx} + P_n y = Q,$$

where $P_1, P_2, \ldots, P_n$, Q are functions of x, or constants; it is of the nth order.

Thus a *first order linear equation* is of the form

$$\frac{dy}{dx} + Py = Q,$$

where P, Q are functions of x or constants. This type of differential equation deserves special attention because an integrating factor, when required and if obtainable, is of a standard form.

Let us assume that the general first order linear equation given above can be made into an exact equation by using the integrating factor R, a function of x. If this is so,

$$R\frac{dy}{dx} + RPy = RQ \tag{1}$$

is an exact equation, and it is apparent from the first term that the L.H.S. of (1) is $\dfrac{d}{dx}(Ry) = R\dfrac{dy}{dx} + y\dfrac{dR}{dx}$. Thus (1) may also be written

$$R\frac{dy}{dx} + y\frac{dR}{dx} = RQ. \tag{2}$$

Equating the second terms on the L.H.S. of (1) and (2),

$$y\frac{dR}{dx} = RPy,$$

$$\therefore \frac{dR}{dx} = RP.$$

Separating the variables,

$$\int \frac{1}{R}\,dR = \int P\,dx,$$

$$\therefore \log_e R = \int P\,dx,$$

$$\therefore R = e^{\int P\,dx}.$$

Thus the required **integrating factor is** $e^{\int P\,dx}$. The initial assumption that an integrating factor exists is therefore justified provided that it is possible to find $\int P\,dx$.

Example 3. *Solve the differential equation* $\dfrac{dy}{dx}+3y=e^{2x}$, *given that* $y=\frac{6}{5}$ *when* $x=0$.

The integrating factor is $e^{\int 3\,dx} = e^{3x}$.

Multiplying each side of the given equation by e^{3x},

$$e^{3x}\frac{dy}{dx}+3e^{3x}y = e^{5x},$$
$$\therefore\ e^{3x}y = \tfrac{1}{5}e^{5x}+A.$$

Therefore the general solution is

$$y = \tfrac{1}{5}e^{2x}+A\,e^{-3x}.$$

But $y=\frac{6}{5}$ when $x=0$, $\quad\therefore\ \frac{6}{5}=\frac{1}{5}+A, \quad\therefore\ A=1,$

Therefore the particular solution is

$$y = \tfrac{1}{5}e^{2x}+e^{-3x}.$$

Example 4. *Solve* $\dfrac{dy}{dx}+y\cot x = \cos x$.

The integrating factor is

$$e^{\int \cot x\,dx} = e^{\log_e \sin x} = \sin x.\dagger$$

Multiplying each side of the given equation by $\sin x$,

$$\sin x\frac{dy}{dx}+y\cos x = \cos x\sin x,$$
$$\therefore\ y\sin x = \tfrac{1}{2}\sin^2 x+A,$$

Therefore the general solution is

$$y = \tfrac{1}{2}\sin x+A\csc x.$$

Qu. 11. Find the general solution of $\dfrac{dy}{dx}+2xy=x$. What is the particular solution given by $y=-\frac{1}{2}$ when $x=0$?

Qu. 12. Show that the equation in Qu. 10 (i) is of the type under discussion, and find the required integrating factor as $e^{\int P\,dx}$.

Qu. 13. Solve: (i) $\dfrac{dy}{dx}-y\tan x = x$, (ii) $\dfrac{dy}{dx}+y+3 = x$.

$\dagger$ Remember that if $\log_e a=b$ then $a=e^b$. It follows that $e^{\log_e a}=e^b=a$.

First order homogeneous equations

17.6. In a homogeneous differential equation all the terms are of the same dimensions. To obtain a clear picture of what is meant by this, suppose x and y to measure units of length. The term x^2y is of dimensions (length)^3, or L^3; the term $\dfrac{(x^2+y^2)^2}{x}$ is of dimensions $\dfrac{L^4}{L} = L^3$. The dimensions of some other terms are given below:

$$\frac{y}{x} \qquad\qquad \frac{L}{L} = L^0$$

$$\frac{y}{x^2} \qquad\qquad \frac{L}{L^2} = L^{-1}$$

$$\frac{dy}{dx} = \lim_{\Delta x \to 0} \frac{\Delta y}{\Delta x} \qquad\qquad \frac{L}{L} = L^0$$

$$\frac{d^2y}{dx^2} = \lim_{\Delta x \to 0} \frac{\Delta \left(\dfrac{dy}{dx}\right)}{\Delta x} \qquad\qquad \frac{L^0}{L} = L^{-1}$$

$$x\frac{dy}{dx} \qquad\qquad L \times L^0 = L^1$$

$$\frac{\left(\dfrac{y^2}{x}\right)^2}{\dfrac{dy}{dx}} \qquad\qquad \frac{L^2}{L^0} = L^2$$

Qu. 14. Pick out that member of each of the following groups of terms and expressions which is not of the same dimensions as the rest:

(i) $xy\dfrac{dy}{dx}$, $\quad y^3\dfrac{d^2y}{dx^2}$, $\quad \left(\dfrac{y}{x}\right)^2$, $\quad x^2+y^2$,

(ii) $(x+y)^2\dfrac{dy}{dx}$, $\quad x^2\left(1+\dfrac{y}{x}\right)$, $\quad \left(\dfrac{dy}{dx}\right)^2 xy$, $\quad \dfrac{d^2y}{dx^2}+xy$,

(iii) $(y+2x)\dfrac{dy}{dx}$, $\quad (y^2-x^2)\dfrac{d^2y}{dx^2}$, $\quad x\surd(x^2+y^2)$, $\quad 2x+\dfrac{y^2}{x}$.

Qu. 15. Which of the following equations are homogeneous?

(i) $x^2\dfrac{dy}{dx} = y^2$,

(ii) $xy\dfrac{dy}{dx} = x^2+y^2$,

(iii) $x^2\dfrac{dy}{dx} = 1+xy$,

(iv) $x^2\dfrac{d^2y}{dx^2} = y\dfrac{dy}{dx}$,

(v) $(x^2-y^2)\dfrac{dy}{dx} = 2xy$,

(vi) $(1+y^2)\dfrac{dy}{dx} = x^2$.

A first order homogeneous equation is of the form

$$P \frac{dy}{dx} = Q.$$

Since $\frac{dy}{dx}$ is of dimensions 0, P and Q are homogeneous functions of x and y of the same dimensions, i.e. of the same degree. The significant point to note is that, if P and Q are of degree n, we may divide each side of the equation by x^n and thereby obtain

$$P' \frac{dy}{dx} = Q',$$

where P' and Q' are functions of $\frac{y}{x}$.

For example, the equation

$$xy \frac{dy}{dx} = x^2 + y^2,$$

when each side is divided by x^2, becomes

$$\frac{y}{x} \frac{dy}{dx} = 1 + \left(\frac{y}{x}\right)^2.$$

This suggests the substitutions

$$\frac{y}{x} = u,$$

and, since $y = ux$, $\qquad \dfrac{dy}{dx} = u + x \dfrac{du}{dx}.$

Example 5. *Solve* $xy \dfrac{dy}{dx} = x^2 + y^2.$

Dividing each side by x^2,

$$\frac{y}{x} \frac{dy}{dx} = 1 + \left(\frac{y}{x}\right)^2.$$

Let $\qquad y = ux, \quad$ then $\quad \dfrac{dy}{dx} = u + x \dfrac{du}{dx}.$

$$\therefore \ u\left(u + x \frac{du}{dx}\right) = 1 + u^2,$$

$$\therefore \ ux \frac{du}{dx} = 1.$$

Separating the variables,

$$\int u \, \mathrm{d}u = \int \frac{1}{x} \, \mathrm{d}x,$$

$$\therefore \quad \tfrac{1}{2}u^2 = \log_e (Bx),$$

$$\therefore \quad \left(\frac{y}{x}\right)^2 = 2 \log_e (Bx),$$

$$\therefore \quad \left(\frac{y}{x}\right)^2 = \log_e (Ax^2), \text{ where } A = B^2.$$

Therefore the general solution is

$$y^2 = x^2 \log_e (Ax^2).$$

Qu. 16. Solve the following equations by the method of Example 5:

(i) $x^2 \dfrac{\mathrm{d}y}{\mathrm{d}x} = y^2 + xy,$ (ii) $x \dfrac{\mathrm{d}y}{\mathrm{d}x} = x - y,$ (iii) $x^2 \dfrac{\mathrm{d}y}{\mathrm{d}x} = 2y^2.$

Qu. 17. Solve the equation $x \dfrac{\mathrm{d}y}{\mathrm{d}x} = 2x + y$ (i) by the method of Example 5, (ii) by the method of Example 4.

Qu. 18. Solve the equations in Qu. 16 (ii) and (iii) *not* using the method of Example 5.

The above questions serve not only to illustrate the method under discussion but also to stress that the types of equations given in this chapter are not all mutually exclusive.

Exercise 17b

1. Solve the following exact differential equations:

(i) $y^2 + 2xy \dfrac{\mathrm{d}y}{\mathrm{d}x} = \dfrac{1}{x^2},$ (ii) $xy^2 + x^2 y \dfrac{\mathrm{d}y}{\mathrm{d}x} = \sec^2 2x,$

(iii) $\log_e y + \dfrac{x}{y} \dfrac{\mathrm{d}y}{\mathrm{d}x} = \sec x \tan x,$ (iv) $(1 - 2x) \, \mathrm{e}^y \dfrac{\mathrm{d}y}{\mathrm{d}x} - 2\mathrm{e}^y = \sec^2 x,$

(v) $2t \, \mathrm{e}^s + t^2 \, \mathrm{e}^s \dfrac{\mathrm{d}s}{\mathrm{d}t} = \sin t + t \cos t,$ (vi) $\mathrm{e}^u r^2 + 2r \, \mathrm{e}^u \dfrac{\mathrm{d}r}{\mathrm{d}u} = -\operatorname{cosec}^2 u.$

2. Find, by inspection, the integrating factors required to make the following differential equations into exact equations, and solve them:

(i) $\sin y + \tfrac{1}{2}x \cos y \dfrac{\mathrm{d}y}{\mathrm{d}x} = 3,$ (ii) $\dfrac{\mathrm{d}y}{\mathrm{d}x} + \dfrac{y}{x} = \dfrac{\mathrm{e}^x}{x},$

(iii) $\dfrac{1}{x} \tan y + \sec^2 y \dfrac{\mathrm{d}y}{\mathrm{d}x} = 2\mathrm{e}^{x^2},$ (iv) $y \, \mathrm{e}^x + y^2 \, \mathrm{e}^x \dfrac{\mathrm{d}x}{\mathrm{d}y} = 1.$

3. Solve the following first order linear equations:

(i) $\dfrac{dy}{dx} + 2y = e^{-2x} \cos x$,   (ii) $\dfrac{1}{t}\dfrac{ds}{dt} = 1 - 2s$,

(iii) $\dfrac{dy}{dx} + (2x+1)y - e^{-x^2} = 0$,   (iv) $\dfrac{dr}{d\theta} + 2r \cot \theta = \operatorname{cosec}^2 \theta$,

(v) $\dfrac{dr}{d\theta} + r \tan \theta = \cos \theta$,   (vi) $x\dfrac{dy}{dx} + 2y = \dfrac{\cos x}{x}$,

(vii) $x\dfrac{dy}{dx} - y = \dfrac{x}{x-1}$,   (viii) $2x\dfrac{dy}{dx} = x - y + 3$,

(ix) $\sin x \dfrac{dy}{dx} + y = \sin^2 x$,   (x) $3y + (x-2)\dfrac{dy}{dx} = \dfrac{2}{x-2}$,

(xi) $x\dfrac{dy}{dx} - y = x^2(\cos x - x \sin x)$,

(xii) $x\dfrac{dy}{dx} + y + xy \cot x = x \operatorname{cosec} x$,

(xiii) $y\dfrac{dx}{dy} = x + 2y$,   (xiv) $(1+x^2)\left(x\dfrac{dy}{dx} - 2y\right) = x^3(1+2x)$.

4. Solve the following homogeneous equations:

(i) $x^2\dfrac{dy}{dx} = 3x^2 + xy$,   (ii) $xy\dfrac{dy}{dx} = x^2 - y^2$,

(iii) $x^2\dfrac{dy}{dx} = x^2 + xy + y^2$,   (iv) $3x^2\dfrac{dy}{dx} = y^2$,

(v) $(x^2+y^2)\dfrac{dy}{dx} = xy$,   (vi) $(4x-y)\dfrac{dy}{dx} = 4x$,

(vii) $x\dfrac{dy}{dx} = y + \surd(x^2-y^2)$,   (viii) $x\dfrac{dy}{dx} = x + 2y$,

(ix) $y\dfrac{dy}{dx} = 2x + y$,   (x) $x^2\dfrac{dy}{dx} = x^2 + y^2$,

(xi) $xy\dfrac{dy}{dx} = y^2 + x\surd(x^2+y^2)$,   (xii) $x(x^2+y^2)\dfrac{dy}{dx} = x^3 + 2x^2y + y^3$.

*5. Solve the equation $\dfrac{dy}{dx} = \dfrac{x-y+2}{x+y}$, reducing it to a homogeneous equation by the change of variables $x = X - 1$, $y = Y + 1$. $\Big($Note that this implies a change of origin to $(-1, 1)$ the point of intersection of the straight lines $x - y + 2 = 0$ and $x + y = 0$; see § 7.6. The new axes are parallel to the old so $\dfrac{dy}{dx} = \dfrac{dY}{dX}.\Big)$

6. Solve the following equations by the method indicated in No. 5:

(i) $\dfrac{dy}{dx} = \dfrac{y-2}{x+y-5}$,   (ii) $2y\dfrac{dy}{dx} = x + y - 3$.

*7. State why the equation $\dfrac{dy}{dx} = \dfrac{y-x+2}{y-x-4}$ may not be reduced to a homogeneous equation by the method of No. 5. Solve it by the change of variable $y - x = z$.

8. Solve the following equations by the method indicated in No. 7:

   (i) $\dfrac{dy}{dx} = \dfrac{2x+y-2}{2x+y+1}$, (ii) $(x+y)\dfrac{dy}{dx} = x+y-2$.

9. Solve the following differential equations:

   (i) $(x+3)\dfrac{dy}{dx} - 2y = (x+3)^3$, (ii) $x^2\dfrac{dy}{dx} = x^2 - xy + y^2$,

   (iii) $\dfrac{dy}{dx} + (y+3)\cot x = e^{-2x}\operatorname{cosec} x$,

   (iv) $\sin y + (x+3)\cos y\,\dfrac{dy}{dx} = \dfrac{1}{x^2}$,

   (v) $(x-4y+2)\dfrac{dy}{dx} = x+y-3$, (vi) $\dfrac{dy}{dx} = y+2+e^{2x}(x+1)$,

   (vii) $2y\log_e y + x\dfrac{dy}{dx} = \dfrac{y}{x}\cot x$,

   (viii) $2\tan\theta\,\dfrac{dr}{d\theta} + (2r+3)\tan^2\theta + 2r = 0$,

   (ix) $x(y-x)\dfrac{dy}{dx} = y(x+y)$,

   (x) $\dfrac{dy}{dx} + y(1+\cot x) = 2x\,e^{-x}\operatorname{cosec} x$,

   (xi) $\dfrac{dy}{dx} = \dfrac{x-y+1}{x-y+3}$.

10. Find the particular solutions of the following differential equations which satisfy the given conditions:

   (i) $(x+1)\dfrac{dy}{dx} - 3y = (x+1)^4$, $y = 16$ when $x = 1$,

   (ii) $\dfrac{du}{d\theta} + u\cot\theta = 2\cos\theta$, $u = 3$ when $\theta = \dfrac{\pi}{2}$,

   (iii) $(x+y)\dfrac{dy}{dx} = x-y$, $y = -2$ when $x = 3$,

   (iv) $(x^2-y^2)\dfrac{dy}{dx} = xy$, $y = 2$ when $x = 4$,

   (v) $x-1+\dfrac{dx}{dt} = e^{-t}\,t^{-2}$, $x = 1$ when $t = 1$,

   (vi) $x(x+2y)\dfrac{dy}{dx} = 4x^2 + xy + 3y^2$, $y = 2$ when $x = 1$,

   (vii) $x\dfrac{dy}{dx} - 3y = x^3\log_e x$, $y = \tfrac{1}{2}$ when $x = 1$.

$\dfrac{\mathrm{d}y}{\mathrm{d}x} = p$ for second order equations

17.7. A general treatment of second order equations is beyond the scope of this book. To the form $\dfrac{\mathrm{d}^2y}{\mathrm{d}x^2}=\mathrm{f}(x)$ we may apply direct integration twice, as also to an equation such as

$$x\,\frac{\mathrm{d}^2y}{\mathrm{d}x^2}+\frac{\mathrm{d}y}{\mathrm{d}x} = 2x,$$

which is exact, giving

$$x\,\frac{\mathrm{d}y}{\mathrm{d}x} = x^2+A \quad \text{and} \quad y = \tfrac{1}{2}x^2+A\log_e x+B.$$

Of wide application to other forms of second order equations is the substitution $\dfrac{\mathrm{d}y}{\mathrm{d}x} = p,$

from which we obtain

$$\frac{\mathrm{d}^2y}{\mathrm{d}x^2} = \frac{\mathrm{d}p}{\mathrm{d}x} = \frac{\mathrm{d}p}{\mathrm{d}y}\cdot\frac{\mathrm{d}y}{\mathrm{d}x} = p\,\frac{\mathrm{d}p}{\mathrm{d}y}.$$

Thus:

(i) the equation $\dfrac{\mathrm{d}^2y}{\mathrm{d}x^2}=\mathrm{f}(y)$ becomes $p\,\dfrac{\mathrm{d}p}{\mathrm{d}y}=\mathrm{f}(y)$,

(ii) an equation containing $\dfrac{\mathrm{d}^2y}{\mathrm{d}x^2}, \dfrac{\mathrm{d}y}{\mathrm{d}x}, y$ but *with x absent*, becomes a *first order* equation containing $p\,\dfrac{\mathrm{d}p}{\mathrm{d}y}, p, y$,

(iii) an equation containing $\dfrac{\mathrm{d}^2y}{\mathrm{d}x^2}, \dfrac{\mathrm{d}y}{\mathrm{d}x}, x$ but *with y absent*, becomes a *first order* equation containing $\dfrac{\mathrm{d}p}{\mathrm{d}x}, p, x$.

Example 6. *Solve* $(1+x^2)\dfrac{\mathrm{d}^2y}{\mathrm{d}x^2} = 2x\dfrac{\mathrm{d}y}{\mathrm{d}x}.$

Let $\dfrac{\mathrm{d}y}{\mathrm{d}x} = p$, and since y is absent, write $\dfrac{\mathrm{d}^2y}{\mathrm{d}x^2}$ as $\dfrac{\mathrm{d}p}{\mathrm{d}x}.$

$$(1+x^2)\,\frac{\mathrm{d}p}{\mathrm{d}x} = 2xp.$$

Separating the variables,

$$\int \frac{1}{p} \, \mathrm{d}p = \int \frac{2x}{1+x^2} \, \mathrm{d}x,$$

$$\therefore \ \log_e p = \log_e \{C(1+x^2)\},$$

$$\therefore \ p = \frac{\mathrm{d}y}{\mathrm{d}x} = C + Cx^2,$$

$$\therefore \ y = Cx + \tfrac{1}{3}Cx^3 + B.$$

Therefore, writing $3A$ for C, the general solution is

$$y = Ax^3 + 3Ax + B.$$

Qu. 19. Solve: (i) $x \dfrac{\mathrm{d}^2 y}{\mathrm{d}x^2} = 2$,   (ii) $\dfrac{\mathrm{d}^2 y}{\mathrm{d}x^2} = x \cos x$,

(iii) $x \dfrac{\mathrm{d}^2 y}{\mathrm{d}x^2} + \dfrac{\mathrm{d}y}{\mathrm{d}x} = 9x^2$,   (iv) $y \dfrac{\mathrm{d}^2 y}{\mathrm{d}x^2} + \left(\dfrac{\mathrm{d}y}{\mathrm{d}x}\right)^2 = \cos x$.

Qu. 20. Solve: (i) $y \dfrac{\mathrm{d}^2 y}{\mathrm{d}x^2} = \left(\dfrac{\mathrm{d}y}{\mathrm{d}x}\right)^2$,

(ii) $(2x-1) \dfrac{\mathrm{d}^2 y}{\mathrm{d}x^2} - 2 \dfrac{\mathrm{d}y}{\mathrm{d}x} = 0$.

Qu. 21. Write the differential equation $2 \dfrac{\mathrm{d}y}{\mathrm{d}x} + x \dfrac{\mathrm{d}^2 y}{\mathrm{d}x^2} = \dfrac{2}{x}$, by means of the substitution $\dfrac{\mathrm{d}y}{\mathrm{d}x} = p$, as a differential equation linear in p, and proceed as in § 17.4.

Simple harmonic motion

17.8. The substitutions mentioned in § 17.7 arise in a less abstract form in mechanics. With the usual notation, the velocity

$$v = \frac{\mathrm{d}x}{\mathrm{d}t} \quad \left(\text{compare with } p = \frac{\mathrm{d}y}{\mathrm{d}x}\right),$$

and the acceleration is

$$\frac{\mathrm{d}^2 x}{\mathrm{d}t^2} = \frac{\mathrm{d}v}{\mathrm{d}t} = \frac{\mathrm{d}v}{\mathrm{d}x} \cdot \frac{\mathrm{d}x}{\mathrm{d}t} = v \frac{\mathrm{d}v}{\mathrm{d}x}.$$

The reader may already appreciate that in dealing with variable forces, the equation of motion, $P = mf$, may be usefully written

$$P = m \frac{\mathrm{d}v}{\mathrm{d}t}, \quad \text{if } P \text{ is a function of } t,$$

or $\qquad\qquad P = mv \dfrac{\mathrm{d}v}{\mathrm{d}x}, \quad \text{if } P \text{ is a function of } x.$

A particular case of motion under the action of a force varying with displacement is Simple Harmonic Motion (S.H.M.).

DEFINITION: *A body moves in Simple Harmonic Motion in a straight line when its acceleration is proportional to its distance from a given point on the line, and is directed always towards that point.*

Before studying this section the reader should have some knowledge of this topic. We shall not confine our attention only to finding the general solution of the typical S.H.M. equation

$$\frac{d^2x}{dt^2} = -n^2x;$$

we must discuss in some detail the constants which arise in the solution. Now the constant n in the above equation is determined by the physical situation which gives rise to S.H.M. For example, if a body of given mass hangs at rest from a given spring attached to a fixed point, and is then displaced vertically and released, it will oscillate in S.H.M.; in this case the mass of the body, and the natural length and elasticity of the spring, together determine the constant n, and the periodic time $\frac{2\pi}{n}$.

But n must not be confused with the two *arbitrary constants* which will arise in the general solution of the above second order differential equation.

(i) Quite independent of the periodic time is an *arbitrary* **amplitude** a, the maximum displacement from the centre of oscillation (dependent in the above example upon how far we displace the body from its equilibrium position before releasing it).

(ii) The general solution of the S.H.M. equation will give the displacement x from the centre of oscillation at time t; here is the second *arbitrary* choice, the instant at which we take t to be zero.

Example 7. *Find the general solution of the Simple Harmonic Motion equation* $\frac{d^2x}{dt^2} = -n^2x$.

$\left[\text{Since } t \text{ is absent, we write } \frac{d^2x}{dt^2} \text{ as } v\,\frac{dv}{dx}. \right]$

$$v\,\frac{dv}{dx} = -n^2x,$$

$$\therefore \ \tfrac{1}{2}v^2 = -\tfrac{1}{2}n^2x^2 + c.$$

[At this stage we prefer to express the arbitrary c in terms of the arbitrary amplitude a.]

If the amplitude is a, $v=0$ when $x=a$,

$$\therefore \quad 0 = -\tfrac{1}{2}n^2a^2 + c \quad \text{whence} \quad c = \tfrac{1}{2}n^2a^2.$$

$$\therefore \quad v^2 = n^2(a^2 - x^2).$$

We must now deal separately with the positive and negative velocities which occur in any position (other than the extreme positions when $x = \pm a$). Thus

$$\frac{dx}{dt} = +n\sqrt{(a^2 - x^2)} \quad \text{or} \quad \frac{dx}{dt} = -n\sqrt{(a^2 - x^2)}.$$

Separating the variables,

$$+ \int \frac{1}{\sqrt{(a^2 - x^2)}}\, dx = \int n\, dt \quad \text{or} \quad - \int \frac{1}{\sqrt{(a^2 - x^2)}}\, dx = \int n\, dt.$$

[Here it is preferable to use the change of variable $x = a \cos u$ on the L.H.S. of each equation, rather than the more usual $x = a \sin u$, since it enables one to handle more easily the remaining arbitrary constant.]
The solution of these equations may be written

$$-\cos^{-1}\frac{x}{a} = nt + \epsilon' \qquad \text{and} \qquad \cos^{-1}\frac{x}{a} = nt + \epsilon,$$

$$\therefore \quad x = a \cos(-nt - \epsilon') \qquad \text{and} \qquad x = a \cos(nt + \epsilon).$$

But $\cos(-\theta) = \cos\theta$, so we may write

$$x = a \cos(nt + \epsilon') \qquad \text{and} \qquad x = a \cos(nt + \epsilon)$$

for motion in the

$$\textit{positive} \qquad \text{and} \qquad \textit{negative}$$

directions respectively.

At an extreme position when $x=a$, and $t=t_1$ say, the motion is changing from positive to negative direction, both solutions are valid, and

$$\cos(nt_1 + \epsilon') = \cos(nt_1 + \epsilon) = 1,$$

$$\therefore \quad nt_1 + \epsilon' \text{ and } nt_1 + \epsilon \text{ are multiples of } 2\pi.$$

$$\therefore \quad \epsilon' = \epsilon + 2k\pi \text{ (where } k \text{ is an integer or zero).}$$

Hence $x = a \cos(nt + \epsilon') = a \cos(nt + \epsilon + 2k\pi) = a \cos(nt + \epsilon)$. Therefore the motion is fully defined by the general solution

$$x = a \cos(nt + \epsilon). \tag{1}$$

Qu. 22. Write down the general solutions of the following equations:

(i) $\dfrac{d^2x}{dt^2} = -4x,$ (ii) $\dfrac{d^2y}{dx^2} + 9y = 0,$ (iii) $\dfrac{d^2y}{dx^2} = -16x.$

Qu. 23. In Example 7 what integrating factor will enable you to obtain the first order equation

$$\left(\frac{dx}{dt}\right)^2 = -n^2x^2 + k?$$

Qu. 24. A Simple Harmonic Motion of amplitude 2 cm has the equation $\dfrac{d^2x}{dt^2} = -\dfrac{9}{4}x.$ Write down the solution of this equation given that $x = 2$ when $t = 0$. Find expressions for $\dfrac{dx}{dt}$ (i) in terms of x, (ii) in terms of t.

Qu. 25. What special form is taken by the general solution $x = a \cos(nt + \epsilon)$ if the motion is timed (i) from an extreme position (i.e. $x = a$ when $t = 0$), (ii) from the centre of oscillation?

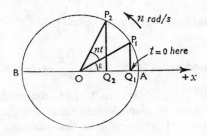

Fig. 17.1

The reader is no doubt familiar with the fact that, if a radius OP of a circle centre O radius a rotates about O with constant angular velocity n rad/sec, and Q is the projection of P on a diameter AB, Q moves with S.H.M. along AB.

Let us take $t = 0$ at Q_1, where $\angle AOP_1 = \epsilon$ radians, and suppose that Q moves directly to position Q_2 in time t. Then $\angle P_1OP_2 = nt$ radians, and $\angle AOP_2 = (nt + \epsilon)$ radians (Fig. 17.1). Thus if x is the displacement of Q from O at time t,

$$x = a \cos(nt + \epsilon).$$

When $t = 0$, $x = a \cos \epsilon$, and so we see the significance of this constant ϵ, which is called the **initial phase**.

Qu. 26. What does the general solution $x = a \cos (nt + \epsilon)$ become if the initial phase is (i) 0, (ii) $-\frac{\pi}{2}$? Illustrate each case with a sketch.

Qu. 27. A Simple Harmonic Motion has amplitude 3 mm. If $t = 0$ when the body is $+1 \cdot 5$ mm from the centre of oscillation, what is the initial phase?

The two arbitrary constants which appear in the general solution (1) are the amplitude a, and the initial phase ϵ. However, the general solution is often given in a form in which these are not explicitly stated.

Expanding the R.H.S. of (1)

$$x = a \cos nt \cos \epsilon - a \sin nt \sin \epsilon,$$

or $\qquad x = A \cos nt + B \sin nt \qquad\qquad (2)$

where $\qquad A = a \cos \epsilon \quad \text{and} \quad B = -a \sin \epsilon.$

In this form we see that the amplitude

$$a = \sqrt{(A^2 + B^2)},$$

and the initial phase $\qquad \epsilon = \tan^{-1} \left(-\frac{B}{A} \right).$

Reduction to the form $\dfrac{\mathrm{d}^2x}{\mathrm{d}t^2} = -n^2x$

17.9. In tackling a problem on S.H.M. the reader may inadvertently choose to measure displacement from a point other than the centre of oscillation. How an equation thus obtained may be reduced to the standard form is illustrated in the following example.

Example 8. *Solve* $\dfrac{\mathrm{d}^2x}{\mathrm{d}t^2} + 9x - 18 = 0.$

This equation may be written

$$\frac{\mathrm{d}^2x}{\mathrm{d}t^2} = -9(x - 2).$$

Let $\qquad\qquad x - 2 = u,$

then $\qquad \dfrac{\mathrm{d}x}{\mathrm{d}t} = \dfrac{\mathrm{d}u}{\mathrm{d}t} \quad \text{and} \quad \dfrac{\mathrm{d}^2x}{\mathrm{d}t^2} = \dfrac{\mathrm{d}^2u}{\mathrm{d}t^2},$

$$\therefore \frac{\mathrm{d}^2u}{\mathrm{d}t^2} = -9u.$$

The general solution of this equation is

$$u = a \cos (3t + \epsilon).$$

But
$$x = u + 2,$$
$$\therefore \ x = a \cos (3t + \epsilon) + 2.$$

Qu. 28. Solve the following equations:

(i) $\dfrac{d^2y}{dx^2} + 4y + 4 = 0$,   (ii) $\dfrac{d^2\theta}{dt^2} + 2\theta - 6 = 0$,   (iii) $\dfrac{d^2x}{dt^2} + \frac{9}{4}t = -1$.

Exercise 17c

1. Solve the following differential equations:

(i) $x \dfrac{d^2y}{dx^2} - 1 = 0$,   (ii) $\cos \theta \dfrac{d^2y}{d\theta^2} - \sin \theta \dfrac{dy}{d\theta} = \cos \theta$,

(iii) $e^x \dfrac{d^2y}{dx^2} = 2$,   (iv) $(2x+1) \dfrac{d^2y}{dx^2} + 2 \dfrac{dy}{dx} = 0$,

(v) $\dfrac{1}{x} \dfrac{d^2y}{dx^2} - \dfrac{1}{x^2} \dfrac{dy}{dx} = e^x$,   (vi) $\sin^2 t \dfrac{d^2x}{dt^2} + 1 = 0$.

2. Use the substitution $\dfrac{dy}{dx} = p$ to solve the following differential equations:

(i) $x \dfrac{d^2y}{dx^2} = 3 \dfrac{dy}{dx}$,   (ii) $\dfrac{d^2y}{dx^2} = 2 \left(\dfrac{dy}{dx} \right)^2$,

(iii) $\dfrac{d^2y}{dx^2} + y \left(\dfrac{dy}{dx} \right)^3 = 0$,   (iv) $(1+x^2) \dfrac{d^2y}{dx^2} + 2x \dfrac{dy}{dx} = 0$,

(v) $(1-x^2) \dfrac{d^2y}{dx^2} = x \dfrac{dy}{dx}$,   (vi) $y \dfrac{d^2y}{dx^2} = 3 \left(\dfrac{dy}{dx} \right)^2$,

(vii) $(3y-2) \dfrac{d^2y}{dx^2} = 3 \left(\dfrac{dy}{dx} \right)^2$,   (viii) $(2y+1) \dfrac{d^2y}{dx^2} = \left(\dfrac{dy}{dx} \right)^2$,

(ix) $y^3 \dfrac{d^2y}{dx^2} + 9 = 0$,   (x) $\dfrac{d^2y}{dx^2} = 8x \left(\dfrac{dy}{dx} \right)^2$,

(xi) $\dfrac{d^2y}{dx^2} = 3 \dfrac{dx}{dy}$,   (xii) $y \dfrac{d^2y}{dx^2} - \left(\dfrac{dy}{dx} \right)^2 - 2 \dfrac{dy}{dx} = 0$.

3. Solve the differential equation $\dfrac{d^2s}{dt^2} + \frac{1}{10} \left(\dfrac{ds}{dt} \right)^2 = 0$, using the substitution

$\dfrac{ds}{dt} = v$, and writing $\dfrac{d^2s}{dt^2}$ (i) as $\dfrac{dv}{dt}$, (ii) as $v \dfrac{dv}{ds}$.

4. Use the substitution $\dfrac{dy}{dx} = p$ to write the following as differential equations linear in p, and proceed as in § 17.5:

(i) $\dfrac{d^2y}{dx^2} + \cot x \dfrac{dy}{dx} = 1$,   (ii) $\dfrac{d^2y}{dx^2} + \dfrac{1}{x+2} \dfrac{dy}{dx} = x$,

(iii) $x \dfrac{d^2y}{dx^2} + 2 \dfrac{dy}{dx} = x \log_e x$.

5. Find the particular solution of the differential equation

$$\frac{dy}{dx}\frac{d^2y}{dx^2}+x = 0$$

which satisfies the condition that y and $\frac{dy}{dx}$ are both zero when $x=1$.

6. Write down the general solution of each of the following differential equations:

(i) $\frac{d^2s}{dt^2} = -25s,$ (ii) $\frac{d^2y}{dx^2}+\frac{9}{4}y = 0,$

(iii) $9\frac{d^2s}{d\theta^2} = -2\theta,$ (iv) $4\frac{d^2y}{dt^2}+3y = 0.$

7. Find the solution of the differential equation

$$16\frac{d^2s}{dt^2}+9s = 0,$$

given that $s=4$ and $\frac{ds}{dt}=0$ when $t=0$.

8. A body moves in a straight line so that when it is x cm from a point O on the line its acceleration is $9x$ cm/s² towards O. Write down the differential equation which describes this motion, and then present a complete solution of it (see Example 7) given that the body is at rest when 2 cm from O, and its distance from O is $+\sqrt{3}$ cm at the instant from which time is measured.

9. A body moves in S.H.M. of amplitude 4 cm and has initial phase $-\frac{\pi}{2}$ s. It takes 1 s to travel 2 cm from the centre of oscillation, O. What was its initial position, and what is its periodic time?

10. A body moving in S.H.M. is timed from an extreme position, and is found to take 2 s to reach a point mid-way between the centre of oscillation and the other end of its path. State the initial phase, and calculate the periodic time.

11. A body moves in a straight line so that it is x m from a fixed point on the line at time t s, where $x = \cos 2t+\sin 2t$. Write this in the form $x = a\cos(nt+\epsilon)$ and state the amplitude, initial phase, and periodic time of the motion.

12. Repeat No. 11 for $x = 3\cos\frac{1}{2}t - 4\sin\frac{1}{2}t$, giving the initial phase correct to three significant figures.

13. The two simple harmonic motions defined by $x = a\cos(nt+\epsilon_1)$ and $x = a\cos(nt+\epsilon_2)$ are said to have a *phase difference* of $\epsilon_1 - \epsilon_2$. Find the phase difference between the following pairs of S.H.M.:

(i) $x = a\cos nt, \quad x = a\sin nt,$

(ii) $x = 2\cos\left(3t+\dfrac{\pi}{6}\right),\quad x = \sqrt{2}\,(\cos 3t - \sin 3t),$

(iii) $x = \dfrac{3\sqrt{2}}{2}(\cos nt - \sin nt),\quad x = \dfrac{3\sqrt{2}}{2}(\cos nt + \sin nt),$

(iv) $x = a\sin nt,\quad x = -a\sin nt,$

(v) $x = 5\cos nt - 5\sqrt{3}\sin nt,\quad x = 5\sqrt{2}\cos nt + 5\sqrt{2}\sin nt.$

14. Solve the following differential equations:

(i) $\dfrac{d^2y}{dx^2} = -4(y+3),$ 　　　　(ii) $2\dfrac{d^2\theta}{dt^2}+9\theta = 3,$

(iii) $3\dfrac{d^2s}{dt^2}+4t = 1,$ 　　　　(iv) $\dfrac{d^2x}{dt^2}+4x+8 = 0,$

given that in (iv), $x = -1$ when $t=0$, and $x = -3$ when $t = \dfrac{\pi}{4}$.

15. Solve the following differential equations:

(i) $\dfrac{d^2y}{dx^2}-2\dfrac{dy}{dx}-2 = 0,$ 　　(ii) $\dfrac{d^2y}{dx^2} = \dfrac{dy}{dx},$

(iii) $\dfrac{d^2x}{dt^2}+x = 0,$ 　　　　(iv) $x\dfrac{d^2y}{dx^2}+\dfrac{dy}{dx} = 0,$

(v) $(3y-1)\dfrac{d^2y}{dx^2} = \left(\dfrac{dy}{dx}\right)^2,$ 　(vi) $\dfrac{d^2y}{dx^2}-\log_e x = 0,$

(vii) $9\dfrac{d^2x}{dt^2}+4x-1 = 0,$ 　　(viii) $2\dfrac{d^2x}{du^2}+9u = 0,$

(ix) $x\dfrac{d^2y}{dx^2}-\dfrac{dy}{dx} = x^2\,e^x.$

Exercise 17d (Miscellaneous)

Solve the following differential equations Nos. 1 to 26:

1. $\cos t\,\dfrac{dx}{dt} = x.$ 　　　**2.** $x\dfrac{dy}{dx}-y = x^2\log_e x.$

3. $\sec x\,\dfrac{d^2y}{dx^2} = e^x.$ 　　**4.** $(1+\cos\theta) = (1-\cos\theta)\dfrac{d\theta}{dr}.$

5. $\log_e(y+1)+\dfrac{x}{y+1}\dfrac{dy}{dx} = \dfrac{1}{x(x+1)}.$ 　**6.** $3\dfrac{d^2y}{dx^2}+4y = 0.$

7. $(2x-1)\dfrac{dy}{dx}+8y = 4(2x-1)^{-2}.$ 　**8.** $y\dfrac{d^2y}{dx^2}+\left(\dfrac{dy}{dx}\right)^2-3\dfrac{dy}{dx} = 0.$

9. $x(x+y)\dfrac{dy}{dx} = x^2+xy-3y^2.$ 　**10.** $u\dfrac{dv}{du} = \log_e u.$

11. $e^x\left(\dfrac{d^2y}{dx^2}+\dfrac{dy}{dx}\right) = 1.$ 　**12.** $(x-1)\dfrac{d^2y}{dx^2} = x\dfrac{dy}{dx}.$

13. $\dfrac{dy}{dx}+y = e^{-x}\sin^{-1}x.$ 　**14.** $\dfrac{dy}{dx}-y\tan x = e^x.$

15. $xy \dfrac{dy}{dx} = (x+y)^2$.

16. $(y-x+6) \dfrac{dy}{dx} = 2(x-4)$.

17. $\dfrac{d^3y}{dx^3} + \dfrac{d^2y}{dx^2} = 1$.

18. $\dfrac{dy}{dx} = \dfrac{x+y}{x+y+1}$.

19. $\dfrac{dy}{dx} = \sqrt{\dfrac{1-y^2}{1-x^2}}$.

20. $y = \sin x \left(1 - \dfrac{dy}{dx}\right)$.

21. $(x-y) \dfrac{dy}{dx} = 2x+y$.

22. $\sin x \dfrac{d^2y}{dx^2} + \dfrac{dy}{dx} = 2 \cos^3 \dfrac{x}{2}$.

23. $(1-x^2) \dfrac{dy}{dx} - xy = 1$.

24. $9 \dfrac{d^2x}{dt^2} + 4x = 0$, given that $x=2$ when $t=0$, and $x=-4$ when $t=\pi$.

25. $y \dfrac{d^2y}{dx^2} + 25 = \left(\dfrac{dy}{dx}\right)^2$, given that $\dfrac{dy}{dx} = 4$ when $y=1$, and $y=\frac{5}{3}$ when $x=0$.

26. $\dfrac{d^2s}{dt^2} + 9(s-1) = 0$, given that $s=2$ when $t=0$, and $\dfrac{d^2s}{dt^2} = -9\sqrt{3}$ when $t = \dfrac{\pi}{6}$.

27. Solve the differential equation

$$xy \frac{dy}{dx} = y^2 + x^2 e^{y/x},$$

by means of the substitution $y = vx$.   O.C.

28. The normal at any point of a curve passes through the point $(1, 1)$. Express this condition in the form of a differential equation, and hence find the equation of the family of curves which satisfy the condition.   N.

29. Solve the differential equation

$$\frac{dy}{dx} + \frac{2xy}{1+x^2} = \cos x,$$

given that $y=2$ when $x=0$.   O.C.

30. Transform the equation

$$\frac{d^2y}{dx^2} + x^2 + y + 2 = 0$$

by the substitution $y = t - x^2$, and hence obtain the general solution of the equation.   L.

31. Given that y satisfies the differential equation

$$\frac{dy}{dx} + 2y \tan x = \sin x,$$

and that $y=1$ when $x = \dfrac{\pi}{3}$, express $\dfrac{dy}{dx}$ in terms of x.   N.

32. By means of the substitution $y = vx$ reduce the differential equation

$$xy \frac{dy}{dx} = y^2 + \sqrt{(x^2 + y^2)}$$

to an equation in v and x. Find the solution, given that $y = 1$ when $x = 1$.   O.C.

33. Integrate the differential equation

$$(x^2 + x) \frac{dy}{dx} + y = x + 1. \qquad\qquad \text{O.C.}$$

34. Transform the equation

$$v \frac{dv}{dx} + \frac{v^2}{2a} = -\mu x,$$

where a and μ are constants, by the substitution $y = v^2$, and hence find the general solution.

If $v = 0$ when $x = a$, show that $v^2 = 2\mu a^2$ when $x = 0$.   L.

35. If $\dfrac{d^2 u}{d\theta^2} + u = 1$ and if $\dfrac{du}{d\theta} = 0$ when $u = 2$, prove that

$$\left(\frac{du}{d\theta} \right)^2 = 2u - u^2.$$

If $u = 0$ when $\theta = 0$, prove also that $u = 1 - \cos \theta$.   O.C.

36. Find the general solution of the differential equation

$$\frac{dy}{dx} + \frac{y}{x} = \tfrac{1}{2} \sin \tfrac{1}{2} x.$$

If y is the solution which takes the value unity when $x = 2\pi$, find the limit of y as x tends to zero.   L.

37. Solve the differential equation

$$(x + 1) \frac{dy}{dx} - xy = e^{2x}. \qquad\qquad \text{O.C.}$$

CHAPTER 18

APPROXIMATIONS—FURTHER EXPANSIONS IN SERIES

Approximation

18.1. To students of elementary mathematics the word *approximation* no doubt conjures up the idea of a 'rough calculation'; but it should also be a reminder that answers may often be relied upon only to a certain degree of accuracy, due to limitations set by data and by the available means of computation.

However, there is a positive aspect of approximation which must be stressed at this stage. In the ever increasing field of application to engineering and complex organization, mathematics often assumes a character less exact, possibly less aesthetically satisfying, than when it is pursued for its own sake. A problem may arise in which many variables or 'parameters' are involved; only when attention is confined to the more significant of these is the problem susceptible to known mathematical techniques, and even then the functions concerned are often only manageable when reduced to an approximate form. It must therefore be appreciated that approximation, far from always implying a sacrifice of accuracy, can provide the means whereby mathematics may be brought to bear on practical problems which would otherwise be out of reach.

The main object of this chapter is to consider new ways of re-writing functions in an approximate form; already we have used the binomial theorem to this end, and in Chapter 10 we saw how e^x and $\log_e (1+x)$ could be expressed as power series in x. We now start by establishing a basic form of approximation, and applying it to numerical examples.

Linear approximation

18.2. Figure 18.1 represents part of the graph of $y = f(x)$. P is a fixed point $(a, f(a))$, PT is the tangent at P, and Q is a variable point on

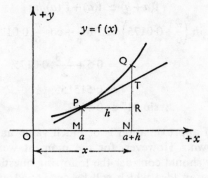

Fig. 18.1

the curve given by $x = a + h$, where h is small. We shall establish an approximate relationship between $f(a)$ and $f(a + h)$.

$$f(a + h) = \text{NQ} \simeq \text{NT}, \quad \text{since } h \text{ is small.}$$

$$\text{NT} = \text{MP} + \text{RT} = f(a) + \text{PR} \tan \angle \text{RPT} = f(a) + f'(a) \, h.\dagger$$

Hence, if h is small,

$$f(a + h) \simeq f(a) + f'(a) \, h.$$

This is called a *linear approximation* since we consider the straight line PT in lieu of the curve PQ. Expressing it another way, when $x \simeq a$, the function $f(x)$ may be expressed in an approximate linear form, since

$$f(x) \simeq f(a) + f'(a) \, (x - a).$$

Example 1. *Find an approximate value of* $\sin 31°$.

[Here we wish to establish an approximate relationship between $\sin 30°$, which we know, and $\sin 31°$. Since we use the derivative of $\sin x$ we work in radians.]

$$31° \simeq \frac{\pi}{6} + 0 \cdot 0175 \text{ radians.}$$

† The fact that $\text{RQ} \simeq \text{RT} = f'(a) \, h$ has been used in P.M.I., §6.6; there it was stated in the form $\varDelta y \simeq \dfrac{\mathrm{d}y}{\mathrm{d}x} \, \varDelta x$.

Since $$f(a+h) \simeq f(a) + f'(a)\, h,$$

$$\sin\left(\frac{\pi}{6} + 0 \!\cdot\! 0175\right) \simeq \sin\frac{\pi}{6} + \cos\frac{\pi}{6}.0 \!\cdot\! 0175,$$

$$= 0 \!\cdot\! 5 + \frac{\sqrt{3}}{2}.0 \!\cdot\! 0175,$$

$$\simeq 0 \!\cdot\! 515155,$$

$$\therefore\ \sin 31° \simeq 0 \!\cdot\! 5152.$$

No method of approximation is of much value unless its degree of accuracy is known. However, for the moment we must avoid this issue; the reader should consider the following questions as just a first step in mastering an idea which will be developed more fully later in the chapter.

Qu. 1. Use the method of Example 1 to find an approximate value of tan 45° 40′, retaining five significant figures. Compare this with the value you find in four-figure tables.

Qu. 2. Assuming that $\cos^{-1} 0 \!\cdot\! 8 \simeq 36°\ 52'$ and $52' \simeq 0 \!\cdot\! 0151$ radians, find an approximation for cos 36°.

Qu. 3. If $x \simeq \dfrac{\pi}{6}$, prove that $\cos x \simeq \frac{1}{12}(\pi + 6\sqrt{3} - 6x)$, using the second form of linear approximation given in § 18.2.

Approximate roots of an equation—Newton's method

18.3. When there is no better direct method of solving an equation, an attempt must be made to find an approximate solution, and early training may suggest a graphical approach. However, there are alternative numerical methods, and one of these, due to Newton, involves the use of the relationship established in § 18.2 to obtain successive approximations to a root of an equation once we have roughly located it.

Suppose that we wish to solve the equation $f(x) = 0$, that we know the graph of $y = f(x)$ to be continuous throughout the range we are considering, and that we have established that $f(a)$ is negative and that $f(a+1)$ is positive. (See Fig. 18.2.) It follows that $f(x)$ is zero for some value of x between a and $a+1$, and therefore there is (at least) one root of the equation $f(x) = 0$ lying between a and $a+1$. Having thus roughly located a root we proceed as follows.

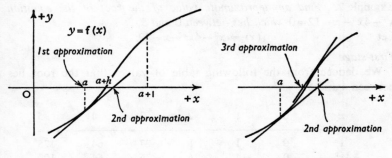

Fig. 18.2   Fig. 18.3

Let the root be $(a+h)$, then

$$f(a+h) = 0. \quad \text{(See Fig. 18.2.)}$$

But from § 18.2, if h is small,

$$f(a+h) \simeq f(a) + f'(a) h,$$

$$\therefore f(a)+f'(a) h \simeq 0,$$

$$\therefore h \simeq -\frac{f(a)}{f'(a)}.$$

This enables us to write down a second approximation to the root as $a-\dfrac{f(a)}{f'(a)}$, and the process may be repeated to obtain successively closer approximations. (See Fig. 18.3.)

We then apply this method to any other root of the equation $f(x) = 0$. However, it must be noted that various factors, such as h being too large, $f'(a)$ being too small, $f''(a)$ being too large, may together or separately cause this method to break down (compare Fig. 18.4 with Fig. 18.3), so it must be used with discretion.

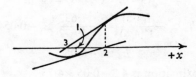

Fig. 18.4

Example 2. *Find an approximate value of the root of the equation* $x^3 - 4x^2 - x - 12 = 0$ *which lies between 0 and 5.*

Let
$$f(x) = x^3 - 4x^2 - x - 12.$$

First stage

We deduce from the following table of values that the root lies between 4 and 5.

x	0	1	2	3	4	5
x^3	0	1	8	27	64	125
$-4x^2$	0	-4	-16	-36	-64	-100
$-x$	0	-1	-2	-3	-4	-5
-12	-12	-12	-12	-12	-12	-12
$f(x)$	-12	-16	-22	-24	-16	8

Also since f(5) is numerically less than f(4) it is reasonable to guess that the root is nearer 5 than 4.

Hence the first approximation is 5.

Second stage

Let the root be $5 + h_1$.

Since
$$f(5 + h_1) = 0,$$
$$\therefore f(5) + f'(5)\, h_1 \simeq 0.$$

$$f'(x) = 3x^2 - 8x - 1$$
$$\therefore f'(5) = 75 - 40 - 1,$$
$$= 34.$$
$$\therefore h_1 \simeq -\frac{f(5)}{f'(5)} = -\frac{8}{34} \simeq -0.2353,$$
$$\therefore h_1 \simeq -0.2.$$

The second approximation is $5 - 0.2 = 4.8$.

Third stage

Let the root be $4.8 + h_2$.

$$f(4.8) + f'(4.8)\, h_2 \simeq 0.$$

$$\therefore h_2 \simeq -\frac{f(4.8)}{f'(4.8)} = -\frac{1.632}{29.72} \simeq -0.05491.$$

$$\therefore h_2 \simeq -0.05.$$

The third approximation is $4.8 - 0.05 = 4.75$.

[Note that at each stage we have added one decimal place to the approximation; this of course does not imply that we expect the answer to be correct to the number of decimal places taken. We now attempt to discover a range of values within which the root lies.]

Accuracy check

$$f(4·75) \simeq 107·17 - 90·25 - 4·75 - 12 = +0·17 > 0.$$

[Since this is positive, and the previous calculations show $f'(x)$ to be positive for this part of the curve, we deduce that the root is less than 4·75.]

$$f(4·74) \simeq 106·50 - 89·87 - 4·74 - 12 = -0·11 < 0.$$

Therefore the root lies between 4·74 and 4·75.

Qu. 4. Show that the equation $x^4 - 3x^3 - 10 = 0$ has a root which lies between 3 and 4. Apply Newton's method twice to determine the root to the same degree of accuracy as in Example 2.

The next question shows how this method can be combined with an initial graphical approach to locate the roots of an equation.

Qu. 5. Solve the equation $\sin \theta = \frac{\theta}{2}$. (First sketch on the same axes the graphs of $y = \sin \theta$ and $y = \frac{\theta}{2}$, to locate the non-zero roots; then apply Newton's method.) State, to three significant figures, the limits of the ranges of values within which the approximate roots must lie.

The reader might like to compare his solution of Qu. 5 with the entirely graphical solution of this equation on p. 331 of P.M.I.

Exercise 18a

1. Use the method of Example 1 to obtain approximations for the following. Retain four decimal places and compare your answers with the values given in four figure tables.

    (i) cosec 61° 30'   (take $\sqrt{3}$ as 1·7321),
    (ii) cot 28° 40'   (take $\sqrt{3}$ as 1·7321),
    (iii) sec 47°   (take $\sqrt{2}$ as 1·4142),
    (iv) cos 67°   (take $\tan^{-1} 2·4$ as 67° 23'),
    (v) tan 54°   (take $\tan^{-1} \frac{4}{3}$ as 53° 8'),
    (vi) sin 61°   (take $\tan^{-1} 1\frac{5}{8}$ as 61° 56'),
    (vii) $e^{1·08}$ (take e as 2·7183),
    (viii) $\log_e 2·001$ (take $\log_e 2$ as 0·6931).

2. Use the fact that if $x \simeq a$, $f(x) \simeq f(a) + f'(a)(x-a)$ to prove that:

   (i) if $x \simeq 0$, $e^x \simeq 1 + x$,

   (ii) if $x \simeq \pi$, $\sin x \simeq \pi - x$,

   (iii) if $x \simeq 2$, $\dfrac{1}{(1+x)^2} \simeq \frac{1}{27}(7 - 2x)$,

   (iv) if $x \simeq \dfrac{\pi}{4}$, $\tan x \simeq 1 - \dfrac{\pi}{2} + 2x$,

   (v) if $x \simeq 7$, $\sqrt{(2+x)} \simeq \frac{1}{6}(x + 11)$,

   (vi) if $x \simeq 1$, $\log_e x \simeq x - 1$.

In Nos. 3 to 8, unless otherwise instructed, state to two decimal places the limits of the range of values within which an approximate root lies.

3. Locate the roots of the equation $x^3 - 10x + 10 = 0$, stating the integral values between which they lie. Apply Newton's method twice to obtain an approximation for the negative root.

4. Find an approximation for the root of the equation $4 + 5x^2 - x^3 = 0$, given that it lies between 0 and 10.

5. Solve the equation $e^x = 2x + 1$, locating the roots by sketching the graphs of $y = e^x$ and $y = 2x + 1$. State the limits of the range of values within which the approximate root lies to two, or three, decimal places, as your table of e^x permits.

6. Show that the smallest positive root of the equation $3 \tan \theta + 4\theta - 6 = 0$ is approximately $\dfrac{\pi}{4}$. Find an approximation to this root, correct to two significant figures.

7. Find an approximation for the positive root of the equation
$$x^4 - 4x^3 - x^2 + 4x - 10 = 0.$$

8. Plot the graphs of $y = \log_e(x+2)$ and $y = 2x + 1$ for values of x between -2 and 0 at intervals of $0 \cdot 1$. Hence estimate the roots of the equation $\log_e(x+2) - 2x - 1 = 0$ to two decimal places. Then use Newton's method to obtain closer approximations, and give to three decimal places the limits of the ranges of values within which the roots lie.

Quadratic approximation

18.4. In Chapter 10 it was established that
$$e^x = 1 + x + \frac{x^2}{2!} + \frac{x^3}{3!} + \frac{x^4}{4!} + \cdots.$$

If x is small, an approximation for e^x may be found by ignoring high powers of x. Thus the linear approximation obtained in Exercise 18a, No. 2(i) is $1 + x$, the first two terms of the above series. Clearly a better approximation would be obtained by taking more terms; let us see how this fits in with the approach being developed in this chapter.

$y = x + 1$ is the tangent to $y = e^x$ at $(0, 1)$. So when $x = 0$, the graphs of the function and of its linear approximation have equal ordinates and equal gradients. If we take a quadratic approximation, $f(x)$, we can further stipulate that the *rate of change of gradient* of $y = e^x$ and $y = f(x)$ are equal when $x = 0$. $y = f(x)$ is a parabola, and this gives a better approximation to the curve $y = e^x$ over a wider range of values of x.

Suppose $$f(x) = c_0 + c_1 x + c_2 x^2.$$

Then for small values of x,

$$e^x \simeq c_0 + c_1 x + c_2 x^2.$$

Differentiating twice,

$$e^x \simeq c_1 + 2c_2 x,$$
and $$e^x \simeq 2c_2.$$

But we have stipulated that, when $x = 0$, these are not approximations but equalities,

$$\therefore \quad c_0 = 1, \quad c_1 = 1, \quad c_2 = \tfrac{1}{2}.$$

Therefore for small values of x

$$e^x \simeq 1 + x + \frac{x^2}{2},$$

and, as expected, we have obtained the first three terms of the series for e^x.

Qu. 6. Sketch with the same axes the graphs of e^x, $1 + x$, $1 + x + \dfrac{x^2}{2}$ from $x = 0$ to $x = 1$ at intervals of $0 \cdot 1$.

Proceeding on the same lines, we now consider in Qu. 7 the function $\log_e x$ when $x \simeq 1$, and investigate how we can obtain an improvement on the linear approximation $\log_e x \simeq x - 1$. (See Exercise 18a, No. 2 (vi).)

Qu. 7. Given that the graphs of $y = \log_e x$ and $y = c_0 + c_1(x - 1) + c_2(x - 1)^2$ have the same ordinate, gradient, and rate of change of gradient when $x = 1$, prove that when $x \simeq 1$,

$$\log_e x \simeq -\tfrac{3}{2} + 2x - \frac{x^2}{2}.$$

The following table gives values of the function $\log_e x$, and of the first and second approximations $x-1$ and $-\frac{3}{2}+2x-\frac{x^2}{2}$, in the vicinity of $x=1$, and the graphs of these functions are shown in Fig. 18.5.

x	0.2	0.3	0.4	0.5	0.6	0.7	0.8	0.9	1
$x-1$	−0.8	−0.7	−0.6	−0.5	−0.4	−0.3	−0.2	−0.1	0
$-\frac{3}{2}+2x-\frac{x^2}{2}$	−1.12	−0.95	−0.78	−0.63	−0.48	−0.345	−0.22	−0.105	0
$\log_e x$	−1.61	−1.20	−0.92	−0.69	−0.51	−0.357	−0.223	−0.1054	0

x	1.1	1.2	1.3	1.4	1.5	1.6	1.7	1.8	1.9	2
$x-1$	0.1	0.2	0.3	0.4	0.5	0.6	0.7	0.8	0.9	1
$-\frac{3}{2}+2x-\frac{x^2}{2}$	0.095	0.18	0.255	0.32	0.38	0.42	0.46	0.48	0.495	0.5
$\log_e x$	0.0953	0.182	0.262	0.34	0.41	0.47	0.53	0.59	0.64	0.69

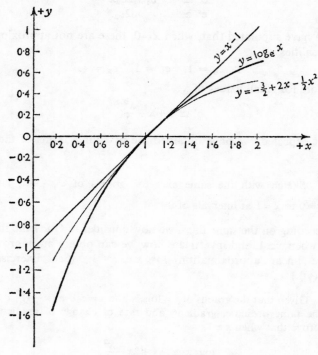

Fig. 18.5

Qu. 8. Given that the graphs of $y = f(x)$ and $y = c_0 + c_1(x - a) + c_2(x - a)^2$ have the same ordinate, gradient, and rate of change of gradient when $x = a$, find c_0, c_1, c_2, and hence give an approximation for $f(x)$ when $x \simeq a$.

Qu. 9. If a is a constant and h is small, re-write the answer to Qu. 8 so as to give an approximation for $f(a + h)$ in ascending powers of h as far as h^2

Taylor's theorem

18.5. Pursuing the ideas of Qu. 7 and Qu. 8 we may reasonably suppose that if we add terms of successively higher powers of $(x - a)$ to an approximation for a function $f(x)$ when $x \simeq a$, determining the co-efficients so that successively higher derivatives of the approximate function are equal to those of $f(x)$ when $x = a$, then we shall obtain ever closer approximations to $f(x)$.

In Chapter 10, e^z and $\log_e(1 + x)$ are expressed as infinite series in ascending powers of x. We shall now assume that if $f(x)$ is any function of x, and a is a constant, then provided that $f(a)$ exists and that successive derivatives of $f(x)$ all have finite values when $x = a$, $f(x)$ may be expressed as an infinite series in ascending powers of $(x - a)$.† In what follows, we assume that it is in order to differentiate an infinite series term by term.

Let
$$f(x) = c_0 + c_1(x - a) + c_2(x - a)^2 + c_3(x - a)^3 + c_4(x - a)^4 + \ldots,$$
then
$$f'(x) = c_1 + 2c_2(x - a) + 3c_3(x - a)^2 + 4c_4(x - a)^3 + \ldots,$$
$$f''(x) = 2!c_2 + 3 \cdot 2c_3(x - a) + 4 \cdot 3c_4(x - a)^2 + \ldots,$$
$$f'''(x) = 3!c_3 + 4 \cdot 3 \cdot 2c_4(x - a) + \ldots,$$
$$f''''(x) = 4!c_4 + \ldots,$$

and putting $x = a$ in each line, we find that

$$c_0 = f(a), \quad c_1 = f'(a), \quad c_2 = \frac{f''(a)}{2!}, \quad c_3 = \frac{f'''(a)}{3!}, \quad c_4 = \frac{f(a)''''}{4!}.$$

Thus

$$f(x) = f(a) + f'(a)(x - a) + \frac{f''(a)}{2!}(x - a)^2 + \frac{f'''(a)}{3!}(x - a)^3$$
$$+ \frac{f''''(a)}{4!}(x - a)^4 + \ldots,$$

† We proved in § 10.2 that the expansion of $\log_e(1 + x)$ in ascending powers of x is valid only if $-1 < x \leqslant +1$; we may therefore expect some limitations on the value of x in certain cases of the general expansions we are about to discuss. Consideration of this is delayed until § 18.9.

or if $x = a+h$,

$$f(x) = f(a+h) = f(a)+f'(a)\,h+\frac{f''(a)}{2!}\,h^2+\frac{f'''(a)}{3!}\,h^3+\frac{f''''(a)}{4!}\,h^4+\ldots$$

This result is a form of **Taylor's theorem** (1716).

Example 3. *Use Taylor's theorem to expand* $\sin\left(\dfrac{\pi}{6}+h\right)$ *in ascending powers of h as far as the term in* h^4.

Let $\quad f(x) = \sin x = \sin\left(\dfrac{\pi}{6}+h\right), \qquad f\left(\dfrac{\pi}{6}\right) = \sin\dfrac{\pi}{6} = \dfrac{1}{2},$

$\qquad f'(x) = \cos x, \qquad\qquad\qquad f'\left(\dfrac{\pi}{6}\right) = \cos\dfrac{\pi}{6} = \dfrac{\sqrt{3}}{2},$

$\qquad f''(x) = -\sin x, \qquad\qquad\quad f''\left(\dfrac{\pi}{6}\right) = -\sin\dfrac{\pi}{6} = -\dfrac{1}{2},$

$\qquad f'''(x) = -\cos x, \qquad\qquad\quad f'''\left(\dfrac{\pi}{6}\right) = -\cos\dfrac{\pi}{6} = -\dfrac{\sqrt{3}}{2},$

$\qquad f''''(x) = \sin x, \qquad\qquad\quad\; f''''\left(\dfrac{\pi}{6}\right) = \sin\dfrac{\pi}{6} = \dfrac{1}{2}.$

By Taylor's theorem,

$$f(a+h) = f(a)+f'(a)\,h+\frac{f''(a)}{2!}\,h^2+\frac{f'''(a)}{3!}\,h^3+\frac{f''''(a)}{4!}\,h^4+\ldots,$$

$$\therefore\;\sin\left(\frac{\pi}{6}+h\right) = \frac{1}{2}+\frac{\sqrt{3}}{2}\,h+\frac{(-\frac{1}{2})}{2!}\,h^2+\frac{(-\sqrt{3}/2)}{3!}\,h^3+\frac{\frac{1}{2}}{4!}\,h^4+\ldots,$$

$$= \frac{1}{2}+\frac{\sqrt{3}}{2}\,h-\frac{1}{4}\,h^2-\frac{\sqrt{3}}{12}\,h^3+\frac{1}{48}\,h^4+\ldots.$$

Qu. 10. Using only the first three terms of the expansion obtained in Example 3, obtain a value for $\sin 31°$ to five significant figures, taking $\sqrt{3}$ as $1{\cdot}7321$ and $1°$ as $0{\cdot}01745$ radians. Compare your answer with that of Example 1. (From five-figure tables $\sin 31° \simeq 0{\cdot}51504$.)

Qu. 11. Use Taylor's theorem to express $\tan\left(\dfrac{\pi}{4}+h\right)$ as a series in ascending powers of h as far as the term in h^3.

Qu. 12. Use Taylor's theorem to find the first four terms in the expansion of $\cos x$ in ascending powers of $(x-\alpha)$, where $\alpha = \tan^{-1}\frac{4}{3}$.

Maclaurin's theorem

18.6. Bearing in mind the relationship $x = a+h$, where a is a constant, and x and h are variable (see Fig. 18.1), we see that there is a special case given by $a=0$, when $x=h$, and either form of Taylor's theorem given in § 18.5 reduces to

$$f(x) = f(0)+f'(0)\ x+\frac{f''(0)}{2!}\ x^2+\frac{f'''(0)}{3!}\ x^3+\frac{f''''(0)}{4!}\ x^4+\ldots.$$

This is a form of **Maclaurin's theorem** (1742).

Example 4. *Use Maclaurin's theorem to expand* $\log_e (1+x)$ *in ascending powers of x as far as the term in x^5.*

$$
\begin{array}{ll}
f(x) = \log_e (1+x), & f(0) = 0, \\
f'(x) = (1+x)^{-1}, & f'(0) = 1, \\
f''(x) = -(1+x)^{-2}, & f''(0) = -1, \\
f'''(x) = 2(1+x)^{-3}, & f'''(0) = 2! \\
f''''(x) = -3.2(1+x)^{-4}, & f''''(0) = -3! \\
f'''''(x) = 4.3.2(1+x)^{-5}, & f'''''(0) = 4!
\end{array}
$$

By Maclaurin's theorem

$$f(x) = f(0)+f'(0)\ x+\frac{f''(0)}{2!}\ x^2+\frac{f'''(0)}{3!}\ x^3+\frac{f''''(0)}{4!}\ x^4+\frac{f'''''(0)}{5!}\ x^5+\ldots,$$

$$\therefore\ \log_e (1+x) = 0+1.x+\frac{(-1)}{2!}\ x^2+\frac{2!}{3!}\ x^3+\frac{(-3!)}{4!}\ x^4+\frac{4!}{5!}\ x^5+\ldots,$$

$$\therefore\ \log_e (1+x) = x-\frac{x^2}{2}+\frac{x^3}{3}-\frac{x^4}{4}+\frac{x^5}{5}-\ldots.$$

Qu. 13. Use Maclaurin's theorem:

(i) to expand e^x in ascending powers of x as far as the x^5 term,

(ii) to show that when x is small, $\sin x \simeq x-\dfrac{x^3}{3!}+\dfrac{x^5}{5!}$,

(iii) to find the first three terms of the expansion of $\cos x$ in ascending powers of x.

Qu. 14. Express $17° 11'$ in radians correct to one significant figure. Use the approximation given in Qu. 13 (ii) to express $\sin 17° 11'$ to four significant figures. Check your answer against the value given in four-figure tables.

Exercise 18b

1. Given that the graphs of $y=\log_e x$ and $y=c_0+c_1(x-2)+c_2(x-2)^2$ have the same ordinate, gradient and rate of change of gradient when $x=2$, determine c_0, c_1, c_2 and deduce an approximation for $\log_e x$ when $x \simeq 2$.

2. Obtain a quadratic approximation for $\sin x$ when $x \simeq \alpha$.

3. Apply Taylor's theorem:

    (i) to expand $\log_e x$ in ascending powers of $(x - e)$ as far as the term in $(x - e)^4$,

    (ii) to expand $\operatorname{cosec} x$ in ascending powers of $\left(x - \dfrac{\pi}{2}\right)$ as far as the term in $\left(x - \dfrac{\pi}{2}\right)^4$.

4. Use Taylor's theorem to expand $\cos\left(\dfrac{\pi}{3} + h\right)$ in ascending powers of h up to the h^3 term. Taking $\sqrt{3}$ as $1 \cdot 7321$ and $5° \, 30'$ as $0 \cdot 095\,99$ radians, find the value of $\cos 54° \, 30'$ to three decimal places.

5. Given that the functions $f(x)$ and $c_0 + c_1 x + c_2 x^2 + c_3 x^3 + c_4 x^4 + \ldots$ have the same value when $x = 0$, and equal successive derivatives when $x = 0$, deduce the first five terms of the Maclaurin expansion of $f(x)$ in ascending powers of x.

6. We have used Maclaurin's theorem to establish the following expansions, which should be memorized:

$$e^x = 1 + x + \frac{x^2}{2!} + \frac{x^3}{3!} + \ldots,$$

$$\log_e (1 + x) = x - \frac{x^2}{2} + \frac{x^3}{3} - \frac{x^4}{4} + \ldots,$$

$$\cos x = 1 - \frac{x^2}{2!} + \frac{x^4}{4!} - \frac{x^6}{6!} + \ldots,$$

$$\sin x = x - \frac{x^3}{3!} + \frac{x^5}{5!} - \frac{x^7}{7!} + \ldots.$$

Write down the first four terms of the expansions of the following in ascending powers of x:

    (i) e^{2x},   (ii) $\log_e (1 - x)$,   (iii) $\cos x^2$,   (iv) $\sin \dfrac{x}{2}$.

7. By subtracting the expansion of $\log_e (1 - x)$ from that of $\log_e (1 + x)$ deduce that

$$\log_e \sqrt{\frac{1 + x}{1 - x}} = x + \frac{x^3}{3} + \frac{x^5}{5} + \frac{x^7}{7} + \ldots.$$

8. Find approximations for the following:

    (i) $e^{0 \cdot 4}$ (correct to five significant figures),

    (ii) $\log_e 1 \cdot 2$ (correct to four significant figures),

    (iii) $\cos 17° \, 11'$ (correct to three significant figures, taking $17° \, 11'$ as $0 \cdot 3$ radian),

    (iv) $\sin 11° \, 28'$ (correct to three significant figures, taking $11° \, 28'$ as $0 \cdot 2$ radian).

9. Apply Maclaurin's theorem directly (see Example 4) to obtain expansions for the following in ascending powers of x up to the given term:

(i) $\sin^2 x$,	(x^4),	(ii) $(1+x)^n$,	(x^3),
(iii) 2^x,	(x^3),	(iv) $\cos^{-1} x$,	(x^3),
(v) $e^x \sin x$,	(x^5),	(vi) $\log_e \{x + \sqrt{(x^2+1)}\}$,	(x^3).

10. If $f(x) = e^x \sin x$ show that $f^n(x) = (\sqrt{2})^n e^x \sin \left(x + \dfrac{n\pi}{4}\right)$, and use this with Maclaurin's theorem to find an expansion for $f(x)$ in ascending powers of x as far as the x^6 term.

11. Find the expansion of $\log_e x$ in ascending powers of $(x-4)$ up to the fourth term:

(i) by writing $\log_e x$ as $\log_e \left\{4\left(1 + \dfrac{x-4}{4}\right)\right\}$ and applying the expansion for $\log_e (1+x)$ given in No. 6,

(ii) by applying Taylor's theorem.

Deduce an approximation for $\log_e 4\cdot02$ correct to four decimal places.

Expansion by integration

18.7. If we wish to expand $f(x)$ in ascending powers of x, and we find that *an expansion for* $f'(x)$ *is known* or is easily obtainable, then the required expansion may be obtained from the latter by integration. This is illustrated graphically by saying that if two curves approximate over a certain range of values of x, then the area under these curves will be approximately equal over that range.   Since, for example,

$$\frac{d}{dx} \log_e (1+x) \text{ is } \frac{1}{1+x},$$

which may be expanded by the binomial theorem, this provides an alternative method of obtaining an expansion for $\log_e (1+x)$.

Example 5. *Expand* $\log_e (1+x)$ *in ascending powers of x as far as the* x^4 *term.*

$$\left[\log_e (1+u)\right]_0^x = \int_0^x \frac{1}{1+u}\, du,$$

$$= \int_0^x (1 - u + u^2 - u^3 + \ldots)\, du, \text{ (provided } -1 < u < 1)$$

$$= \left[u - \tfrac{1}{2}u^2 + \tfrac{1}{3}u^3 - \tfrac{1}{4}u^4 + \ldots\right]_0^x,$$

$$\therefore \log_e (1+x) = x - \tfrac{1}{2}x^2 + \tfrac{1}{3}x^3 - \tfrac{1}{4}x^4 + \ldots.$$

Qu. 15. (i) Use Maclaurin's theorem to obtain the coefficients in the expansion of $\tan^{-1} x$ in ascending powers of x, up to the x^4 term.

(ii) Now obtain this expansion up to the x^7 term by using the fact that

$$\left[\tan^{-1} u\right]_0^x = \int_0^x \frac{1}{1+u^2}\, du.$$

Miscellaneous methods

18.8. Qu. 15 brings out the value of the integration method; this is just one way of avoiding the laborious differentiation sometimes involved in the direct application of Maclaurin's theorem. It is also useful to bear in mind the following possibilities:

(i) the use of a known approximation together with a known expansion (see Example 6),

(ii) the use of the product of known, or more easily obtained, expansions (see Qu. 17).

Example 6. *Expand $\sec x$ in ascending powers of x as far as the x^6 term.*

$$\sec x = \frac{1}{\cos x} \simeq \left\{1 - \frac{x^2}{2!} + \frac{x^4}{4!} - \frac{x^6}{6!}\right\}^{-1},$$

$$= \left\{1 - \left(\frac{x^2}{2} - \frac{x^4}{24} + \frac{x^6}{720}\right)\right\}^{-1},$$

$$= 1 + \left(\frac{x^2}{2} - \frac{x^4}{24} + \frac{x^6}{720}\right) + \left(\frac{x^2}{2} - \frac{x^4}{24} + \frac{x^6}{720}\right)^2$$

$$+ \left(\frac{x^2}{2} - \frac{x^4}{24} + \frac{x^6}{720}\right)^3 + \ldots,$$

$$= 1 + \frac{x^2}{2} - \frac{x^4}{24} + \frac{x^6}{720} + \frac{x^4}{4} - \frac{x^6}{24} + \ldots + \frac{x^6}{8} + \ldots$$

$$\therefore \sec x = 1 + \tfrac{1}{2}x^2 + \tfrac{5}{24}x^4 + \tfrac{61}{720}x^6 + \ldots.$$

Qu. 16. Use the expansions

$$e^z = 1 + x + \frac{x^2}{2!} + \frac{x^3}{3!} + \frac{x^4}{4!} + \ldots,$$

and

$$\sin x = x - \frac{x^3}{3!} + \ldots,$$

to express $e^{\sin z}$ in ascending powers of x as far as the x^4 term.

Qu. 17. Expand $\dfrac{\cos x}{\sqrt{(1-x)}}$ in ascending powers of x as far as the x^4 term, by considering the product of the expansions of $\cos x$ and of $(1-x)^{-\frac{1}{2}}$.

Validity of expansions

18.9.   So far we have avoided the issue that some of the expansions obtained in this chapter may be valid only for certain values of x.

For example, the binomial theorem only enables us to expand $(1-x)^{-1}$ as the infinite series $1+x+x^2+\ldots$ provided that $-1<x<1$. As a reminder of why this is so we may employ an even more elementary method of expansion; by long division

$$\frac{1}{1-x} = 1+x+\frac{x^2}{1-x} = 1+x+x^2+\frac{x^3}{1-x} = 1+x+x^2+x^3+\frac{x^4}{1-x} \text{ etc.}$$

Only if $-1<x<1$† is it true that

$$\left|\frac{x^4}{1-x}\right| < \left|\frac{x^3}{1-x}\right| < \left|\frac{x^2}{1-x}\right|,$$

and hence that $1+x$, $1+x+x^2$, $1+x+x^2+x^3$ are progressively better approximations to $\frac{1}{1-x}$ since the error involved is progressively decreasing in size; taking the approximation to n terms, the error is $\frac{x^n}{1-x}$. If we let $n \to \infty$, we assume that $x^n \to 0$ if $-1<x<1$, and so the error $\to 0$; it follows that we may make the approximation

$$1+x+x^2+\ldots+x^{n-1}$$

as near as we please to $\frac{1}{1-x}$ by taking n sufficiently large.   Expressing this in other words, we may say that the infinite series $1+x+x^2+\ldots$ *converges* to the sum $\frac{1}{1-x}$ provided that $-1<x<1$.

The expansion

$$\log_e (1+x) = x-\frac{x^2}{2}+\frac{x^3}{3}-\frac{x^4}{4}+\frac{x^5}{5}-\ldots$$

is valid only if $-1<x\leqslant+1$. (This is proved in § 10.2.) Fig. 18.6 shows the graphs of $y=\log_e (1+x)$ together with those of the successive approximate functions

(i) $y = x,$ (ii) $y = x-\frac{x^2}{2},$

(iii) $y = x-\frac{x^2}{2}+\frac{x^3}{3},$ (iv) $y = x-\frac{x^2}{2}+\frac{x^3}{3}-\frac{x^4}{4}.$

† The consequences of taking values of x outside this range are discussed in P.M.I., § 11.4.

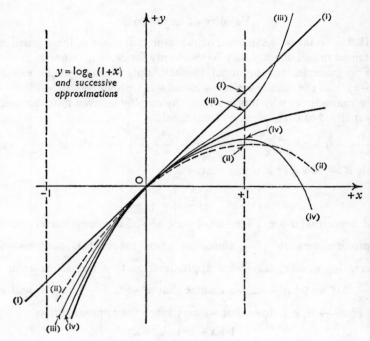

$y = \log_e (1+x)$
and successive
approximations

Fig. 18.6

$x = -1$ is cut by all the latter, but is an asymptote to $y = \log_e (1+x)$, which illustrates the condition $-1 < x$; it can be seen that the values of the successive approximate functions approach that of $\log_e (1+x)$ for positive values of x up to *and including* $+1$, but thereafter diverge rapidly from it, hence the condition $x \leqslant 1$.

An expansion obtained by the use of Maclaurin's theorem is valid only if the series is convergent but general consideration of this matter must be delayed until a later stage. The reader should learn the following conditions for validity, and should always quote them, or derived conditions, whenever they apply.

Expansion	*Condition*
$(1+x)^n$	$-1 < x < 1$†
$\log_e (1+x)$	$-1 < x \leqslant 1$
$\tan^{-1} x$	$-1 \leqslant x \leqslant 1$

† Also valid for $x = 1$ if $n > -1$, and for $x = -1$ if $n > 0$; but these refinements are ignored elsewhere in this book.

Expansions of e^x, $\sin x$, $\cos x$ are valid for all values of x.

Qu. 18. For what ranges of values of x may the following be expanded as infinite series in ascending powers of x?

(i) $\dfrac{1}{2+x}$, (ii) $\log_e (1-x)$, (iii) $\tan^{-1} (1+x)$,

(iv) e^{3x}, (v) $\log_e (1+2x)$, (vi) $\log_e (1+x^2)$,

(vii) $\dfrac{\sqrt{(2+x)}}{1+3x}$, (viii) $\log_e \dfrac{1+x}{1-x}$, (ix) $\dfrac{\log_e (1+x)}{1+x}$.

Qu. 19. What are the conditions that the following may be expanded as an infinite series in ascending powers of $\dfrac{1}{x}$?

$$\text{(i) } \sqrt{\left(1+\frac{2}{x}\right)}, \quad \text{(ii) } \log_e \left(1+\frac{1}{x}\right).$$

Qu. 20. State the conditions for validity of the following expansions:

(i) $\log_e x$ in ascending powers of $(x-2)$,

(ii) $\sin x \sqrt{(1+x)}$ in ascending powers of $\left(x-\dfrac{\pi}{2}\right)$.

Rate of convergence

18.10. For practical purposes we wish to know not only that a particular expansion is valid, but also that it converges sufficiently rapidly for the value of the variable we are considering; in other words, if we are to be able to obtain a satisfactory approximation by considering reasonably few terms, these must decrease rapidly in size.

Now
$$\tan^{-1} x = x - \frac{x^3}{3} + \frac{x^5}{5} - \frac{x^7}{7} + \ldots$$

is valid when $-1 \leqslant x \leqslant 1$, and if we put $x=1$,

$$\tan^{-1} 1 = \frac{\pi}{4} = 1 - \tfrac{1}{3} + \tfrac{1}{5} - \tfrac{1}{7} + \ldots ;$$

this provides a method of calculating π, but it is not a very good one since the series converges very slowly; a better one is given in Exercise 18c, No. 5. See also Exercise 12b, No. 29.

To take another example,

$$\sin x = x - \frac{x^3}{3!} + \frac{x^5}{5!} - \frac{x^7}{7!} + \ldots,$$

is valid for all values of x, but only converges rapidly when x is small. If we are concerned with large values of x, where $x \simeq a$, it is necessary to use a Taylor expansion in ascending powers of $(x-a)$.

As a third example let us consider the expansion

$$\log_e (1+x) = x - \frac{x^2}{2} + \frac{x^3}{3} - \frac{x^4}{4} + \ldots$$

which is valid when $-1 < x \leqslant 1$. Putting $x = 1$, we obtain

$$\log_e 2 = 1 - \tfrac{1}{2} + \tfrac{1}{3} - \tfrac{1}{4} + \ldots;$$

this is an extreme case of a fruitless application of an expansion, since somewhere in the order of 10 000 terms are needed to produce a value of $\log_e 2$ correct to four decimal places! A rather more economical method of evaluating $\log_e 2$ is given in Exercise 18c. Meanwhile the reader should consider two less contrasting methods of evaluating $\log_e 1 \cdot 5$, which nevertheless stress the practical value of rapid convergence.

Qu. 21. Obtain approximations for $\log_e 1 \cdot 5$ by the following methods, and compare them with the value given in four-figure tables:

(i) use the first five terms of the expansion of $\log_e (1 + x)$ in ascending powers of x, putting $x = 0 \cdot 5$,

(ii) find a value of x for which $\dfrac{1+x}{1-x} = 1 \cdot 5$, and substitute this value in the first three terms of the expansion of $\log_e \dfrac{1+x}{1-x}$ in ascending powers of x.

Exercise 18c

1. Use the method of integration given in Example 5 to obtain expansions of the following in ascending powers of x up to the given term:
   (i) $\sin^{-1} x$ (x^7 term),
   (ii) $\log_e (\sec x + \tan x)$ (x^5 term),
   (iii) $\tan^{-1} (x+1)$ (x^5 term),
   (iv) $\cos^{-1} x$ (x^5 term).

2. Make use of known expansions to obtain expansions of the following in ascending powers of x up to the given term:
   (i) $\tfrac{1}{2}(e^x + e^{-x})$ (x^6 term),
   (ii) $x \operatorname{cosec} x$ (x^4 term),
   (iii) $\cos^3 x$ (x^4 term),
   (iv) $\tan x$ (x^5 term),
   (v) $\log_e \dfrac{\sin x}{x}$ (x^4 term),
   (vi) $\log_e (1 + \sin x)$ (x^4 term),
   (vii) $\log_e (1 + e^x)$ (x^4 term).

3. Verify the following expansions, and state any limitations on the value of x required for the validity of Nos. (i) to (v):

(i) $e^x \cos x = 1 + x - \tfrac{1}{3}x^3 - \tfrac{1}{6}x^4 + \ldots$,

(ii) $\dfrac{1}{e^x(1-x)} = 1 + \tfrac{1}{2}x^2 + \tfrac{1}{3}x^3 + \ldots$,

(iii) $\dfrac{\cos x}{(1-x)^2} = 1 + 2x + \tfrac{5}{2}x^2 + 3x^3 + \tfrac{85}{24}x^4 + \ldots$,

(iv) $\dfrac{e^{\sin x}}{\sqrt{(4+x^2)}} = \tfrac{1}{2} + \tfrac{1}{4}x + \tfrac{3}{16}x^2 - \tfrac{1}{16}x^3 - \tfrac{21}{256}x^4 + \ldots$,

(v) $\dfrac{\log_e(1+x)}{\tan^{-1}x} = 1 - \tfrac{1}{2}x + \tfrac{2}{3}x^2 - \tfrac{5}{12}x^3 + \tfrac{2}{9}x^4 + \ldots$,

(vi) $\dfrac{x(1-x)}{\log_e(1+x)} = 1 - \tfrac{1}{2}x - \tfrac{7}{12}x^2 + \tfrac{1}{8}x^3 + \ldots$; is this valid for $0 < x \leqslant 1$?

4. By substituting $x = \dfrac{1}{2m+1}$ in the expansion of $\log_e \dfrac{1+x}{1-x}$, show that

$$\log_e \frac{m+1}{m} = 2\left\{\frac{1}{2m+1} + \frac{1}{3(2m+1)^3} + \frac{1}{5(2m+1)^5} + \ldots\right\}.$$

Use four terms of this expansion to find approximate values for $\log_e 1.5$, $\log_e 2$, $\log_e 3$, and compare them with those given in tables.

5. Show that $\dfrac{\pi}{4} = \tan^{-1}\tfrac{1}{2} + \tan^{-1}\tfrac{1}{3}$, and use this relationship to obtain an approximation for π correct to four significant figures.

6. Use Taylor's theorem to obtain an approximation for $\sin 131° 28'$ correct to four significant figures. (Assume $11° 28' \simeq 0.2$ radian.)

7. Verify the following expansion of $\dfrac{\cos x}{1-x}$ in ascending powers of $(x-a)$, and state the range of values of x for which it is valid:

$$\frac{\cos x}{1-x} = \frac{\cos a}{1-a} + \{\cos a - (1-a)\sin a\}\frac{x-a}{(1-a)^2}$$

$$+ \left\{\cos a - (1-a)\sin a - \tfrac{1}{2}(1-a)^2\cos a\right\}\frac{(x-a)^2}{(1-a)^3} + \ldots.$$

8. Obtain the expansion of $\sin^{-1}(\tfrac{1}{2} - x)$ in ascending powers of x as far as the x^3 term.

9. If $f(x) = e^{x\sqrt{3}}\cos x$, obtain the expansion of $f(x)$ in ascending powers of x as far as the x^6 term. (Establish an expression for $f^n(x)$.)

Exercise 18d (Miscellaneous)

1. Express $\dfrac{10(x+1)}{(x+3)(x^2+1)}$ in partial fractions. Hence obtain the expansion of the given function in ascending powers of x as far as the term in x^3, stating the necessary restrictions on the value of x.   O.C.

2. Show that

$$\log_e (1 + x + x^2) = \int \frac{dx}{1-x} - 3 \int \frac{x^2 \, dx}{1-x^3} + c,$$

where c is an arbitrary constant. By expanding the integrands as far as the terms in x^5, find the first six terms of the series for

$$\log_e (1 + x + x^2)$$

for small values of x.

N.

3. Prove that

$$\frac{x}{1 + x + \sqrt{(2x+1)}} = \frac{1}{x} [1 + x - \sqrt{(2x+1)}]$$

and show that, if x is small, the expression is approximately equal to $\frac{1}{2}x(1-x)$.

O.C.

4. Use Maclaurin's theorem to show that, if x^5 and higher powers of x are neglected,

$$\log_e \{x + \sqrt{(1+x^2)}\} = x - \frac{1}{6}x^3.$$

O.C.

5. State Taylor's theorem for the expansion of $f(a+h)$ in a series of ascending powers of h. Prove that the first four terms in the Taylor expansion of $\tan^{-1} (1+h)$ are

$$\tfrac{1}{4}\pi + \tfrac{1}{2}h - \tfrac{1}{4}h^2 + \tfrac{1}{12}h^3.$$

O.C.

6. Express

$$E \equiv \frac{2x^2 + 7}{(x+2)^2(x-3)}$$

in partial fractions. Hence, if x is so large that $\dfrac{1}{x^5}$ can be neglected, prove that

$$E = \frac{1}{x^4} (2x^3 - 2x^2 + 25x - 17).$$

O.C.

7. Write down the expansions of e^z and e^{-z}. The limit of

$$\frac{e^{2z} - e^{-2z} - 4x}{f(x)},$$

where $f(x)$ is a polynomial, is 8 as $x \to 0$; show that the term of lowest degree in the polynomial $f(x)$ is $\frac{1}{3}x^3$.

O.C.

8. Expand

$$E \equiv \log_e \left(\frac{2-x}{1-x} \right)$$

in ascending powers of x up to x^3, stating the necessary conditions for your expansion. Evaluate E when $x = \frac{1}{3}$ and hence find $\log_e 3$ to three places of decimals, given that $\log_e 2 = 0.6931$.

O.C.

9. Using Maclaurin's theorem, expand $x \tan (\tfrac{1}{4}\pi - x)$ in ascending powers of x as far as the term containing x^4.

L.

10. Prove that, if x is small so that x^6 and higher powers of x may be neglected, then

$$\frac{e^{2x} - e^{-2x}}{e^{2x} + e^{-2x}} = 2x - \tfrac{8}{3}x^3 + \tfrac{64}{15}x^5.$$

O.C.

11. Find the coefficient of x^r in the expansion of $(1 + 3x)\, e^{-3x}$ as a series of ascending powers of x.

O.C.

12. By means of Taylor's theorem, or otherwise, prove that when $(x - a)$ is small the expression $\sqrt{(x + 3a)} - \sqrt{(5a - x)}$ is represented approximately by

$$\frac{1}{2a^{\frac{1}{2}}}\,(x - a) + \frac{1}{256a^{\frac{5}{2}}}\,(x - a)^3.$$

O.C.

13. Prove, when $x > 1$, that

$$2 \log_e x - \log_e (x + 1) - \log_e (x - 1) = \frac{1}{x^2} + \frac{1}{2x^4} + \frac{1}{3x^6} + \ldots,$$

and state why the condition $x > 1$ is necessary.

O.C.

14. Expand the function given by

$$y = \log_e \left\{ \frac{(2 - x)^2}{4 - 4x} \right\}$$

in a series of ascending powers of x as far as x^4, stating the limitations on the value of x, and giving the coefficient of x^n. Prove that, up to x^4,

$$y - \frac{x^2}{4}\,e^x = \frac{3x^4}{32}.$$

O.C.

15. If a is an approximation to a root of the equation $\mathrm{f}(x) = 0$, prove that in general a closer approximation to the root is

$$a - \frac{\mathrm{f}(a)}{\mathrm{f}'(a)}.$$

A root of the equation $2x^3 + 5 \tan^2 x = 6$ is known to lie close to $\dfrac{\pi}{4}$.

Obtain the value of this root correct to three significant figures. [Take $\pi = 3.142$.]

O.C.

16. A root of the equation

$$\sin^3 \tfrac{1}{2}x^\circ + \cos x^\circ - \tfrac{249}{400} = 0$$

is close to 60. Find the value of the root, correct to 0.1. [Take $\pi = 3.142$.]

O.C.

17. Given that a root of the equation

$$(x^2+9)^{3/2} + 8x^2 + x - 258 = 0$$

is close to 4, find the value of the root correct to three significant figures.

O.C.

18. If $y = e^{4x} \cos 3x$, prove that

$$\frac{dy}{dx} = 5e^{4x} \cos(3x + \alpha),$$

where $\tan \alpha = \frac{3}{4}$.   Use Maclaurin's theorem to find the expansion of y in ascending powers of x as far as the term in x^3.

O.C.

19. Express the function E given by

$$E = \frac{x+3}{(2x+1)(1+x^2)}$$

in partial fractions.   Hence prove that, if x is so large that $\frac{1}{x^4}$ can be neglected, then

$$E = \frac{5+2x}{4x^3}.$$

O.C.

20. Assuming the expansion of $\log_e(1+x)$ in ascending powers of x, prove that

$$\log_e \sqrt{\left(\frac{1+x}{1-x}\right)} = x + \frac{x^3}{3} + \frac{x^5}{5} + \dots,$$

and deduce that, when $0 \leqslant \theta \leqslant \frac{1}{4}\pi$,

$$\sin 2\theta + \tfrac{1}{3}\sin^3 2\theta + \tfrac{1}{5}\sin^5 2\theta + \dots = 2(\tan \theta + \tfrac{1}{3}\tan^3 \theta + \tfrac{1}{5}\tan^5 \theta + \dots).$$

O.C.

21. If $\log_e y = xy$ find the values of $\frac{dy}{dx}$ and $\frac{d^2y}{dx^2}$ when $x = 0$ and hence show that the Maclaurin expansion of y in powers of x is

$$y = 1 + x + \tfrac{3}{2}x^2 + \dots.$$

Find also the expansion of x in powers of $(y-1)$ up to the term in $(y-1)^2$.

L.

22. Prove that

$$\log_e\left(\frac{x+1}{x}\right) = 2\left[\frac{1}{2x+1} + \frac{1}{3(2x+1)^3} + \frac{1}{5(2x+1)^5} + \dots\right],$$

stating the range of values of x for which the series is valid.

O.C.

23. If $a = b(1 + h)$, where h is small and $a > b > 0$, expand $\dfrac{2(a - b)}{(a + b)}$ as a series in ascending powers of h.

Show that, when a is nearly equal to b, $\log_e \dfrac{a}{b}$ differs from $\dfrac{2(a - b)}{(a + b)}$ by approximately $\dfrac{1}{12}\left(\dfrac{a - b}{b}\right)^3$.   N.

24. Prove that if $a^2 > 2$ then

$$\sqrt{(a^2 - 2)} = \sqrt{(a^2 - 1)}\left[1 - \frac{1}{2(a^2 - 1)} - \frac{1}{8(a^2 - 1)^2} - \frac{1}{16(a^2 - 1)^3} \cdots\right].$$

Use this result when $a^2 = 51$ to show that $\frac{140}{99}$ is a close approximation to the value of $\sqrt{2}$.   O.C.

25. Show that

$$\log_e \sqrt{x} = \frac{x - 1}{x + 1} + \frac{1}{3}\left(\frac{x - 1}{x + 1}\right)^3 + \frac{1}{5}\left(\frac{x - 1}{x + 1}\right)^5 + \dots,$$

and state for what range of values of x the expansion is valid.   L.

26. Find the expansion of $\log_e\left(1 + \dfrac{1}{n}\right)$ in ascending powers of $\dfrac{1}{2n + 1}$ and state the range of values of n for which the expansion is valid.   N.

CHAPTER 19

SOME NUMERICAL METHODS†

Differentiation

19.1. A cyclist who, together with his cycle, has mass 75 kg, accelerates on the level from rest to a speed of 40 km/h. The estimated speed at two-second intervals is given in the following table:

Time t s	0	2	4	6	8	10	12	14	16	18	20
Speed v km/h	0	3·4	10·0	17·6	23·0	26·8	30·0	32·7	35·3	37·8	40·0

Can we estimate, from this data, the acceleration at a number of instants during the motion?

It would be possible to plot a graph, draw tangents at intervals and estimate their gradients; but in this chapter we shall be concerned with methods which do not entail drawing, so we shall look for an alternative method.

The velocity-time graph has been sketched in Fig. 19.1, and Fig. 19.2 (not drawn to scale) represents an enlargement of part of the graph.

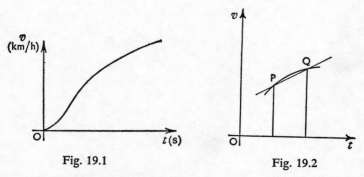

Fig. 19.1   Fig. 19.2

† It is strongly recommended that the questions in the text should be answered when reading each section.

Suppose that the tops of ordinates at intervals of two seconds are joined, as PQ in Fig. 19.2, then it may be seen that somewhere in the interval the gradient of the curve is equal to the gradient of PQ; and we shall not be very far wrong if we take the gradient of PQ to be equal to the gradient of the curve at the *mid-point* of the interval. This has been done in the table that follows. The increase in v is denoted by $[v]$ so that the acceleration a (in km per hour per second) at the *mid-point* of each interval is estimated to be $\frac{1}{2}[v]$.

t (s)	v km/h	$[v]$ km/h	$a = \frac{1}{2}[v]$ km/h per s
0	0		
		3·4	1·7
2	3·4		
		6·6	3·3
4	10·0		
		7·6	3·8
6	17.6		
		5·4	2·7
8	23·0		
		3·8	1·9
10	26·8		
		3·2	1·6
12	30·0		
		2·7	1·3(5)
14	32·7		
		2·6	1·3
16	35·3		
		2·5	1·2(5)
18	37·8		
		2·2	1·1
20	40·0		

Qu. 1.   Convert the acceleration in the table above into m/s². Estimate the net force in N used to accelerate the cycle (i.e. the excess of the total force applied over the resistances) at two-second intervals during the motion. Sketch a graph of the force against time.

Qu. 2.   The distance travelled by a particle in twenty seconds of its motion is given by the following table:

Time (s)	0	2	4	6	8	10	12	14	16	18	20
Distance (m)	0	30	75	136	208	269	316	347	365	376	381

Estimate the velocity of the particle at intervals of two seconds and, from these values, estimate the acceleration after 2, 4, ..., 18 seconds. Draw the velocity-time and acceleration-time graphs.

The trapezium rule

19.2. In the example taken in § 19.1, we could draw an accurate graph and estimate the distance travelled during the motion by counting squares. We now turn to consider how this could be estimated without going through this rather tedious process.

An estimate of the area under a curve could be obtained by drawing in ordinates (see Fig. 19.3), joining the tops of adjacent ones and

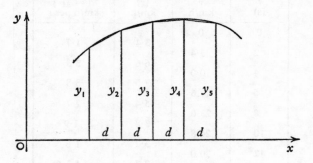

Fig. 19.3

calculating the areas of the trapeziums obtained. Here we have taken five ordinates y_1, y_2, y_3, y_4, y_5 at *equal intervals* of d. The area estimated by this method will be

$$\frac{y_1+y_2}{2}d+\frac{y_2+y_3}{2}d+\frac{y_3+y_4}{2}d+\frac{y_4+y_5}{2}d$$
$$= \tfrac{1}{2}d(y_1+2y_2+2y_3+2y_4+y_5). \qquad (1)$$

This is the trapezium rule for five ordinates.

Qu. 3. Use the trapezium rule to estimate the area, from $x=0.2$ to $x=1$, under the curve given by

x	0·20	0·40	0·60	0·80	1·00
y	0·24	0·56	0·96	1·44	2·00

Given that the equation of the curve is $y=x^2+x$, check your answer by integration.

Qu. 4. Find expressions similar to (1) for (i) eight, (ii) nine ordinates. Now express the trapezium rule in words.

Qu. 5. Estimate the area under the curve given by the following table. Beware of the catch!

x	0	10	15	20	25
y	7	9	11	12	10

Another way of looking at the above expression (1) for the area is to take $a = 4d$ so that a is the total interval along the x-axis. In this case the area is estimated to be

$$a\left(\frac{y_1 + 2y_2 + 2y_3 + 2y_4 + y_5}{8}\right) \qquad (2)$$

where the expression in brackets appears as the average height of the curve, with a total of eight ordinates (y_2, y_3, y_4 counted twice) divided by 8.

Qu. 6. Obtain the expressions equivalent to (2) for (i) eight ordinates, (ii) n ordinates.

The following example has been chosen to illustrate the accuracy of the trapezium rule. We shall compare the answer with that obtained by another rule later.

Example 1. *Use the trapezium rule to estimate the area under the curve* $y = \frac{1}{x}$ *from* $x = 1$ *to* $x = 2$.

To begin with, let us take six ordinates.

x	1·0	1·2	1·4	1·6	1·8	2·0
y	1	0·8333	0·7143	0·6250	0·5556	0·5

$$y_1 = 1\cdot0000 \qquad y_2 = 0\cdot8333\dagger$$
$$y_6 = 0\cdot5000 \qquad y_3 = 0\cdot7143$$
$$\overline{} \qquad y_4 = 0\cdot6250$$
$$1\cdot5000 \qquad y_5 = 0\cdot5556$$

$$2\cdot7282$$
$$\times 2$$

$$5\cdot4564 \longleftarrow \qquad 5\cdot4564$$

$$6\cdot9564$$

$$\tfrac{1}{2}d = 0\cdot1.$$

$$\therefore \text{ estimated area} = 0\cdot6956.$$

Now by integration the area is

$$\int_1^2 \frac{1}{x}\,\mathrm{d}x = \Big[\log_e x\Big]_1^2,$$
$$= \log_e 2,$$
$$= 0\cdot6931.$$

Qu. 7. Repeat the calculation of Example 1 but with eleven ordinates instead of six.

Qu. 8. Use the trapezium rule to find the distance travelled by the cyclist in § 19.1 before he reaches a speed of 40 km/h.

Simpson's rule

19.3. It will have been clear from Fig. 19.3 that the trapezium rule will not be very accurate for curves like the one illustrated. If, on the other hand, we were to join the tops of the ordinates by a smooth curve, we might expect to get a better estimate. The question then arises as to what curve to use—and there are a number of possibilities. But if we take three ordinates we can find a parabola in the form

$$y = ax^2 + bx + c$$

to pass through the three corresponding points.

† The symbol $\simeq$ has not been used in this chapter.

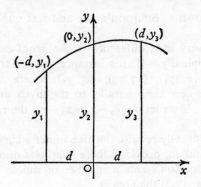

Fig. 19.4

Given a curve with three ordinates y_1, y_2, y_3 at *equal* intervals of d apart, take the y-axis along the middle ordinate and the x-axis through its foot as in Fig. 19.4.

Let

$$y = ax^2 + bx + c$$

be the *parabola* through the points $(-d, y_1)$, $(0, y_2)$, (d, y_3); its equation is therefore satisfied by their coordinates.

$$\therefore \quad y_1 = ad^2 - bd + c,$$
$$y_2 = \qquad\qquad c,$$
$$y_3 = ad^2 + bd + c.$$

The area under the parabola is

$$\int_{-d}^{d} (ax^2 + bx + c)\,dx = \left[\frac{ax^3}{3} + \frac{bx^2}{2} + cx\right]_{-d}^{d},$$
$$= \tfrac{2}{3}ad^3 + 2cd.$$

(Note that we do not need to find the equation of the parabola because we can express this area in terms of the data y_1, y_2, y_3, d.)

Now
$$y_1 + y_3 - 2y_2 = 2ad^2,$$
$$\therefore \quad y_1 + 4y_2 + y_3 = 2ad^2 + 6c.$$
$$\therefore \quad \tfrac{1}{3}d(y_1 + 4y_2 + y_3) = \tfrac{2}{3}ad^3 + 2cd.$$

So an approximation for the area under the given curve is

$$\tfrac{1}{3}d(y_1 + 4y_2 + y_3)$$

This result is known as Simpson's rule and was published by Thomas Simpson in 1743.

NOTE: it makes very little difference to the proof exactly what points we are given originally. If, for instance, we are told that the curve passes through (x_1, y_1), (x_2, y_2), (x_3, y_3), where $x_2 = \frac{1}{2}(x_1 + x_3)$, we can at once take new axes, parallel to the given ones, with the new origin at $(x_2, 0)$. If we let $d = x_3 - x_2 = x_2 - x_1$, the rest of the proof is as above.

In practice we usually require the area under a curve with more than three ordinates and so, provided there is an *odd* number of ordinates, we may apply Simpson's rule a number of times. Thus with seven ordinates (see Fig. 19.5) the area is

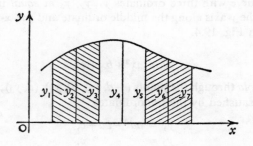

Fig. 19.5

$$\tfrac{1}{3}d(y_1 + 4y_2 + y_3) + \tfrac{1}{3}d(y_3 + 4y_4 + y_5) + \tfrac{1}{3}d(y_5 + 4y_6 + y_7)$$
$$= \tfrac{1}{3}d(y_1 + 4y_2 + 2y_3 + 4y_4 + 2y_5 + 4y_6 + y_7).$$

Qu. 9. Find similar expressions for the area with (i) five, (ii) nine ordinates. Now express Simpson's rule for an odd number of ordinates in words.

The next example is the same as Example 1. This is so that the reader may compare the accuracy of Simpson's rule and the trapezium rule for this case.

Example 2. *Use Simpson's rule to find an approximation for the area under the curve $y = \dfrac{1}{x}$ between $x = 1$ and $x = 2$.*

Five ordinates have been used.

x	1	1·25	1·5	1·75	2
y	1	0·8000	0·6667	0·5714	0·5

$$y_1 = 1\cdot0000 \qquad y_3 = 0\cdot6667 \qquad y_2 = 0\cdot8000$$
$$y_5 = 0\cdot5000 \qquad\qquad \times 2 \qquad\qquad y_4 = 0\cdot5714$$

$$
\begin{array}{ccc}
1\cdot5000 & 1\cdot3334 & 1\cdot3714 \\
1\cdot3334 & & \times 4 \\
5\cdot4856 & & \\
\hline
8\cdot3190 & & 5\cdot4856 \\
\end{array}
$$

$$\tfrac{1}{3}d = \tfrac{1}{12}; \quad \therefore \text{ the area} = \frac{8\cdot3190}{12} = 0\cdot6933.$$

This is a nearer value for $\log_e 2$ than the result obtained with the trapezium rule using eleven ordinates (see Qu. 7).

Qu. 10. Evaluate approximately $\displaystyle\int_1^2 \frac{1}{x}\,dx$ using Simpson's rule with eleven ordinates.

Qu. 11. Repeat Qu. 3, p. 410, using Simpson's rule.

The next example has been chosen to show how a complicated calculation may be set out.

Example 3. *Find the x-coordinate of the centroid of the area bounded by $y = \sqrt{(\sin x)}$, $x = \tfrac{1}{2}\pi$ and the x-axis, using Simpson's rule with five ordinates.*

The centroid $\bar{x}$ is found by

$$\bar{x} \int_0^{\frac{1}{2}\pi} y\,dx = \int_0^{\frac{1}{2}\pi} xy\,dx.$$

First find
$$\int_0^{\frac{1}{2}\pi} y\,dx = \int_0^{\frac{1}{2}\pi} \sqrt{(\sin x)}\,dx$$

We use
$$\int_0^{\frac{1}{2}\pi} y \, dx = \tfrac{1}{3}d(y_1 + 4y_2 + 2y_3 + 4y_4 + y_5).$$

x	0	$\frac{1}{8}\pi$	$\frac{1}{4}\pi$	$\frac{3}{8}\pi$	$\frac{1}{2}\pi$
†sin x	0	0·3827	0·7071	0·9239	1
$\sqrt{(\sin x)}$	0	0·6187	0·8409	0·9611	1

$$
\begin{array}{lll}
y_1 = 0\cdot0000 & y_2 = 0\cdot6187 & y_3 = 0\cdot8409 \\
y_5 = 1\cdot0000 & y_4 = 0\cdot9611 & 2 \\
\hline
1\cdot0000 & 1\cdot5798 & 1\cdot6818 \\
6\cdot3192 & 4 & \\
1\cdot6818 & \hline & \\
\hline & 6\cdot3192 & \\
9\cdot0010 & &
\end{array}
$$

$$\tfrac{1}{3}d = 0\cdot1309; \quad \therefore \int_0^{\frac{1}{2}\pi} \sqrt{(\sin x)} \, dx = 0\cdot1309 \times 9\cdot0010.$$

We now require
$$\int_0^{\frac{1}{2}\pi} xy \, dx = \int_0^{\frac{1}{2}\pi} x\sqrt{(\sin x)} \, dx.$$

We have already found values for $\sqrt{(\sin x)}$.   Writing x in decimal form,

x	0	0·3927	0·7854	1·1781	1·5708
$\sqrt{(\sin x)}$	0	0·6187	0·8409	0·9611	1·0000
‡ $x\sqrt{(\sin x)}$	0	0·2430	0·6604	1·1323	1·5708

Evaluating $\int_0^{\frac{1}{2}\pi} x\sqrt{(\sin x)} \, dx$, we treat $x\sqrt{(\sin x)}$ as y and use the

† The tables use degrees, so we need the sines of 0°, 22½°, 45°, 67½°, 90°.

‡ These values were calculated on a machine.   If none is available write:

x	0	$\frac{1}{8}\pi$	$\frac{1}{4}\pi$	$\frac{3}{8}\pi$	$\frac{1}{2}\pi$
$\sqrt{(\sin x)}$	0	0·6187	0·8409	0·9611	1·0000
$x\sqrt{(\sin x)}$	0	0·0773π	0·2102π	0·3604π	0·5000π

and retain the factor π until the final stage.

same formula as before: $\frac{1}{3}d(y_1 + 4y_2 + 2y_3 + 4y_4 + y_5)$.

$y_1 = 0 \cdot 0000$	$y_2 = 0 \cdot 2430$	$y_3 = 0 \cdot 6604$
$y_5 = 1 \cdot 5708$	$y_4 = 1 \cdot 1323$	2
$1 \cdot 5708$	$1 \cdot 3753$	$1 \cdot 3208$
$5 \cdot 5012$	4	
$1 \cdot 3208$		
	$5 \cdot 5012$	
$8 \cdot 3928$		

$$\frac{1}{3}d = 0 \cdot 1309; \qquad \therefore \int_0^{\frac{1}{2}\pi} x\sqrt{(\sin x)}\,dx = 0 \cdot 1309 \times 8 \cdot 3928.$$

Now $\quad \bar{x}\displaystyle\int_0^{\frac{1}{2}\pi} \sqrt{(\sin x)}\,dx = \int_0^{\frac{1}{2}\pi} x\sqrt{(\sin x)}\,dx.$

$$\therefore \bar{x} \cdot 0 \cdot 1309 \times 9 \cdot 0010 = 0 \cdot 1309 \times 8 \cdot 3928.$$

$$\therefore \bar{x} = 8 \cdot 3928 \div 9 \cdot 0010.$$

Therefore the x-coordinate of the centroid is approximately $0 \cdot 93$.

Exercise 19a

1. A cyclist travels a distance of 200 m from rest to rest in 14 s. The distance covered at intervals of two seconds is given below:

Time (s)	0	2	4	6	8	10	12	14
Distance (m)	0	2	9	21	34	46	57	61

From a velocity-time graph, estimate the greatest acceleration during this motion.

2. The distance moved by a body starting from rest is given by the following table:

Time (s)	0	1	2	3	4	5	6
Distance (m)	0	5	17	34	57	84	115

Plot a velocity-time graph and estimate the speed after 6 s.

3. The total distance travelled from rest by a motor car of mass 950 kg at intervals of 2½ s from rest is given as follows:

Time (s)	0	2½	5	7½	10	12½	15	17½	20
Distance (m)	0	23	107	233	386	558	748	952	1168

Find the speed at intervals of 2½ s during the motion; estimate the net accelerating force acting on the car 5 s after the start and deduce the power being used on acceleration at this instant.

4. Estimate the area enclosed by the curve in Fig. 19.6. Take eight intervals.

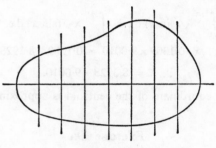

Fig. 19.6

5. The area in square centimetres of the cross-section of a model boat 28 cm long at intervals of 3·5 cm is as follows:

0	11·5	15·3	16·3	16·2	13·4	9·3	4·9	0

Find the volume of the boat.

6. A jug of circular cross-section is 16 cm high inside and its internal diameter is measured at equal intervals from the bottom:

Height (cm)	0	4	8	12	
Diameter (cm)	10·2	13·8	15·3	9·3	9·9

What volume of liquid will the jug hold if filled to the brim?

7. A vase is in the form of a solid of revolution with its axis vertical. 100 cm³ of water are poured in at a time and the depth of water measured as below:

Volume (cm³)	100	200	300	400	500	600	700	800
Depth (cm)	4·6	7·2	9·6	11·6	13·5	15·6	18·4	25·1

Use this data to estimate the area of the cross-section of the surface of the water when the vessel contains 50, 150, . . ., 750 cm³. Given that the internal radius of the base is 2·3 cm, draw the section of the vase made by a vertical plane through the axis. [The best-fitting curve will not pass through all the points you plot.]

8. The force, acting in a constant direction, on a body of mass 2 kg moving freely from rest is given in the table below:

Time (s)	0	5	10	15	20	25	30
Force (N)	7	8·5	11	13	13·5	10·5	5

Find the velocity of the body after 30 s.

9. A body of mass 686 kg, moving freely from rest, is acted on by a variable force as below:

Distance (m)	0	20	40	60	80	100	120
Force (N)	588	490	333	118	− 176	− 392	− 529

Find the velocity of the body after it has travelled 120 m.

10. (x_1, y_1), (x_2, y_2), (x_3, y_3), where $x_2 = \frac{1}{2}(x_1 + x_3)$, are three points on the parabola $y = ax^2 + bx + c$. Prove that the area under the curve between the lines $x - x_1 = 0$, $x - x_3 = 0$ is equal to $\frac{1}{3}(x_2 - x_1)(y_1 + 4y_2 + y_3)$.

Use this formula to find the area between the parabola $y = x(10 - x)$ and the x-axis. Check your answer by integration.

11. Evaluate $\int_0^1 e^{-x^2}\,dx$ by Simpson's rule taking ten intervals.

12. Using tables where necessary, calculate the value of

$$\int_{0·1}^{0·5} e^{-x}\,dx$$

(i) by direct integration,

(ii) by Simpson's rule, using 5 ordinates spaced at intervals of 1/10 unit.

(Give your answers to four places of decimals.)   N.

13. By means of Simpson's rule and taking unit intervals of x from $x=8$ to $x=12$, find approximately the area enclosed by the curve $y=\log_{10} x$, the lines $x=8$ and $x=12$, and the x-axis. Deduce the average value of $\log_{10} x$ between $x=8$ and $x=12$.   N.

14. The coordinates of three points on the curve $y=A+Bx+Cx^2$ are (x_1, y_1), (x_2, y_2) and (x_3, y_3), where $x_2=\frac{1}{2}(x_3+x_1)$. Prove that the area under the curve between the lines $x=x_1$ and $x=x_3$ is equal to

$$\tfrac{1}{6}(x_3-x_1)(y_1+4y_2+y_3).$$

Deduce Simpson's rule for five ordinates.

Using five ordinates and four-figure tables, apply Simpson's rule to evaluate the integral $4\int_0^1 \dfrac{dx}{1+x^2}$ and thus to find a value for π correct to three places of decimals.   O.C.

15. The coordinates of three points on the curve $y=ax^3+bx^2+cx+d$ are (x_1, y_1), (x_2, y_2), (x_3, y_3). Prove that, if $x_2-x_1=x_3-x_2=h$, the area under the curve between the lines $x=x_1$, $x=x_3$ is $\frac{1}{3}h(y_1+4y_2+y_3)$.

Find the area between the curve $y=x(x-2)^2$ and the x-axis by means of Simpson's rule with three ordinates. Use integration to check that your answer is exact.

16. Show that the area under the curve $y=1/x$, from $x=n-1$ to $x=n+1$, is $\log_e \{(n+1)/(n-1)\}$, provided $n>1$.

By applying Simpson's rule to this area, deduce that, approximately,

$$\log_e \frac{n+1}{n-1} = \frac{1}{3}\left(\frac{1}{n-1}+\frac{4}{n}+\frac{1}{n+1}\right),$$

and show that the error in this approximation is $4/(15n^5)$, when higher powers of $1/n$ are neglected.   N.

17. The part of the curve $y=+x^{1/2}(x-1)^{1/4}$ between the lines $x=1$ and $x=3$ is rotated through four right angles about the axis of x. Prove that the volume enclosed is $44\pi\sqrt{2}/15$.

Calculate the approximate volume directly from the integral using Simpson's rule with ordinates at $x=1$, $1\frac{1}{2}$, 2, $2\frac{1}{2}$ and 3.   O.C.

18. Prove that the area enclosed by the curve $y=\tan x$, the x-axis, and $x=\frac{1}{3}\pi$ is $\log_e 2$.

If the point $(\bar{x}, \bar{y})$ is the centroid of this area, prove that

$$\bar{y} = (3\sqrt{3}-\pi)/6\log_e 2$$

and calculate $\bar{x}$ by means of Simpson's rule with ordinates at intervals of $\frac{1}{12}\pi$. [$\tan \frac{1}{12}\pi=2-\sqrt{3}$.]   O.C.

A numerical method of solving equations

19.4. The reader will be familiar with using the difference columns in four-figure tables. For example, finding $\log_{10} 7\cdot463$, we find $\log_{10} 7\cdot460 = 0\cdot8727$ and, since there is a 2 under the 3 in the difference columns, we find $\log_{10} 7\cdot463 = 0\cdot8729$. This process is one of *interpolation* and the difference columns save us a great deal of trouble. Using the same book of tables, the author wanted to find the value of $e^{1\cdot257}$. Now, from the tables,

$$e^{1\cdot25} = 3\cdot4903, \qquad e^{1\cdot26} = 3\cdot5254.$$

The difference between these is $0\cdot0351$, so we may estimate the value of $e^{1\cdot257}$ as

$$3\cdot4903 + \tfrac{7}{10} \times 0\cdot0351 = 3\cdot4903 + 0\cdot02457 = 3\cdot5149,$$

but the method does not give any indication of the accuracy of the answer. In fact, using seven-figure tables it is found that $3\cdot5148$ is a better approximation.

In § 18.3, p. 386, we discussed Newton's method of obtaining approximate solutions of equations and now we shall consider another simple method using proportional parts as in the last paragraph.

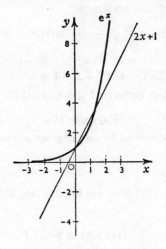

Fig. 19.7

Consider the equation $e^x = 2x + 1$. We can see from the sketch graph of the functions e^x, $2x + 1$ (Fig. 19.7) that there is a root of the equation between $x = 1$ and $x = 1 \cdot 5$. Let $f(x) \equiv e^x - 2x - 1$.

Stage 1.

$$f(1) = e - 3 = -0 \cdot 2817.$$

(Note that $f(x) > 0$ when x is greater than the larger root, so try a value of x greater than 1.)

$$f(1 \cdot 5) = e^{1 \cdot 5} - 4 = 0 \cdot 4817.$$

So we expect the root to be approximately

$$1 + \frac{0 \cdot 2817}{0 \cdot 4817 + 0 \cdot 2817} \times 0 \cdot 5 = 1 \cdot 2.$$

Stage 2.

$$f(1 \cdot 2) = e^{1 \cdot 2} - 3 \cdot 4 = -0 \cdot 0799.$$
$$f(1 \cdot 3) = e^{1 \cdot 3} - 3 \cdot 6 = 0 \cdot 0693.$$

Therefore a better approximation is

$$1 \cdot 2 + \frac{0 \cdot 0799}{0 \cdot 0693 + 0 \cdot 0799} \times 0 \cdot 1 = 1 \cdot 25.$$

Stage 3.

$$f(1 \cdot 25) = e^{1 \cdot 25} - 3 \cdot 5 = -0 \cdot 0097.$$
$$f(1 \cdot 26) = e^{1 \cdot 26} - 3 \cdot 52 = 0 \cdot 0054.$$

Therefore a better approximation is

$$1 \cdot 25 + \frac{0 \cdot 0097}{0 \cdot 0054 + 0 \cdot 0097} \times 0 \cdot 01 = 1 \cdot 256.$$

$$f(1 \cdot 256) = e^{1 \cdot 256} - 3 \cdot 512 = -0 \cdot 00065.$$
$$f(1 \cdot 257) = e^{1 \cdot 257} - 3 \cdot 514 = 0 \cdot 00086.$$

Therefore this root lies between $1 \cdot 256$ and $1 \cdot 257$.

Exercise 19b

Solve the following equations by any appropriate method.

1. $x^3 = x + 1$.
2. $e^x = 5x$.
3. $x^4 - 2x - 1 = 0$.
4. $x^3 - 3x^2 - 1 = 0$. [For hint, see Ex. 8d, No. 30.]
5. $x = 3 \log_e x$.
6. $e^x = x^2 + 2$.
7. $x^3 - 6x^2 + 11x - 8 = 0$. [See hint in No. 4.]
8. $2 + \log_e x = x$.
9. $10^x = x^{10}$.

10. A cuboid has volume 100 cm³, surface area 150 cm², and its length is twice its breadth. What are its dimensions?

11. When the height of water in a hemispherical bowl is h, the volume of water in the bowl is $\pi(rh^2 - \frac{1}{3}h^3)$, where r is the radius of the bowl. Find the height of the water when half the volume of the bowl is filled.

12. If I pay £100 on January 1st for fifteen consecutive years and draw £2100 on January 1st of the next year, what rate of compound interest do I receive?

13. A donkey is tied by a rope to a point on the circumference of a circular field of radius r. If the donkey is to be allowed to graze half the area of the field, how long should the rope be?

*14. Square roots may be found by the following method. If b is an approximate root of the equation $x^2 = a$, a second approximation is given by

$$\frac{\dfrac{a}{b}+b}{2}$$

(i) Take $a = 17$, $b = 4$ and hence estimate $\sqrt{17}$. (Two divisions will give this to four places of decimals.)

(ii) Show that, if $b = \sqrt{a}+h$, where h is so small that the cube and higher powers of $h/\sqrt{a}$ may be neglected,

$$\frac{\dfrac{a}{b}+b}{2} \simeq \sqrt{a}+\frac{h^2}{2\sqrt{a}}.$$

Finite differences

19.5. Look at Tables 1, 2:

TABLE 1

x	0		1		2		3		4		5		6
x^2	0		1		4		9		16		25		36
First difference		1		3		5		7		9		11	
Second difference			2		2		2		2		2		
Third difference				0		0		0		0			

TABLE 2

x	0		1		2		3		4		5		6
x^3	0		1		8		27		64		125		216
First difference		1		7		19		37		61		91	
Second difference			6		12		18		24		30		
Third difference				6		6		6		6			
Fourth difference					0		0		0				

The differences are, in fact, *increases*. The first difference is the increase in the value of the function from one value to the one on the right; the second difference is the increase in the first difference, and so on. We see that the **third** differences of x^2 are zero and that the **fourth** differences of x^3 are zero as far as we have taken them. So we might make two guesses:

(i) that these results will hold if we extend the range of values of x, using intervals of 1 unit;

that the $(n+1)$th differences of x^n are zero, taking intervals in x of 1 unit.

But for the moment we shall not prove these.

Qu. 12. For the same values of x, verify that the fifth differences of x^4 are zero.

Qu. 13. Tabulate x^2 and x^3 at intervals of $\frac{1}{2}$ between -3 and $+3$. Are the third and fourth differences zero again? [Remember that the differences are, in fact, increases, so that the signs must be retained.]

If you have worked through Qu. 13, you should have found that, with intervals of $\frac{1}{2}$ in x, the third differences of the function x^2 are again zero. So let us try a general interval of h in x. This has been done in Table 3, but the rows and columns have been interchanged to make the table clearer.

TABLE 3

x	x^2	First difference	Second difference	Third difference
a	a^2			
		$2ah + h^2$		
$a+h$	$a^2 + 2ah + h^2$		$2h^2$	
		$2ah + 3h^2$		0
$a+2h$	$a^2 + 4ah + 4h^2$		$2h^2$	
		$2ah + 5h^2$		
$a+3h$	$a^2 + 6ah + 9h^2$			

Qu. 14. Find expressions for x^3 when $x = a$, $a+h$, $a+2h$, $a+3h$, $a+4h$ and calculate the differences up to the fourth.

Now we shall show how the method of differences can be used in connection with polynomials both for checking and calculating values. But first note the following so that we can apply the work above:

(i) if a is a constant, the $(n+1)$th differences of ax^n are zero, since all the differences will be a times the corresponding differences of x^n.

(ii) the differences of $a_0 + a_1 x + \ldots + a_{n-1}x^{n-1} + a_n x^n$ will be the sum of the differences of $a_0, a_1 x, \ldots, a_{n-1}x^{n-1}, a_n x^n$, therefore the $(n+1)$th differences will vanish.

As an example of using the method of differences as a check, the values of $2x^2 - x + 3$ have been calculated at unit intervals from -3 to $+3$ (working not shown) and the results tabulated (see Table 4).

TABLE 4

x	-3	-2	-1	0	1	2	3
$2x^2-x+3$	24	13	6	3	4	9	18
First difference		-11	-7	-3	1	5	9
Second difference			4	4	4	4	4
Third difference				0	0	0	0

As the third differences vanish, we can feel reasonably confident about the calculated values of the function.

Qu. 15. Find the values of $3x^2 - 4x + 1$ for $x = -3, -2, -1, 0, 1, 2, 3$ and check your results by the method of differences.

Qu. 16. Repeat Qu. 15 for $x^3 - 10x + 1$.

Example 4. *Tabulate the function $2x^2 - x + 3$ at intervals of $\frac{1}{2}$ from $x = -2$ to $x = +2$.*

First stage

Calculate a few easy values of the function (three for a quadratic):

x	0	$\frac{1}{2}$	1
$2x^2-x+3$	3	3	4
First difference		0	1
Second difference		1	

The third differences are zero for a quadratic, so the second differences are equal. We therefore enter the other first differences getting:

Second stage

x	-2	$-1\frac{1}{2}$	-1	$-\frac{1}{2}$	0	$\frac{1}{2}$	1	$1\frac{1}{2}$	2
$2x^2-x+3$	13	9	6	4	3	3	4	6	9
First difference		-4	-3	-2	-1	0	1	2	3

As a check, work out the values of the function at the ends of the range.

Example 5. *The values of a function tabulated at equal intervals are* 25, a, 11, 8, 9, 16. *Estimate the value of a.*

With five values given, it is possible to determine the five coefficients of a function $a_0 + a_1 x + a_2 x^2 + a_3 x^3 + a_4 x^4$ which would have the values 25, 11, 8, 9, 16 when, say, $x = 0, 2, 3, 4, 5$ (the values of x were chosen arbitrarily subject to the restriction on the intervals in the data). But the fifth differences of a quartic are zero, so it is easier to proceed as follows:

Function	25		a		11			8	9	16
First difference		$a-25$		$11-a$			-3		1	7
Second difference			$-2a+36$		$a-14$			4		6
Third difference				$3a-50$		$-a+18$			2	
Fourth difference					$-4a+68$		$a-16$			
Fifth difference						$5a-84$				

Equating the fifth differences to zero,
$$5a - 84 = 0$$
$$\therefore a = 16 \cdot 8.$$

In fact, the function from which the data were calculated was $2^x + (x-4)^2$, taking $x = -1, 1, 2, 3, 4$. Substituting $x = 0$, $2^x + (x-4)^2 = 17$. Since the function is not a polynomial we cannot expect perfect agreement.

Difference notation

19.6. The reader may have noticed a certain similarity between the successive differences and the result of differentiating a function. For instance in Table 1, p. 423, the first differences are the values of $d(x^2)/dx$ when $x = \frac{1}{2}, 1\frac{1}{2}, \ldots, 5\frac{1}{2}$; the second differences are 2, which is the second derivative of x^2. Before jumping to conclusions, however, note that the first differences in Table 2, p. 423, are *not* the values of $d(x^3)/dx$ when $x = \frac{1}{2}, 1\frac{1}{2}, \ldots, 5\frac{1}{2}$. However, the similarity is worth noting.

Qu. 17. Find the quadratic function whose values for $x = \frac{1}{2}, 1\frac{1}{2}, \ldots, 5\frac{1}{2}$ are the first differences in Table 2, p. 423.

The symbol Δy has been used for an increase in y and, as that is what we have been concerned with in the last section (although y was not used), it is natural to use it here too. Further, its use leads to very neat symbolic expressions for some of the results.

Let y_0, y_1, y_2, y_3, y_4 be values of a function y at *equal* intervals, then we write the increase in y as Δy.

By Δy_0, we mean the increase starting from y_0, i.e. $y_1 - y_0$.

By Δy_1, we mean the increase starting from y_1, i.e. $y_2 - y_1$. Similarly for Δy_2, Δy_3. (See Table 5.)

TABLE 5

y	y_0	y_1	y_2	y_3	y_4
$\Delta y \;\Big\{$	$\begin{matrix} y_1 - y_0 \\ = \Delta y_0 \end{matrix}$	$\begin{matrix} y_2 - y_1 \\ = \Delta y_1 \end{matrix}$	$\begin{matrix} y_3 - y_2 \\ = \Delta y_2 \end{matrix}$	$\begin{matrix} y_4 - y_3 \\ = \Delta y_3 \end{matrix}$	

The increase in Δy may be written $\Delta(\Delta y)$ and this is abbreviated to $\Delta^2 y$. (Note that Δ is *not* an algebraic symbol, so we cannot assume that it follows any of the laws of algebra.) In the same way as before, $\Delta(\Delta y_0)$ means the increase starting from Δy_0, so that

$$\begin{aligned} \Delta(\Delta y_0) &= \Delta y_1 - \Delta y_0, \\ &= (y_2 - y_1) - (y_1 - y_0), \\ &= y_2 - 2y_1 + y_0. \end{aligned}$$

Hence we write $\qquad \Delta^2 y_0 = y_2 - 2y_1 + y_0.$

Now examine Table 6 to see how it has been made up. The third differences $\Delta(\Delta^2 y)$ are written $\Delta^3 y$, and the fourth difference $\Delta(\Delta^3 y)$ is written $\Delta^4 y$ for short.

TABLE 6

y	Δy	$\Delta^2 y$	$\Delta^3 y$	$\Delta^4 y$
y_0				
	Δy_0			
y_1		$\Delta^2 y_0$		
	Δy_1		$\Delta^3 y_0$	
y_2		$\Delta^2 y_1$		$\Delta^4 y_0$
	Δy_2		$\Delta^3 y_1$	
y_3		$\Delta^2 y_2$		
	Δy_3			
y_4				

TABLE 7

y	Δy	$\Delta^2 y$	$\Delta^3 y$	$\Delta^4 y$
y_0				
	$y_1 - y_0$			
y_1		$y_2 - 2y_1 + y_0$		
	$y_2 - y_1$		$y_3 - 3y_2 + 3y_1 - y_0$	
y_2		$y_3 - 2y_2 + y_1$		$y_4 - 4y_3 + 6y_2 - 4y_1 + y_0$
	$y_3 - y_2$		$y_4 - 3y_3 + 3y_2 - y_3$	
y_3		$y_4 - 2y_3 + y_2$		
	$y_4 - y_3$			
y_4				

In Table 7, the differences have been expressed in terms of $y_0, y_1, \ldots, y_4$. We see, for example, that

$$\Delta y_0 = y_1 - y_0,$$
$$\Delta^2 y_0 = y_2 - 2y_1 + y_0,$$
$$\Delta^3 y_0 = y_3 - 3y_2 + 3y_1 - y_0,$$
$$\Delta^4 y_0 = y_4 - 4y_3 + 6y_2 - 4y_1 + y_0.$$

As the coefficients are those found in Pascal's triangle (P.M.I., p. 238) we might expect to find that

$$\Delta^5 y_0 = y_5 - 5y_4 + 10y_3 - 10y_2 + 5y_1 - y_0$$

and in general that

$$\Delta^n y_0 = y_n - \binom{n}{1} y_{n-1} + \ldots + (-1)^r \binom{n}{r} y_{n-r} + \ldots + (-1)^n y_0,$$

where

$$\binom{n}{r} = \frac{n!}{(n-r)! \, r!}.$$

This may be proved by induction, see Exercise 19c, No. 15. Note that (since the numbering of the y's could start with another value) similar expressions hold for $\Delta^n y_1$, $\Delta^n y_2$, etc.

This result can assist someone checking calculated values of a function. Suppose that *one* value of the function has an error of ϵ. The greatest coefficients in the expressions for

	Δy	$\Delta^2 y$	$\Delta^3 y$	$\Delta^4 y$	$\Delta^5 y$	$\Delta^6 y$
are	1	2	3	6	10	20

so that an error of ϵ can produce in the differences errors of up to

ϵ	2ϵ	3ϵ	6ϵ	10ϵ	20ϵ

Thus the errors in successive differences will *increase*.

There is a deliberate mistake in the values of e^x in Table 8. Decimal points have been left out of the differences:

TABLE 8

x	e^x	$\Delta(e^x)$	$\Delta^2(e^x)$	$\Delta^3(e^x)$	$\Delta^4(e^x)$
3·0	20·086				
3·1	22·198	2112	223		
3·2	24·533	2335	245	22	4
3·3	27·113	2580	271	26	3
3·4	29·964	2851	300	29	103
3·5	33·115	3151	432	132	−398
3·6	36·698	3583	166	−266	605
3·7	40·447	3749	505	339	−397
3·8	44·701	4254	447	−58	106
3·9	49·402	4701	495	48	3
4·0	54·598	5196	546	51	7
4·1	60·340	5742	604	58	
4·2	66·686	6346			

An examination of the table suggests that the entries under $\Delta^4(e^x)$ should be single figures. The largest entry in this column is 605, so that the error probably will be found opposite this. In fact, the value 36·698 should read 36·598.

Qu. 18. Correct Table 8. [There are only 2, 3, 4, 5 entries wrong in the columns of $\Delta(e^x)$, $\Delta^2(e^x)$, $\Delta^3(e^x)$, $\Delta^4(e^x)$, respectively.]

Example 6. *A function is given at equal intervals by* 0, 41, 79, y_3, 146. *Estimate the value of* y_3.

Write   0,   41,   79, y_3,   146.
respectively as y_0, y_1, y_2, y_3, y_4.

Given four values of y, it is possible to find a cubic function of x in the form

$$y = ax^3 + bx^2 + cx + d$$

which takes these values, since we could form four equations for the four unknowns a, b, c, d. But as this function is cubic, its fourth differences will vanish.

$$\Delta^4 y_0 = y_4 - 4y_3 + 6y_2 - 4y_1 + y_0.$$

Writing $\Delta^4 y_0 = 0$,

$$y_4 - 4y_3 + 6y_2 - 4y_1 + y_0 = 0.$$

$$\therefore \ 146 - 4y_3 + 474 - 164 + 0 = 0.$$

$$\therefore \ y_3 = 114.$$

In fact the data were obtained from tables using the formula

$$y = 100 \log_{10} x, \quad x = 1, 1{\cdot}1, 1{\cdot}2, 1{\cdot}4.$$

Tables give $y_3 = 100 \log_{10} 1{\cdot}3 = 114$ to three significant figures.

Qu. 19. A function is given at equal intervals by 82, y_1, 13, 11, 17. Estimate the value of y_1.

Evaluate the function $2^x + 3^{(4-x)}$ for $x = 0, 1, \ldots, 4$ and compare these values with those above.

If you have worked through Qu. 19, you will have seen that there is a considerable discrepancy between y_1 and the corresponding value calculated from the function $2^x + 3^{(4-x)}$. Why should this be so? Because, with four values given, we can only determine a comparatively simple function—a cubic—to take these values. There is, in fact, no limit to the number of functions which can be constructed to take these values, some of which would depart much further from the calculated value of y_1. In general, better results can be expected when the function is increasing, or decreasing, throughout the range under consideration.

The last example is not a sort which occurs very often in practice. It is more likely that we should know the values of a function at equal intervals and want to find the value of the function in between these. This, for example, is what happens when functions are being tabulated in, say, four-figure tables. A certain number of values of the function are calculated and the rest are found by interpolation. One method of interpolation will now be explained.

First we shall try to express some values of a function in terms of y_0 and the differences Δy_0, $\Delta^2 y_0$, etc.

$$\Delta y_0 = y_1 - y_0.$$
$$\therefore \ y_1 = \Delta y_0 + y_0.$$

$$\Delta^2 y_0 = y_2 - 2y_1 + y_0.$$
$$\therefore \ \Delta^2 y_0 = y_2 - 2(\Delta y_0 + y_0) + y_0.$$
$$\therefore \ y_2 = \Delta^2 y_0 + 2\Delta y_0 + y_0.$$

Qu. 20. Show, as above, that

$$y_3 = \Delta^3 y_0 + 3\,\Delta^2 y_0 + 3\,\Delta y_0 + y_0.$$

The reason for the coefficients from Pascal's triangle appearing again may be seen by examining Tables 9 and 10. In these the two-tailed arrows indicate that the expression at the head of an arrow is the sum of the expressions at its tails. If Table 10 is continued to include $\Delta^4 y_0$

TABLE 9

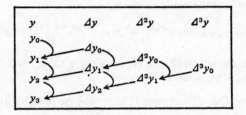

TABLE 10

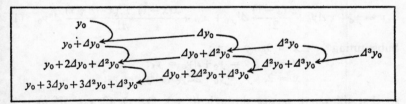

and $\Delta^5 y_0$, it will readily be seen how the coefficients of Pascal's triangle occur in each column. By induction (see Exercise 19c, No. 16) it may be shown that

$$y_n = y_0 + n\,\Delta y_0 + \frac{n(n-1)}{2!}\,\Delta^2 y_0 + \frac{n(n-1)(n-2)}{3!}\,\Delta^3 y_0 + \ldots + \Delta^n y_0.$$

But this is what we should obtain if we treated Δ as an algebraic symbol and expanded $(1+\Delta)^n y_0$ in ascending powers of Δ. So for brevity we write

$$y_n = (1+\Delta)^n y_0.$$

We now have an expression which may readily be used (if this turns out to be permissible) to find intermediate values of the function; for instance, we might hope that the value of y mid-way between y_0 and y_1 would be

$$y_{\frac{1}{2}} = (1+\Delta)^{\frac{1}{2}}y_0.$$

and evaluated as

$$y_{\frac{1}{2}} = y_0 + \tfrac{1}{2}\Delta y_0 + \frac{(\frac{1}{2})(-\frac{1}{2})}{2!}\Delta^2 y_0 + \frac{(\frac{1}{2})(-\frac{1}{2})(-\frac{3}{2})}{3!}\Delta^3 y_0 + \ldots,$$
$$= y_0 + \tfrac{1}{2}\Delta y_0 - \tfrac{1}{8}\Delta^2 y_0 + \tfrac{1}{16}\Delta^3 y_0 + \ldots.$$

Qu. 21. Show that the cubic curve

$$y = y_0 + x\,\Delta y_0 + \frac{x(x-1)}{2!}\Delta^2 y_0 + \frac{x(x-1)(x-2)}{3!}\Delta^3 y_0$$

passes through the points $(0, y_0)$, $(1, y_1)$, $(2, y_2)$, $(3, y_3)$ and find y in terms of y_0, Δy_0, $\Delta^2 y_0$, $\Delta^3 y_0$ when $x = \frac{1}{2}$.

The result of Qu. 21 suggests that it may be possible to write down the equation of a curve through the points $(0, y_0)$, $(1, y_1)$, $\ldots$, (n, y_n). Consider the curve

$$y = y_0 + x\,\Delta y_0 + \frac{x(x-1)}{2!}\Delta^2 y_0 + \ldots + \frac{x(x-1)\ldots(x-n+1)}{n!}\Delta^n y_0. \quad (1)$$

Substituting
$$x = 0, \quad y = y_0;$$
$$x = 1, \quad y = y_0 + \Delta y_0 = y_1;$$
$$x = 2, \quad y = y_0 + 2\Delta y_0 + \Delta^2 y_0 = y_2;$$

so that the curve passes through $(0, y_0)$, $(1, y_1)$, $(2, y_2)$. If m is a positive integer $(m \leqslant n)$, substituting $x = m$,

$$y = y_0 + m\,\Delta y_0 + \binom{m}{2}\Delta^2 y_0 + \ldots + \frac{m(m-1)\ldots(m-r+1)}{r!}\Delta^r y_0 + \ldots$$
$$+ \frac{m(m-1)\ldots 1}{m!}\Delta^m y_0,$$

any subsequent terms containing a factor of 0.

So, when $x = m$,

$$y = y_0 + \binom{m}{1}\Delta y_0 + \ldots + \binom{m}{r}\Delta^r y_0 + \ldots + \Delta^m y_0 = y_m.$$

Therefore the curve passes through $(0, y_0)$, $(1, y_1)$, $\ldots$, (n, y_n).

So if we have a curve which passes through the points $(0, y_0)$, $(1, y_1)$,

$\ldots$, (n, y_n), the ordinates at intermediate values of x will be given approximately by equation (1). The equation of the curve may be remembered as the first $n+1$ terms of the expansion of

$$y = (1+\Delta)^x y_0.$$

(If we have n values of y given, we can only find the first $n-1$ differences and subsequent terms of the expansion are ignored.)

This formula, which is called Newton's forward-difference formula, is of limited value but is of use (i) when only a few values of the function are known, or (ii) when the series converges rapidly.

Example 7. *The following values of a function are given*:

x	0	1	2	3	4
y	1000	1158	1282	1366	1409

Estimate the value of the function when $x = 3 \cdot 6$.

The differences are found below:

x	y	Δy	$\Delta^2 y$	$\Delta^3 y$	$\Delta^4 y$
0	1000				
1	1158	158			
2	1282	124	-34	-6	
3	1366	84	-40	-1	5
4	1409	43	-41		

By Newton's method, when $x = 3 \cdot 6$

$$y = 1000 + 3 \cdot 6 \times 158 + \frac{3 \cdot 6 \times 2 \cdot 6}{2!}(-34)$$

$$+ \frac{3 \cdot 6 \times 2 \cdot 6 \times 1 \cdot 6}{3!}(-6) + \frac{3 \cdot 6 \times 2 \cdot 6 \times 1 \cdot 6 \times 0 \cdot 6}{4!} \times 5$$

$$= 1000 + 568 \cdot 8 - 159 \cdot 1 - 15 \cdot 0 + 1 \cdot 9 = 1396 \cdot 6.$$

Our estimate for y when $x = 3 \cdot 6$ is 1397.

[The function given was $y = 1000 (\sin 10x° + \cos 10x°)$, and tables give $y = 1396 \cdot 8$ when $x = 3 \cdot 6$.]

Qu. 22. Estimate y by straight line interpolation when $x = 3 \cdot 6$ using the data of Example 7.

Qu. 23. With the data of Example 7, take y_0, y_1, y_2, y_3, y_4 as 1409, 1366, 1282, 1158, 1000, respectively, and hence find a value for y when $x = 3 \cdot 6$.

The following exercise has been constructed for those who do not have access to desk calculating machines. With such comparatively simple numbers the methods are easily learned but their advantages become more apparent when less easy numbers and the use of machines are combined.

Exercise 19c

1. Find the values of 30^2, 31^2, 32^2 and use the method of differences to find the values of 33^2, 34^2, ..., 40^2.

2. Find the values of the functions $n(n-1)/(2!)$, $n(n-1)(n-2)/(3!)$ for $n = 1, 2, \ldots, 10$ by the method of differences. Check your results by calculating the values for $n = 10$ directly from the formulae.

3. A quadratic function has been calculated at equal intervals. Check the values given and correct any mistake: 42, 25, 14, 7, 4, 5, 10.

4. Find the values of the function $4x^3 - 3x + 1$ when $x = -\frac{1}{2}$, 0, $\frac{1}{2}$, 1 and obtain the values of the function when $x = -2$, $-1\frac{1}{2}$, -1, $1\frac{1}{2}$, 2 by the method of differences.

5. It is known that a quadratic function takes the values 3, -7, 6 when $x = 1, 2, 3$ respectively. Find the value of the function when $x = 5, 6, 7$.

6. A function has values at equal intervals $1 \cdot 00$, $1 \cdot 54$, $3 \cdot 76$, a, $27 \cdot 31$. Estimate the value of a.

7. A function has values given in the following table:

x	0	1	2	3	4
y	7·9	9·2	10·4	11·6	12·8

Estimate its value when $x = 5$.

8. There is an error in the function tabulated at equal intervals below. Find and correct it.

9091   9009   8929   8850   8727   8696

9. Use the equation $y = y_0 + x \, \Delta y_0 + \frac{1}{2}x(x-1) \, \Delta^2 y_0$ to obtain Simpson's rule.

10. Given that $y = 10^4\{f(x) - 1\}$ has the values in the table below, find the value of y when $x = \frac{1}{2}$ and deduce the value of $f(\frac{1}{2})$.

x	0	1	2	3	4
y	0	50	201	453	811

11. Given the values below, estimate the value of (i) $\log_{10} 1 \cdot 01$, (ii) $\log_{10} 1 \cdot 07$.

x	$1 \cdot 00$	$1 \cdot 02$	$1 \cdot 04$	$1 \cdot 06$	$1 \cdot 08$
$10^4 \log_{10} x$	0	86	170	253	334

12. A function has the values shown in the table below:

x	0	1	2	3	4	5
y	564	632	703	778	857	941

Estimate the value of y when (i) $x = \frac{1}{2}$, (ii) $x = 2\frac{1}{2}$.

13. The following values of $\sin x°$ can be calculated without using series. Estimate the values of $\sin 10°$, $\sin 20°$ and compare the results with those in your tables.

x	0	$7\frac{1}{2}$	15	$22\frac{1}{2}$	30	$37\frac{1}{2}$
$\sin x°$	0	$0 \cdot 1305$	$0 \cdot 2588$	$0 \cdot 3827$	$0 \cdot 5000$	$0 \cdot 6088$

14. Prove by induction that the $(n+1)$th differences of x^n are zero.

15. Prove by induction that

$$\Delta^n y_0 = y_n - \binom{n}{1} y_{n-1} + \ldots + (-1)^r \binom{n}{r} y_{n-r} + \ldots + (-1)^n y_0,$$

where

$$\binom{n}{r} = \frac{n!}{(n-r)! \, r!}.$$

16. Prove by induction that

$$y_n = y_0 + \binom{n}{1} \Delta y_0 + \ldots + \binom{n}{r} \Delta^r y_0 + \ldots + \Delta^n y_0.$$

CHAPTER 20

HYPERBOLIC FUNCTIONS

Hyperbolic cosine and sine

20.1. We shall begin by defining two new functions, the hyperbolic cosine and the hyperbolic sine. No attempt to explain the reason for adopting these definitions will be given at present because a knowledge of complex numbers is needed if the reason is to be fully appreciated. The reader will, however, very soon find some strong similarities between the hyperbolic functions and the familiar trigonometrical functions which, to save confusion, are often referred to as the *circular functions*. These similarities would not, by themselves, justify the inclusion of a study of the hyperbolic functions in this book: they are being introduced because they will very quickly extend the reader's powers of integration, and the reader may begin to need them in mechanics. But first we shall study the functions themselves. They were introduced by J. H. Lambert in a paper read in 1768.

DEFINITIONS: *the hyperbolic cosine of x*

$$\cosh x = \tfrac{1}{2}(e^x + e^{-x}),$$

and the hyperbolic sine of x

$$\sinh x = \tfrac{1}{2}(e^x - e^{-x}).$$

$\cosh x$ is pronounced as it is spelled. $\sinh x$ may be pronounced 'sinch x' (or shine x).

First we sketch their graphs. Starting with the graph of e^x in Fig. 20.1 (i), that of e^{-x} has been shown dotted in the same figure. $\cosh x$ is half the sum of these two functions (see Fig. 20.1(ii)) and $\sinh x$ is half the difference (see Fig. 20.1 (ii)). The two graphs are distinct in the first quadrant but they approach so close that, when they are sketched to this scale, the lines run together.

In general, the properties of hyperbolic functions are easily proved and this will be left to the reader to do in Exercise 20a. We shall first prove one important identity.

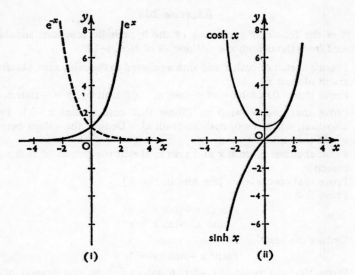

Fig. 20.1

Example 1. *Prove the identity*

$$\cosh^2 x - \sinh^2 x = 1.$$

From the definitions of $\cosh x$ and $\sinh x$,

$$\cosh^2 x - \sinh^2 x = \{\tfrac{1}{2}(e^x + e^{-x})\}^2 - \{\tfrac{1}{2}(e^x - e^{-x})\}^2,$$
$$= \tfrac{1}{4}(e^{2x} + 2 + e^{-2x}) - \tfrac{1}{4}(e^{2x} - 2 + e^{-2x}),$$
$$= \tfrac{1}{4}e^{2x} + \tfrac{1}{2} + \tfrac{1}{4}e^{-2x} - \tfrac{1}{4}e^{2x} + \tfrac{1}{2} - \tfrac{1}{4}e^{-2x},$$

$$\therefore \cosh^2 x - \sinh^2 x = 1.$$

DEFINITIONS: *the hyperbolic tangent, cotangent, secant, cosecant are defined as follows:*

$$\tanh x = \frac{\sinh x}{\cosh x}, \qquad \coth x = \frac{1}{\tanh x},$$
$$\operatorname{sech} x = \frac{1}{\cosh x}, \qquad \operatorname{cosech} x = \frac{1}{\sinh x}.$$

Exercise 20a

Most of the following properties of the hyperbolic functions should be deduced from the definitions. Work all of Nos. 1–14.

1. From a sketch of $\cosh x$ and $\sinh x$ referred to the same axes, sketch the graph of $\tanh x$.

2. Prove that: (i) $\cosh(-x) = \cosh x$, (ii) $\sinh(-x) = -\sinh x$.

3. Prove that $\cosh x > \sinh x$. [Show that $\cosh x - \sinh x > 0$.] Prove also that, when $x < 0$, $\cosh x > |\sinh x|$. Deduce the values between which $\tanh x$ lies.

4. From sketches of $\cosh x$ and $\sinh x$, sketch the graphs of $\operatorname{sech} x$ and $\operatorname{cosech} x$.

5. Prove that $\cosh x > 1$. [See hint in No. 3.]

6. Prove that
$$\cosh x + \sinh x = e^x,$$
$$\cosh x - \sinh x = e^{-x}.$$

Deduce the identity
$$\cosh^2 x - \sinh^2 x = 1.$$

7. Prove that the point $(a \cosh t, \ b \sinh t)$ lies on one branch of the hyperbola
$$\frac{x^2}{a^2} - \frac{y^2}{b^2} = 1.$$

[Hence the name hyperbolic functions. Use the result of No. 6.]

8. Prove that $\sinh 2x = 2 \sinh x \cosh x$.

9. Prove that
$$\cosh 2x = \cosh^2 x + \sinh^2 x,$$
$$= 2 \cosh^2 x - 1,$$
$$= 1 + 2 \sinh^2 x.$$

10. Use the results of Nos. 8, 9 to show that
$$\tanh 2x = 2 \tanh x / (1 + \tanh^2 x).$$

11. Prove that
$$\operatorname{sech}^2 x = 1 - \tanh^2 x,$$
$$\operatorname{cosech}^2 x = 1 - \coth^2 x.$$

[Use the identity connecting $\cosh x$ and $\sinh x$.]

12. Prove that
$$\cosh(A + B) = \cosh A \cosh B + \sinh A \sinh B.$$

Deduce a similar expression for $\cosh(A - B)$.

13. Prove that
$$\sinh(A + B) = \sinh A \cosh B + \cosh A \sinh B.$$

Deduce a similar expression for $\sinh(A - B)$.

14. Use the results of Nos. 12, 13 to find expressions for $\tanh(A+B)$, $\tanh(A-B)$ in terms of $\tanh A$, $\tanh B$.

15. Solve the equation $8 \cosh x + 17 \sinh x = 20$.

16. Find the condition that the equation $a \cosh x + b \sinh x = c$ should have equal roots.

17. If $a > b > 0$, prove that

$$b < \frac{a\,e^z + b\,e^{-z}}{e^z + e^{-z}} < a.$$

18. If $|a| < |b|$, prove that the equation $a \cosh x + b \sinh x = 0$ has one and only one root.

19. Prove that

$$\sinh 3\theta = 3 \sinh \theta + 4 \sinh^3 \theta.$$

20. Prove that

$$\cosh^2 x \sin^2 x - \sinh^2 x \cos^2 x = \tfrac{1}{2}(1 - \cosh 2x \cos 2x).$$

Osborn's rule

20.2. The reader will have noticed a striking similarity between the identities connecting hyperbolic functions and those connecting the corresponding circular functions. In fact the standard identities are in the same form except that certain signs are changed. Osborn's rule provides a simple way of remembering these changes of signs. The rule is to change the sign of any term containing the *square* of a sine (or cosecant, tangent, or cotangent, because these all include a sine by implication: $\operatorname{cosec} x = 1/\sin x$, $\tan x = (\sin x)/(\cos x)$, $\cot x = (\cos x)/(\sin x)$.) For instance,

$$\sin 2x = 2 \sin x \cos x, \qquad \sinh 2x = 2 \sinh x \cosh x.$$
$$\cos 2x = \cos^2 x - \sin^2 x, \qquad \cosh 2x = \cosh^2 x + \sinh^2 x.$$
$$\tan 2x = \frac{2 \tan x}{1 - \tan^2 x}, \qquad \tanh 2x = \frac{2 \tanh x}{1 + \tanh^2 x}.$$

Qu. 1. Write down the identities connecting hyperbolic functions corresponding to:

    (i) $\sin A + \sin B = 2 \sin \tfrac{1}{2}(A+B) \cos \tfrac{1}{2}(A-B)$,

    (ii) $\cos A + \cos B = 2 \cos \tfrac{1}{2}(A+B) \cos \tfrac{1}{2}(A-B)$,

    (iii) $\cos A - \cos B = -2 \sin \tfrac{1}{2}(A+B) \sin \tfrac{1}{2}(A-B)$,

    (iv) $\sec^2 \theta = 1 + \tan^2 \theta$,   (v) $\operatorname{cosec}^2 \theta = 1 + \cot^2 \theta$,

    (vi) $\cos 3\theta = 4 \cos^3 \theta - 3 \cos \theta$,

    (vii) $\tan 3\theta = (3 \tan \theta - \tan^3 \theta)/(1 - 3 \tan^2 \theta)$.

WARNING: Osborn's rule holds for the standard trigonometrical identities but it must not be applied indiscriminately; for instance, application to

$$\frac{\cos 2A}{\cos A + \sin A} = \cos A - \sin A$$

and

$$\frac{\cos A}{\sin B} - \frac{\sin A}{\cos B} = \frac{2 \cos (A+B)}{\sin 2B}$$

leads to incorrect results.   Further it *cannot* be relied upon as an aid to remembering calculus formulae.   It is to these that we now turn.

Derivatives of hyperbolic functions

20.3.   The derivatives of cosh x and sinh x are most easily obtained by starting from the definitions of these functions.

$$\frac{d}{dx}(\cosh x) = \frac{d}{dx}\{\tfrac{1}{2}(e^x + e^{-x})\},$$

$$= \tfrac{1}{2}(e^x - e^{-x}).$$

$$\therefore \frac{d}{dx}(\cosh x) = \sinh x.$$

Similarly, $\dfrac{d}{dx}(\sinh x) = \cosh x.$

The derivatives of the other hyperbolic functions are easily obtained by first expressing them in terms of cosh x, sinh x.

Qu. 2.   Remembering that

$$\frac{dy}{dx} = \frac{dy}{dt} \cdot \frac{dt}{dx},$$

write down the derivatives of:

(i)   cosh $2x$,   (ii)   sinh $\tfrac{1}{2}x$,   (iii)   3 cosh $\tfrac{1}{3}x$,

(iv) $\tfrac{1}{4}$ sinh $4x$,   (v)   sinh$^2 x$,   (vi)   cosh$^3 2x$.

Inverse hyperbolic functions

20.4.   In the following exercise use will be made of the inverse functions cosh$^{-1} x$, sinh$^{-1} x$, tanh$^{-1} x$. The reader may remember how in § 12.4, p. 254, we distinguished between Cos$^{-1} x$, a many-valued expression, and cos$^{-1} x$, which is uniquely defined. A similar convention will be used here.

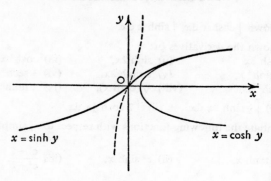

$x = \sinh y$

$x = \cosh y$

Fig. 20.2

Figure 20.2 shows the graphs of $x = \cosh y$, $x = \sinh y$, $x = \tanh y$. (Compare Fig. 20.1, p. 437, and note how x, y have been interchanged in the equations. Once again the graphs are so close that they have run together when sketched to this scale.) This shows that:

if $x \geqslant 1$, there are two values of y (coincident if $x = 1$) such that $x = \cosh y$,

for any value of x, there is one value of y such that $x = \sinh y$,

and, if $-1 < x < 1$, there is one value of y such that $x = \tanh y$.

Hence we define the inverse hyperbolic functions as follows:

if $\cosh y = x$ $(x \geqslant 1)$, then $y = \cosh^{-1} x$, where $\cosh^{-1} x$ is positive or zero, while $\mathrm{Cosh}^{-1} x$ may be positive, negative or zero;

if $\sinh y = x$, then $y = \sinh^{-1} x$;

if $\tanh y = x$ $(-1 < x < 1)$, then $y = \tanh^{-1} x$.

Exercise 20b

1. Prove the following results by first expressing the functions concerned in terms of $\cosh x$ and $\sinh x$:

(i) $\dfrac{\mathrm{d}}{\mathrm{d}x} (\tanh x) = \mathrm{sech}^2 x$,

(ii) $\dfrac{\mathrm{d}}{\mathrm{d}x} (\coth x) = -\mathrm{cosech}^2 x$,

(iii) $\dfrac{\mathrm{d}}{\mathrm{d}x} (\mathrm{sech}\ x) = -\mathrm{sech}\ x \tanh x$,

(iv) $\dfrac{\mathrm{d}}{\mathrm{d}x} (\mathrm{cosech}\ x) = -\mathrm{cosech}\ x \coth x$.

2. Write down $\int \cosh x \, dx$, $\int \sinh x \, dx$.

3. Write down the derivatives of:
   (i) $\cosh 3x$,   (ii) $\sinh 2x$,   (iii) $\cosh^2 x$,
   (iv) $2 \sinh^3 x$,   (v) $3 \tanh 2x$,   (vi) $\frac{1}{2} \operatorname{sech}^2 x$,
   (vii) $\sinh^2 3x$,   (viii) $\sqrt{(\coth x)}$, (ix) $2 \tanh^2 \frac{1}{2}x$.

4. Find: (i) $\int \frac{1}{2} \sinh 3x \, dx$,   (ii) $\int 2 \cosh \frac{1}{3}x \, dx$.

5. Differentiate the following functions with respect to x, simplifying your answers:
   (i) $\log_e \tanh x$,   (ii) $e^x \sinh x$,   (iii) $\dfrac{e^x - 1}{e^x + 1}$.

6. Find:
   (i) $\int \operatorname{sech}^2 2x \, dx$,   (ii) $\int \dfrac{\sinh x}{\cosh^2 x} \, dx$.

7. Find the minimum value of $5 \cosh x + 3 \sinh x$.

8. Prove that
   (i) $\dfrac{d}{dx}(\cosh^{-1} x) = \dfrac{1}{\sqrt{(x^2 - 1)}}$,   (ii) $\dfrac{d}{dx}(\tanh^{-1} x) = \dfrac{1}{1 - x^2}$,

   and find an expression for $\dfrac{d}{dx}(\sinh^{-1} x)$.

   [HINT: see the method of p. 234.]

9. Prove that $\dfrac{d}{dx}\{\tan^{-1}(e^x)\} = \frac{1}{2} \operatorname{sech} x$.

10. Find $\dfrac{d}{dx}\left[\log_e\{x + \sqrt{(1 + x^2)}\} - \sinh^{-1} x\right]$.

11. Find: (i) $\int \cosh 2x \sinh 3x \, dx$,   (ii) $\int \cosh x \cosh 3x \, dx$.

12. Find the distance from the y-axis of the centroid of the area formed by $y = \sinh x$, $x - 1 = 0$ and the x-axis.

13. Find the equations of the tangent and normal to the hyperbola
$$b^2 x^2 - a^2 y^2 = a^2 b^2$$
at the point $(a \cosh \theta, b \sinh \theta)$. If the tangent meets the y-axis at T and the normal meets the x-axis at N, find the locus of the mid-point of NT.

14. If $y = A \cosh 2x + B \sinh 3x$, find an equation connecting $\dfrac{d^2 y}{dx^2}$, $\dfrac{dy}{dx}$, y which does not contain A, B.

15. Prove that
$$\cosh x = 1 + \frac{x^2}{2!} + \frac{x^4}{4!} + \ldots + \frac{x^{2n}}{(2n)!} + \ldots,$$
$$\sinh x = x + \frac{x^3}{3!} + \frac{x^5}{5!} + \ldots + \frac{x^{2n+1}}{(2n+1)!} + \ldots,$$

and obtain the first three non-zero terms of the expansion in ascending powers of x of tanh x.

16. Expand $\tanh^{-1} x$ as a series of ascending powers of x.   Express the sum of this series as a logarithm.

17. If $\tanh^{-1} x = y$, show that $x = (e^{2y} - 1)/(e^{2y} + 1)$.   Hence express y as a logarithm.

18. Investigate the stationary values of $\cosh 3x - 12 \cosh x$.

Integration

20.5.   As was mentioned at the beginning of this chapter, the chief purpose in including hyperbolic functions in this book is to extend the reader's powers of integration.   So far, we have integrated functions in the forms

$$\frac{1}{\sqrt{(a^2 - x^2)}}, \qquad \frac{1}{x^2 + a^2}.$$

but not those in the forms

$$\frac{1}{\sqrt{(x^2 - a^2)}}, \qquad \frac{1}{\sqrt{(x^2 + a^2)}}.$$

We have seen (p. 12) that the identity

$$\cos^2 \theta + \sin^2 \theta = 1$$

helps us to eliminate the square root sign in $1/\sqrt{(a^2 - x^2)}$ and we may expect the identity

$$\cosh^2 \theta - \sinh^2 \theta = 1$$

to assist us with corresponding integrals.

Qu. 3.   What substitution using hyperbolic functions would eliminate the square root sign in the following?

(i) $\int \dfrac{1}{\sqrt{(x^2 - a^2)}} \, dx,$ (ii) $\int \dfrac{1}{\sqrt{(x^2 + a^2)}} \, dx.$

★Qu. 4.   Show that

(i) $\int \dfrac{1}{\sqrt{(a^2 - x^2)}} \, dx = \sin^{-1} \dfrac{x}{a} + c,$ $\int \dfrac{1}{\sqrt{(a^2 + x^2)}} \, dx = \sinh^{-1} \dfrac{x}{a} + c,$

(ii) $\int \dfrac{1}{\sqrt{(x^2 - a^2)}} \, dx = \cosh^{-1} \dfrac{x}{a} + c.$

There is little difficulty in remembering these results and confusion between them is easily avoided by thinking, 'What substitution would eliminate the square root.'   In fact, by taking this thought and doing some side-work, the reader can dispense with memorizing these as formulae. Those who do memorize them should note that the results in Qu. 6 are slightly and inconveniently different.

Qu. 5. What substitution would enable us to write the denominator of the following integrand as a square?

(i) $\int \dfrac{1}{a^2+x^2}\,dx,$ (ii) $\int \dfrac{1}{a^2-x^2}\,dx.$

⋆**Qu. 6.** Show that

$$\int \frac{1}{a^2+x^2}\,dx = \frac{1}{a}\tan^{-1}\frac{x}{a}+c, \qquad \int \frac{1}{a^2-x^2}\,dx = \frac{1}{a}\tanh^{-1}\frac{x}{a}+c.$$

The last two questions between them raise a number of points of interest:

(1) Those who like to memorize formulae should note that these results contain a factor $1/a$ in front of the $\tan^{-1}(x/a)$ and $\tanh^{-1}(x/a)$. See also Qu. 7.

(2) $\int \dfrac{1}{a^2-x^2}\,dx$ is usually found by first expressing $\dfrac{1}{a^2-x^2}$ in partial fractions. See Qu. 8.

(3) $x=a\tan\theta$, $x=a\tanh\theta$ are not the only possible substitutions by which to express a^2+x^2, a^2-x^2 respectively as squares, but if the reader wonders why these particular substitutions have been favoured, he can easily try the others for himself!

Qu. 7. If x, a have dimensions of length L, write down the dimensions of both sides of the formulae below. $\left[\text{Since } \int y\,dx \text{ is the limit of } \sum y\varDelta x \text{ as}\right.$ $\varDelta x\to 0$, we take dx to have the same dimensions as $\varDelta x$, i.e. L.$\Big]$

(i) $\int \dfrac{1}{\sqrt{(a^2-x^2)}}\,dx = \sin^{-1}\dfrac{x}{a}+c,$ (ii) $\int \dfrac{1}{a^2+x^2}\,dx = \dfrac{1}{a}\tan^{-1}\dfrac{x}{a}+c.$

Qu. 8. Use partial fractions to show that

$$\int \frac{1}{a^2-x^2}\,dx = \frac{1}{2a}\log_e \frac{a+x}{a-x}+k.$$

What conclusion can be drawn from a comparison of this result and the second result in Qu. 6?

⋆**Qu. 9.** (A repeat of Exercise 20b, No. 17.) If $\tanh^{-1}x=y$, show that $x=(e^{2y}-1)/(e^{2y}+1)$. Hence prove that

$$\tanh^{-1}x = \tfrac{1}{2}\log_e \frac{1+x}{1-x}.$$

The result of Qu. 9 suggests that it may be possible to express $\cosh^{-1} x$, $\sinh^{-1} x$ in the terms of logarithms. We begin with $\sinh^{-1} x$ because (see Fig. 20.1 (ii)) there is only one value of the function $\sinh y$ for any given value of y.

Let $$y = \sinh^{-1} x,$$
then by the definition of the function (see p. 441),

$$\sinh y = x. \tag{1}$$

But $$\cosh^2 y = 1 + \sinh^2 y,$$
$$\therefore \cosh y = \sqrt{(1 + x^2)}. \tag{2}$$

[$\cosh y > 0$, so the negative square root does not give a real value of y.]

Now $$\cosh y + \sinh y = \tfrac{1}{2}(e^y + e^{-y}) + \tfrac{1}{2}(e^y - e^{-y}),$$
$$= e^y.$$

But from (1), (2),

$$\cosh y + \sinh y = x + \sqrt{(1 + x^2)}.$$
$$\therefore e^y = x + \sqrt{(1 + x^2)}.$$
$$\therefore y = \log_e \{x + \sqrt{(1 + x^2)}\}.$$

That is $$\sinh^{-1} x = \log_e \{x + \sqrt{(1 + x^2)}\}.$$

An expression for $\cosh^{-1} x$ may be obtained in a similar manner.

Let $$y = \cosh^{-1} x,$$
then $$\cosh y = x.$$

$$\sinh^2 y = \cosh^2 y - 1.$$
$$\therefore \sinh y = \pm \sqrt{(x^2 - 1)}.$$

Now $$e^y = \cosh y + \sinh y = x \pm \sqrt{(x^2 - 1)}.$$
$$\therefore \cosh^{-1} x = y = \log_e \{x \pm \sqrt{(x^2 - 1)}\}.$$

But from Fig. 20.2, p. 441, it appears that the two values of $\cosh^{-1} x$ obtained above are equal and opposite, i.e.

$$\log_e \{x + \sqrt{(x^2 - 1)}\} + \log_e \{x - \sqrt{(x^2 - 1)}\} = 0.$$

To prove this, write the left-hand side as

$$\log_e \{x^2 - (x^2 - 1)\} = \log_e 1 = 0.$$

Therefore we may write

$$\cosh^{-1} x = \pm \log_e \{x + \sqrt{(x^2 - 1)}\}.$$

and the principal value of the function

$$\cosh^{-1} x = \log_e \{x + \sqrt{(x^2 - 1)}\}.$$

Qu. 10.   Find, as accurately as your tables allow, the values of: (i) $\sinh^{-1} 1$, (ii) $\cosh^{-1} 2$, (iii) $\sinh^{-1} 0.58$.

Once the reader has grasped the forms which require the substitution of a hyperbolic function, the integrations in Exercise 20c should present no new difficulty. Only in exceptional cases as, for instance, in Example 5, is the treatment of hyperbolic functions completely different from the treatment of circular functions. If the reader is unable to integrate any particular function in Exercise 20c, he should refer back to Chapters 1 and 13 for help. The following examples illustrate how a knowledge of integrating with circular functions helps with the present work.

Example 2.   *Find* $\displaystyle\int \frac{1}{\sqrt{(x^2+2x+10)}}\ dx.$

First complete the square:

$$x^2+2x+10 = (x+1)^2+9.$$

[The substitution $x+1 = 3\sinh\theta$ makes $(x+1)^2+9 = 9\cosh^2\theta.$]

$$\int \frac{1}{\sqrt{(x^2+2x+10)}}\ dx = \int \frac{1}{\sqrt{(x^2+2x+10)}}\ \frac{dx}{d\theta}\ d\theta,$$

$$= \int \frac{1}{3\cosh\theta}\ 3\cosh\theta\ d\theta, \qquad \text{Let } x+1 = 3\sinh\theta.$$
$$\frac{dx}{d\theta} = 3\cosh\theta.$$

$$= \int 1\ d\theta = \theta + c.$$

$$\therefore \int \frac{1}{\sqrt{(x^2+2x+10)}}\ dx = \sinh^{-1}\frac{x+1}{3}+c.$$

Example 3.   *Evaluate* $\displaystyle\int_2^3 \cosh^{-1} x\ dx.$

$$\left[\int \cos^{-1} x\ dx \text{ we integrate by parts as } \int 1.\cos^{-1} x\ dx.\right]$$

$$\int_2^3 1.\cosh^{-1} x\ dx = \left[\ x\cosh^{-1} x\ \right]_2^3 - \int_2^3 x\,\frac{1}{\sqrt{(x^2-1)}}\ dx.$$

$$= 3\cosh^{-1} 3 - 2\cosh^{-1} 2 - \left[\ \sqrt{(x^2-1)}\ \right]_2^3.$$

$$3 \cosh^{-1} 3 = \quad 3 \log_e (3 + \sqrt{8}) = \quad 3 \log_e 5 \cdot 828 = \quad 5 \cdot 288$$
$$2 \cosh^{-1} 2 = -2 \log_e (2 + \sqrt{3}) = -2 \log_e 3 \cdot 732 = -2 \cdot 634$$
$$-\sqrt{8} = -2 \cdot 828$$
$$+\sqrt{3} = \quad 1 \cdot 732$$
$$\overline{ 1 \cdot 558}$$

$$\therefore \int_2^3 \cosh^{-1} x \, dx = 1 \cdot 56, \quad \text{to 3 sig. fig.}$$

Example 4. *Find* $\int \sinh^3 \theta \, d\theta$.

$$\int \sinh^3 \theta \, d\theta = \int (\cosh^2 \theta - 1) \sinh \theta \, d\theta,$$
$$= \tfrac{1}{3} \cosh^3 \theta - \cosh \theta + c.$$

Qu. 11. Find $\int \sinh^3 \theta \, d\theta$ by means of the identity analogous to

$$\sin 3A = 3 \sin A - 4 \sin^3 A.$$

Example 5. *Integrate* sech x *with respect to* x.

[The method of p. 274 cannot be used here because $\cosh x$ is *not* $\sinh (x + \tfrac{1}{2}\pi)$. We might guess an integral but we can go back to the definition of $\cosh x$.]

$$\int \text{sech } x \, dx = \int \frac{2}{e^x + e^{-x}} \, dx, \qquad \text{Let } e^x = \tan \theta.$$
$$= \int \frac{2e^x}{e^{2x} + 1} \frac{dx}{d\theta} \, d\theta, \qquad e^x \frac{dx}{d\theta} = \sec^2 \theta.$$
$$= \int \frac{2 \sec^2 \theta}{\sec^2 \theta} \, d\theta, \qquad e^{2x} + 1 = \tan^2 \theta + 1$$
$$= 2\theta + c. \qquad\qquad = \sec^2 \theta.$$

$$\therefore \int \text{sech } x \, dx = 2 \tan^{-1} (e^x) + c.$$

Exercise 20c

Integrate with respect to x:

1. $\dfrac{1}{\sqrt{(x^2 + 9)}}$.

2. $\dfrac{3}{\sqrt{\{1 - (x - 2)^2\}}}$.

3. $\dfrac{2}{\sqrt{(4x^2 - 1)}}$.

4. $\dfrac{1}{\sqrt{(4x + x^2)}}$.

5. $\dfrac{1}{x^2 + x + 1}$.

6. $\dfrac{1}{\sqrt{(x^2 - 6x + 10)}}$.

7. $\dfrac{1}{\sqrt{(4x^2 + x)}}$.

8. $\dfrac{1}{\sqrt{(3x - 4x^2)}}$.

Evaluate:

9. $\displaystyle\int_0^1 \dfrac{1}{\sqrt{(x^2 + 4)}} \, dx$.

10. $\displaystyle\int_{1\frac{1}{2}}^2 \dfrac{2}{\sqrt{(4x^2 - 9)}} \, dx$.

11. $\displaystyle\int_1^2 \dfrac{1}{\sqrt{(x^2 + 4x + 5)}} \, dx$.

12. $\displaystyle\int_0^{\sqrt{3}} \dfrac{1}{\sqrt{(2x^2 + 3)}} \, dx$.

Integrate with respect to x:

13. $\cosh^2 x$.

14. $\cosh^3 x$.

15. $\sinh^4 x$.

16. $\tanh^2 x$.

17. $\tanh x$.

18. $\coth^3 x$.

19. $\tanh^4 x$.

20. $\operatorname{cosech} x$.

21. $\sinh^2 x \cosh^3 x$.

22. $\operatorname{sech}^3 x \tanh x$.

23. $2 \sinh 3x \cosh 5x$.

24. $\cosh 3x \cosh x$.

25. $\sinh^{-1} x$.

26. $\tanh^{-1} x$.

27. $x \cosh 2x$.

28. $x^2 \sinh x$.

29. $e^x \cosh x$.

30. $\dfrac{1}{x^2} \sinh \dfrac{1}{x}$.

31. $\dfrac{\sinh x + 2 \cosh x}{\sinh x + \cosh x}$.

32. $\dfrac{\cosh x}{3 \sinh x - 2 \cosh x}$.

33. $\dfrac{x + 1}{\sqrt{(x^2 - 1)}}$.

34. $\dfrac{x - 1}{\sqrt{(x^2 + 1)}}$.

35. $\sqrt{\dfrac{x - 1}{x + 1}}$.

36. $\sqrt{(x^2 - 4)}$.

Evaluate:

37. $\displaystyle\int_0^1 \sqrt{(x^2 + 1)} \, dx$.

38. $\displaystyle\int_{\frac{1}{2}}^1 \dfrac{x - 1}{\sqrt{(4x^2 - 1)}} \, dx$.

39. $\displaystyle\int_3^4 x \cosh x \, dx$.

40. $\displaystyle\int_0^1 \dfrac{2x - 3}{x^2 + 1} \, dx$.

41. If $a > 0$ and the equation $ax^2 + bx + c = 0$ has two real roots, find

$$\int \frac{1}{\sqrt{(ax^2 + bx + c)}} \, dx.$$

42. If $\cosh y = x$, use the definition of the hyperbolic cosine to find a quadratic equation for e^y and hence obtain an expression for $\cosh^{-1} x$.

Also show by a similar method that

$$\sinh^{-1} x = \log_e\{x + \sqrt{(1 + x^2)}\}.$$

43. Find

$$\int \sqrt{(x^2 + a^2)} \, dx \quad \text{and} \quad \int \sqrt{(x^2 - a^2)} \, dx.$$

Hence obtain

$$\int \sqrt{(x^2 + 2px + q)}\, dx,$$

distinguishing between the different cases.

Exercise 20d (Miscellaneous)

1. Define the functions $\cosh x$ and $\sinh x$. From your definitions prove that:
   (i) $\sinh(x+y) = \sinh x \cosh y + \cosh x \sinh y$.
   (ii) $\cosh 2x = 2\cosh^2 x - 1$.

2. Solve the equations:
   (i) $2\cosh x + \sinh x = 2$,
   (ii) $2\sinh^2 x + 8 = 7\cosh x$.

3. The position of a particle at time t is given by $x = a\cosh t$, $y = b\sinh t$. Show that the acceleration is proportional to the distance of the particle from a fixed point and is in the direction of the line from the fixed point to the particle. Find also the least speed of the particle.

4. If $y = A\cosh px + B\sinh qx$, where A, B, p, q are constants, find the simplest differential equation satisfied by y which does not contain A and B. [Hint: differentiate four times.]

5. (i) Differentiate with respect to x:
   $$\text{(a) } \tanh^{-1}(2x), \qquad \text{(b) } \cosh^{-1}\sqrt{(x^2+1)}.$$
   (ii) If $x = a\cosh\theta$, $y = b\tanh\theta$, find $\dfrac{d^2y}{dx^2}$ in terms of a, b, θ.

6. Find the area between the rectangular hyperbola $x^2 - y^2 = a^2$ and the latus rectum $x = \sqrt{2}a$.

7. Through the point $(a\cosh\theta,\ b\sinh\theta)$ on the hyperbola
   $$b^2x^2 - a^2y^2 = a^2b^2,$$
   a chord is drawn parallel to the minor axis. Find the area of the segment so formed and the position of its centroid.

8. A tangent is drawn from the origin to touch the curve $y = \cosh x$ in the first quadrant. Find the coordinates of the point of contact correct to three significant figures.

9. An arc AB of the hyperbola $b^2x^2 - a^2y^2 = a^2b^2$ is cut off by $x = 2a$. Perpendiculars from A, B meet the y-axis at D, C. The area bounded by the arc AB and the lines BC, CD, DA is rotated once about the y-axis. Find the volume of the solid generated.

10. (i) Expand in a series of ascending powers of x and give the term in x^r:
    $$\text{(a) } \sinh 2x, \qquad \text{(b) } \sinh^2 x.$$

(ii) Write down the series for cosh x in ascending powers of x and deduce the series for sech x as far as the term in x^4.

11. Find the sum to infinity of the series:

(i) $\dfrac{x}{3!} + \dfrac{x^2}{5!} + \ldots + \dfrac{x^{2n-1}}{(2n+1)!} + \ldots,$

(ii) $\dfrac{4}{2!} + \dfrac{16}{4!} + \dfrac{64}{6!} + \ldots.$

12. Define the hyperbolic functions sinh x, cosh x and tanh x, and show that

$$\sinh^{-1} x = \log_e \{x + \surd(1+x^2)\}.$$

If $\tan \tfrac{1}{2}x = \tan \alpha \tanh \beta$, prove that

$$\tan x = \frac{\sin 2\alpha \sinh 2\beta}{1 + \cos 2\alpha \cosh 2\beta},$$

and that

$$\sin x = \frac{\sin 2\alpha \sinh 2\beta}{\cos 2\alpha + \cosh 2\beta}.$$

                                                                 L.

13. Sketch the curve $y^2(a^2 + x^2) = a^4$. Find (i) the area bounded by the y-axis, the curve and $x = a$, (ii) the position of the centroid of this area, (iii) the volume of the solid generated when this area is rotated about the x-axis.

14. Integrate with respect to x:

(i) $\sinh x \operatorname{sech}^3 x$,   (ii) $\dfrac{1}{\cosh^2 x\,(1 + \tanh x)}$,   (iii) $\dfrac{x+a}{\surd(x^2 + a^2)}$.

15. Integrate with respect to x:

(i) $\dfrac{1}{x^2 - x + 1}$,   (ii) $\cosh^4 x$

and evaluate:

(iii) $\displaystyle\int_2^6 \frac{1}{\surd(4x + x^2)}\,\mathrm{d}x$,   (iv) $\displaystyle\int_0^1 x \cosh x \,\mathrm{d}x.$

16. Find:

(i) $\displaystyle\int \surd(x^2 - a^2)\,\mathrm{d}x$,   (ii) $\displaystyle\int \frac{x^2}{\surd(x^2 - a^2)}\,\mathrm{d}x$,

(iii) $\displaystyle\int x^2 \surd(x^2 - a^2)\,\mathrm{d}x$.

17. If a, b, c are constants such that $b^2 \neq 4ac$, determine constants P, Q, R such that

$$\frac{\mathrm{d}}{\mathrm{d}x}\left(\frac{Px + Q}{ax^2 + bx + c}\right) = \frac{1}{(ax^2 + bx + c)^2} - \frac{R}{ax^2 + bx + c}.$$

Evaluate the integrals

$$\int_0^1 \frac{1}{(x^2 + 4x + 1)^2}\,\mathrm{d}x \quad \text{and} \quad \int_0^1 \frac{1}{(x^2 + 4x + 4)^2}\,\mathrm{d}x. \qquad \textbf{L.}$$

18. Use the identity
$$2 \sinh A \sinh B = \cosh (A+B) - \cosh (A-B)$$
to sum to n terms the series
$$\sinh x + \sinh (x+y) + \sinh (x+2y) + \ldots.$$
Also find the sum to n terms of the series
$$\cosh x + \cosh (x+y) + \cosh (x+2y) + \ldots.$$

19. Find the value of $\int_0^\infty \operatorname{sech} \theta \, \mathrm{d}\theta$.

Show that if $I_m = \int_0^\infty \operatorname{sech}^n \theta \, \mathrm{d}\theta$,
$$(m-1)I_m = (m-2)I_{m-2}.$$
Hence find the value of $\int_0^\infty \operatorname{sech}^7 \theta \, \mathrm{d}\theta$.

20. Show that if $I_n = \int \cosh^n x \, \mathrm{d}x$,
$$nI_n = \cosh^{n-1} x \sinh x + (n-1)I_{n-2}.$$
Hence find $\int \cosh^6 x \, \mathrm{d}x$.

21. (i) Show that the general solution of the differential equation
$$\frac{\mathrm{d}^2 y}{\mathrm{d}x^2} - n^2 y = 0$$
may be written
$$y = A \cosh nx + B \sinh nx.$$

(ii) A chain hanging under gravity lies approximately in a curve (called a catenary) which satisfies the differential equation
$$c \frac{\mathrm{d}^2 y}{\mathrm{d}x^2} = \sqrt{\left\{ 1 + \left(\frac{\mathrm{d}y}{\mathrm{d}x} \right)^2 \right\}}.$$
Show that the equation of the curve may be written
$$y + B = c \cosh \left(\frac{x+A}{c} \right).$$

22. Show that, if $|x| < 1$,
$$\sinh^{-1} x = x - \frac{x^3}{6} + \ldots + (-1)^r \frac{1.3 \ldots (2r-1)}{(2r+1)2^r \, r!} x^{2r+1} + \ldots.$$

CHAPTER 21

SOME GEOMETRICAL APPLICATIONS OF CALCULUS†

Area of a sector

21.1. A good slogan for a reader who finds the formulae in this chapter difficult to remember, or prefers to work them out for himself, is, 'When in doubt, differentiate'. By differentiation we mean, in this instance, take a small increment—which is the fundamental step in differentiating a new function. With this approach we now work out an expression for areas of sectors and closed curves in polar coordinates.

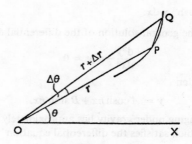

Fig. 21.1

In Fig. 21.1, the radius vectors OP, OQ are r, $r+\Delta r$; the angles between them and the fixed line OX are θ, $\theta+\Delta\theta$. If $\Delta\theta$ is small, the area of sector OPQ is approximately equal to the area of triangle OPQ.

$$\therefore \text{ sector OPQ} \simeq \tfrac{1}{2}r(r+\Delta r)\sin\Delta\theta.$$

But

$$\sin\Delta\theta = \Delta\theta - \frac{(\Delta\theta)^3}{3!} + \dots,$$

† The reader should work all the Questions in the text.

and $\Delta r \sin \Delta \theta$ is small compared with $\Delta \theta$ so, correct to the term in $\Delta \theta$,

$$\text{sector OPQ} = \tfrac{1}{2} r^2 \, \Delta \theta.$$

(Here it is assumed that the difference between the sector OPQ and triangle OPQ is small compared to $\Delta \theta$.)

Summing for all the elements in the sector concerned and proceeding to the limit,

area of sector $= \displaystyle\int_{\alpha}^{\beta} \tfrac{1}{2} r^2 \, \mathbf{d\theta},$

where α, β are the values of θ corresponding to the bounding radius vectors of the sector.

Qu. 1. Sketch the curve given by $r = a$. What does the integral $\displaystyle\int_{\alpha}^{\beta} \tfrac{1}{2} r^2 \, \mathrm{d}\theta$ represent in this case?

Qu. 2. Sketch the cardioid $r = a(1 + \cos \theta)$ and find the area enclosed by it.

Qu. 3. Find the area swept out by the radius vector of the equiangular spiral $r = a \, e^{k\theta}$ as θ increases from $-\pi$ to π. Show this area on a sketch.

Qu. 4. Sketch the trefoil $r = a \sin 3\theta$ and find the area of one of its loops.

Qu. 5. Regarding the limacon $r = 1 + 2 \cos \theta$ as having a small loop contained within a larger one, find the area of the larger loop.

★**Qu. 6.** If x and y are functions of a parameter t, show that

$$\int \tfrac{1}{2} r^2 \, \mathrm{d}\theta = \int \tfrac{1}{2} \left(x \frac{\mathrm{d}y}{\mathrm{d}t} - y \frac{\mathrm{d}x}{\mathrm{d}t} \right) \mathrm{d}t.$$

★**Qu. 7.** The vertices of a triangle are

$$O(0, 0), \quad P(x, y), \quad Q(x + \Delta x, y + \Delta y).$$

Show that the area of the triangle is $\tfrac{1}{2}(x \, \Delta y - y \, \Delta x)$. Hence show that the area of a sector may be found from the expression

$$\int_{t_1}^{t_2} \tfrac{1}{2} \left(x \frac{\mathrm{d}y}{\mathrm{d}t} - y \frac{\mathrm{d}x}{\mathrm{d}t} \right) \mathrm{d}t,$$

where t_1, t_2 are the values of t corresponding to the bounding radius vectors of the sector.

Qu. 8. Find the area enclosed by the ellipse $x = a \cos \theta$, $y = b \sin \theta$.

Qu. 9. Find the area enclosed by the loop of the curve given by $x = t^2 - 4$, $y = t^3 - 4t$.

Qu. 10. Find the area of one loop of the curve given by $x = \sin \theta$, $y = \sin 2\theta$. Why does the formulae of give Qu. 7 a negative answer?

Qu. 11. Find the area between the cycloid $x = a(\theta - \sin \theta)$, $y = a(1 - \cos \theta)$ and the portion of the x-axis between the points determined by $\theta = 0$ and $\theta = 2\pi$.

Length of a curve

21.2. The reader may be aware of a gap in his knowledge about curves: that he does not even know how to find the length of an arc of a curve given by such a simple equation as $y = x^2$. An approximation could be found by marking a number of points on the arc and finding the sum of the lengths of the chords. We should *expect* this sum to approach the length of the curve as we increased the number of chords and decreased their lengths.

At this stage it is advisable to pause to remark that we have only an intuitive idea of the length of a curve, based on our experience of string, measuring tapes and other flexible material which can be measured or graduated when it is placed against a ruler. Another thing to note is that we have no proof that the limit of the sum of the lengths of the chords would be the same if the points were marked off on the arc in different ways. The reader may find such questions discussed if he goes on to study higher mathematics.

Our object is to find expressions for s, the arc length of a curve, in terms of the coordinates in which the equations of curves are commonly written. The most important of these, for our present purposes, are the Cartesian coordinates x, y.

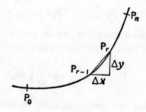

Fig. 21.2

Suppose an arc of a curve is divided into n parts by points $P_0, P_1, \ldots, P_n$. (See Fig. 21.2.) We shall assume that the

sum of the lengths of chords → length of arc

as the lengths of the chords → 0. If $\varDelta x$, $\varDelta y$ are the increments in x, y from P_{r-1} to P_r,

$$P_{r-1}P_r{}^2 = (\varDelta x)^2 + (\varDelta y)^2. \tag{1}$$

For an equation in the form $y = f(x)$, it will be convenient to work in terms of x. We therefore rewrite (1) as

$$P_{r-1}P_r{}^2 = \left\{1 + \left(\frac{\varDelta y}{\varDelta x}\right)^2\right\}(\varDelta x)^2.$$

$$\therefore P_{r-1}P_r = \sqrt{\left\{1 + \left(\frac{\varDelta y}{\varDelta x}\right)^2\right\}}\,\varDelta x.$$

Summing for all the chords and proceeding to the limit,

$$\textbf{length of arc } s = \int_a^b \sqrt{\left\{1 + \left(\frac{dy}{dx}\right)^2\right\}}\,dx,$$

where a, b are the values of x corresponding to the ends of the arc. From this it follows that the arc length to a variable point on the curve is given by

$$s = \int_a^z \sqrt{\left\{1 + \left(\frac{dy}{dx}\right)^2\right\}}\,dx. \tag{2}$$

$$\therefore \frac{ds}{dx} = \sqrt{\left\{1 + \left(\frac{dy}{dx}\right)^2\right\}},$$

a result that will be used later in this chapter.

The reader is not recommended to memorize the formulae in this chapter but, rather, to remember how they were obtained and to work them out for himself when they are wanted. In this respect, it is easier to treat $P_{r-1}P_r$ as if it were $\varDelta s$, the increment in arc length. The triangle in Fig. 21.3 helps us to remember the expression for arc length and to work out relationships such as $\frac{dy}{ds} = \sin \psi$, where ψ is the angle the tangent to the curve makes with the x-axis. Some convention is needed to specify the direction in which s is measured. We take the integrand of (2) to

Fig. 21.3

be positive; this fits in with the convention that the square root sign denotes the positive square root. (See P.M. I, p. 165.)

Qu. 12. (i) Express $\tan \psi$ as a derivative.

(ii) Express $\sec \psi$ as a derivative.

(iii) Use the identity $\sec^2 \psi = 1 + \tan^2 \psi$ to express the derivative in (ii) in terms of the derivative in (i).

(iv) Draw diagrams to show both ψ and the direction in which s is measured when $\dfrac{dy}{dx}$ is (a) positive, (b) negative, and s is given by (2), p. 455.

Qu. 13. Find the length of the arc in the first quadrant of $y = 2x^{3/2}$ from $x = 0$ to $x = \frac{1}{3}$.

Qu. 14. Find the length of the arc of $y = \log_e \sec x$ from $x = -\frac{1}{6}\pi$ to $x = \frac{1}{6}\pi$.

Qu. 15. Find the length of the arc of the parabola $y = x^2$ bounded by the line $y - 2 = 0$.

Arc length: parametric equations

21.3. Now suppose that we were given a curve in the form $x = f(t)$, $y = g(t)$ as, for example, $x = at^2$, $y = 2at$. It would then be more convenient in finding the length of an arc to have an integral with respect to t.

From (1), (p. 455),

$$P_{r-1}P_r{}^2 = \left\{ \left(\frac{\Delta x}{\Delta t} \right)^2 + \left(\frac{\Delta y}{\Delta t} \right)^2 \right\} (\Delta t)^2,$$

$$\therefore \ P_{r-1}P_r = \sqrt{\left\{ \left(\frac{\Delta x}{\Delta t} \right)^2 + \left(\frac{\Delta y}{\Delta t} \right)^2 \right\}} \, \Delta t.$$

Hence
$$s = \int_{t_1}^{t_2} \sqrt{\left\{ \left(\frac{dx}{dt} \right)^2 + \left(\frac{dy}{dt} \right)^2 \right\}} \, dt,$$

where t_1, t_2 are the values of t corresponding to the ends of the arc. Care should be taken to ensure that the integrand is positive throughout the range of integration. This applies particularly to Qu. 18.

Qu. 16. Find the length of the arc from $\theta = 0$ to $\theta = \alpha$ of the curve given by $x = a \cos \theta$, $y = a \sin \theta$. What is this curve?

Qu. 17. Find an expression for the distance measured along the curve from the origin to any point on the locus $x = at^2$, $y = at^3$.

Qu. 18. Sketch the astroid given by $x = a \cos^3 t$, $y = a \sin^3 t$ and find the length of its circumference.

Qu. 19. Sketch the arc of the cycloid $x = a(\theta - \sin \theta)$, $y = a(1 - \cos \theta)$ from $\theta = 0$ to $\theta = 2\pi$. Find its length.

Arc length: polar equations

21.4.

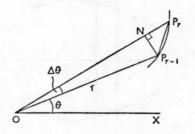

Fig. 21.4

*Qu. 20. In Fig. 21.4, $P_{r-1}N$ is the perpendicular from P_{r-1} on to a neighbouring radius OP_r of a curve given in polar coordinates.

(i) Find approximations for NP_r, NP_{r-1} in terms of r, θ;

(ii) Obtain the expression for arc length

$$s = \int_\alpha^\beta \sqrt{\left\{r^2 + \left(\frac{dr}{d\theta}\right)^2\right\}}\, d\theta,$$

where α, β are the values of θ corresponding to the ends of the arc.

Qu. 21. Find the length of the equiangular spiral $r = a\,e^{k\theta}$ from $\theta = 0$ to $\theta = 2\pi$.

Qu. 22. Find the length of the spiral of Archimedes $r = a\theta$ from $\theta = 0$ to $\theta = \pi$.

Qu. 23. What is the length of the circumference of the cardioid

$$r = a(1 + \cos\,\theta)?$$

[Make sure the integrand is positive.]

Area of surface of revolution

21.5. In considering areas of surfaces, we come up against a difficulty straight away. How can we measure the area of a curved surface? With a cylinder or cone, the surface can be 'developed', i.e. laid out on a plane surface, but with other figures, such as a sphere, this cannot be done. It is beyond the scope of this book to define, in mathematical terms, what is meant by the area of a curved surface and we shall assume that the following method of finding an expression for the area of a surface of revolution is valid.

Divide the surface with n planes $p_0, p_1, \ldots, p_n$ perpendicular to the axis of revolution, here the x-axis. (See Fig. 21.5.) We shall find an

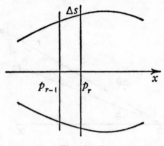

Fig. 21.5

approximation for the area between planes p_{r-1}, p_r. If the distance between the planes is small, the area of the surface between the planes is approximately a frustum of a cone with radii of the ends y, $y+\Delta y$ and with slant height Δs. Hence its area is given by (see Qu. 5, p. 319)

$$\pi\{y+(y+\Delta y)\}\,\Delta s = \pi(2y+\Delta y)\,\Delta s.$$

Summing and proceeding to the limit,

area of surface of revolution $= \int 2\pi y \, \mathrm{d}s$.

The limits of the integral will depend on the substitution used. See Qu. 24–29.

Qu. 24. Find the curved surface area of the frustum formed by rotating the segment of the line $y = 2x+3$ between $x = 1$ and $x = 3$ about the x-axis. $\left[\text{First show that } \dfrac{\mathrm{d}s}{\mathrm{d}x} = \sqrt{5} \quad \text{and evaluate} \quad \int_1^3 2\pi y \, \dfrac{\mathrm{d}s}{\mathrm{d}x} \, \mathrm{d}x.\right]$

Qu. 25. Show that for the circle $y^2 = a^2 - x^2$

$$\frac{\mathrm{d}s}{\mathrm{d}x} = \frac{a}{\sqrt{(a^2-x^2)}}.$$

Hence obtain the surface area of (i) a sphere, (ii) a section of a sphere bounded by two parallel planes at a distance h apart.

Qu. 26. The parabola $x = at^2$, $y = 2at$ is cut by the latus rectum and the arc so formed is rotated through two right angles about the axis of the parabola. Show that $\dfrac{\mathrm{d}s}{\mathrm{d}t} = 2a(t^2+1)^{1/2}$ and find the area of the surface of revolution.

Qu. 27.   The astroid $x = a \cos^3 t$, $y = a \sin^3 t$ is rotated through two right angles about the x-axis.   Find the area of the surface of revolution.

Qu. 28.   The part of the rectangular hyperbola $x^2 - y^2 = a^2$ which lies between $y = b$ and $y = -b$ is rotated through two right angles about the y-axis.   Find the area of the hyperboloid of one sheet so formed.

Qu. 29.   The cardioid $r = a(1 + \cos \theta)$ is rotated through two right angles about its axis of symmetry.   Use the expression

$$\int 2\pi y \, \frac{\mathrm{d}s}{\mathrm{d}\theta} \, \mathrm{d}\theta$$

to find the surface of revolution. $\left[\text{First show that } \dfrac{\mathrm{d}s}{\mathrm{d}\theta} = 2a \cos \tfrac{1}{2}\theta.\right]$

Pappus' theorems

21.6.   The two following theorems are due to Pappus, who lived at Alexandria about the end of the third century A.D.   Note that they both involve the distance moved by a centroid.

First consider the solid generated when an area is rotated about the x-axis.   We consider only areas which do not cut the x-axis, although the axis may form part of the boundary.

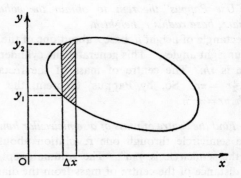

Fig. 21.6

Think of a lamina of area A and of uniform density ρ per unit area.   Take an element of area, as shown in Fig. 21.6.   Its mass is then approximately $\rho(y_2 - y_1) \, \varDelta x$ and its centroid at $\{x, \tfrac{1}{2}(y_2 + y_1)\}$. Therefore the moment about the x-axis is

$$\rho(y_2 - y_1) \, \varDelta x \cdot \tfrac{1}{2}(y_2 + y_1) = \tfrac{1}{2}\rho(y_2{}^2 - y_1{}^2) \, \varDelta x.$$

If $\bar{y}$ is the centre of mass of the lamina, taking moments about the x-axis,

$$\rho A . \bar{y} = \int_a^b \tfrac{1}{2}\rho(y_2{}^2 - y_1{}^2)\,\mathrm{d}x,$$

where a, b are the extreme values of x.

If we multiply each side by $\dfrac{2\pi}{\rho}$, we obtain

$$A . 2\pi\bar{y} = \int_a^b \pi(y_2{}^2 - y_1{}^2)\,\mathrm{d}x, \qquad (1)$$

where the R.H.S. is the volume of the solid generated when the lamina is rotated through 2π rad about the x-axis. Hence we may write

$$area \times distance\ moved\ by\ centroid = volume\ of\ revolution. \qquad (2)$$

Note that, although we have considered the special case where the area is rotated through an angle 2π, we might have written (1) as

$$A . \alpha\bar{y} = \int_a^b \tfrac{1}{2}\alpha(y_2{}^2 - y_1{}^2)\,\mathrm{d}x$$

for a general rotation α, so that (2) remains valid.

Example 1. *Use Pappus' theorem to obtain the volume of a right circular cylinder, base radius r, height h.*

Rotate a rectangle of height h, base r about one of the sides of length h through four right angles. This generates the cylinder. The area of the rectangle is rh. The centre of mass of the rectangle moves a distance $2\pi . \tfrac{1}{2}r = \pi r$. So, by Pappus' theorem, the volume of the cylinder is $rh . \pi r = \pi r^2 h$.

Example 2. *Find the centre of mass of a semicircular lamina of radius a.*

Rotate the semicircle through one revolution about its diameter. The area of the semicircle is $\tfrac{1}{2}\pi a^2$. The volume swept out is $4\pi a^2/3$. Let $\bar{y}$ be the distance of the centre of mass from the diameter then, by Pappus' theorem,

$$\tfrac{1}{2}\pi a^2 \times 2\pi\bar{y} = \tfrac{4}{3}\pi a^3.$$

$$\therefore \bar{y} = \frac{4a}{3\pi}.$$

Therefore the centre of mass of the semicircular lamina is $4a/(3\pi)$ from the bounding diameter.

Now consider the area of the surface of revolution swept out when an arc is rotated about the x-axis. We consider only arcs which do not cut the axis.

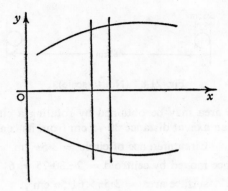

Fig. 21.7

Think of the arc as having a density σ per unit length. An element of length of arc has a mass $\sigma \, \Delta s$ with moment $y . \sigma \, \Delta s$ about the x-axis. If the total length of the arc is s and if its centre of mass is at a distance $\bar{y}$ from the x-axis, taking moments about the x-axis,

$$\sigma s . \bar{y} = \int y\sigma \, ds,$$

the integral being evaluated between the appropriate limits.

$$\therefore \ s . 2\pi \bar{y} = \int 2\pi y \, ds.$$

But this integral represents the area of the surface of revolution so that we may write

length of arc × distance moved by centre of mass
= area of surface of revolution.

If the reader can remember that both of Pappus' theorems involve the distance moved by a centre of mass, the idea of dimensions will help him to work out what the formulae must be, e.g. to find a volume $[L^3]$, the distance $[L]$ moved by the centre of mass must be multiplied by an area $[L^2]$.

Example 3. *An inflated inner tube of a bicycle tyre has a section (through a plane of symmetry) as shown in Fig. 21.8. Find the surface area of the tube.*

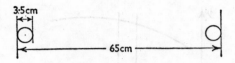

3·5cm

65cm

Fig. 21.8   (Not to scale).

The surface area may be obtained by rotating a circle of diameter 3·5 cm about an axis at distance 30·75 cm from its centre.

$$\text{Circumference of circle} = 3 \cdot 5\pi \text{ cm.}$$

$$\text{Distance moved by centroid} = 2\pi.30 \cdot 75 = 61 \cdot 5\pi \text{ cm.}$$

$$\therefore \text{ surface area} = 3 \cdot 5\pi \times 61 \cdot 5\pi \text{ cm}^2,$$

$$= 215 \cdot 25\pi^2 \text{ cm}^2 \ (\simeq 2120 \text{ cm}^2).$$

Qu. 30.   A reel of tape is in the form of a hollow cylinder, external diameter 8·5 cm, internal diameter 6·5 cm, height 2 cm. Use Pappus' theorem to find the volume of tape and check your answer by an elementary method.

Qu. 31.   A triangle has a right angle included by sides r, h; the hypotenuse is of length l. By considering the cone formed by rotating the triangle about the side h, obtain the formulae for the volume and curved surface area of a cone.

Qu. 32.   Three sides of a rectangle are r, h, r. By considering the total surface area of the cylinder obtained by rotating these lines about the fourth side of the rectangle, find the distance of their centre of mass from the fourth side.

Qu. 33.   Find the centre of mass of a semi-circle of uniform wire.

Qu. 34.   An anchor ring is formed by rotating a circle of radius a about an axis in the plane of the circle at distance b from the centre. Find (i) the volume, (ii) the surface area of the ring.

Qu. 35.   A trapezium PQRS is right-angled at P and Q, PQ $= h$, QR $= b$, RS $= l$, SP $= a$. It is rotated about PQ to form the frustum of a cone.

(i) Obtain the formula for the curved surface area of a frustum of a cone in terms of a, b, l.†

† It should be pointed out that Pappus' theorem was deduced from this result. However, Qu. 35 provides an easy method of deriving the formula in case it is forgotten.

(ii) Use the formula for the volume of a frustum of a cone (page 319) to find the distance of the centre of mass from PQ.

Qu. 36. An arc of a circle, radius a, is rotated about the diameter perpendicular to the axis of symmetry of the arc. If the arc subtends an angle 2α at the centre of the circle, find the surface area generated by the arc. Hence find the distance of the centre of mass of the arc from the centre of the circle.

Curvature

21.7. Any user of the roads, especially in Great Britain, will be familiar with their bends and corners. Cyclists and drivers have a particular interest in the sharpness of the turns because they can easily lead to skids under bad road conditions or excessive speed; but how many could say how sharp a particular bend was?

There are two ways in which such a question might be answered. One way is to compare a bend with that of a circular arc of some radius; another is to state the rate of turning and we shall start with this.

How, then, can a rate of turning be measured? The word 'turning' suggests an angle but Fig. 21.9 shows that an angle by itself is not

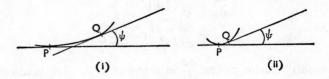

(I)   (II)

Fig. 21.9

enough: between two points P and Q on two curves the tangent rotates through the same angle but one is clearly more bent than the other. This suggests that we might take into account the distance moved along the curve. For a *constant* rate of turning we shall take

$$\frac{\text{angle turned through}}{\text{distance along curve}}$$

as a measure of the curvature. For a variable rate of turning, the average curvature in a small displacement along the curve would be $\frac{\Delta\psi}{\Delta s}$ and in the limit we should obtain as a measure of curvature

$$\kappa = \frac{d\psi}{ds}.$$

The *radius of curvature* ρ of a curve is defined by $\rho = \dfrac{1}{\kappa}$. Thus

$$\rho = \frac{1}{\kappa} = \frac{ds}{d\psi}.$$

There should be no difficulty in remembering which way up this derivative is written because ρ is a length and so has ds in the numerator.

Qu. 37. Show that, for a circle, ρ is equal to the radius of the circle.

Qu. 38. Find the curvature of the cycloid $s = 4a \sin \psi$ at the points where $\psi = 0$.

Curvature: Cartesian coordinates

Our next aim is to find an expression for curvature in terms of cartesian coordinates x, y. We therefore shall try to work with x as the independent variable:

$$\kappa = \frac{d\psi}{ds} = \frac{d\psi}{dx} \Big/ \frac{ds}{dx}.$$

But from p. 455,

$$\frac{ds}{dx} = \sqrt{\left\{1 + \left(\frac{dy}{dx}\right)^2\right\}},$$

$$\therefore \kappa = \frac{d\psi}{dx} \Big/ \sqrt{\left\{1 + \left(\frac{dy}{dx}\right)^2\right\}}. \qquad (1)$$

So it only remains to find $\dfrac{d\psi}{dx}$ in terms of x, y. The gradient of a curve is

$$\tan \psi = \frac{dy}{dx}.$$

Differentiating with respect to x,

$$\sec^2 \psi \, \frac{d\psi}{dx} = \frac{d^2y}{dx^2}.$$

But $\sec^2 \psi = 1 + \tan^2 \psi$,

$$\therefore \left\{1 + \left(\frac{dy}{dx}\right)^2\right\} \frac{d\psi}{dx} = \frac{d^2y}{dx^2}.$$

$$\therefore \frac{d\psi}{dx} = \frac{d^2y}{dx^2} \Big/ \left\{1 + \left(\frac{dy}{dx}\right)^2\right\}.$$

So from (1),

$$\kappa = \frac{d^2y}{dx^2} \Big/ \sqrt{\left\{1 + \left(\frac{dy}{dx}\right)^2\right\}^3}.$$

Qu. 39.   Find the curvature of the parabola $y = x^2$ (i) at (1, 1), (ii) at the origin.

★Qu. 40.   If the coordinates x, y of a point on a curve are given in terms of a parameter t, use the equation

$$\kappa = \frac{d\psi}{dt} \Big/ \frac{ds}{dt}$$

to show that

$$\kappa = \left(\frac{dx}{dt} \frac{d^2y}{dt^2} - \frac{dy}{dt} \frac{d^2x}{dt^2} \right) \Big/ \sqrt{ \left\{ \left(\frac{dx}{dt} \right)^2 + \left(\frac{dy}{dt} \right)^2 \right\}^3 }.$$

Qu. 41.   Find the curvature of the parabola $y^2 = 4ax$ at $(at^2, 2at)$, (i) by treating y as a function of x, (ii) by the formula of Qu. 40.

Qu. 42.   Find the least curvature of the cycloid

$$x = a(\theta - \sin \theta), \quad y = a(1 - \cos \theta).$$

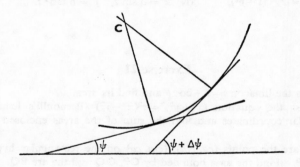

Fig. 21.10

Qu. 43.   If tangents to a curve make angles ψ and $\psi + \varDelta\psi$ with the x-axis (see Fig. 21.10), find the angle between the corresponding normals.   Let the normals intersect at C.   Find the limiting distance of C from the curve (measured along one normal) as $\varDelta\psi \to 0$.   Take s as the arc length.

This point C is called the *centre of curvature*.   Note that: (i) it is on a normal at a distance ρ from the curve on the concave side, (ii) it is the limiting position of the point of intersection of neighbouring normals.   The circle with centre C and radius ρ is called the *circle of curvature*.

Qu. 44.   Find the equation of the circle of curvature at the point (c, c) on the rectangular hyperbola $xy = c^2$.

Qu. 45. The equation of the normal at $(at^2, 2at)$ to the parabola $y^2 = 4ax$ is

$$xt + y - at^3 - 2at = 0.$$

Treating this as a cubic in t, show that the condition that it should have two equal roots is

$$27ay^2 = 4(x - 2a)^3.$$

What is this the locus of?

Qu. 46. A curve touches the x-axis at the origin. Write down the first non-zero term in the expansion of y in ascending powers of x. Show that the radius of curvature at the origin is the limiting value of $x^2/2y$ as $x, y \to 0$. This is known as Newton's formula.

Qu. 47. Use Newton's formula to find the radius of curvature at the origin for

(i) $y = x^2$,
(ii) $y = x^2/(1 - x^2)$,
(iii) $x = 2y^2/(1 + y^2)$,
(iv) $x = a \sin t, \quad y = b \tan^2 t$.

Exercise 21

1. Sketch the limaçon $r = 2 + \cos \theta$ and find its area.
2. Express the equation $(x^2 + y^2)^2 = a^2(x^2 - y^2)$ (Bernoulli's lemniscate) in polar coordinates and find the sum of the areas enclosed by the loops.
3. P, Q are the points $(ca, c/a)$, $(c/a, ca)$ on the rectangular hyperbola $xy = c^2$. Find the area bounded by OP, OQ and the arc PQ.
4. An arc AB of a circle, radius a, subtends an angle 2α at the centre O. The sector OAB is rotated through four right angles about the diameter parallel to AB. Write down the area of the surface of revolution generated by the perimeter of the sector and hence find the distance from O of the centroid of the perimeter.
5. $P(at^2, 2at)$ is a point on the parabola $y^2 = 4ax$ and S is the focus. Show that the area bounded by the parabola, its axis and the line PS is $\frac{1}{3}a^2(3t + t^3)$.
6. Find the area enclosed by the astroid

$$x = a \cos^3 t, \qquad y = a \sin^3 t.$$

7. A cylindrical hole of radius r is made in a sphere of radius a so that the axis of the cylinder coincides with a diameter of the sphere. Find the volume that remains. [HINT: with the diameter as y-axis, take an element of area $2y \varDelta x$ and use Pappus' theorem.]

8. Find the centre of curvature C of the ellipse

$$b^2x^2 + a^2y^2 = a^2b^2 \quad \text{at } (a, 0).$$

Show that the circle with centre C which passes through $(a, 0)$ has no other point of intersection, real or imaginary, with the ellipse.

9. Find the length of the arc of the catenary $y = c \cosh (x/c)$ from the vertex $(0, c)$ to any point on the curve.

Find also the area generated when this arc is rotated through four right angles about the y-axis.

10. Sketch the curve $x = e^t \cos t$, $y = e^t \sin t$ from $t = -\pi$ to $t = +\pi$. Find the area enclosed by this arc and the segment of the straight line joining its end points.

11. Sketch the curve $y^2 = (x-1)^2(2-x)$. If the area enclosed by the loop is rotated through four right angles about the y-axis, find the volume of the solid generated.

12. For the cardioid $r = a(1 + \cos \theta)$ find, in any order, (i) its area, (ii) the distance from the x-axis of the centroid of the half of the cardioid above the axis, (iii) the volume of the solid generated when this area is rotated through one revolution about the x-axis.

13. A circle of radius a rolls touching externally a fixed circle of radius $2a$. Show that the locus of a point on the smaller circle (a two-cusped epicycloid) may be written

$$x = 3a \cos \theta - a \cos 3\theta, \qquad y = 3a \sin \theta - a \sin 3\theta.$$

Find the length of this curve and also its area.

14. A three-cusped hypocycloid is traced out by a point of a circle of radius a which rolls touching internally a fixed circle of radius $3a$. Show that the locus may be written

$$x = 2a \cos \theta + a \cos 2\theta, \qquad y = 2a \sin \theta - a \sin 2\theta.$$

   (i) Find the curvature at $(-a, 0)$.
   (ii) Find the length of the curve.
   (iii) Find the area generated when the arc from $\theta = 0$ to $\theta = 2\pi/3$ is rotated through four right angles about the x-axis.

15. Show that

$$x \frac{dy}{dt} - y \frac{dx}{dt} = x^2 \frac{d}{dt} \left(\frac{y}{x} \right).$$

Hence, or otherwise, find the area of the loop of the curve

$$x = 3t/(1 + t^3), \qquad y = 3t^2/(1 + t^3).$$

What is the Cartesian equation of this locus?

16. A flanged wheel of radius b rolls along a straight rail with its centre a distance a vertically above the rails. Show that the locus of a point on the circumference of the wheel may be written:

$$x = a\theta - b \sin \theta, \qquad y = a - b \cos \theta,$$

(a subtrochoid) and find the area of a loop correct to three significant figures when $a = 1$, $b = 2$.

17. A circle of radius a rolls on a fixed straight line. Show that the locus of a point P on the circumference of the circle may be written in the form:

$$x = a(\theta - \sin \theta), \qquad y = a(1 - \cos \theta)$$

(a cycloid).

Show that, in general, if C is the centre of curvature corresponding to P, then the mid-point of PC is the point of contact of the line and the circle.

18. Express the equation of the normal at P$(ct, c/t)$ to the rectangular hyperbola $xy = c^2$ as a quartic in t.

If this equation has a repeated root, derive from it another (cubic) equation which is satisfied by the same root. Hence find the coordinates of the centre of curvature C at $(ct, c/t)$.

Verify that PC is equal in length to the radius of curvature at P.

19. Find the length of arc from the point $(1, 0)$ to any point on the curve $y = \log_e x$.

If the arc is rotated through four right angles about the y-axis, find the area of the surface of revolution generated.

20. The arc of the rectangular hyperbola $x^2 - y^2 = a^2$ bounded by the chord $x = a \cosh \theta$ is rotated through two right angles about the x-axis. Find the area of the surface generated.

CHAPTER 22

COMPLEX NUMBERS

22.1. To those who have not used them, complex numbers have about them an air of mystery which, although it may be understandable, is unjustified. It is hoped that this rather long introduction to the chapter will help the reader to see complex numbers in perspective—as the culmination of a series of stages in which the concept of 'number' is extended. However, before we start on this, it is well to point out that complex numbers are extremely useful to the mathematician in the development of the calculus and algebra. Complex numbers are frequently used, too, in the mathematical theory of electricity and in hydrodynamics and aerodynamics. Engineers have such frequent need of them that their notation is slightly different—this will be noted later.

Extending the concept of number

22.2. To the infant, a 'number' is one of the natural numbers: 1, 2, 3, 4, ... Later on he meets fractions and his concept of 'number' is extended to include these. Often he does not find this process easy, especially as he has to learn how to add, subtract, multiply and divide two fractions.

The next stage in the development of this idea of number is usually to introduce directed numbers: that is to say, numbers (including fractions) with signs + or −, and which correspond to points on a line:

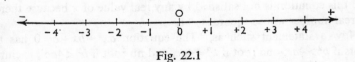

Fig. 22.1

However, there are some points on the line which do not correspond to any integer or fraction; for instance, the points at a distance $\sqrt{2}$ from the origin O. Another way of putting this is to say that there are some numbers, such as $\sqrt{2}$, which cannot be expressed in the form a/b.

469

where a, b are integers.   (See Exercise 22a, No. 20.)   And so the pupil in the early stages of the secondary school meets *irrational numbers* for the first time: *rational* numbers are those which can be expressed in the form a/b, where a, b are integers (positive or negative)—*irrational* numbers cannot be written in this way, e.g. $\sqrt{2}$, $\sqrt{3}$, $\sqrt{5}$, e, π.   With the concept of number extended to include irrational numbers, we may now say that, to every point on the line in Fig. 22.1 (considered extended indefinitely in each direction), there corresponds a number (rational or irrational); and to every rational or irrational number there corresponds a point on the line.   Clearly it would be convenient to have a word to describe the numbers which are either rational or irrational: they are called *real numbers*.

Numbers as roots of equations

22.3.   A linear equation of the form

$$ax + b = 0, \quad (a, b \text{ rational}, a \neq 0)$$

always has a rational root (see Qu. 1).

Qu. 1.   Prove the above statement.   Let $a = p/q$, $b = r/s$ where p, q, r, s are integers.

A quadratic equation of the form

$$ax^2 + bx + c = 0 \quad (a, b, c \text{ rational})$$

does not always have rational roots, for instance, the roots of

$$x^2 + 2x - 1 = 0$$

are $-1 \pm \sqrt{2}$.   Nor are the roots always real, e.g. if

$$x^2 - 4x + 5 = 0,$$
$$(x-2)^2 = -1,$$

and this equation is not satisfied by any real value of x because there is no real number whose square is -1.

Now a statement such as, 'The equation $ax^2 + bx + c = 0$ has two roots if $b^2 > 4ac$, one root if $b^2 = 4ac$, and no root if $b^2 < 4ac$', is clumsy and it would be preferable if we could make a more general statement such as, 'The equation $ax^2 + bx + c = 0$ has two roots'.   We have seen that this cannot be done if we restrict ourselves to real numbers so, in a search for generality, we are led to extend the concept of 'number'.

Going back to the equation

$$x^2 - 4x + 5 = 0,$$
$$(x-2)^2 = -1,$$

so that $\qquad x - 2 = \pm\sqrt{(-1)}.$

For brevity we write $\sqrt{(-1)} = i$† so that

$$x = 2 \pm i.$$

This means that our extended system of 'numbers' is going to include numbers such as $2 \pm i$ if all quadratic equations are to have two roots in that system. Of course, $2 \pm i$ are not numbers in the former sense of real numbers—they force us to use the word 'number' in a wider sense than before.

Now consider the equation

$$ax^2 + bx + c = 0 \quad (a, b, c \text{ real}).$$

By completing the square it is found that

$$x = \{-b \pm \sqrt{(b^2 - 4ac)}\}/2a.$$

If $b^2 > 4ac$, the roots are real and distinct.

If $b^2 < 4ac$, $\qquad x = \{-b \pm i\sqrt{(4ac - b^2)}\}/2a.$

so that the roots are in the form $p + iq$ (p, q real).

If $b^2 = 4ac$, we say that the equation has a repeated root.

We may now state with complete generality that the quadratic equation has two roots.

An expression in the form $a + ib$, where a, b are real and $i^2 = -1$, is called a *complex number*. This notation was introduced by Euler in 1748.

Since it was necessary to introduce complex numbers in order to include the roots of all quadratic equations, it might be thought that yet further types of number would be necessary in order to cover the roots of equations of higher degree. However, this is not so: it can be proved that an equation of the nth degree has exactly n complex roots (possibly repeated), some of which may be in the form $a + i0$. The proof of the theorem is beyond the scope of this book.

† Engineers prefer to write $\sqrt{(-1)} = j$ so as to avoid confusion with i which they use as a symbol for electric current.

Manipulation of complex numbers

22.4. When a child first meets fractions, he has to learn how to add, subtract, multiply, and divide with them. So now we are faced with the problem of manipulating complex numbers. The operations addition, subtraction, multiplication, and division which we have used so far are concerned with real numbers, hence it is necessary to define what we mean by these operations with regard to complex numbers. It is easiest for us to define these operations by saying that we shall use the usual laws of algebra together with the relation $i^2 = -1$. Thus

$$
\begin{aligned}
(a+ib)+(c+id) &= (a+c)+i(b+d), \\
(a+ib)-(c+id) &= (a-c)+i(b-d), \\
(a+ib)\times(c+id) &= ac+aid+ibc+i^2bd, \\
&= (ac-bd)+i(ad+bc).
\end{aligned}
$$

At this stage it is worth comparing the corresponding operations with real numbers in the form $a+\sqrt{(2)}b$ (a, b rational):

$$
\begin{aligned}
(a+b\sqrt{2})+(c+d\sqrt{2}) &= (a+c)+\sqrt{2}(b+d), \\
(a+b\sqrt{2})-(c+d\sqrt{2}) &= (a-c)+\sqrt{2}(b-d), \\
(a+b\sqrt{2})\times(c+d\sqrt{2}) &= ac+ad\sqrt{2}+bc\sqrt{2}+2bd, \\
&= (ac+2bd)+\sqrt{2}(ad+bc).
\end{aligned}
$$

This helps us to find a way of expressing $\dfrac{a+ib}{c+id}$ in the form $p+iq$. The reader will probably recall the corresponding process with $\dfrac{a+b\sqrt{2}}{c+d\sqrt{2}}$. The method is to multiply numerator and denominator in such a way that the new denominator involves a difference of two squares:

$$
\begin{aligned}
\frac{a+b\sqrt{2}}{c+d\sqrt{2}}\times\frac{c-d\sqrt{2}}{c-d\sqrt{2}} &= \frac{(ac-2bd)+\sqrt{2}(bc-ad)}{c^2-2d^2}, \\
&= \frac{ac-2bd}{c^2-2d^2}+\sqrt{2}\,\frac{bc-ad}{c^2-2d^2}.
\end{aligned}
$$

Similarly, the expression $\dfrac{a+ib}{c+id}$ may be expressed in the form $p+iq$ by multiplying numerator and denominator by $c-id$ because

$$
(c+id)\times(c-id) = c^2-i^2d^2 = c^2+d^2.
$$

DEFINITION: *Two complex numbers in the form $x+iy$, $x-iy$ are called* conjugate *complex numbers.*

Qu. 2. Express $\dfrac{2+3i}{1+i}$ in the form $p+iq$ (p, q real). [Multiply numerator and denominator by $1-i$.]

Do not attempt to memorize expressions for the sum, difference, product, and quotient of two complex numbers: simply use the usual laws of algebra, together with the relation $i^2 = -1$.

Exercise 22a

Simplify:

1. (i) i^3, (ii) i^4, (iii) i^5, (iv) i^6, (v) $\dfrac{1}{i^2}$, (vi) $\dfrac{1}{i}$, (vii) $\dfrac{1}{i^3}$.

2. (i) $(3+i)+(1+2i)$, (ii) $(5-3i)+(4+3i)$,
(iii) $(2-3i)-(1+2i)$, (iv) $(1+i)-(1-i)$.

3. (i) $(2+3i)(4+5i)$, (ii) $(2-i)(3+2i)$,
(iii) $(1+i)(1-i)$, (iv) $(3+4i)(3-4i)$,
(v) $(u+iv)(u-iv)$, (vi) $(x+2iy)(2x+iy)$,
(vii) $i(2p+3iq)$, (viii) $(p+2iq)(p-2iq)$.

4. Express with real denominators:

(i) $\dfrac{1-i}{1+i}$, (ii) $\dfrac{1}{2-3i}$, (iii) $\dfrac{3i-2}{1+2i}$, (iv) $\dfrac{5+4i}{5-4i}$,

(v) $\dfrac{1}{x+iy}$, (vi) $\dfrac{1}{x-iy}$, (vii) $\dfrac{1}{2+3i}+\dfrac{1}{2-3i}$.

Simplify the expressions in Nos. 5 to 7:

5. (i) $(2+3i)^2$, (ii) $(4-5i)^2$, (iii) $(x+iy)^2$,

6. (i) $(1+i)^3$, (ii) $(1-i)^3$, (iii) $\dfrac{1}{(1+i)^3}$.

7. (i) $(\cos\theta+i\sin\theta)(\cos\theta-i\sin\theta)$,
(ii) $(\cos\theta+i\sin\theta)^2$,
(iii) $(\cos\theta+i\sin\theta)(\cos\phi+i\sin\phi)$.

8. Express with real denominators:

(i) $\dfrac{1}{\cos\theta-i\sin\theta}$, (ii) $\dfrac{1}{1-\cos\theta-i\sin\theta}$,

(iii) $\dfrac{1}{1+\cos\theta-i\sin\theta}$, (iv) $\dfrac{\cos\theta+i\sin\theta}{\cos\phi-i\sin\phi}$.

9. Show that
$$(\cos\theta+i\sin\theta)^3 = \cos3\theta+i\sin3\theta,$$
and express $(\cos\theta-i\sin\theta)^3$ in a similar form.

10. Prove by induction that if n is a positive integer,
$$(\cos\theta+i\sin\theta)^n = \cos n\theta+i\sin n\theta$$
and that
$$(\cos\theta-i\sin\theta)^n = \cos n\theta-i\sin n\theta.$$

11. Simplify: (i) $(a+ib)^6 - (a-ib)^6$,
           (ii) $(a+ib)^6 + (a-ib)^6$,
           (iii) $(a+ib)^6 (a-ib)^6$.

12. Find the roots of the following equations:
  (i) $x^2 - 6x + 10 = 0$,   (ii) $x^2 + 12x + 40 = 0$,
  (iii) $2x^2 + 3x + 2 = 0$,   (iv) $3x^2 - 5x + 7 = 0$.

13. If the equation $ax^2 + bx + c = 0$ (a, b, c real) has complex roots. Show that the two roots are conjugate (see p. 472). Show also that the sum and product of the roots are real.

14. Show with the aid of a graph that the cubic equation

$$ax^3 + bx^2 + cx + d = 0 \quad (a, b, c, d \text{ real})$$

either has three real roots (possibly repeated) or has one real root and two conjugate complex roots.

15. Find the sum of the series:
  (i) $1 + i + i^2 + \ldots + i^n$,
  (ii) $1 - 2i + 4i^2 - 8i^3 + \ldots + (-2i)^n$.

16. If

$$S = 1 + 2ix + 3i^2 x^2 + \ldots + n(ix)^{n-1},$$

find $S(1 - ix)$ and hence find an expression for S with a real denominator.

17. Express $\dfrac{2x}{x^2 + a^2}$ in partial fractions in the form

$$\frac{A}{x - ia} + \frac{B}{x + ia}.$$

Find a similar expression for $\dfrac{2a}{x^2 + a^2}$.

18. Express $\dfrac{1}{x^3 - 1}$ as the sum of three partial fractions with linear denominators.

19. We have not extended the idea of integration to cover complex numbers but, nevertheless, find out what happens when the expressions $2x/(x^2 + a^2)$, $2a/(x^2 + a^2)$ are integrated treating i as an ordinary algebraic constant except that $i^2 = -1$. Use the expressions obtained in No. 17.

*20. To prove that $\sqrt{2}$ is not rational.

Suppose that $\sqrt{2} = m/n$ where m, n are integers and the fraction has been reduced to its lowest terms by cancelling all common factors. Prove that (i) m^2 is even, (ii) m is even.

Write $m = 2l$, where l is an integer. Obtain an equation connecting l^2, n^2. Prove that (i) n^2 is even, (ii) n is even. This contradicts the hypothesis that the fraction m/n is in its lowest terms.

Complex numbers as ordered pairs

22.5. To see how a satisfactory definition of complex numbers can be given, consider the problem of defining rational numbers in terms of the integers.

Note: (i) a rational number is formed from a *pair* of integers, e.g. $\frac{2}{3}, \frac{7}{5}, \frac{4}{1}$ (the last of which is commonly abbreviated to 4).

(ii) the position of the integers is important because in general

$$\frac{a}{b} \neq \frac{b}{a}.$$

We therefore say that a rational number is an *ordered pair* of integers— but this, by itself, is not enough. To complete the definition, we must say how numbers of this type are to be added, subtracted, multiplied and divided.

We know that for rational numbers

$$\frac{a}{b} + \frac{c}{d} = \frac{ad + bc}{bd},$$

but this is by no means the only possible way of defining addition of the ordered pairs $\frac{a}{b}, \frac{c}{d}$. For instance, it would be much simpler to define addition by the rule:

$$\frac{a}{b} + \frac{c}{d} = \frac{a + c}{b + d}.$$

As to multiplication, with rational numbers,

$$\frac{a}{b} \times \frac{c}{d} = \frac{ac}{bd},$$

but multiplication of the ordered pairs $\frac{a}{b}, \frac{c}{d}$ might have been defined by the rule:

$$\frac{a}{b} \times \frac{c}{d} = \frac{ac - bd}{ad + bc}.$$

We need not go through the process of defining subtraction and division: the point to note about defining the various operations on ordered pairs is that the properties of the numbers so defined will depend on the rules chosen.

Now consider complex numbers. We have seen that a complex number involves a *pair* of real numbers and that the *order* of the pair is important because in general $a+ib \neq b+ia$. We therefore define a complex number as an ordered pair of real numbers which we shall write as $[a, b]$. The fundamental operations of addition and multiplication are defined by the rules:

$$[a, b] + [c, d] = [a+c, b+d],$$
$$[a, b] \times [c, d] = [ac-bd, ad+bc].$$

Subtraction and division are defined in terms of addition and multiplication thus, for any type of number,

$$p-q \text{ is the number } x \text{ such that } q+x = p \text{ and}$$
$$p \div q \text{ is the number } y \text{ such that } q \times y = p.$$

Now

$$[c, d] + [a-c, b-d] = [a, b]$$
$$\therefore \quad [a, b] - [c, d] = [a-c, b-d].$$

Qu. 3. Use the definition of division above to show that:

(i) for the rational numbers $\dfrac{a}{b}, \dfrac{c}{d}, \quad \dfrac{a}{b} \div \dfrac{c}{d} = \dfrac{ad}{bc}$.

(ii) for the complex numbers $[a, b]$, $[c, d]$,

$$[a, b] \div [c, d] = \left(\frac{ac+bd}{c^2+d^2}, \frac{bc-ad}{c^2+d^2} \right).$$

Qu. 4. Note that to every real number a there corresponds a unique complex number $[a, 0]$. Find, from the definitions of the four operations on complex numbers:

(i) $[a, 0] + [c, 0]$,　　　　　　(ii) $[a, 0] \times [c, 0]$,

(iii) $[a, 0] - [c, 0]$,　　　　　　(iv) $[a, 0] \div [c, 0]$.

Compare these results with the corresponding operations on the real numbers a, c.

The definition of a complex number as an ordered pair was first given by Hamilton in 1835.

The Argand diagram

22.6. The last section was written to show the reader that complex numbers can be put on a satisfactory logical basis. However, manipulation of complex numbers is most easily carried out as before: the ordered pair notation is simply a device for defining these numbers without reference to $\sqrt{(-1)}$. We could write $\sqrt{(-1)}$ as the ordered pair [0, 1] but this would be rather clumsy and it is easier to write $\sqrt{(-1)} = i$.

Qu. 5. Prove from the definition of multiplication of complex numbers that

$$[0, 1] \times [0, 1] = [-1, 0].$$

Although the idea of an ordered pair may appear to some readers to have been a digression, it leads us to the next step in our treatment of the subject. The Argand diagram is named after J. R. Argand, who published his work on the graphical representation of complex numbers in 1806.

Corresponding to every complex number $[x, y]$ or $x + iy$, there is a point (x, y) in the plane with axes OX, OY; and corresponding to any point (x, y) in the plane, there is a complex number $x + iy$. (Here it is worth comparing the equivalent situation with real numbers. Corresponding to every real number x there is a point on a line OX. What is less easy to prove is that corresponding to every point on OX there is a real number.) At first this correspondence between complex numbers and points on the plane may appear to be rather obvious and not very useful, but in fact it proves to be a considerable importance to the theory of complex numbers.

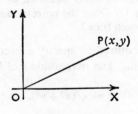

Fig. 22.2

The value of this correspondence is increased by the fact that with every point $P(x, y)$ in the plane there is associated a *radius vector* OP. (See Fig. 22.2.) This means that corresponding to every complex number $x + iy$ there is a radius vector OP where P is (x, y). Further, corresponding to every radius vector OP in the plane there is complex number $x + iy$ where P is (x, y).

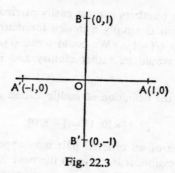

Fig. 22.3

Look at Fig. 22.3. The points A, B, A′, B′ are respectively (1, 0), (0, 1), (−1, 0), (0, −1). Corresponding to:

OA	there is the complex number	$1 + 0i$	or	1
OB	there is the complex number	$0 + 1i$	or	i
OA′	there is the complex number	$-1 + 0i$	or	-1
OB′	there is the complex number	$0 + (-1)i$	or	$-i$

Looking down the right-hand side of the last four lines, each number is equal to the previous one multiplied by i. Meanwhile, the corresponding radius vector has rotated in the positive (anticlockwise) sense through one right angle. Would the same thing happen if any complex number were multiplied by i?

Qu. 6. Find the complex numbers obtained by multiplying $x + iy$ once, twice and three times by i. Does the corresponding radius vector rotate through one right angle each time?

Two quantities are required to specify a vector through the origin: magnitude and direction. The magnitude r of OP (Fig. 22.4) presents no difficulty

$$r = \sqrt{(x^2 + y^2)}.$$

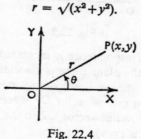

Fig. 22.4

This quantity is called the *modulus* of the complex number $x+iy$. 'The modulus of $x+iy$' is abbreviated to $|x+iy|$ hence

$$|x+iy| = \sqrt{(x^2+y^2)}.$$

Qu. 7. Write down the moduli of:

  (i) $3+4i$,   (ii) $-i$,   (iii) $\cos\theta+i\sin\theta$,

(iv) $\frac{1}{2}-\frac{1}{2}\sqrt{3}i$,   (v) -3,   (vi) $1+i$.

The direction specifying the radius vector OP is not quite so easy to deal with because there are infinitely many positive and negative angles which would do.

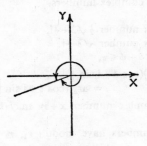

Fig. 22.5

The problem of which angle to choose is well illustrated by a radius vector in the third quadrant. (Fig. 22.5.) It is simply a matter of convention whether we take the positive reflex angle or the negative obtuse angle. In fact the numerically smaller angle is used. The angle between the radius vector OP and the positive x-axis is called the *argument* of the complex number $x+iy$. This is abbreviated to Arg $(x+iy)$ and has, as we have said before, infinitely many values. The value uniquely specified by the above convention is called the *principal value* of the argument and is written arg $(x+iy)$, so that

$$-\pi < \text{arg } (x+iy) \leqslant \pi.$$

[In some textbooks, the argument is called the amplitude but this term is less acceptable because of possible confusion with the amplitude of a current, motion, or wave.]

Qu. 8. Find the principal values of the arguments of:

  (i) $\cos\frac{1}{4}\pi+i\sin\frac{1}{4}\pi$,   (ii) 1,   (iii) -1,   (iv) $1-i$,

(v) $\frac{1}{2}+\frac{1}{2}\sqrt{3}i$,   (vi) $\cos\frac{7}{3}\pi+i\sin\frac{7}{3}\pi$. (vii) $\cos\frac{2}{3}\pi-i\sin\frac{2}{3}\pi$.

Exercise 22b

1. Mark on the Argand diagram the radius vectors corresponding to:
   (i) $1+i$,   (ii) $-3+2i$,   (iii) $-3-2i$,
   (iv) $3-4i$,   (v) $-4+3i$,   (vi) $\cos \frac{1}{3}\pi + i \sin \frac{1}{3}\pi$,
   (vii) $-\frac{1}{2}\sqrt{3} + \frac{1}{2}i$,   (viii) $\cos \pi + i \sin \pi$.
   Write down the moduli of these complex numbers and give the principal values of their arguments.

2. Simplify: $(1+i)^2$, $(1+i)^3$, $(1+i)^4$.
   Draw in the Argand diagram the radius vectors corresponding to $1+i$, $(1+i)^2$, $(1+i)^3$, $(1+i)^4$. Find the principal values of the arguments of these complex numbers.

3. Repeat No. 2:
   (i) for the complex number $\frac{1}{2}\sqrt{3} + \frac{1}{2}i$,
   (ii) for the complex number $\sqrt{3} + i$.

4. Show that, if $-\pi < \theta + \phi \leqslant \pi$,
   $$\arg \{(\cos \theta + i \sin \theta)(\cos \phi + i \sin \phi)\}$$
   $$= \arg (\cos \theta + i \sin \theta) + \arg (\cos \phi + i \sin \phi).$$

*5. Show that the complex number $x + iy$ may be written in the form $r(\cos \theta + i \sin \theta)$.
   Two complex numbers have moduli r_1, r_2 and arguments θ_1, θ_2. Show that:
   (i) the modulus of the product is $r_1 r_2$,
   (ii) one value of the argument of the product is $\theta_1 + \theta_2$.

6. The principal values of the arguments of two complex numbers are θ_1, θ_2. Draw diagrams to illustrate the cases when the principal value of the argument of the product of the two complex numbers is:
   (i) $\theta_1 + \theta_2 - 2\pi$,   (ii) $\theta_1 + \theta_2 + 2\pi$,   (iii) $\theta_1 + \theta_2$.
   Are there any other cases?

7. Use the result of No. 5 to express in the form $r(\cos \theta + i \sin \theta)$:
   (i) $(\cos \frac{1}{4}\pi + i \sin \frac{1}{4}\pi)(\cos \frac{1}{6}\pi + i \sin \frac{1}{6}\pi)$,
   (ii) $(\cos \frac{1}{4}\pi + i \sin \frac{1}{4}\pi)(\cos \frac{3}{4}\pi + i \sin \frac{3}{4}\pi)$,
   (iii) $(\frac{1}{2} + \frac{1}{2}\sqrt{3}i)(\frac{1}{2}\sqrt{3} + \frac{1}{2}i)$,
   (iv) $(1+i)^2$,
   (v) $(1 - \sqrt{3}i)(1+i)$,
   (vi) $(-1 + \sqrt{3}i)^2$.

8. Express in the form $r(\cos \theta + i \sin \theta)$, where θ is the principal value of the argument:
   (i) $(\cos \frac{3}{4}\pi + i \sin \frac{3}{4}\pi)^2$,   (ii) $\{\cos (-\frac{2}{3}\pi) + i \sin (-\frac{2}{3}\pi)\}^2$,
   (iii) $(-1+i)(1 - \sqrt{3}i)$,   (iv) $(1 + \sqrt{3}i)(-\sqrt{3} + i)$.

9. Show on the Argand diagram the radius vectors corresponding to $(\cos \frac{1}{6}\pi + i \sin \frac{1}{6}\pi)^n$ when $n = 0, 1, 2, \ldots, 11, 12$.

10. Repeat No. 9 for

$$(\cos \tfrac{2}{5}\pi + i \sin \tfrac{2}{5}\pi)^n, \quad n = 0, 1, \ldots, 5.$$

11. The conjugate of $x + iy$ is $x - iy$ and for brevity we write $z = x + iy$, $\bar{z} = x - iy$. Show that

$$z\bar{z} = |x + iy|^2 = |x - iy|^2.$$

What are the values of $|z\bar{z}|$, $\left|\dfrac{z}{\bar{z}}\right|$, $\arg (z\bar{z})$?

In the remainder of the exercise we shall refer to the points (x_1, y_1), (x_2, y_2), (x_3, y_3) as P_1, P_2, P_3 and the corresponding complex numbers $x_1 + iy_1$, $x_2 + iy_2$, $x_3 + iy_3$ will be called z_1, z_2, z_3.

★12. Show that the radius vector corresponding to $z_1 + z_2$ is a diagonal of the parallelogram defined by OP_1, OP_2. (See Fig. 22.6.)

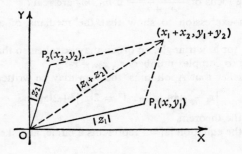

Fig. 22.6

Deduce that $|z_1 + z_2| \leqslant |z_1| + |z_2|$.
Under what circumstances is $|z_1 + z_2| = |z_1| + |z_2|$?

13. Show that

$$|z_1 - z_2| \leqslant |z_1| + |z_2|,$$
$$|z_1 - z_2| \geqslant \big||z_1| - |z_2|\big|.$$

[Show that, in Fig. 22.6, $P_1 P_2 = |z_1 - z_2|$.]

14. If $z_1 = 2 + i$, $z_2 = 1 + 3i$, show on the Argand diagram the radius vectors corresponding to

(i) $z_1 + z_2$,   (ii) $z_1 - z_2$,   (iii) $z_2 - z_1$,   (iv) $z_1 z_2$.

Arrange in order of magnitude:

$$|z_1|, \quad |z_2|, \quad |z_1 + z_2|, \quad |z_1 - z_2|, \quad |z_1 z_2|.$$

15. Repeat No. 14 for

$$z_1 = \tfrac{1}{2} + \tfrac{1}{2}i, \qquad z_2 = -\tfrac{1}{4} - \tfrac{3}{4}i.$$

16. Find interpretations, in terms of the Argand diagram, of:

    (i) $\dfrac{z_1 + z_2}{2}$,   (ii) $\dfrac{z_1 + z_2 + z_3}{3}$,   (iii) $\dfrac{m_1 z_1 + m_2 z_2}{m_1 + m_2}$.

17. Find the conditions that the triangle $P_1 O P_2$ should be:
    (i) isosceles with $OP_1 = OP_2$,
    (ii) right-angled at O.

    A radius vector, starting at OX, rotates in the positive sense, through positions OP_1, OP_2, OP_3 in order. Show that, if

    $$z_3 - z_2 = (z_2 - z_1)(\cos \tfrac{1}{2}\pi + i \sin \tfrac{1}{2}\pi),$$

    then the isosceles triangle $P_1 P_2 P_3$ is right-angled at P_2.

18. What is the locus of P as λ varies through real values from $-\infty$ to $+\infty$ if $z = 1 + i + \lambda(1 + 2i)$?
    Find the minimum value of $|z|$.

19. What is the locus of $\dfrac{m_1 z_1 + m_2 z_2}{m_1 + m_2}$ if m_1, m_2 are real?

    Use this expression to show that the medians of a triangle are concurrent.

20. Show that, for any triangle, an origin may be taken so that the vertices correspond to complex numbers z_1, z_2, $-z_2$.

    Show further that Apollonius' theorem may be written in the form

    $$|z_1 - z_2|^2 + |z_1 + z_2|^2 = 2|z_1|^2 + 2|z_2|^2$$

    and prove the theorem.

21. Show that the equation $|z| = 1$ represents a circle, and sketch the locus.

*22. Show that

    $$|z - z_1| = \sqrt{\{(x - x_1)^2 + (y - y_1)^2\}}.$$

    This means that we may interpret $|z - z_1|$ as the distance between the points corresponding to z, z_1.

23. If $c = a + ib$, express the equation $|z - c| = r$ as an equation in x, y, a, b, r. What is the locus?

24. Shade the areas on the Argand diagram represented by the inequalities:

    (i) $|z| < 1$,   (ii) $|z - 3| < 1$,   (iii) $|z + 5| < 2$.

25. If z satisfies both the inequalities $|z| < 4$, $|z - 2| > 1$, shade the area on the Argand diagram in which (x, y) must lie.

26. What locus is represented by the equations:

    (i) $|z - a| = x + a$,   (ii) $|z - ae| + |z + ae| = 2a$?

27. Express in modulus notation the equations of the circles:
    (i) centre O, radius 2,   (ii) centre (1, 0), radius 1,
    (iii) centre $(-a, 0)$, radius b.

28. Write in modulus notation the equation of the circles:

(i) $(x-3)^2+y^2 = 2$,　　(ii) $x^2+y^2+6x-4y-12 = 0$.

29. We define $\dfrac{dz}{dt}$ by the relationship

$$\frac{dz}{dt} = \frac{dx}{dt}+i\frac{dy}{dt}$$

and write this $\dot{z}=\dot{x}+i\dot{y}$. Express $\dot{z}$ in terms of r, θ, $\dot{r}$, $\dot{\theta}$ where $x=r\cos\theta$, $y=r\sin\theta$ and deduce the radial and transverse components of the velocity of a particle at the point given in polar coordinates (r, θ). Also find the radial and transverse components of acceleration at this point.

De Moivre's theorem

22.7. In Exercise 22a, No. 9, the reader was introduced to what is known as De Moivre's theorem. This theorem was published by De Moivre in 1730, but the substance of it had appeared in the posthumous publication of Cotes in 1722. The theorem states that, for any rational value of n, one value of $(\cos\theta+i\sin\theta)^n$ is given by

$$(\cos\theta+i\sin\theta)^n = \cos n\theta+i\sin n\theta.$$

The reason for saying *one* value of $(\cos\theta+i\sin\theta)^n$ is that there is more than one value for expressions such as $(\cos\theta+i\sin\theta)^{3/2}$: we return to this in § 22.8. A proof of the theorem may proceed as follows:

Stage 1. Prove that

$$(\cos\theta+i\sin\theta)(\cos\phi+i\sin\phi) = \cos(\theta+\phi)+i\sin(\theta+\phi).$$

[Expand the left-hand side.]

Stage 2. Use induction to prove the theorem for positive integral values of n.

Stage 3. Use the identity $(\cos\theta+i\sin\theta)(\cos\theta-i\sin\theta)=1$ to show that:

(i) $(\cos\theta+i\sin\theta)^{-1}=\cos\theta-i\sin\theta=\cos(-\theta)+i\sin(-\theta)$.

(ii) if $n=-m$, where m is a positive integer,

$$(\cos\theta+i\sin\theta)^n = \cos n\theta+i\sin n\theta.$$

In accordance with the usual laws of algebra we take

$$(\cos\theta+i\sin\theta)^0 = 1.$$

Stage 4. If $n=\dfrac{1}{q}$, where q is an integer (positive or negative, but not zero), show that *one* value of $(\cos\theta+\text{i}\sin\theta)^n$ is $\cos\dfrac{1}{q}\theta+\text{i}\sin\dfrac{1}{q}\theta$ by finding the value of $\left(\cos\dfrac{1}{q}\theta+\text{i}\sin\dfrac{1}{q}\theta\right)^q$.

Stage 5. If n is a rational number say $\dfrac{p}{q}$, where p, q are integers, we have by Stage 4 one value of

$$(\cos\theta+\text{i}\sin\theta)^{\frac{1}{q}} = \cos\dfrac{1}{q}\theta+\text{i}\sin\dfrac{1}{q}\theta,$$

$$(\cos\theta+\text{i}\sin\theta)^{\frac{p}{q}} = \left(\cos\dfrac{1}{q}\theta+\text{i}\sin\dfrac{1}{q}\theta\right)^p,$$

$$= \cos\dfrac{p}{q}\theta+\text{i}\sin\dfrac{p}{q}\theta.$$

Qu. 9. Express in the form $x+\text{i}y$:
  (i) $(\cos\theta+\text{i}\sin\theta)^5$, (ii) $1/(\cos\theta+\text{i}\sin\theta)^2$,
  (iii) $(\cos\theta-\text{i}\sin\theta)^{-3}$, (iv) $(\cos\theta+\text{i}\sin\theta)^2(\cos\theta+\text{i}\sin\theta)^3$,
  (v) $\dfrac{\cos\theta+\text{i}\sin\theta}{\cos\phi+\text{i}\sin\phi}$, (vi) $\dfrac{\cos\theta+\text{i}\sin\theta}{\cos\phi-\text{i}\sin\phi}$,

Qu. 10. Find one value of:
  (i) $\sqrt{(\cos2\theta+\text{i}\sin2\theta)}$, (ii) $\sqrt[3]{(\cos2\pi+\text{i}\sin2\pi)}$,
  (iii) $\sqrt{(\cos\theta+\text{i}\sin\theta)^3}$.

Complex roots of unity

22.8. In § 22.3, p. 471, we referred to the theorem that an equation of the nth degree has n roots. This means that the equation

$$z^3-1 = 0$$

has three roots: one of them is 1 but what are the others? Qu. 11 gives one method of finding out.

Qu. 11. Use the identity $z^3-1=(z-1)(z^2+z+1)$ to find the three cube roots of unity.

Still, the method of Qu. 11 would not work for, say, the roots of $z^7-1=0$, and the reader may well have been wondering what this has got to do with De Moivre's theorem. Now, we can express 1 as a complex number in infinitely many ways:

$$\ldots \cos(-2\pi) + i \sin(-2\pi), \quad \cos 0 + i \sin 0,$$
$$\cos 2\pi + i \sin 2\pi, \quad \cos 4\pi + i \sin 4\pi, \ldots$$

or, in general $\qquad \cos 2k\pi + i \sin 2k\pi$

where k is an integer.

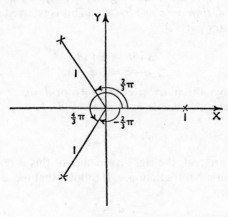

Fig. 22.7

By De Moivre's theorem, one value of

$$\sqrt[3]{(\cos\theta + i\sin\theta)} = (\cos\theta + i\sin\theta)^{\frac{1}{3}} = \cos\tfrac{1}{3}\theta + i\sin\tfrac{1}{3}\theta.$$

Therefore values of $\sqrt[3]{1}$ are given by

$$\ldots \cos\left(-\frac{2\pi}{3}\right) + i\sin\left(\frac{2\pi}{3}\right), \quad \cos\frac{0}{3} + i\sin\frac{0}{3},$$
$$\cos\frac{2\pi}{3} + i\sin\frac{2\pi}{3}, \quad \cos\frac{4\pi}{3} + i\sin\frac{4\pi}{3}, \ldots$$

or in general

$$\cos\frac{2k\pi}{3} + i\sin\frac{2k\pi}{3}. \quad \text{(See Fig. 22.7.)}$$

Qu. 12. Show that the expression $\cos\frac{2}{3}k\pi + i\sin\frac{2}{3}k\pi$ represents the same complex number when k is replaced by (i) $k+3$, (ii) $k+3m$, (k, m integral).

Qu. 13. By writing $-1 = \cos\pi + i\sin\pi$, and in two other ways, find the cube roots of -1.

Qu. 14. Writing $\omega = \cos \frac{2}{3}\pi + i \sin \frac{2}{3}\pi$, show that the cube roots of 1 are 1, ω, ω^2. Prove that $1 + \omega + \omega^2 = 0$ in two different ways. [For a hint, see Qu. 11, p. 484.] Show also, that ω is the square of ω^2.

Real and imaginary parts

22.9. One advantage of using complex numbers is that sometimes it is possible to do two pieces of working simultaneously! The possibility rests on the following result: *If $a + ib = c + id$, where a, b, c, d are real, then $a = c$ and $b = d$.* [This is referred to as 'equating real and imaginary parts'.]

Proof:

$$a + ib = c + id.$$
$$\therefore\ a - c = i(d - b).$$

We now have two alternatives: either $d \neq b$ or $d = b$.

Assuming that $d \neq b$, divide both sides by $d - b$.

$$\therefore\ i = \frac{a - c}{d - b}.$$

Since a, b, c, d are real, the right-hand side of this is real; but i is not, and so we obtain a contradiction. It follows that the assumption that $d \neq b$ is incorrect.

$$\therefore\ d = b.$$
$$\therefore\ a - c = i.0.$$
$$\therefore\ a = c.$$

Example 1. *Prove that* $\tan 3\theta = \dfrac{3 \tan \theta - \tan^3 \theta}{1 - 3 \tan^2 \theta}$.

By De Moivre's theorem,

$$\cos 3\theta + i \sin 3\theta = (\cos \theta + i \sin \theta)^3.$$

The right-hand side may be written

$$\cos^3 \theta + 3i \cos^2 \theta \sin \theta - 3 \cos \theta \sin^2 \theta - i \sin^3 \theta.$$

$$\therefore\ \cos 3\theta + i \sin 3\theta = \cos^3 \theta - 3 \cos \theta \sin^2 \theta + i(3 \cos^2 \theta \sin \theta - \sin^3 \theta).$$

Equating real and imaginary parts,

$$\cos 3\theta = \cos^3 \theta - 3 \cos \theta \sin^2 \theta,$$
$$\sin 3\theta = 3 \cos^2 \theta \sin \theta - \sin^3 \theta.$$

By division, $\quad \dfrac{\sin 3\theta}{\cos 3\theta} = \dfrac{3 \cos^2 \theta \sin \theta - \sin^3 \theta}{\cos^3 \theta - 3 \cos \theta \sin^2 \theta}.$

Dividing numerator and denominator of the right-hand side by $\cos^3 \theta$,

$$\tan 3\theta = \frac{\sin 3\theta}{\cos 3\theta} = \frac{3 \tan \theta - \tan^3 \theta}{1 - 3 \tan^2 \theta}.$$

Qu. 15. Show that

$$\tan 4\theta = (4 \tan \theta - 4 \tan^3 \theta)/(1 - 6 \tan^2 \theta + \tan^4 \theta).$$

Exercise 22c

Simplify:

1. $(\cos \theta + i \sin \theta)^3 (\cos 2\theta + i \sin 2\theta)$.
2. $(\cos \theta + i \sin \theta)^2 (\cos \theta - i \sin \theta)^{-2}$.
3. $\dfrac{\cos 4\theta + i \sin 4\theta}{(\cos \theta + i \sin \theta)^3}$.
4. $\dfrac{(\cos 2\theta + i \sin 2\theta)^3}{(\cos \theta + i \sin \theta)^5}$.
5. $\dfrac{\cos 5\theta + i \sin 5\theta}{(\cos \theta - i \sin \theta)^3}$.
6. $\dfrac{\cos \theta + i \sin \theta}{\cos 2\theta - i \sin 2\theta}$.
7. $(\cos \phi + i \sin \phi)^2 (\cos \theta + i \sin \theta)^3$.
8. $(\cos 2\theta + i \sin 2\theta)^3 (\cos 3\phi - i \sin 3\phi)^2$.
9. By writing 1 in the form $\cos \theta + i \sin \theta$, where $\theta = -2\pi$, 0, 2π, 4π, find the fourth roots of unity.

   What are the fourth roots of -1?
*10. By writing 1 in the form $\cos 2k\pi + i \sin 2k\pi$ $(k = -2, -1, 0, 1, 2)$, find the fifth roots of unity.

   Show that, if z, $\bar{z}$ are conjugate complex numbers, the expansion of $(x - \bar{z})(x - z)$ is a quadratic in x with real coefficients. Hence write $x^5 - 1$ in the form $(x - 1)q_1 q_2$, where q_1, q_2 are quadratic expressions with real coefficients.
11. Find the sixth roots of 1 and show the corresponding vectors on the Argand diagram. Deduce the real quadratic factors of $x^4 + x^2 + 1$.
12. Show that the nth roots of -1, together with the nth roots of 1, form a complete set of $2n$th roots of 1. How can the $2n$th roots of -1 be deduced from these? Illustrate your answer by showing the corresponding vectors on the Argand diagram for the case $n = 5$.
13. If α is a seventh root of unity, other than 1, show that the other roots are α^2, α^3, α^4, α^5, α^6, 1. Show further that

$$1 + \alpha + \alpha^2 + \alpha^3 + \alpha^4 + \alpha^5 + \alpha^6 = 0.$$

   Do analogous properties hold for all other nth roots of unity?
14. Find the real factors of:
    (i) $x^7 - 1$,   (ii) $x^5 + 1$,   (iii) $x^{2n} - 1$.

★15. If $z = \cos\theta + i\sin\theta$, show that:

$$\frac{1}{z} = \cos\theta - i\sin\theta, \qquad z^3 = \cos 3\theta + i\sin 3\theta,$$

$$\frac{1}{z^3} = \cos 3\theta - i\sin 3\theta.$$

Show further that

$$\left(z + \frac{1}{z}\right)^3 = 8\cos^3\theta,$$

and, by expanding $\left(z + \dfrac{1}{z}\right)^3$, prove that

$$2\cos 3\theta + 6\cos\theta = 8\cos^3\theta.$$

Hence express $\cos 3\theta$ in terms of powers of $\cos\theta$.

16. With the notation of No. 15, show that $\left(z - \dfrac{1}{z}\right)^3 = -8i\sin^3\theta$ and hence express $\sin 3\theta$ in terms of $\sin\theta$.

17. Use the method of No. 15 to prove that

$$\cos^4\theta = \tfrac{1}{8}(\cos 4\theta + 4\cos 2\theta + 3),$$

and express $\cos 4\theta$ in terms of $\cos\theta$.

$$\left[\text{Expand } \left(z + \frac{1}{z}\right)^4.\right]$$

18. Prove that

$$\sin^5\theta = \tfrac{1}{16}(\sin 5\theta - 5\sin 3\theta + 10\sin\theta).$$

19. Prove that

$$\cos^6\theta = \tfrac{1}{32}(\cos 6\theta + 6\cos 4\theta + 15\cos 2\theta + 10).$$

20. Show that

$$\tan 5\theta = \frac{5\tan\theta - 10\tan^3\theta + \tan^5\theta}{1 - 10\tan^2\theta + 5\tan^4\theta}.$$

[Use the method of Example 1, p. 486.]

21. Find expressions in terms of $\tan\theta$ for:

    (i) $\tan 6\theta$,   (ii) $\tan 2n\theta$,   (iii) $\tan(2n+1)\theta$.

22. Show that

$$\tan 4\theta = \frac{4t - 4t^3}{1 - 6t^2 + t^4},$$

where $t = \tan\theta$.

Hence find the roots of the equation

$$t^4 + 4t^3 - 6t^2 - 4t + 1 = 0$$

correct to three significant figures.

23. Solve the equation

$$t^5 - 10t^4 - 10t^3 + 20t^2 + 5t - 2 = 0$$

correct to three significant figures.

24. If $w = u + iv$, $z = x + iy$, and $w = z^3$, express u, v in terms of x, y.

25. Find $(a + ib) \div (c + id)$ by equating real and imaginary parts in the equation

$$(c + id)(p + iq) = a + ib.$$

26. (i) Find the square roots of $-5 + 12i$ by equating real and imaginary parts of $(a + ib)^2 = -5 + 12i$.

(ii) Find the square roots of i
  (a) by the method above
  (b) by using De Moivre's theorem.

★27. Let

$$C \equiv 1 + \cos\theta + \cos 2\theta + \ldots + \cos(n-1)\theta,$$
$$S \equiv \sin\theta + \sin 2\theta + \ldots + \sin(n-1)\theta,$$
$$Z = C + iS, \qquad z = \cos\theta + i\sin\theta.$$

(i) Show that $Z = (1 - z^n)/(1 - z)$.

(ii) Express Z with a real denominator.

(iii) Deduce expressions for C, S by equating real and imaginary parts.

28. Sum the series

$$C \equiv 1 + a\cos\theta + a^2\cos 2\theta + \ldots + a^n\cos n\theta,$$
$$S \equiv a\sin\theta + a^2\sin 2\theta + \ldots + a^n\sin n\theta.$$

29. Examine the answers to Nos. 27 (iii), and 28. Show how the series in these two questions could be summed by multiplying both sides of the equations by some expression.

30. Find the sum to infinity of the series

$$\cos\theta\cos\theta + \cos^2\theta\cos 2\theta + \cos^3\theta\cos 3\theta + \ldots.$$

31. Find the sum to infinity of the series

$$\sin\theta + \tfrac{1}{2}\sin(\theta + \phi) + \tfrac{1}{4}\sin(\theta + 2\phi) + \ldots.$$

e^z, $\cos z$, $\sin z$

22.10. In this section it is intended to give the reader a glimpse of the way in which it is possible to extend the above functions to cover complex variables.

First consider the function e^x, where x is real. We have to find some property of the function which we can use to define what we mean by e^z where z is complex. The most obvious way to start is to go back to our definition of e^x but unfortunately this does not lend itself to an extension to complex numbers; so we have to find some other property of e^x. Now, e^x can be expanded as a power series:

$$e^x = 1 + x + \frac{x^2}{2!} + \ldots + \frac{x^n}{n!} + \ldots$$

and we can readily give e^z a meaning by *defining* it by the series

$$e^z = 1 + z + \frac{z^2}{2!} + \ldots + \frac{z^n}{n!} + \ldots.$$

We know that the series for e^x is convergent for all real values of x, and we shall assume that the series for e^z is convergent for complex values of z, but the reader should be aware that this does not follow automatically.

The definition of e^z by a series poses another problem: will the usual laws of indices still hold? In particular, we must satisfy ourselves that

$$e^w \times e^z = e^{w+z}.$$

This is something which has to be deduced from our definition of e^z: that is, it is necessary to show that

$$\left(1 + w + \frac{w^2}{2!} + \ldots + \frac{w^n}{n!} + \ldots\right)\left(1 + z + \frac{z^2}{2!} + \ldots + \frac{z^n}{n!} + \ldots\right)$$
$$= 1 + (w+z) + \frac{(w+z)^2}{2!} + \ldots + \frac{(w+z)^n}{n!} + \ldots.$$

The conditions under which infinite series may be multiplied are outside the scope of this book and so a proof will not be given here, but the reader should work Qu. 16.

Qu. 16. Show that term-by-term multiplication of the first few terms of the series for e^w, e^z give the first few terms of the series for e^{w+z}.

Also find the terms of degree n in the product and show that they reduce to $(w+z)^n/n!$

Now let us see what happens if we write $z = x + iy$ in the expression e^z.

$$e^z = e^{x+iy} = e^x \times e^{iy}.$$

e^x is real (and familiar), so we shall examine the function e^{iy}.

$$e^{iy} = 1 + iy + \frac{i^2 y^2}{2!} + \frac{i^3 y^3}{3!} + \frac{i^4 y^4}{4!} + \ldots + \frac{i^n y^n}{n!} + \ldots,$$
$$= 1 \qquad -\frac{y^2}{2!} \qquad +\frac{y^4}{4!} - \ldots$$
$$+ i\left(y \qquad -\frac{y^3}{3!} + \ldots\right).$$

But
$$1 - \frac{y^2}{2!} + \frac{y^4}{4!} - \ldots = \cos y$$

and
$$y - \frac{y^3}{3!} + \frac{y^5}{5!} - \ldots = \sin y$$

so that
$$e^{iy} = \cos y + i \sin y.$$

A result equivalent to this, discovered by Cotes, was published in 1722, after his death.

Qu. 17. Use the results
$$(\cos \theta + i \sin \theta)(\cos \theta - i \sin \theta) = 1$$

and
$$e^w \times e^z = e^{w+z}$$

to show that
$$e^{-i\theta} = \cos \theta - i \sin \theta.$$

NOTE. In most texts the reader will find the series
$$1 + z + \frac{z^2}{2!} + \ldots + \frac{z^n}{n!} + \ldots$$

denoted by exp z.

We are now in a position to turn to the problem of assigning meanings to the functions $\sin z$, $\cos z$, where z is complex.

From above, $e^{i\theta} = \cos \theta + i \sin \theta,$
and from Qu. 17, $e^{-i\theta} = \cos \theta - i \sin \theta.$

Hence $\cos \theta = \dfrac{e^{i\theta} + e^{-i\theta}}{2},$

$$\sin \theta = \frac{e^{i\theta} - e^{-i\theta}}{2i}.$$

Since e^z has been defined for complex z, we may use these last two equations to define $\cos z$ and $\sin z$:

$$\cos z = \frac{e^{iz} + e^{-iz}}{2}, \tag{1}$$

$$\sin z = \frac{e^{iz} - e^{-iz}}{2i}. \tag{2}$$

These results were given by Euler in 1748, twenty years before Lambert introduced the hyperbolic functions.

The hyperbolic functions $\cosh x$, $\sinh x$ were defined for real values of x on p. 436:

$$\cosh x = \tfrac{1}{2}(e^x + e^{-x}), \qquad \sinh x = \tfrac{1}{2}(e^x - e^{-x}).$$

These definitions can be used to define the functions $\cosh z$, $\sinh z$ of a complex variable z:

$$\cosh z = \tfrac{1}{2}(e^z + e^{-z}), \qquad \sinh z = \tfrac{1}{2}(e^z - e^{-z}).$$

Replacing z by iz,

$$\cosh iz = \tfrac{1}{2}(e^{iz} + e^{-iz}), \qquad \sinh iz = \tfrac{1}{2}(e^{iz} - e^{-iz}), \tag{3}$$

and so from (1), (2), (3) above we obtain the following relations connecting circular and hyperbolic functions:

$$\cosh iz = \cos z, \qquad \sinh iz = i \sin z.$$

Qu. 18. Confirm these last relationships for real x by replacing z by x in equations (3) and expressing e^{ix}, e^{-ix} in terms of $\cos x$, $\sin x$.

Qu. 19. Express $\cosh z$, $\sinh z$ in terms of the corresponding circular functions.

Qu. 20. Use the series

$$e^w = 1 + w + \frac{w^2}{2!} + \ldots + \frac{w^n}{n!} + \ldots$$

to show that

$$\cos z = 1 - \frac{z^2}{2!} + \frac{z^4}{4!} - \ldots,$$

$$\sin z = z - \frac{z^3}{3!} + \frac{z^5}{5!} - \ldots.$$

Much of the earlier part of this book was concerned with identities connecting sines and cosines: will they still hold for $\cos z$ and $\sin z$ when z is complex?

Qu. 21. Deduce from equations (1) and (2) above that:
(i) $\cos^2 z + \sin^2 z = 1$,   (ii) $\cos(-z) = \cos z$,
(iii) $\sin(-z) = -\sin z$,   (iv) $\cos(w+z) = \cos w \cos z - \sin w \sin z$,
(v) $\sin(w+z) = \sin w \cos z + \cos w \sin z$.

The other trigonometrical identities follow from these (if you cannot satisfy yourself about this, turn back to where they were proved for real numbers); and so trigonometrical functions of a complex variable may be manipulated by the same identities as those for a real variable.

Qu. 22. A function of a complex variable is said to be periodic with period p if $f(z+p) = f(z)$ for all z. Show that $\cos z$, $\sin z$ have period 2π.

Exercise 22d (Miscellaneous)

1. Express in the form $a + ib$:

   (i) $(x + iy)^6$, (ii) $\dfrac{1}{3\cos\theta + 2i\sin\theta - 1}$, (iii) $\sqrt{\dfrac{\cos\theta + i\sin\theta}{\cos 2\theta - i\sin 2\theta}}$,

   (iv) $\dfrac{z+1}{z-1}$ where $z = x + iy$.

2. Express in the form $a + ib$:

   (i) $\dfrac{(\cos\theta + i\sin\theta)^3}{(\cos 2\theta + i\sin 2\theta)^2}$, (ii) $\dfrac{(\cos 2\theta + i\sin 2\theta)^2}{(\cos\theta - i\sin\theta)^4}$,

   (iii) $1 - \text{cis}\,\theta + \text{cis}\,2\theta - \ldots + (-1)^{n-1}\,\text{cis}\,(n-1)\theta$,

   where $\text{cis}\,r\theta = \cos r\theta + i\sin r\theta$.

3. P_1, P_2 are points on the Argand diagram corresponding to the complex numbers z_1, z_2. Show that the mid-point of $P_1 P_2$ corresponds to $\frac{1}{2}(z_1 + z_2)$. Hence prove that the mid-points of the sides of a plane quadrilateral are the vertices of a parallelogram.

4. What is the locus given by

$$z\bar{z} + 2(z + \bar{z}) = 0$$

where $z = x + iy$ and $\bar{z}$ is its conjugate?

   Express as an equation connecting z, $\bar{z}$ the condition that (x, y) should lie on the circle centre $(2, -1)$ radius 3.

5. (i) Prove that $\cos^5\theta = \frac{1}{16}(\cos 5\theta + 5\cos 3\theta + 10\cos\theta)$
   and find an expression for $\sin^4\theta$ in terms of $\cos 2\theta$, $\cos 4\theta$.

   (ii) Express $\tan(\theta_1 + \theta_2 + \theta_3 + \theta_4)$ in terms of $\tan\theta_1$, $\tan\theta_2$, $\tan\theta_3$, $\tan\theta_4$.

6. (i) Express -1 in the form $\cos\theta + i\sin\theta$. Hence obtain the three linear factors of $z^3 + 1$.

   (ii) Find the real factors of $x^5 - a^5$.

7. Show that, if a quadratic equation is satisfied by a complex number, then the conjugate complex number is also a root of the equation.

   Hence find the quadratic equation satisfied by the complex number $2 - 3i$.

   Find the four roots of the equation $z^4 - 3z^3 + 4z^2 - 3z + 1 = 0$.

8. (i) Solve the equation $z^4 - 6z^2 + 25 = 0$.

   (ii) Given that one root of the equation

$$z^4 - 6z^3 + 23z^2 - 34z + 26 = 0$$

   is $1 + i$, find the others.

9. If $z + \dfrac{1}{z} = -1$, prove that $z^5 + \dfrac{1}{z^5} = -1$ and find the value of $z^{11} + \dfrac{1}{z^{11}}$.

10. Prove that, if $a + ib = c + id$ where a, b, c, d are real, then $a = c$, $b = d$.

Find the sum to infinity of $1 - \dfrac{1}{2} \cos\,\theta + \dfrac{1}{4} \cos 2\theta - \dfrac{1}{8} \cos 3\theta + \ldots$.

11. Find the sum of the series

$$1 + 2 \cos 2\theta + 4 \cos 4\theta + \ldots + 2^n \cos 2n\theta.$$

12. Prove by induction that, for positive integral values of n,

$$(\cos\,\theta + i \sin\,\theta)^n = \cos n\theta + i \sin n\theta.$$

Hence show that the identity also holds for negative integral values of n.

Find the sum to infinity of the series

$$\sin\,\theta \sin\,\theta + \sin^2\,\theta \sin 2\theta + \sin^3\,\theta \sin 3\theta + \ldots.$$

In Nos. 13 to 18 x, y, u, v are real numbers, z, w are the complex numbers $x + iy$, $u + iv$.

13. If w, z are connected by the equation $w = z + 1$, find the locus of (u, v) when:

(i) $x + 1 = 0$, (ii) $x + y - 1 = 0$, (iii) $|z + 1| = 1$, (iv) $|z - 3| = 2$.

14. If $w = 3z$, find the locus of (u, v) when:

(i) $x - 2y = 0$, (ii) $y - 1 = 0$, (iii) $|z| = 1$, (iv) $|z - 2| = 2$.

15. Prove that $|z^2| = |z|^2$ and that one value of Arg $(z^2) = 2$ arg z.

The point (x, y) moves once round each of the circles (i) $|z| = 1$, (ii) $|z| = 2$ in a counter-clockwise sense. Describe the corresponding motions of (u, v) if $w = z^2$.

16. If $w = z^2$, find the locus of (u, v) if the locus of (x, y) is: (i) $x^2 - y^2 = 1$, (ii) $xy = 1$.

17. The point (x, y) moves once round the circle $|z| = 1$ in a counter-clockwise sense. Describe the corresponding motion of (u, v) if $w = 1/z$.

Also describe the motion of (u, v) if (x, y) moves round the circle $|z| = 2$ in a counter-clockwise sense.

18. The point (x, y) describes in a counter-clockwise sense the boundary of the quadrant defined by:

$$y = 0,\ 0 \leqslant x \leqslant 2; \qquad |z| = 2,\ 0 \leqslant \arg z \leqslant \tfrac{1}{2}\pi;$$

$$x = 0,\ 2 \geqslant y \geqslant 0.$$

Describe, with the aid of diagrams the motion of (u, v) if:

(i) $w = z^2$, (ii) $w = \dfrac{1}{z}$.

19. Express $\dfrac{1}{1-2x\cos\theta+x^2}$ in partial fractions and hence find the expansion of this expression in ascending powers of x.

20. Define $\cos z$, $\sin z$ when z is complex. Prove, directly from your definitions, that:
  (i) $\sin 2z = 2\sin z\cos z$,   (ii) $\cos 2z = \cos^2 z - \sin^2 z$,
  (iii) $\cos(w-z) = \cos w\cos z + \sin w\sin z$,
  (iv) $\cos w + \cos z = 2\cos\frac{1}{2}(w+z)\cos\frac{1}{2}(w-z)$.

21. Express $\cos z$, $\sin z$ in terms of the hyperbolic cosine and sine. Deduce identities connecting hyperbolic functions corresponding to those in No. 20.

22. Define $\cosh z$, $\sinh z$ when z is complex. Deduce from your definitions the identities:
  (i) $\cosh^2 z - \sinh^2 z = 1$,
  (ii) $\sinh(w+z) = \sinh w\cosh z + \cosh w\sinh z$.

    Express $\cosh z$, $\sinh z$ in terms of the circular cosine and sine, and deduce identities corresponding to the two above.

23. (i) Write down the values of $|z|$ and $\arg z$, where $z = x + iy$. Illustrate by means of an Argand diagram.

    The numbers c and p are given, c being real and p being complex, with $p = a + ib$; $\bar{z}$ and $\bar{p}$ denote the conjugates of z and p respectively. Prove that, if

$$z\bar{z} - \bar{p}z - p\bar{z} + c = 0,$$

then the point on the Argand diagram which represents z lies on a certain circle whose centre and radius should be determined.

(ii) Prove that, if

$$x + iy = \frac{1}{\lambda + i\mu},$$

then the points on the Argand diagram defined by making λ constant lie on a circle, and the points defined by making μ a constant also lie on a circle.

    Prove also that, whatever be the values of the constants, the centres of the two systems of circles obtained lie on two fixed perpendicular lines.   O.C.

24. The coordinates (x, y) of a point P are expressible in terms of real variables u and v by the formula

$$x + iy = (u + iv)^2.$$

Prove that the locus of P is a parabola (i) when u varies and v is constant, and also (ii) when v varies and u is constant. Prove also that all the parabolas have a common focus and a common axis.

Prove that through a given point (x_0, y_0) there pass two parabolas, one of each system $u = $ constant, $v = $ constant, which cut at right angles.

O.C.

25. (i) Complex numbers z_1 and z_2 are given by the formulae

$$z_1 = R_1 + i\omega L, \qquad z_2 = R_2 - \frac{i}{\omega C},$$

and z is given by the formula

$$\frac{1}{z} = \frac{1}{z_1} + \frac{1}{z_2}.$$

Find the value of ω for which z is a real number.

(ii) Use De Moivre's theorem to prove that, if

$$2 \cos \theta = x + \frac{1}{x},$$

then

$$2 \cos n\theta = x^n + \frac{1}{x^n}.$$

Hence, or otherwise, solve the equation

$$5x^4 - 11x^3 + 16x^2 - 11x + 5 = 0.$$

O.C.

26. Show how to represent the sum and difference of two complex numbers in an Argand diagram.

If $s_n = z_1 + z_2 + \ldots + z_n$, where $z_1, z_2, \ldots, z_n$ are given complex numbers, prove that

$$|s_n| \leqslant |z_1| + |z_2| + \ldots + |z_n|.$$

Show that the equation $z^3 + 27 = 0$ has one real root and two complex roots. Give the modulus and argument of each root and show the roots in an Argand diagram.

Express $(2 + i)^2(1 - i)/(1 + 3i)$ in the form $a + ib$.

O.C.

27. Prove that, if u, v, x, y are real numbers connected by the relation

$$u + iv = e^{x + iy},$$

then

$$u = e^x \cos y, \qquad v = e^x \sin y.$$

The variable point (x, y) describes, in a counter-clockwise sense, the boundary of the semi-infinite strip defined by the three straight lines

$$y = \tfrac{1}{4}\pi, \quad x \leqslant 0; \qquad x = 0; \qquad y = \tfrac{1}{2}\pi, \quad x \leqslant 0.$$

Illustrate the corresponding locus of the point (u, v) by a sketch, giving a careful explanation.

O.C.

ANSWERS

CHAPTER 1

page 2.

Qu. 1. (i) $16x(2x^2+3)^3$, (ii) $\dfrac{x-1}{\sqrt{(x^2-2x+1)}}$,

(iii) $-4(2x-1)^{-3}$, (iv) $4\cos(4x-7)$,

(v) $3\tan^2 x \sec^2 x$, (vi) $-6\cos 3x \sin 3x$.

Qu. 2. (i) $\frac{1}{6}(x^2+1)^3+c$, (ii) $\frac{1}{10}(2x+1)^5+c$,

(iii) $\frac{1}{7}x^7+\frac{3}{5}x^5+x^3+x+c$, (iv) $-\frac{1}{6}\cos 3x+c$,

(v) $\frac{2}{9}(x^3+1)^{3/2}+c$, (vi) $\frac{1}{2}\tan^2 x+c$.

Qu. 3. (i) $\frac{1}{3}\cos^3 x-\cos x+c$, (ii) $\sin x-\frac{2}{3}\sin^3 x+\frac{1}{5}\sin^5 x+c$.

Qu. 4. (i) $\frac{1}{3}\sin^3 x-\frac{1}{5}\sin^5 x+c$, (ii) $\frac{1}{5}\cos^5 x-\frac{1}{3}\cos^3 x+c$.

Qu. 5. $\frac{1}{3}\sec^3 x-\sec x+c$.

page 3. **Exercise 1a**

1. (i) $30x(5x^2-1)^2$, (ii) $\dfrac{2-8x}{(2x^2-x+3)^3}$,

(iii) $\frac{2}{3}x(x^2+4)^{-2/3}$, (iv) $-5\csc^2 5x$,

(v) $-5\sin(5x-1)$, (vi) $\frac{1}{3}\sin\frac{2}{3}x$,

(vii) $\dfrac{1}{2\sqrt{x}}\sec^2\sqrt{x}$, (viii) $4\sec^2 2x \tan 2x$,

(ix) $-\frac{1}{2}\cot x\sqrt{\csc x}$.

2. (i) $\frac{1}{12}(x^2-3)^6+c$, (ii) $\frac{1}{18}(3x-1)^6+c$,

(iii) $\frac{1}{4}x^4+\frac{4}{3}x^3+2x^2+c$, (iv) $-\frac{1}{2}(x^2+1)^{-1}+c$,

(v) $-\frac{1}{4}(x^2+2x-5)^{-2}+c$, (vi) $\frac{1}{3}(x^2-3x+7)^3+c$,

(vii) $-\frac{1}{8}(4x^2-7)^{-1}+c$, (viii) $\frac{2}{9}(3x^2-5)^{3/2}+c$,

(ix) $\frac{1}{7}x^7+\frac{1}{2}x^4+x+c$, (x) $\frac{2}{3}\sqrt{(x^3-3x)}+c$,

(xi) $-\frac{1}{2}(2x^2-4x+1)^{-1/2}+c$, (xii) $\frac{8}{7}x^7-\frac{12}{5}x^5+2x^3-x+c$.

3. (i) $\sin 3x+c$, (ii) $-\frac{1}{2}\cos(2x+3)+c$,

(iii) $\frac{1}{2}\sin^2 x+c$, (iv) $\frac{1}{6}\sin 2x+c$,

(v) $-\frac{1}{9}\cos^3 3x+c$, (vi) $\frac{1}{3}\tan^3 x+c$,

(vii) $\frac{1}{5}\sec^5 x+c$, (viii) $\frac{2}{3}\sin^{3/2} x+c$,

(ix) $-\frac{1}{2}\cot x^2+c$, (x) $2\sin\sqrt{x}+c$,

(xi) $-\frac{1}{3}\csc^3 x+c$.

4. (i) $\sin x-\frac{1}{3}\sin^3 x+c$, (ii) $2\sin\dfrac{x}{2}-\frac{4}{3}\sin^3\dfrac{x}{2}+\frac{2}{5}\sin^5\dfrac{x}{2}+c$

(iii) $\frac{1}{6}\cos^3 2x-\frac{1}{2}\cos 2x+c$, (iv) $\frac{1}{2}\sin(2x+1)-\frac{1}{6}\sin^3(2x+1)+c$,

page 4.

4. (v) $-\frac{1}{3}\cos^3 x+\frac{2}{5}\cos^5 x-\frac{1}{7}\cos^7 x+c,$ (vi) $\frac{1}{4}\sin^4 x-\frac{1}{6}\sin^6 x+c,$

   (vii) $\tan x+\frac{1}{3}\tan^3 x+c,$ (viii) $\operatorname{cosec} x-\frac{1}{3}\operatorname{cosec}^3 x+c,$

   (ix) $\sec x-\frac{2}{3}\sec^3 x+\frac{1}{5}\sec^5 x+c.$

5. (i) $\frac{1}{4}\sec^4 x+c,$ (ii) $\frac{1}{2}\tan^2 x+\frac{1}{4}\tan^4 x+k.$

6. $A+\frac{1}{2}\sin^2 x,\quad B-\frac{1}{2}\cos^2 x,\quad C-\frac{1}{4}\cos 2x.$

7. (i) $\frac{1}{2}(1-\cos x),$ (ii) $\frac{1}{2}(1+\cos 6x).$

8. (i) $\frac{1}{2}x+\frac{1}{4}\sin 2x+c,$ (ii) $\frac{1}{2}x-\frac{1}{2}\sin x+c,$

   (iii) $\frac{1}{2}x+\frac{1}{12}\sin 6x+c.$

9. $\frac{1}{4}-\frac{1}{2}\cos 2x+\frac{1}{4}\cos^2 2x,\quad \frac{1}{2}(1+\cos 4x).$

10. $\frac{3}{8}x+\frac{1}{4}\sin 2x+\frac{1}{32}\sin 4x+c.$

11. (i) $\frac{1}{2}x-\frac{1}{4}\sin 2x+c,$ (ii) $\frac{1}{2}x+\frac{3}{4}\sin\frac{2}{3}x+c,$

   (iii) $\frac{3}{8}x-\frac{1}{8}\sin 4x+\frac{1}{64}\sin 8x+c,$ (iv) $\frac{3}{8}x+\frac{1}{4}\sin x+\frac{1}{16}\sin 2x+c.$

13. (i) $2\sqrt{2}\sin\frac{x}{2}+c,$ (ii) $-\dfrac{1}{\sqrt{2}}\operatorname{cosec} x+c,$

   (iii) $\frac{1}{2}\sin^4 x+c,$ (iv) $-\frac{8}{3}\cos^3\frac{x}{2}+c.$

14. (i) $2\sin 2x\cos x,$ (ii) $\sin 5x+\sin x,$ (iii) $-\frac{1}{10}\cos 5x-\frac{1}{2}\cos x+c.$

15. (i) $\frac{1}{4}\cos 2x-\frac{1}{8}\cos 4x+c,$ (ii) $\frac{1}{2}\sin 2x+\sin x+c,$

   (iii) $\frac{1}{6}\sin 3x-\frac{1}{10}\sin 5x+c.$

Qu. 6. $\frac{1}{12}\sin^3 4x+c.$ **Qu. 7.** $-\cos x+\frac{2}{3}\cos^3 x-\frac{1}{5}\cos^5 x+c.$

Qu. 8. (i) $\frac{1}{15}(2x+1)^{3/2}(3x-1)+c,$ (ii) $\frac{1}{15}(2x+1)^{3/2}(3x-1)+c,$

   (iii) $\frac{1}{504}(3x-2)^7(21x+2)+c.$

page 7.   **Exercise 1b**

1. (i) $\frac{1}{20}(4x-1)^{3/2}(6x+1)+c,$ (ii) $\frac{2}{375}(5x+2)^{3/2}(15x-4)+c,$

   (iii) $\frac{1}{224}(2x-1)^7(14x+1)+c,$ (iv) $\frac{2}{3}(x+4)\sqrt{(x-2)}+c,$

   (v) $\frac{1}{36}(x-1)^6(5x+13)+c,$ (vi) $\frac{1}{168}(x-2)^6(21x^2+156x+304)+c,$

   (vii) $\dfrac{x^2-4x+8}{x-2}+c,$ (viii) $\frac{1}{5}(x-6)\sqrt{(2x+3)}+c.$

3. (i) $\frac{2}{135}(3x-4)^{3/2}(9x+8)+c,\quad \frac{1}{6}(3x^2-4)^{3/2}+c,$

   (ii) $\frac{1}{14}(x^2+5)^7+c,\quad \frac{1}{56}(x+5)^7(7x-5)+c,$

   (iii) $\frac{2}{3}(x+2)\sqrt{(x-1)}+c,\quad \sqrt{(x^2-1)}+c.$

4. (i) $\frac{1}{6}(2x^2+1)^{3/2}+c,$ (ii) $-\frac{1}{2}(x^3-x+4)^{-2}+c,$

   (iii) $\frac{2}{15}(2x-1)^{3/2}(3x+1)+c,$ (iv) $\frac{1}{2}\sin 2x-\frac{1}{6}\sin^3 2x+c,$

   (v) $-\frac{2}{3}(\cos x)^{3/2}+c,$ (vi) $-\frac{1}{3}\cot^3 x+c,$ (vii) $\frac{1}{16}(4x^2-1)^4+c,$

   (viii) $\frac{1}{2}\sqrt{(2x^2-5)}+c,$ (ix) $-2(8+x)\sqrt{(4-x)}+c,$

   (x) $-2\cos\sqrt{x}+c,$ (xi) $2\sqrt{\sec x}+c,$

   (xii) $\frac{2}{15}\cos^3\frac{x}{2}\left(3\cos^2\frac{x}{2}-5\right)+c,$ (xiii) $\sec x+c,$

   (xiv) $2\sqrt{(1+\sec x)}+c,$ (xv) $\frac{4}{3}(1+\sqrt{x})^{3/2}+c.$

page 9.   **Exercise 1c**

1. (i) $\frac{26}{15}$,   (ii) $\frac{1}{30}$,   (iii) $\sqrt{3}-\frac{2}{3}$,   (iv) $-\frac{7}{20}$,   (v) $\frac{67}{48}$.

2. (i) $\frac{7}{36}$,   (ii) $\frac{8}{15}$,   (iii) $\frac{1}{6}(4-\sqrt{2})$.

3. (i) $1-\frac{1}{2}\sqrt{3}$, (ii) $\frac{256}{15}$,   (iii) $-\frac{1}{10}$,   (iv) $\frac{4}{3}$,   (v) $\frac{23}{108000}$,

   (vi) 24·3,   (vii) $\frac{4}{3}$,   (viii) $\frac{74}{27}$,   (ix) $\frac{2}{3}$.

4. $2\sqrt{2}-\sqrt{3}$.   **5.** $\frac{9}{8}$.   **6.** $\frac{1}{4}\pi^2$.   **7.** $\frac{1}{8}\pi$.

Qu. 9.   (i) $45°$. $\frac{\pi}{4}$ rad,   (ii) $30°$, $\frac{\pi}{6}$ rad,   (iii) $45°$, $\frac{\pi}{4}$ rad,

       (iv) $60°$, $\frac{\pi}{3}$ rad,   (v) $30°$, $\frac{\pi}{6}$ rad,   (vi) $0°$,   0 rad,

       (vii) $60°$, $\frac{\pi}{3}$ rad,   (viii) $30°$, $\frac{\pi}{6}$ rad,   (ix) $30°$, $\frac{\pi}{6}$ rad,

       (x) $60°$, $\frac{\pi}{3}$ rad,   (xi) $90°$, $\frac{\pi}{2}$ rad.

Qu. 10.   (i) 0·3491,   (ii) 1·2217,   (iii) 0·1803(5),
       (iv) 0·9003,   (v) 1·5019.

Qu. 11.   (i) $57° 18'$,   (ii) $1° 43'$,   (iii) $71° 37'$,
       (iv) $40° 58'$,   (v) $36°$.

Qu. 12.   (i) 1·2869,   (ii) 0·9273,   (iii) 0·7841.

Qu. 13.   (i) $3\cos u$, $\sin^{-1}\frac{x}{3}$,   (ii) $\cos u$, $\sin^{-1}5x$,

       (iii) $2\cos u$, $\sin^{-1}\frac{3x}{2}$,   (iv) $\sqrt{7}\cos u$, $\sin^{-1}\frac{x}{\sqrt{7}}$,

       (v) $\cos u$, $\sin^{-1}\sqrt{3}x$,   (vi) $\sqrt{3}\cos u$, $\sin^{-1}\sqrt{\frac{2}{3}}x$.

Qu. 14.   (i) $\sin^{-1}\frac{x}{2}+c$,   (ii) $\frac{1}{\sqrt{3}}\sin^{-1}\sqrt{3}x+c$,   (iii) $\frac{1}{3}\sin^{-1}\frac{3x}{4}+c$.

Qu. 15. $-\cos^{-1}\frac{x}{2}+c = -\cos^{-1}\frac{x}{2}+\frac{\pi}{2}+k = \sin^{-1}\frac{x}{2}+k$.

Qu. 16. $\tan^{-1}x+c$.

Qu. 17.   (i) $9\sec^2 u$, $\tan^{-1}\frac{x}{3}$,   (ii) $\sec^2 u$, $\tan^{-1}2x$,

       (iii) $25\sec^2 u$, $\tan^{-1}\frac{3x}{5}$,   (iv) $3\sec^2 u$, $\tan^{-1}\frac{x}{\sqrt{3}}$,

       (v) $\sec^2 u$, $\tan^{-1}\sqrt{5}x$,   (vi) $7\sec^2 u$, $\tan^{-1}\sqrt{\frac{3}{7}}x$.

Qu. 18.   (i) $\frac{1}{2}\tan^{-1}\frac{x}{2}+c$,   (ii) $\frac{1}{4}\tan^{-1}4x+c$,   (iii) $\frac{1}{2\sqrt{3}}\tan^{-1}\frac{2x}{\sqrt{3}}+c$.

page 14. **Exercise 1d**

1. (i) $45°$, $\frac{\pi}{4}$ rad,   (ii) $45°$, $\frac{\pi}{4}$ rad,   (iii) $17° \, 19'$, $\frac{\pi\sqrt{3}}{18}$ rad,

   (iv) $60°$, $\frac{\pi}{3}$ rad,   (v) $51° \, 58'$, $\frac{\pi\sqrt{3}}{6}$ rad,   (vi) $15°$, $\frac{\pi}{12}$ rad,

   (vii) $67\frac{1}{2}°$, $\frac{3\pi}{8}$ rad,   (viii) $15°$, $\frac{\pi}{12}$ rad.

2. (i) $0{\cdot}5585$,   (ii) $1{\cdot}0533$,   (iii) $0{\cdot}0992$,   (iv) $4{\cdot}1062$.

3. (i) $114° \, 36'$,   (ii) $4° \, 35'$,   (iii) $78° \, 2'$,   (iv) $30°$.

4. (i) $0{\cdot}9273(5)$,   (ii) $0{\cdot}5882$,   (iii) $1{\cdot}1170$.

5. (i) $4\cos u$, $\sin^{-1}\dfrac{x}{4}$,   (ii) $\cos u$, $\sin^{-1} 3x$,

   (iii) $3\cos u$, $\sin^{-1}\frac{2}{3}x$,   (iv) $\sqrt{10}\cos u$, $\sin^{-1}\dfrac{x}{\sqrt{10}}$,

   (v) $\cos u$, $\sin^{-1}\sqrt{6}x$,   (vi) $\sqrt{5}\cos u$, $\sin^{-1}\sqrt{\frac{3}{5}}x$.

6. (i) $\sin^{-1}\dfrac{x}{5}+c$,   (ii) $\frac{1}{2}\sin^{-1} 2x+c$,   (iii) $\frac{1}{3}\sin^{-1}\dfrac{3x}{2}+c$,

   (iv) $\sin^{-1}\dfrac{x}{\sqrt{3}}+c$,   (v) $\dfrac{1}{\sqrt{7}}\sin^{-1}\sqrt{7}x+c$,   (vi) $\dfrac{1}{\sqrt{3}}\sin^{-1}\sqrt{\frac{3}{2}}x+c$.

7. (i) $16\sec^2 u$, $\tan^{-1}\dfrac{x}{4}$,   (ii) $\sec^2 u$, $\tan^{-1} 3x$,

   (iii) $4\sec^2 u$, $\tan^{-1}\dfrac{\sqrt{3}}{2}x$,   (iv) $2\sec^2 u$, $\tan^{-1}\dfrac{x}{\sqrt{2}}$,

   (v) $\sec^2 u$, $\tan^{-1}\sqrt{3}x$,   (vi) $5\sec^2 u$, $\tan^{-1}\sqrt{\frac{2}{5}}x$.

8. (i) $\frac{1}{5}\tan^{-1}\dfrac{x}{5}+c$,   (ii) $\frac{1}{6}\tan^{-1} 6x+c$,

   (iii) $\dfrac{1}{4\sqrt{3}}\tan^{-1}\dfrac{\sqrt{3}}{4}x+c$,   (iv) $\dfrac{1}{\sqrt{5}}\tan^{-1}\dfrac{x}{\sqrt{5}}+c$,

   (v) $\dfrac{1}{\sqrt{6}}\tan^{-1}\sqrt{6}x+c$,   (vi) $\dfrac{1}{\sqrt{30}}\tan^{-1}\sqrt{\frac{10}{3}}x+c$.

9. (i) $\dfrac{1}{3\sqrt{2}}\tan^{-1}\dfrac{\sqrt{2}}{3}x+c$,   (ii) $\dfrac{3}{\sqrt{5}}\sin^{-1}\dfrac{\sqrt{5}}{2}x+c$,

   (iii) $\dfrac{1}{\sqrt{2}}\sin^{-1}\sqrt{\frac{2}{3}}x+c$,   (iv) $\dfrac{2}{\sqrt{15}}\tan^{-1}\sqrt{\frac{5}{3}}x+c$.

10. (i) $\frac{1}{6}\pi$,   (ii) $\frac{1}{4}\pi$,   (iii) π,   (iv) $\frac{1}{12}\pi$,   (v) $\frac{1}{18}\pi$,   (vi) $\frac{1}{6}\pi$.

11a. (i) $\sin^{-1}\dfrac{x}{3}+c$,   (ii) $-\cos^{-1}\dfrac{x}{3}+c=-\cos^{-1}\dfrac{x}{3}+\dfrac{\pi}{2}+k=\sin^{-1}\dfrac{x}{3}+k$.

11b. (i) $\frac{1}{6}\pi$,   (ii) $\frac{1}{3}\pi$.

12. (i) $\sin^{-1}\dfrac{x+1}{2}+c$,   (ii) $\frac{1}{3}\tan^{-1}\dfrac{x-3}{3}+c$.

13. (i) $\frac{1}{5}\tan^{-1}\dfrac{x+3}{5}+c$,   (ii) $\sin^{-1}\dfrac{x-1}{2}+c$,

page 15.

13. (iii) $\dfrac{1}{\sqrt{15}} \tan^{-1} \dfrac{(x-2)\sqrt{3}}{\sqrt{5}} + c$,   (iv) $\dfrac{1}{\sqrt{3}} \sin^{-1} \dfrac{x+1}{\sqrt{3}} + c$.

14a. (i) $(x-3)^2 + 7$,   (ii) $3(x-2)^2 + 2$,   (iii) $2(x-1)^2 + 3$.

14b. (i) $\tfrac{1}{2} \tan^{-1} \dfrac{x-1}{2} + c$,   (ii) $\dfrac{1}{3\sqrt{2}} \tan^{-1} \dfrac{(x+1)\sqrt{2}}{3} + c$,

(iii) $\tfrac{1}{3} \tan^{-1} \dfrac{x-2}{3} + c$,   (iv) $\dfrac{1}{2\sqrt{3}} \tan^{-1} \dfrac{2(x-1)}{\sqrt{3}} + c$.

15a. (i) $4 - (x+1)^2$,   (ii) $9 - (2-x)^2$,   (iii) $7\tfrac{1}{2} - 2(x - \tfrac{1}{2})^2$.

15b. (i) $\sin^{-1} \dfrac{x+1}{2} + c$,   (ii) $\tfrac{1}{2} \sin^{-1} \dfrac{2}{\sqrt{5}}(x-1) + c$,

(iii) $\sin^{-1} \dfrac{x-2}{4} + c$,   (iv) $\dfrac{1}{\sqrt{2}} \sin^{-1} \dfrac{(x-3)\sqrt{2}}{3} + c$.

16. (i) $\tfrac{1}{4}\pi$,   (ii) $\tfrac{1}{2}\pi$.

17. (i) $3 \sin^{-1} x + \sqrt{(1-x^2)} + c$,   (ii) $3 \sin^{-1} \dfrac{x}{2} - 2\sqrt{(4-x^2)} + c$.

18. (i) $\tfrac{1}{2} \sin^{-1} x - \tfrac{1}{2}x\sqrt{(1-x^2)} + c$,   (ii) $\dfrac{1}{54} \tan^{-1} \dfrac{x}{3} + \dfrac{x}{18(9+x^2)} + c$,

(iii) $\tfrac{1}{2} \sin^{-1} \dfrac{x^2}{2} + c$.

19. (i) $\dfrac{x}{\sqrt{(1-9x^2)}} + c$,   (ii) $-\dfrac{1}{x}\sqrt{(1-x^2)} + c$,   (iii) $\sec^{-1} x + c$.

Qu. 19. (i) 0·8660,   (ii) 0·4795,   (iii) 2·5736,
(iv) −0·9900,   (v) 1·0415.

Qu. 20. (i) $\tfrac{1}{2}\pi$,   (ii) $-\tfrac{1}{4}\pi$,   (iii) $\tfrac{1}{6}\pi$,   (iv) 1·1072.

Qu. 22. (i) 0·2527,   (ii) $\tfrac{1}{12}\pi$,   (iii) 0·1676,   (iv) 1·696,
(v) $\tfrac{1}{6}\pi$.

Qu. 23. 0·271.

Qu. 24. (i) $x = \tfrac{1}{2}y + \tfrac{1}{2}\sqrt{3}\sqrt{(1-y^2)}$,   (ii) $x = \dfrac{2y}{1-y^2}$.

page 19.   **Exercise 1e**

1. 0·9659,   (ii) 0·9975,   (iii) 0·0808,   (iv) 1·0538.

3. (i) 1·249,   (ii) $\dfrac{\sqrt{3}}{3}\pi$,   (iii) $\dfrac{\pi}{6}$,   (iv) 1·955,   (v) 1·247.

4. (i) $x = \dfrac{1+y}{1-y}$,   (ii) $x = \tfrac{1}{2}\sqrt{3}y + \tfrac{1}{2}\sqrt{(1-y^2)}$.

5. (i) $\dfrac{\pi}{6}$,   (ii) 0·325.

6. (i) $\dfrac{\pi}{4}$,   (ii) 0·322.

page 19.

7.　(i) 10, 0, -10, 0, 10 m.　　　(ii) After $\dfrac{\pi}{3}$ s.

(iii) After $\frac{2}{3}\pi$ s.

8.　(i) 0, 5, 0, -5, 0 m,　　　(ii) $2\sin^{-1}\frac{3}{5} \simeq 1\cdot29$ s.

9.　(i) $0\cdot464$,　　(ii) $0\cdot927$,　　(iii) $0\cdot0183$,　　(iv) $0\cdot0947$,

(v) $0\cdot168$,　　(vi) $0\cdot300$.

page 20.　　　　　　　　　　　**Exercise 1f**

1. $\frac{2}{9}(x^3-1)^{3/2}+c$.

2. $-\frac{1}{4}(x^2-1)^{-2}+c$.

3. $-\frac{1}{6}\cos^3 2x+c$.

4. $2\sqrt{\tan x}+c$.

5. $\frac{1}{4}\sin 4x-\frac{1}{12}\sin^3 4x+c$.

6. $\frac{1}{2}x-\frac{3}{4}\sin\frac{2}{3}x+c$.

7. $\frac{3}{8}x+\frac{1}{8}\sin 4x+\frac{1}{64}\sin 8x+c$.

8. $-2\sqrt{2}\cos\dfrac{x}{2}+c$.

9. $-3\cos^4\dfrac{x}{6}+c$.

10. $\frac{1}{10}\sin 5x+\frac{1}{2}\sin x+c$.

11. $\frac{1}{270}(15x+7)(3x-7)^5+c$.

12. $\frac{2}{3}(x-10)\sqrt{(5+x)}+c$.

13. $\dfrac{1}{\sqrt{5}}\sin^{-1}\sqrt{\dfrac{5}{6}}\,x+c$.

14. $\dfrac{1}{\sqrt{8}}\tan^{-1}\sqrt{8}x+c$.

15. $\sin^{-1}\dfrac{x+2}{3}+c$.

16. $\dfrac{1}{\sqrt{6}}\tan^{-1}\sqrt{\dfrac{3}{2}}\,(x+1)+c$.

17. $\sin^{-1}\dfrac{x}{\sqrt{5}}-\sqrt{(5-x^2)}+c$.

18. $\frac{1}{2}x\sqrt{(9-x^2)}+\frac{9}{2}\sin^{-1}\dfrac{x}{3}+c$.

19. $\dfrac{x}{4\sqrt{(4-x^2)}}+c$.

20. $-\dfrac{\sqrt{(16-x^2)}}{16x}+c$.

21. (i) $-\sqrt{(1-x^2)}+c$,　　(ii) $2\tan^{-1}x+c$,

(iii) $-\frac{1}{3}(1-x^2)^{3/2}+c$,　(iv) $2\sin^{-1}x+c$,　(v) $2\sin^{-1}x+\sqrt{(1-x^2)}+c$.

22. $\frac{2}{27}(3x+20)\sqrt{(3x-1)}+c$.

23. $\frac{1}{6}(x^2+2)^{3/2}+c$.

24. $-\dfrac{\sqrt{(1-x^2)}}{x}+c$.

25. $\frac{2}{5}\cos^5\dfrac{x}{2}-\frac{2}{3}\cos^3\dfrac{x}{2}+c$.

26. $-\frac{1}{3}\sqrt{(4-x^2)}+c$.

27. $3\sin^{-1}\dfrac{x}{6}+c$.

28. $\frac{1}{168}(x+3)^6(21x^2-66x+61)+c$.

29. $2\sin^{-1}\dfrac{x}{2}-\frac{1}{2}x\sqrt{(4-x^2)}+c$.

30. $2\sin^{-1}\dfrac{x}{3}-\sqrt{(9-x^2)}+c$.

31. $\dfrac{1}{2\sqrt{3}}\tan^{-1}\dfrac{\sqrt{3}}{2}(x-2)+c$.

32. $\frac{1}{2}x-\frac{5}{4}\sin\frac{2}{5}x+c$.

33. $\frac{2}{3}\sqrt{2}\sin\frac{3}{2}x+c$.

34. $2\sin^{-1}\frac{1}{2}x+\frac{1}{2}x\sqrt{(4-x^2)}+c$.

35. $\frac{1}{5}\sec^5 x+c$.

36. $\frac{1}{9}\sin^{-1}\frac{3}{2}x^3+c$.

37. $-\frac{1}{2}\cos 2x+\frac{1}{3}\cos^3 2x-\frac{1}{10}\cos^5 2x+c$.

38. $\frac{1}{2}\tan^{-1}x+\dfrac{x}{2(1+x^2)}+c$.

39. $\frac{1}{2}\cos x-\frac{1}{8}\cos 4x+c$.

page 21.

40. $\dfrac{1}{\sqrt{2}}\sec x + c.$ **41.** $\tfrac{1}{2}\tan^{-1}x^2 + c.$

42. $\tfrac{1}{2}\sin x\sqrt{(1-2\sin^2 x)} + \dfrac{1}{2\sqrt{2}}\sin^{-1}(\sqrt{2}\sin x) + c.$

43. $\tfrac{1}{2}\tan^{-1}(2\tan x) + c.$ **44.** $\tfrac{1}{3}\sec^{-1}\dfrac{x}{3} + c.$

45. $\sqrt{\dfrac{(1+x)}{(1-x)}} + c.$ **46.** $\tfrac{3672}{125}.$ **47.** $\tfrac{1}{2}.$ **48.** 1.

49. $\tfrac{3}{4}\pi.$ **50.** $\dfrac{\sqrt{3}}{4}.$ **51.** $2 - \sqrt{2}.$

52. $\tfrac{1}{2}(\sec^{-1}\sqrt{3} - \sec^{-1}\sqrt{2}) \simeq 0.085.$ **53.** $-2592.$ **54.** $\dfrac{\pi}{2\sqrt{2}}.$

CHAPTER 2

page 23.

Qu. 1. The larger a, the larger the gradient.

Qu. 2. The reflection of $y = 2^x$ in the y-axis.

Qu. 3. (i) 0.7×2^x, 1.1×3^x, (ii) 1.4, 1.9. **Qu. 4.** 0.7×2^x. **Qu. 5.** 1.08.

Qu. 6. (i) $30x^2(2x^3+1)^4$, (ii) $6x^2\cos(2x^3)$, (iii) $6x^2 e^{2x^3}$,

(iv) $2y\,e^{y^2}\dfrac{dy}{dx}$, (v) $-2x\,e^{-x^2}$, (vi) $\sec^2 x\,.\,e^{\tan x}$,

(vii) $\dfrac{1}{2\sqrt{x}}\,e^{\sqrt{x}}$, (viii) $\cos y\,.\,e^{\sin y}\dfrac{dy}{dx}$.

Qu. 7. (i) $-\tfrac{1}{2}(x^2+1)^{-1} + c$, (ii) $-\tfrac{1}{2}\cos x^2 + c$, (iii) $\tfrac{1}{2}e^{x^2} + c$,

(iv) $-e^{\cos x} + c$, (v) $6\,e^{x/3} + c$, (vi) $\tfrac{3}{2}e^{2x} + c$,

(vii) $\tfrac{1}{12}e^{3x^2} + c$, (viii) $-\tfrac{1}{2}e^{\cot 2x} + c$.

page 27.   **Exercise 2a**

4. 1.1×3^x.

5. (i) $4\,e^x$, (ii) $3\,e^{3x}$, (iii) $2\,e^{2x+1}$,

(iv) $4x\,e^{2x^2}$, (v) $-2\,e^{-2x}$, (vi) $3\,e^{3y}\dfrac{dy}{dx}$,

(vii) $2x\,e^{x^2+3}$, (viii) $-2x^{-3}\,e^{x^{-2}}$, (ix) $-5x^{-2}e^{5/x}$,

(x) $\tfrac{1}{3}x^{-2/3}\,e^{x^{1/3}}$, (xi) $2ax\,e^{ax^2+b}$, (xii) $\dfrac{1}{2\sqrt{t}}\,e^{\sqrt{t}}\dfrac{dt}{dx}$.

6. (i) $-e^{\cos x}\sin x$, (ii) $e^{\sec x}\sec x\tan x$,

(iii) $e^{3\tan y}\,3\sec^2 y\dfrac{dy}{dx}$, (iv) $2\,e^{\sin 2x}\cos 2x$,

(v) $e^{-\cot x}\csc^2 x$, (vi) $-2\,e^{\csc^2 x}\csc^2 x\cot x$,

page 28.

6. (vii) $-\dfrac{\sin x}{2\sqrt{\cos x}}\, e^{\sqrt{\cos x}}$,

(viii) $ab\, e^{a\,\sin bx}\cos bx$,

(ix) $3\, e^{\sin 3t}\cos 3t\, \dfrac{dt}{dx}$,

(x) $2x\, e^{\tan x^2}\sec^2 x^2$.

7. (i) $\dfrac{x}{\sqrt{(x^2+1)}}\, e^{\sqrt{(x^2+1)}}$,

(ii) $\dfrac{2x}{(1-x^2)^2}\, e^{(1-x^2)-1}$,

(iii) $4\, e^{\sin^2 4x}\sin 8x$,

(iv) $2x\, e^{\tan(x^2+1)}\sec^2(x^2+1)$,

(v) $6\, e^{\sec^2 3x}\sec^2 3x\tan 3x$,

(vi) $e^{-\operatorname{cosec} x}\operatorname{cosec} x\cot x$,

(vii) $2x^{-3}\, e^{-x-2}$,

(viii) $(\sin x + x\cos x)\, e^{x\,\sin x}$,

(ix) $e^{xy}\left(y + x\,\dfrac{dy}{dx}\right)$,

(x) e^{x+e^x}.

8. (i) $e^x(x^2+2x)$,

(ii) $\dfrac{(x-1)\, e^x}{x^2}$,

(iii) $\tfrac{1}{2}\, e^{\sin x}(1 + x\cos x)$,

(iv) $e^x\operatorname{cosec} x\,(2x - \cot x)$,

(v) $e^x\operatorname{cosec} x\,(1 - \cot x)$,

(vi) $-e^{-x}\, x^{-2}(x\sin x + x\cos x + \cos x)$,

(vii) $(1+x)\, e^{x+xe^x}$,

(viii) $e^{ax}\sec bx\,(a + b\tan bx)$,

(ix) $e^{ax}\operatorname{cosec} bx\,(a - b\cot bx)$,

(x) $n\, e^x\tan^{n-1} e^x\cdot\sec^2 e^x$,

(xi) $2\, e^x\cos x$.

9. (i) $6\, e^{x/2} + c$,

(ii) $-e^{-x} + c$,

(iii) $3\, e^{x/3} + c$,

(iv) $\tfrac{2}{3}\, e^{3x-1} + c$,

(v) $\tfrac{1}{4}\, e^{x^2} + c$,

(vi) $-\tfrac{1}{3}\, e^{-x^3} + c$,

(vii) $-e^{\cos x} + c$,

(viii) $e^{\tan x} + c$,

(ix) $-e^{\cot x} + c$,

(x) $-e^{1/x} + c$,

(xi) $-\tfrac{1}{2}\, e^{x-2} + c$,

(xii) $2\, e^{\sqrt{x}} + c$,

(xiii) $e^{\sin^2 x} + c$,

(xiv) $\tfrac{1}{2}\, e^{\sin 2x} + c$,

(xv) $\dfrac{-1}{2\, e^{x^2}} + c$.

10. $y = e^a(x - a + 1)$, $y - e^2 x + e^2 = 0$. **12.** $\tfrac{1}{2}\pi(e^2 - 1)$. **13.** $0\cdot 56$.

14. $e^x(1+x)$, $e^x(x-1) + c$. **15.** Minimum $-\dfrac{1}{e}$ when $x = -1$, $e^2 y + x + 4 = 0$.

20. $25\, e^{4x}\cos(3x + 2\alpha)$, $125\, e^{4x}\cos(3x + 3\alpha)$.

21. $13^n\, e^{5x}\sin(12x + n\beta)$.

23. (ii) $-\dfrac{1}{\pi}$, $\dfrac{1}{e\pi}$, $\dfrac{-1}{e^2\pi}$, $\dfrac{1}{e^3\pi}$, $\dfrac{-1}{e^4\pi}$,

(iii) minimum when $t = 2\cdot 83$, maximum when $t = 5\cdot 98$,

(iv) $-0\cdot 6065$, $0\cdot 2231$.

Qu. 8. (i) $\log_{10}(100a^2 b^{-\frac{1}{3}})$,

(ii) $\log_c\dfrac{B(1+x)}{1-x}$.

Qu. 9. (i) $\log_c 2 + \log_c a$,

(ii) $2\log_c a$,

(iii) $-\log_c a$,

(iv) $\log_c 2 - \log_c a$,

(v) $\tfrac{1}{2}\log_c a$,

(vi) $\log_c a - \log_c 2$,

(vii) $-2\log_c a$,

(viii) $-\log_c 2 - \log_c a$.

Qu. 10. (i) $x = \tfrac{3}{2}$,

(ii) $x = 6\cdot 02$.

Qu. 11. (ii) $4\cdot 61$.

Qu. 12. $0\cdot 693$.

Qu. 13. (i) $1\cdot 0986$,

(ii) $2\cdot 7550$,

(iii) $\bar{4}\cdot 1640$,

(iv) $6\cdot 7580$.

page 32.   **Exercise 2b**

1. (i) $\log_{10} \dfrac{a^3}{50}$,   (ii) $\log_e \dfrac{x^2 e^3}{3}$,

    (iii) $\log_e \dfrac{(x-3)^4}{(x-2)^3}$,   (iv) $\log_e \{k \sqrt{(1-y^2)}\}$.

2. (i) $\log_e a + \log_e 3$,   (ii) $3 \log_e a$,   (iii) $\log_e a - \log_e 3$,
    (iv) $-3 \log_e a$,   (v) $\log_e 3 - \log_e a$,
    (vi) $-\log_e 3 - \log_e a$,   (vii) $\frac{1}{3} \log_e a$.

3. (i) $\log_e \cos x - \log_e \sin x$,   (ii) $2 \log_e \sin x - 2 \log_e \cos x$,
    (iii) $\log_e (x-2) + \log_e (x+2)$,   (iv) $\frac{1}{2} \log_e (x+1) - \frac{1}{2} \log_e (x-1)$,
    (v) $\log_e 3 + 2 \log_e \sin x$.

4. (i) $a = 100$,   (ii) $y = 100$ or $y = \frac{1}{100}$.

5. (i) $x = \frac{2}{5}$,   (ii) $x = 1 \cdot 26$,   (iii) $x = 0 \cdot 672$,   (iv) $x = 1 \cdot 82$.

6. $1 \cdot 10$.

7. (i) $1 \cdot 7817$,   (ii) $3 \cdot 2312$,   (iii) $1 \cdot 2467$,   (iv) $5 \cdot 7045$.

8. (i) $x = 1$ or $x = -2$,   (ii) $x = 2$ or $x = 3$.

9. (i) $a = 10$ or $a = 100$,   (ii) $c = 1 \cdot 48$ or $c = 0 \cdot 677$.

10. (i) $x = +\frac{2}{9}$,   (ii) $x = +\frac{1}{2}$.

11. (i) $x = 1$, $y = 10$ or $x = 10$, $y = 1$,
     (ii) $x = 2 \cdot 34$, $y = 0 \cdot 66$ or $x = 0 \cdot 66$, $y = 2 \cdot 34$.

12. $0 \cdot 5 \%$.

Qu. 14.   (i) $10x(x^2-2)^4$,   (ii) $-2x \operatorname{cosec} x^2 \cot x^2$,

    (iii) $2x\, e^{x^2}$,   (iv) $\dfrac{2x}{x^2-2}$,   (v) $2 \cot x$,   (vi) $2x \cot x^2$.

Qu. 15. $\dfrac{3x+2}{2x(x+1)}$.

Qu. 16.   (i) $\dfrac{1}{x}$,   (ii) $\dfrac{1}{x}$,   (iii) $\dfrac{3}{3x+1}$,   (iv) $\dfrac{1}{y} \dfrac{dy}{dx}$,

    (v) $\dfrac{3}{x}$,   (vi) $\dfrac{3x^2}{x^3-2}$,   (vii) $\dfrac{3}{x-1}$,   (viii) $\dfrac{1}{t} \dfrac{dt}{dx}$,

    (ix) $\cot x$,   (x) $-3 \tan 3x$,   (xi) $-3 \tan x$,

    (xii) $6 \cot 3x$,   (xiii) $\dfrac{x}{x^2-1}$,   (xiv) $\dfrac{1+x}{x(1-x)}$.

Qu. 17. $4^x \log_e 4$, $16 \log_e 4$.

Qu. 18. (i) $0 \cdot 6931$,   (ii) $1 \cdot 0986$.

Qu. 19. (i) $10^x \log_e 10$,   (ii) $2^{3x} \log_e 64$,
    (iii) $x\, 3^{x^2} \log_e 9$,   (iv) $\cos x . 10^{\sin x} \log_e 10$.

Qu. 20. $a^x \log_e a$.   **Qu. 21.** $5^x \log_e 5$, $\dfrac{5^x}{\log_e 5} + c$.

page 35.

Qu. 22. $x\,2^{x^2}\log_e 4,\; \dfrac{2^{x^2}}{\log_e 4}+c.$

Qu. 23.   (i) $\dfrac{3^{2x}}{\log_e 9}+c,$ (ii) $\tfrac{1}{3}\,e^{x^3}+c,$ (iii) $\dfrac{2^{\tan x}}{\log_e 2}+c.$

page 35.   **Exercise 2c**

1.   (i) $\dfrac{1}{x},$ (ii) $\dfrac{4}{x},$ (iii) $\dfrac{2}{2x-3},$ (iv) $\dfrac{1}{y}\dfrac{dy}{dx},$

    (v) $\dfrac{1}{x-1},$ (vi) $\dfrac{4}{x},$ (vii) $\dfrac{2x}{x^2-1},$ (viii) $\dfrac{2}{x},$

    (ix) $\dfrac{6}{x},$ (x) $\dfrac{2}{x+1},$ (xi) $\dfrac{3}{t}\dfrac{dt}{dx},$ (xii) $-\dfrac{1}{x},$

    (xiii) $\dfrac{1}{x},$ (xiv) $\dfrac{1}{2x},$ (xv) $-\dfrac{1}{x},$ (xvi) $-\dfrac{1}{x},$

    (xvii) $-\dfrac{2}{x},$ (xviii) $\dfrac{1}{x\log_e 10},$ (xix) $-\dfrac{3}{t}\dfrac{dt}{dx},$ (xx) $\dfrac{1}{3x}.$

2.   (i) $-\tan x,$ (ii) $2\cot x,$ (iii) $6\,\text{cosec}\,6x,$
    (iv) $-6\tan 2x,$ (v) $-4\,\text{cosec}\,2x,$ (vi) $-4\tan 2x,$
    (vii) $\text{cosec}\,x,$ (viii) $\tan x,$ (ix) $\sec x,$
    (x) $-2x\cot x^2,$ (xi) $2\sec 2x.$

3.   (i) $\dfrac{1}{x^2-1},$ (ii) $\dfrac{2x^2-1}{x(x^2-1)},$ (iii) $\dfrac{3x-5}{2(x^2-1)},$ (iv) $\dfrac{1}{\sqrt{(x^2-1)}}.$

4.   (i) $\dfrac{1}{t}\dfrac{dt}{dx},$ (ii) $1+\log_e x,$ (iii) $x+2x\log_e x,$

    (iv) $\dfrac{1}{x^2}(1-\log_e x),$ (v) $\log_e y+\dfrac{x}{y}\dfrac{dy}{dx},$ (vi) $\dfrac{y}{x}+\dfrac{dy}{dx}\log_e x,$

    (vii) $\dfrac{1-2\log_e x}{x^3},$ (viii) $\dfrac{\log_e x-1}{(\log_e x)^2},$ (ix) $\dfrac{2}{x}\log_e x,$

    (x) $\dfrac{1}{x\log_e x},$ (xi) $\cos x.$

5.   (i) $5^x\log_e 5,$ (ii) $x\,2^{x^2}\log_e 4,$ (iii) $\tfrac{2}{3}\,3^{2x}\log_e 3,$

    (iv) $-\text{cosec}\,x\cot x.10^{\text{cosec}\,x}\log_e 10,$ (v) $\dfrac{1}{2\sqrt{x}}.3^{\sqrt{x}}.\log_e 3,$

    (vi) $1,$ (vii) $\dfrac{1}{x}\,2^{\log_e x}\log_e 2,$ (viii) $-\dfrac{1}{x^2}\,2^{1/x}\log_e 2.$

6.   (i) $3^x\log_e 3,\; \dfrac{3^x}{\log_e 3}+c,$ (ii) $x\,2^{x^2}\log_e 4,\; \dfrac{2^{x^2}}{\log_e 4}+c.$

7.   (i) $\dfrac{1}{\log_e 10}\,10^x+c,$ (ii) $\dfrac{2^{3x}}{\log_e 8}+c,$

    (iii) $\dfrac{3^{x^2}}{\log_e 9}+c,$ (iv) $-\dfrac{2^{\cos x}}{\log_e 2}+c.$

page 36.

8. $1 + \log_e x$, $x(\log_e x - 1) + c$.

9. $2^x(1 + x \log_e 2)$, $\dfrac{x \, 2^x}{\log_e 2} - \dfrac{2^x}{(\log_e 2)^2} + c$.

10. (i) $\dfrac{1}{x-2}$,   (ii) $\dfrac{1}{x-2}$.

Qu. 24.   (i) $2 \log_e x + c$,   (ii) $\frac{1}{3} \log_e x + c$,

(iii) $\frac{1}{3} \log_e (3x - 2) + c$,   (iv) $\frac{1}{3} \log_e (x - 2) + c$.

Qu. 25.   (i) $\frac{1}{2} \log_e (2x + 3) + c$,   (ii) $\log_e \dfrac{1}{1-x} + c$.

Qu. 26. $\log_e 8$.   **Qu. 27.**   (i) $c = \log_e A$,   (ii) $c = \log_e (k\sqrt 2)$.

Qu. 28.   (i) $\dfrac{1}{3(2 - x^3)} + c$,   (ii) $\frac{1}{3} \sin x^3 + c$,

(iii) $\frac{1}{3} e^{x^3} + c$,   (iv) $\log_e \{k \sqrt[3]{(x^3 - 2)}\}$,

(v) $\log_e \{k \sqrt{(x^2 - 2x)}\}$,   (vi) $\log_e \dfrac{k}{3 - x^2}$,

(vii) $\log_e (k \sin x)$.

Qu. 29. $x + \log_e (x - 1) + c$.

Qu. 30.   (i) $\log_e \frac{3}{4}$,   (ii) $-\log_e 2$.   **Qu. 31.** $-\log_e 2$.

Qu. 32.   No.   **Qu. 33.** $\dfrac{1}{x-3}$, $\dfrac{1}{x-3}$.

Qu. 34.   (i) $\log_e \frac{3}{2}$,   (ii) $-\log_e 5$.   **Qu. 35.** $-\log_e 3$.

page 42.   **Exercise 2d**

1.   (i) $\log_e (kx^{1/4})$,   (ii) $\log_e (kx^5)$,   (iii) $\log_e \{k\sqrt{(2x - 3)}\}$,

(iv) $\log_e \{k\sqrt{(x + 4)}\}$,   (v) $\log_e \{k(3 - 2x)^{-1/2}\}$,   (vi) $\log_e \{k(1 - x^2)^{-1/2}\}$,

(vii) $\log_e \{k(x^2 - 1)^{3/2}\}$,   (viii) $\log_e \{k(x^2 + x - 2)\}$,

(ix) $\log_e \{k\sqrt[3]{(3x^2 - 9x + 4)}\}$,   (x) $x - \log_e \{k(x + 2)^2\}$,

(xi) $\frac{3}{2}x - \log_e \{k(2x + 3)^{9/4}\}$,   (xii) $-2x - \log_e \{k(3 - x)^6\}$,

(xiii) $-x - \log_e \{k(2 - x)\}$,   (xiv) $-2x - \log_e \{k(x - 4)^5\}$,

(xv) $\log_e (k \sec x)$,   (xvi) $\log_e \left(k \sin^2 \dfrac{x}{2}\right)$,

(xvii) $\log_e \{k\sqrt{\sin (2x + 1)}\}$,   (xviii) $\log_e \left(k \cos^3 \dfrac{x}{3}\right)$,

(xix) $\log_e \{k(x - \sin^2 x)\}$,   (xx) $\log_e \{k(\sin x + \cos x)\}$,

(xxi) $\log_e \{k(x + \tan x)\}$.

2. (ii) $\dfrac{2}{2x - 1}$, $\dfrac{2}{2x - 1}$,   (iii) $\frac{1}{2} \log_e 3$, $-\frac{1}{2} \log_e 5$.

3. (i) $\log_e \frac{4}{3}$,   (ii) $\log_e 2$.

4. (i) $\dfrac{1}{3 - x}$, $\dfrac{1}{3 - x}$,   (ii) $-\log_e 1.5$,   (iii) $\log_e 1.5$.

page 43.

5. (i) $\log_e 2$, (ii) $\frac{1}{3} \log_e 2$, (iii) $-\log_e 2$,
    (iv) $\frac{1}{2} \log_e 5 - \log_e 3$, (v) $\frac{1}{2} \log_e 3$, (vi) $-\frac{1}{2} \log_e 2$,
    (vii) $\log_e 7$, (viii) $2 + \log_e 4$, (ix) $2 + \log_e 4$,
    (x) $-\frac{1}{2} - \log_e \frac{3}{2}$, (xi) $\frac{1}{2} \log_e \frac{4}{3}$, (xii) $\frac{1}{2} \log_e 2$,
    (xiii) $\frac{1}{2} \log_e 3$.

page 43. **Exercise 2e**

1. (i) $\dfrac{\log_{10} b}{\log_{10} a}$, $\overline{1}\cdot 36$, $0\cdot 07$, (ii) $x = 4$ or $x = \sqrt{2}$.

2. $x = 0\cdot 178$. 3. $x = 3$ or $x = 1$.

4. (i) $x = 7\cdot 13$, (ii) $x = 0\cdot 304$, (iii) $x = 0\cdot 1$.

5. $y = 100 x^{-\frac{3}{2}}$. 6. $\overline{2}\cdot 7960$. 7. $1\cdot 40$, $0\cdot 716$.

8. (i) $x = -2$, (ii) $x = \frac{1}{3}$, (iii) $x = 2000$ or $0\cdot 50$.

9. (i) $x = 1\cdot 7$, (ii) $2\cdot 0$.

10. (i) $a = 0\cdot 693$, $b = 0\cdot 366$, (ii) $x = 2\cdot 32$.

11. $a = 2\cdot 4$, $b = 1\cdot 6$, $n = -0\cdot 631$.

14. $a = 3$, $b = 4$. (i) $\pi - \tan^{-1} \frac{4}{3} \simeq 2\cdot 21$, (ii) $\tan^{-1} \frac{4}{3} \simeq 0\cdot 93$.

15. Minimum 1 when $x = 0$, maximum $\dfrac{2}{\sqrt{e}}$ when $x = \pm \dfrac{1}{\sqrt{2}}$.

16. $\frac{3}{4}\pi$, $\frac{7}{4}\pi$, $\frac{11}{4}\pi$.

17. (i) $\dfrac{-7 \sin x}{(3 + 4 \cos x)(4 + 3 \cos x)}$. 18. $k = \frac{3}{4}$. 19. $\log_e (2 + \sqrt{5}) \simeq 1\cdot 44$.

20. Minimum 0, maximum $4e^{-2}$, $ey = x$.

CHAPTER 3

page 46.

Qu. 1. (i) $\dfrac{3 - x}{1 - x^2}$, (ii) $\dfrac{(x+2)(x-1)}{(x^2+1)(x+1)}$, (iii) $\dfrac{3x^2 - x + 4}{(x-1)^2(x+1)}$.

Qu. 2. (i) $\dfrac{1}{x-2} - \dfrac{1}{x+2}$, (ii) $\dfrac{1}{2(1-x)} + \dfrac{1}{2(1+x)}$,

  (iii) $\frac{1}{2} - \frac{1}{3}$, (iv) $\dfrac{1}{n} - \dfrac{1}{n+1}$.

Qu. 3. (i) $12A - 3B + 4C = 17$, $6A - 4B + 3C = 5$, $10A - 15B + 6C = -1$,
  (ii) $A = 2$, $B = 1$, $C = -1$.

Qu. 4. (i) $A = 1$, $B = -3$, $C = 4$, (ii) $A = 2$, $B = 1$, $C = -3$,
  (iii) $A = 5$, $B = -1$.

Qu. 5. $A = 3$, $B = -2$, $C = -1$.

Qu. 6. (i) $A = -\frac{2}{3}$, $B = \frac{2}{3}$, $C = 1$, (ii) No.

page 49. **Exercise 3a**

1. (i) $\dfrac{x-12}{(x+3)(x-2)}$, (ii) $\dfrac{7-3x-5x^2}{(x+2)^2(3x-1)}$,

(iii) $\dfrac{2-4x-3x^2}{(2+3x^2)(1-x)}$, (iv) $\dfrac{-x^3+6x^2-7x+6}{(x^2+1)(x-1)^2}$.

2. (i) $\dfrac{1}{3-x}-\dfrac{1}{3+x}$, (ii) $\dfrac{1}{2(a-b)}+\dfrac{1}{2(a+b)}$,

(iii) $\frac{1}{5}-\frac{1}{6}$, (iv) $\dfrac{1}{1-p}+\dfrac{1}{p}$.

3. (i) $A=3,\ B=7$,
(ii) $A=1,\ B=-1,\ C=2$,
(iii) $A=2,\ B=-1,\ C=-3$,
(iv) $A=1,\ B=-2,\ C=3$.

4. (i) $A=1,\ B=-1$, (ii) $A=2,\ B=1$,
(iii) $A=2,\ B=-1,\ C=3$,
(iv) $A=1,\ B=-2,\ C=3$.

5. (i) $A=3,\ B=-\frac{1}{2},\ C=-\frac{1}{2}$, (ii) No,
(iii) $A=2,\ B=3,\ C=1$, (iv) No,
(v) $A=1,\ B=3,\ C=2,\ D=-1$,
(vi) $A=1,\ B=-1,\ C=-1$,
(vii) $A=2,\ B=-1,\ C=2,\ D=3$.

6. $A=1,\ B=1,\ C=1$. (i) $(x+1)(x^2-x+1)$,
(ii) $(x-2)(x^2+2x+4)$, (iii) $(x+3)(x^2-3x+9)$,
(iv) $(2x-3)(4x^2+6x+9)$, (v) $(3x+5)(9x^2-15x+25)$.

7. $x(x-1)(x-2)+3x(x-1)+x+1$. **8.** $a=60,\ b=25$.

9. $\alpha+\beta=-\dfrac{b}{a}$, $\alpha\beta=\dfrac{c}{a}$.

10. $\alpha+\beta+\gamma=-\dfrac{q}{p}$, $\beta\gamma+\gamma\alpha+\alpha\beta=\dfrac{r}{p}$, $\alpha\beta\gamma=-\dfrac{s}{p}$.

Qu. 7. (i) $\dfrac{1}{x-3}-\dfrac{1}{x+3}$, (ii) $\dfrac{1}{2(2-x)}-\dfrac{1}{2(2+x)}$,

(iii) $\dfrac{1}{x-2}-\dfrac{2}{3x-5}$, (iv) $\dfrac{1}{2(x+1)}-\dfrac{1}{x+2}+\dfrac{1}{2(x-3)}$,

(v) $\dfrac{2}{1+2x}-\dfrac{1}{2-x}$.

Qu. 8. (i) $\dfrac{1}{1-x}+\dfrac{2+x}{4+x^2}$, (ii) $\dfrac{1}{x+1}-\dfrac{2x-1}{2x^2+x+3}$,

(iii) $\dfrac{1}{x+1}+\dfrac{1}{x-2}-\dfrac{2}{x+2}$, (iv) $\dfrac{1}{2-x}+\dfrac{x}{3+x^2}$.

Qu. 9. (i) $\dfrac{1}{x+3}-\dfrac{2}{(x+3)^2}$, (ii) $\dfrac{5}{x-2}-\dfrac{3}{x-1}-\dfrac{4}{(x-1)^2}$.

Qu. 10. $A=3,\ B=-2,\ C=1,\ D=5$.

page 55.

Qu. 11. (i) $x + \dfrac{x+2}{(x-1)(x+3)}$, (ii) $3 + \dfrac{x-1}{(x-2)(x+1)}$.

Qu. 12. (i) $1 + \dfrac{2}{x+1} - \dfrac{1}{x-2}$, (ii) $x - 1 - \dfrac{3}{4(x-2)} + \dfrac{3}{4(x+2)}$.

page 55. **Exercise 3b**

1. (i) $\dfrac{2}{x+3} - \dfrac{1}{x-4}$, (ii) $\dfrac{1}{2(5-x)} - \dfrac{1}{2(5+x)}$,

   (iii) $\dfrac{4}{x+1} + \dfrac{2}{x-2} - \dfrac{3}{x-3}$, (iv) $\dfrac{3}{x-1} - \dfrac{1}{x} + \dfrac{2}{x+1}$,

   (v) $\dfrac{1}{x+2} + \dfrac{2}{2x+1} - \dfrac{2}{3x+2}$, (vi) $2x - 1 + \dfrac{1}{x+3} - \dfrac{3}{x-2}$.

2. (i) $\dfrac{2}{x-3} + \dfrac{3x-1}{x^2+4}$, (ii) $\dfrac{2}{x+1} - \dfrac{1}{x^2+2}$,

   (iii) $\dfrac{1}{x-1} + \dfrac{2x}{x^2+5}$, (iv) $\dfrac{3}{2x-3} + \dfrac{1-3x}{2x^2+1}$,

   (v) $\dfrac{3}{x-3} - \dfrac{2}{x+3} - \dfrac{1}{x+5}$, (vi) $2 + \dfrac{5}{x-3} + \dfrac{x}{x^2+1}$.

3. (i) $\dfrac{1}{x-2} - \dfrac{3}{(x-2)^2}$, (ii) $\dfrac{1}{x-1} - \dfrac{1}{x+2} + \dfrac{2}{(x+2)^2}$,

   (iii) $\dfrac{23}{4(3x+1)} - \dfrac{1}{4(x+1)} - \dfrac{7}{2(x+1)^2}$, (iv) $x + \dfrac{1}{x+2} + \dfrac{2}{x-1} + \dfrac{1}{(x-1)^2}$.

4. (i) $\dfrac{1}{x-2} + \dfrac{2}{x+1} - \dfrac{3}{(x+1)^2} + \dfrac{1}{(x+1)^3}$,

   (ii) $\dfrac{3}{x-1} - \dfrac{1}{(x-1)^3} - \dfrac{3}{x+1} - \dfrac{2}{(x+1)^2}$.

5. (i) $x + 2 + \dfrac{1}{x-3} - \dfrac{2}{x+3}$, (ii) $3 - \dfrac{2}{x-1} - \dfrac{1}{x+2}$,

   (iii) $2x - \dfrac{6}{x-2} + \dfrac{12}{x^2+3}$, (iv) $x - 2 - \dfrac{6}{x+1} + \dfrac{2}{(x+1)^2} + \dfrac{3}{x}$.

6. $\dfrac{1}{6(x+2)} - \dfrac{7}{2x} + \dfrac{10}{3(x-1)}$. 7. $\dfrac{3}{2x^2} - \dfrac{3}{4x} + \dfrac{3}{4(x+2)}$.

8. $2x + 4 - \dfrac{1}{3(x-2)} - \dfrac{5x+61}{3(x^2+5)}$. 9. $\dfrac{5}{3+x} + \dfrac{2}{4-x} - \dfrac{3}{4+x}$.

10. $\dfrac{1}{x-1} - \dfrac{x}{x^2+x+1}$. 11. $\dfrac{2}{(2x+1)^2} - \dfrac{5}{2x+1} + \dfrac{3}{x-3}$.

12. $\dfrac{13}{29(2x-5)} - \dfrac{5}{29(3x+7)}$. 13. $\dfrac{3}{x-1} - \dfrac{3}{x} - \dfrac{2}{x^2} - \dfrac{1}{(x-1)^2}$.

14. $\dfrac{2}{5(5x-2)} - \dfrac{10x+1}{5(25x^2+10x+4)}$.

page 56.

15. $\dfrac{2}{x} - \dfrac{2x}{x^2+3} + \dfrac{2-5x}{(x^2+3)^2}.$ **16.** $\dfrac{1}{x^2+2} - \dfrac{1}{x^2+3}.$

17. $\dfrac{\sqrt 3}{36(x-\sqrt 3)} - \dfrac{\sqrt 3}{36(x+\sqrt 3)} - \dfrac{1}{6(x^2+3)}.$

Qu. 14. (i) $\frac{1}{6}\log_e(x-3) - \frac{1}{6}\log_e(x+3)+c,$

(ii) $\frac{1}{2}\log_e(2x-3) - \frac{5}{2}(2x-3)^{-1}+c.$

Qu. 15. $-\frac{1}{2}\log_e(2-x) - \frac{1}{2}\log_e(2+x)+c = \log_e \dfrac{k}{\sqrt{(4-x^2)}}.$

Qu. 16. No. **Qu. 17.** (i) $\log_e \frac{2}{3}\frac{5}{2},$ (ii) $\log_e 3.$

page 59. **Exercise 3c**

1. $\dfrac{1}{n} - \dfrac{1}{n+2}.$ **2.** $\dfrac{2}{n-1} - \dfrac{3}{n} + \dfrac{1}{n+1}.$

3. (i) $\dfrac{n+4}{n(n+1)(n+2)},$ (ii) $\dfrac{3}{2} - \dfrac{n+3}{(n+1)(n+2)},$ (iii) $1\frac{1}{2}.$

4. 2.

5. (i) $\dfrac{11}{18} - \dfrac{3n^2+12n+11}{3(n+1)(n+2)(n+3)},$ (ii) $\dfrac{n}{4(n+1)},$

(iii) $\dfrac{n}{9(n+1)},$ (iv) $\dfrac{3}{16} - \dfrac{2n+3}{8(n+1)(n+2)},$

(v) $\dfrac{11}{96} - \dfrac{1}{8(n+1)} - \dfrac{1}{8(n+2)} + \dfrac{1}{8(n+3)} + \dfrac{1}{8(n+4)},$

(vi) $\dfrac{1}{6} - \dfrac{n+2}{(n+3)(n+4)}.$

6. (i) $\dfrac{2n}{2n+1},$ (ii) $\dfrac{1}{12} - \dfrac{1}{4(2n+1)(2n+3)}.$

(iii) $\dfrac{5}{24} - \dfrac{4n+5}{8(2n+1)(2n+3)}.$

7. (i) $\frac{1}{2}\log_e \dfrac{k(x-2)}{x},$ (ii) $\frac{1}{17}\log_e \dfrac{k(5x-2)}{x+3},$

(iii) $\log_e \dfrac{kx}{3x+1} - \dfrac{2}{x},$ (iv) $\log_e \dfrac{k}{\sqrt{(16-x^2)}},$

(v) $\frac{1}{6}\log_e \dfrac{k(x-5)}{x+1},$ (vi) $\log_e \{k(x^2-4x-5)^{1/2}\},$

(vii) $\log_e \{k(x+2)\sqrt{(x^2+3)}\},$ (viii) $\log_e \dfrac{k(3+x)^2(2-x)}{(4-x)^3},$

(ix) $2\log_e \{k(2x+1)\} - \frac{1}{2}\log_e \{(x-3)(x+3)^3\},$

(x) $\frac{1}{3}\log_e \dfrac{k(x-2)}{x+1} - \dfrac{4}{x-2},$

(xi) $\log_e \{k(2x+1)^{1/2}\} - \frac{4}{3}\tan^{-1}\dfrac{x}{3},$

page 60.

7. (xii) $\log_e \{k(3x+2)^{\frac{1}{3}}\} - \frac{1}{6}\log_e (9x^2-6x+4)$,

(xiii) $\frac{1}{2}x^2 + 3x + \log_e \dfrac{k(x-5)^2}{x+2}$,

(xiv) $\frac{1}{4}\log_e \{k(x-3)\} - \frac{1}{8}\log_e (1+4x^2) - \frac{3}{2}\tan^{-1} 2x$.

8. (i) $\tan^{-1} x + c$,   (ii) $\log_e \{k(1+x^2)^{\frac{1}{2}}\}$,

(iii) $\tan^{-1} x + \log_e \{k(1+x^2)^{\frac{1}{2}}\}$,

(iv) $\log_e \left(k\sqrt{\dfrac{1+x}{1-x}} \right)$,   (v) $\log_e \dfrac{k}{\sqrt{(1-x^2)}}$,

(vi) $c - \sqrt{(1-x^2)}$,   (vii) $\sin^{-1} x + c$,

(viii) $\sin^{-1} x - \sqrt{(1-x^2)} + c$,   (ix) $-\log_e \{k(1-x)\}$,

(x) $-\log_e \{k(x-1)\}$,   (xi) $x - \log_e (1+x) + c$,

(xii) $\dfrac{1}{1-x} + c$,   (xiii) $\dfrac{1}{1-x} + \log_e (1-x) + c$.

9. (i) $\log_e \frac{4}{3} \simeq 0\cdot 29$,   (ii) $\frac{1}{2}\log_e 2 + \frac{1}{4}\pi \simeq 1\cdot 1$,   (iii) $\log_e \frac{45}{64} \simeq -0\cdot 35$,

(iv) $-3\log_e 2 - \frac{1}{2}\log_e 3 \simeq -2\cdot 6$.

10. $\dfrac{\pi}{2}$, $(2+2\log_e 2,\ 0)$.

CHAPTER 4

page 63.

Qu. 1. (i) $(-1)^r(r+1)x^r$,   (ii) $3^r x^r$,   (iii) $(\frac{1}{2})^{r+1}(r+1)(r+2)x^r$,

(iv) $\frac{1}{6}(-1)^r(r+1)(r+2)(r+3)x^r$.

Qu. 2. (i) $\dfrac{20!(21-2r)}{(21-r)!\,r!}\, x^r$,   (ii) $\dfrac{10!(33-5r)(-1)^r x^r}{(11-r)!\,r!}$.

Qu. 3. (i) $(-1)^{r-1}(3r+1)x^r$.

Qu. 4. (i) $\dfrac{1}{x} - \dfrac{1}{x^2} + \dfrac{1}{x^3} - \ldots + \dfrac{(-1)^{r+1}}{x^r} + \ldots$, $|x| > 1$.

(ii) $\dfrac{1}{x^2} - \dfrac{4}{x^3} + \dfrac{12}{x^4} - \ldots + \dfrac{(r-1)(-1)^r 2^{r-2}}{x^r} + \ldots$, $|x| > 2$.

(iii) $\dfrac{1}{9x^2} - \dfrac{2}{27x^3} + \dfrac{1}{27x^4} - \ldots + \dfrac{(-1)^r(r-1)}{3^r x^r} + \ldots$, $|x| > \dfrac{1}{3}$.

(iv) $\dfrac{3}{x^2} + \dfrac{9}{x^3} + \dfrac{21}{x^4} + \ldots + \dfrac{3(2^{r-1}-1)}{x^r} + \ldots$, $|x| > 2$.

(v) $\dfrac{1}{x} - \dfrac{1}{x^3} + \dfrac{1}{x^5} - \ldots + \dfrac{(-1)^r}{x^{2r+1}} + \ldots$, $|x| > 1$.

page 66.   **Exercise 4a**

1. (i) $1 - 3x + 9x^2 - \ldots + (-1)^r 3^r x^r + \ldots$, $|x| < \frac{1}{3}$.

(ii) $1 + 2x + 4x^2 + \ldots + 2^r x^r + \ldots$, $|x| < \frac{1}{2}$.

(iii) $1 - 2x + 3x^2 - \ldots + (-1)^r(r+1)x^r + \ldots$, $|x| < 1$.

page 66.

1. (iv) $1 + x + \frac{3}{4}x^2 + \ldots + \frac{(r+1)}{2^r}x^r + \ldots$, $|x| < 2$.

   (v) $1 - 3x + 6x^2 - \ldots + \frac{1}{2}(r+1)(r+2)(-1)^r x^r + \ldots$, $|x| < 1$.

   (vi) $\frac{1}{2} - \frac{1}{4}x + \frac{1}{8}x^2 - \ldots + (-1)^r x^r/2^{r+1} + \ldots$, $|x| < 2$.

   (vii) $\frac{1}{9} + \frac{2}{27}x + \frac{1}{27}x^2 + \ldots + (r+1)x^r/3^{r+2} + \ldots$, $|x| < 3$.

   (viii) $\frac{1}{8} + \frac{9}{16}x + \frac{27}{16}x^2 + \ldots + (r+1)(r+2)3^r x^r/2^{r+4} + \ldots$, $|x| < \frac{2}{3}$.

   (ix) $1 + \frac{1}{2}x - \frac{1}{8}x^2 + \ldots + (-1)^{r-1}\frac{1.3\ldots(2r-3)}{2^r r!}x^r + \ldots$, $|x| < 1$.

2. (i) $\frac{1}{1-x} + \frac{2}{1+2x}$, $3 - 3x + 9x^2 + \ldots + \{1 - (-2)^{r+1}\}x^r + \ldots$, $|x| < \frac{1}{2}$.

   (ii) $\frac{1}{1+x} - \frac{1}{2+x}$, $\frac{1}{2} - \frac{3}{4}x + \frac{7}{8}x^2 + \ldots + (-1)^r\{1 - (\frac{1}{2})^{r+1}\}x^r + \ldots$,

   $|x| < 1$.

   (iii) $\frac{1}{x+1} - \frac{2}{(x+1)^2}$, $-1 + 3x - 5x^2 + \ldots + (-1)^{r+1}(2r+1)x^r + \ldots$,

   $|x| < 1$.

   (iv) $\frac{3}{1-3x} + \frac{2}{1+2x}$, $5 + 5x + 35x^2 + \ldots + \{3^{r+1} + (-1)^r 2^{r+1}\}x^r + \ldots$,

   $|x| < \frac{1}{3}$.

   (v) $\frac{1}{x-2} + \frac{5}{(x-2)^2}$, $\frac{3}{4} + x + \frac{13}{16}x^2 + \ldots + (5r+3)x^r/2^{r+2} + \ldots$,

   $|x| < 2$.

   (vi) $\frac{3}{2(x-1)} - \frac{1}{2(x+1)}$, $-2 - x - 2x^2 + \ldots - \{3 + (-1)^r\}x^r/2 + \ldots$,

   $|x| < 1$.

3. (i) $1 - x^2 + x^4 - \ldots + (-1)^r x^{2r} + \ldots$, $|x| < 1$.

   (ii) $x + x^3 + x^5 + \ldots + x^{2r+1} + \ldots$, $|x| < 1$.

   (iii) $1 - 2x + 2x^2 - \ldots + 2(-1)^r x^r + \ldots$, $|x| < 1$.

   (iv) $1 - 9x + 35x^2 - \ldots + \frac{10!(-1)^r(11-2r)}{(11-r)!r!}x^r + \ldots$, all x.

   (v) $\frac{4}{3} - \frac{16}{9}x + \frac{52}{27}x^2 - \ldots + (-1)^r 2\left(1 - \frac{1}{3^{r+1}}\right)x^r + \ldots$, $|x| < 1$.

   (vi) $-\frac{5}{3} - \frac{28}{9}x - \frac{110}{27}x^2 - \ldots - (6 - 13.2^r/3^{r+1})x^r + \ldots$, $|x| < 1$.

   (vii) $-\frac{7}{2} + \frac{19}{4}x - \frac{57}{8}x^2 + \ldots - \{(\frac{1}{2})^{r+1} + (2r+3)(-1)^r\}x^r + \ldots$, $|x| < 1$.

4. (i) $\frac{1}{x} - \frac{2}{x^2} + \frac{4}{x^3} - \ldots + \frac{(-1)^{r-1}2^{r-1}}{x^r} + \ldots$, $|x| > 2$.

   (ii) $-\frac{1}{x^3} - \frac{9}{x^4} - \frac{54}{x^5} - \ldots - \frac{1}{2}(r-2)(r-1)\frac{3^{r-3}}{x^r} + \ldots$, $|x| > 3$.

   (iii) $\frac{1}{4x^2} + \frac{1}{4x^3} + \frac{3}{16x^4} + \ldots + \frac{r-1}{2^r x^r} + \ldots$, $|x| > \frac{1}{2}$.

   (iv) $1 + \frac{1}{x} - \frac{1}{x^2} + \ldots + \frac{(-1)^{r+1}}{x^r} + \ldots$, $|x| > 1$.

page 67.

4. (v) $\dfrac{1}{x}-\dfrac{5}{x^2}+\dfrac{16}{x^3}-\ldots+(-1)^{r-1}(3r-1)\dfrac{2^{r-2}}{x^r}+\ldots,\ |x|>2.$

(vi) $\dfrac{1}{x^2}+\dfrac{5}{x^3}+\dfrac{19}{x^4}+\ldots+\dfrac{3^{r-1}-2^{r-1}}{x^r}+\ldots,\ |x|>3.$

(vii) $\dfrac{2}{x}+\dfrac{6}{x^3}-\dfrac{12}{x^4}+\ldots+\dfrac{3\{1+(-1)^{r-1}3^{r-2}\}}{2x^r}+\ldots,\ |x|>3$

(viii) $-\left\{\dfrac{2}{x}+\dfrac{2}{x^3}+\dfrac{2}{x^5}+\ldots+\dfrac{2}{x^{2r+1}}+\ldots\right\},\ |x|>1.$

(ix) $-\dfrac{1}{x^3}-\dfrac{1}{x^4}-\dfrac{1}{x^7}-\dfrac{1}{x^8}-\ldots-\dfrac{1}{x^{4r-1}}-\dfrac{1}{x^{4r}}+\ldots,\ |x|>1.$

5. $x^{1/2}-x^{-1/2}-\tfrac{1}{2}x^{-3/2}$; $1{\cdot}414\ 2.$

6. $1{\cdot}259\ 92.$ **7.** $2{\cdot}009\ 93.$

8. $3{\cdot}014\ 963.$ **9.** $0{\cdot}009\ 920.$

10. $0{\cdot}242\ 5.$ **11.** $1+2x+\tfrac{5}{2}x^2.$

12. $1-x-x^2+3x^3.$ **13.** $1-4x+6x^2+4x^3.$

14. $1+2x-\tfrac{1}{2}x^2.$ **15.** $1-\tfrac{5}{3}x+\tfrac{14}{9}x^2-\tfrac{130}{81}x^3.$

16. $-1-\tfrac{1}{2}x+\tfrac{1}{4}x^2.$ **17.** $x-x^2-\tfrac{2}{3}x^3.$

18. $\pm 5\%.$ **19.** $\dfrac{86400}{86400-x}$ s, 17 s.

20. $21(\cdot 6)$ s.   **21.** $2x+y-z.$

22. $\tfrac{1}{2}(p-q-r).$ **23.** $\tfrac{2}{3}.$

24. $1/\sqrt{(1+2x)},\ |x|<\tfrac{1}{2}.$ **25.** $1/(1-2x)^2,\ |x|<\tfrac{1}{2}.$

26. $\tfrac{9}{16}.$ **27.** $\sqrt[3]{\tfrac{3}{2}}.$

28. $1/(1+2x)^3,\ |x|<\tfrac{1}{2}.$ **29.** $(1-2x)^{1/2},\ |x|<\tfrac{1}{2}.$

30. $\sqrt[3]{4}.$ **31.** $\sqrt[3]{(1+3x)^2},\ |x|<\tfrac{1}{3}.$

32. $\sqrt{2}.$ **33.** $\tfrac{1}{4}\sqrt[3]{25}.$

34. $\sqrt{\tfrac{5}{3}}.$ **35.** $5\sqrt{\tfrac{5}{3}}-6$

36. $9\sqrt[3]{\tfrac{3}{2}}-10.$ **37.** $3\sqrt{\tfrac{5}{3}}-4.$

38. $x^2+2px+q=0.$

Qu. 5. $792.4^7.3^5.$

Qu. 6. (i) $^{n+2}C_{r+1}=c_{r+1}+2c_r+c_{r-1},\ (1\leqslant r\leqslant n-1);$

(ii) $^{n+2}C_{r+2}=c_{r+2}+2c_{r+1}+c_r,\ (0\leqslant r\leqslant n-2).$

page 72.   **Exercise 4b**

1. (i) 15360,   (ii) 20412,   (iii) $792.4^7.3^5=2^{17}.3^7.11,$

(iv) $15504.2^5.5^{15}$ or $38760.2^6.5^{14}=2^9.3.5^{15}.17.19,$

(v) $330.2^4/3^4=2^5.5.11/3^3,$ (vi) $126.3^5.2^4=2^5.3^7.7,$

(vii) $11^{11}/12^{12},$ (viii) $3.5^5/7^7.$

2. (i) $66.2^2.9^{10}=2^3.3^{21}.11,$ (ii) $210(\tfrac{1}{2})^4(\tfrac{2}{3})^6=\tfrac{280}{243},$

(iii) $56(\tfrac{4}{3})^3(\tfrac{5}{2})^5=2^4.5^5.7/3^3,$ (iv) $3^{18}.2^{-8}.5^{-10}.$

page 73.   **Exercise 4c**

1. $\dfrac{3}{1-x} - \dfrac{2}{1+x^2}$, $1 + 3x + 5x^2 + 3x^3$.

2. $A = 1$, $B = -1$, $C = 1$, $2x + 2x^2 + 2x^5$; $|x| < 1$.

3. (i) 323,   (ii) $1 - x - x^2 - \tfrac{5}{3}x^3$; 1·71.

4. $\dfrac{1}{x} - \dfrac{1}{x^2} + \dfrac{4}{x^3} - \dfrac{1}{x+2}$, $-7 - 14y - 29y^2$; $|y| < 1$.

5. (i) $\dfrac{3}{3x-1} - \dfrac{1}{x+1} + \dfrac{1}{(x+1)^2}$, $n(-1)^n - 3^{n+1}$;

   (ii) $1 - \tfrac{1}{3000} - \tfrac{1}{9}10^{-6}$, 3·332 222.

6. $1 + \tfrac{1}{2}x - \tfrac{1}{8}x^2$, $1 - \tfrac{1}{2}x + \tfrac{3}{8}x^2$.

7. 0·000 103 2.

9. (i) $2p + 4q - r$,

   (ii) (a) $1 + 4x + 9x^2 + 16x^3$, $|x| < 1$; (b) $-\dfrac{1}{x^2} - \dfrac{4}{x^3} - \dfrac{9}{x^4} - \dfrac{16}{x^5}$, $|x| > 1$.

10. (a) 10 201 810 000,   (b) $1 + 2x + 3x^2 + \ldots + (n+1)x^n + \ldots$.

11. $1 - \tfrac{1}{2}y + \tfrac{3}{8}y^2 - \tfrac{5}{16}y^3$;

   $c_1 = \cos\theta$, $c_2 = \tfrac{1}{2}(3\cos^2\theta - 1)$, $c_3 = \tfrac{1}{2}\cos\theta(5\cos^2\theta - 3)$.

12. (ii) $1 + \tfrac{1}{2}x - \tfrac{1}{8}x^2 + \tfrac{1}{16}x^3$,

   (iii) $8n$, $-(8n+4)$.

13. $1 + \tfrac{1}{2}x - \tfrac{1}{8}x^2 + \tfrac{1}{16}x^3 - \tfrac{5}{128}x^4$.

15. $\dfrac{4}{9(2x+1)} - \dfrac{2}{9(x+2)} - \dfrac{1}{3(x+2)^2}$,

   $\tfrac{1}{4} - \tfrac{3}{4}x + \tfrac{27}{16}x^2 - \tfrac{7}{2}x^3$, $\tfrac{4}{9}(-2)^n - \dfrac{3n+7}{36}(-\tfrac{1}{2})^n$, $-\tfrac{1}{8}$.

16. $1 - x - \tfrac{1}{2}x^2 - \ldots - 1.3.5\ldots(2n-3)x^n/n! - \ldots$, 3·316 66...; correct to 4 places of decimals.

18. (a) $\tfrac{21}{16}$,   (b) $1 + x - x^2 + \tfrac{5}{3}x^3$, 1·009 90.

19. (i) 10, 29.   (ii) $(3n)!/\{(2n-r)!(n+r)!\}$.

22. Tenth.

23. (a) $r > \tfrac{1}{2}(n+1)$,   (b) 1·221 3.

24. $n = 6$, $a_3 = 10$.

25. (a) $q = \tfrac{1}{2}p(p-1)$, $p = -3$, $q = 6$, $a_6 = 16$.

CHAPTER 5

page 80.
Qu. 1. (i) $\tfrac{1}{6}$,   (ii) $\tfrac{1}{13}$,   (iii) $\tfrac{4}{11}$.

Qu. 2. $\tfrac{5}{42}$.

page 81.   **Exercise 5a**

1. (i) $\tfrac{1}{4}$,   (ii) $\tfrac{1}{13}$,   (iii) $\tfrac{3}{13}$.

2. (i) $\tfrac{1}{6}$,   (ii) $\tfrac{1}{2}$,   (iii) $\tfrac{1}{3}$.

page 82.

3. (i) $\frac{5}{36}$, (ii) $\frac{1}{18}$, (iii) $\frac{1}{36}$, (iv) $\frac{5}{6}$.

4. (i) $\frac{1}{216}$, (ii) $\frac{1}{36}$, (iii) $\frac{5}{72}$.

5. False. **6.** $\frac{1}{4}$.

7. $4/^{52}C_{13} \simeq 6 \cdot 30 \times 10^{-12}$. **8.** $\frac{3}{7}$.

9. $\frac{23}{25}$. **10.** $\frac{1}{2}$. **11.** $\frac{3}{4}$.

12. $^{20}C_{13}/^{52}C_{13} \simeq 1 \cdot 22 \times 10^{-7}$. **13.** $\frac{3}{4}$.

14. $\frac{4}{19}$. **15.** $\dfrac{4(39!)^2}{26!\,52!} \simeq 0 \cdot 0512$.

16. $\frac{5}{12}$. **17.** (i) 0, (ii) $\frac{6}{35}$.

18. $\frac{28}{111}$. **19.** (i) $\frac{2}{5}$, (ii) $\frac{41}{125}$.

Qu. 3. (i), (iv), (vi), (vii).

Qu. 4. (i) $\frac{1}{4}$, (ii) $\frac{1}{2}$; $\frac{3}{4}$.

Qu. 5. (i) $\frac{3}{10}$, (ii) $\frac{1}{5}$, (iii) $\frac{1}{10}$, (iv) $\frac{2}{5}$, (v) $\frac{3}{10}$.

15 satisfies both A and B, so they are not mutually exclusive.

Qu. 6. $\frac{3}{8}$.

Qu. 7. $\dfrac{5n-2}{2(4n-1)}$.

Qu. 8. 15/128.

page 87. **Exercise 5b**

1. (i) $\frac{1}{64}$, (ii) $\frac{9}{64}$, (iii) $\frac{27}{64}$.

2. (i) $\frac{1}{216}$, (ii) $\frac{1}{36}$.

3. (i) $\frac{1}{729}$, (ii) $\frac{512}{729}$, (iii) $\frac{217}{729}$.

4. 0·23 **5.** 17.

6. 2197/83 300, 1/649 740.

7. (i) $(\frac{4}{9})^6$, (ii) $\frac{52}{81}$.

8. (i) $\frac{1}{11}$, (ii) $\frac{53}{66}$. **9.** A.

10. $2^{10}/3^7$. **11.** 3:4.

12. $\frac{25}{27}$.

13. $\frac{1}{64}, \frac{6}{64}, \frac{15}{64}, \frac{20}{64}, \frac{15}{64}, \frac{6}{64}, \frac{1}{64}$.

14. $2^5/3^5$. **15.** $\frac{3}{4}$.

16. $\frac{5}{7}$. **17.** $\frac{81}{400}, \frac{3}{11}, \frac{6}{11}$.

18. 1225 : 1296 : 1260.

page 89. **Exercise 5c**

1. (i) 40 320, (ii) 4320. **2.** (ii) 1260.

3. $n^2(n-1)$, 3. **4.** (i) 1260, (ii) $\frac{2}{15}$.

5. (i) 15 600, (ii) 17 576, (iii) 6300.

6. (i) 24!/12!, (ii) $\frac{1}{3}$. **7.** 120, $\frac{1}{2}$.

8. 12!/(5! 4! 2!), 9! 4!/12! **9.** (i) $2(n-2)!$ (ii) $6(n-3)!$

10. 857/1105. **11.** 9/220, 7/12.

12. (i) 3^{10}, (ii) 3^3, (iii) 280.

13. 10/81, 20/243. **14.** (a) 1/63, (b) 74/315.

page 91.

15. (i) 1/221, (ii) 40/243.

16. $49 \cdot \dfrac{19^{19}}{20^{20}}$, $1 - \dfrac{49}{3} \cdot \dfrac{19^{19}}{20^{20}} - \dfrac{902}{3} \cdot \dfrac{9^{18}}{10^{20}}$.

17. (i) $^{20}C_{15}3^{15} \cdot 2^5/5^{20}$, (ii) $\{3^{20} + {}^{20}C_{15}3^{15} \cdot 2^5 + {}^{20}C_{10}3^{10} \cdot 2^{10}\}/5^{20}$.

18. 105/512.   19. 3/16, 8/47.

20. (a) 179:1 against,   (b) 8:7 against.

21. (a) $1 - (19/20)^{12}$,   (b) $b!w!/(b+w)!$

22. (a) (i) 1/22, (ii) 3/11,   (b) (i) 23/100, (ii) 379/1800.

23. (a) 2/15,   (b) 11/24.

24. 2/3.   25. $(p+m)!/(p!m!)$, $\frac{1}{4}n^2(n+1)^2$.

CHAPTER 6

page 96.

Qu. 3. (i) 46·4°,   (ii) 87·3°.

Qu. 4. 44·9°.   **Qu. 7.** 67·4°.

page 99.   **Exercise 6a**

1. (i) 50·1°, (ii) 51·5°, (iii) 36·9°.

2. (i) 67·4°, (ii) 71·6°, (iii) 28·1°.

3. (i) 22·03 cm, (ii) 65·2°.   4. 29·4°.

5. 97·9°.   6. 54·7°.

7. $1/\sqrt{3}$, (ii) $1/\sqrt{5}$.   8. 6·53 cm, 54·7°, 70·5°.

9. $\tan^{-1}(1/\sqrt{2})$.   10. 22·2°.   11. 75·2°.

12. $\sqrt{\cdot\frac{2}{3}}$.   13. 28·1°.   14. 80·4°.

15. $\tan^{-1}\dfrac{c\sqrt{(a^2+b^2)}}{ab}$.   16. $\dfrac{2\sqrt{3}}{\sqrt{133}}$.

page 103.   **Exercise 6b**

12. $S \cos^{-1}\dfrac{5\tan\alpha}{4}$ E or W.

Qu. 10. 42.   **Qu. 12.** BC′D′A.   **Qu. 14.** ∠ABC′.

Qu. 15. A plane bisecting, and at right angles to their join.

Qu. 16. A line through the circumcentre of the triangle of which the points are the vertices, and perpendicular to their plane.

page 113.   **Exercise 6c**

5. A plane parallel to the first two planes.

9. VC, BC.   19. A plane ∥ both lines equidistant from each.

Qu. 23. 53·1°.

page 118. **Exercise 6d**

3. $\dfrac{a}{2\sqrt{6}}$.

4. $\dfrac{a}{2}, \dfrac{a\sqrt{3}}{2}$.

6. $\dfrac{b^2\sqrt{3}}{2\sqrt{(3b^2-a^2)}}$.

7. $\dfrac{h^2+r^2}{2h}$.

11. (i) $\sqrt{\frac{119}{8}}$ cm,   (ii) $57\cdot2°$.

12. (i) $8\cdot0$ cm,   (ii) $65\cdot4°$.

13. $\dfrac{a}{3\sqrt{2}}$.

16. $46\cdot7°$.

17. (i) $39\cdot7°$,   (ii) $39\cdot7°$.

19. $22\cdot6°$.

20. $69\cdot4°, 35\cdot8°$.

22. $a\sqrt{\frac{2}{3}}$.

23. Sphere.

26. $97\cdot8°$.

27. $\dfrac{a}{2}\ \sqrt{(\sec^2 54° + \tan^2 54° \tan^2 \alpha)}$ cm, $\tan^{-1}(\sin 54° \tan \alpha)$.

28. (i) $60\cdot8°$,   (ii) $54\cdot6°$.

29. (i) $36\cdot9°, 56\cdot3°, 36\cdot0°$,   (ii) $100\cdot0°$.

30. $45\cdot9°$.   **32.** $35\cdot3°$.   **35.** $\sqrt{\{\frac{1}{2}(b^2+c^2-a^2)\}}$.

CHAPTER 7

page 123.

Qu. 1. (i) Yes,   (ii) No.

Qu. 2. (i) No,   (ii) Yes.

Qu. 3. No.

Qu. 4. Not necessarily.

page 126. **Exercise 7a**

1. (i) $x < \frac{1}{2}, x > 2$.   (ii) $\frac{1}{2} < x < 2$.

2. (i) $x < -2, x > -\frac{1}{2}$.   (ii) $-2 < x < -\frac{1}{2}$.

3. $-2 < x < 3$.

4. $x > 2\frac{1}{2}, x < -2\frac{1}{3}$.

5. $-3 < x < 2\frac{1}{2}$.

6. $x < -2, x > 2\frac{1}{2}$.

7. $x < -1\frac{1}{4}, x > 1$.

8. $\frac{3}{5} < x < 7$.

9. $1 < x < 2, 3\frac{1}{4} < x < 3\frac{3}{4}$.

10. $x < -1, 1 < x < 2, x > 3$.

11. $\frac{3}{4}$.   **12.** -1.

13. $\frac{7}{8}$.   **14.** $-\frac{11}{12}$.

15. $x < 0, x > 1; |y| > \frac{1}{2}$.

16. $|x| > 2, |y| > \sqrt{3}$.

17. $x < -1, 0 < x < 1$.

18. $|x| > 2, |y| > \sqrt{(-2+2\sqrt{2})}$.

19. Discontinuities when $x = 2, 3$.

20. Discontinuities when $x = -1\frac{1}{2}, \frac{1}{2}$.

21. $y < -\frac{1}{2}, y > 4\frac{1}{2}$.

22. $y < 0, y > \frac{4}{3}$.

23. $x \geqslant 0$.

24. $x \leqslant 0$.

25. $x \leqslant -3, x \geqslant -\frac{1}{3}$.

26. $x \leqslant -9, x \geqslant -\frac{1}{3}$.

27. $|x| \leqslant \sqrt{2}$.

28. $-4 \leqslant x \leqslant 6$.

31. $(2x-3y)^2 + (x-2)^2 > 0$, unless $x=2, y=1\frac{1}{3}$.

32. 1.   **34.** $(3, -1)$.

page 133. **Exercise 7b**

In Nos. 10 to 15, *y cannot* lie in the following ranges:

10. $-4 < y < 0$. **11.** $\frac{1}{6} < y < 1\frac{1}{2}$.

12. $y < 0$, $y > 1$. **13.** $y < -\frac{1}{2}$, $y > 4\frac{1}{2}$.

14. $y > 1$. **15.** $\frac{2}{3}(1-\sqrt{10}) < y < \frac{2}{3}(1+\sqrt{10})$.

16. $(2, 4)$ min.

17. $\left(-6-2\sqrt{10}, (11+2\sqrt{10})/18\right)$ min., $\left(-6+2\sqrt{10}, (11-2\sqrt{10})/18\right)$ max.

18. $\left((-1-\sqrt{7})/2, (-23-8\sqrt{7})/9\right)$ max.,
$\left((-1+\sqrt{7})/2, (-23+8\sqrt{7})/9\right)$ min.

19. $(-\frac{1}{3}, 4\frac{1}{2})$ max.

Qu. 5. (i) a, b, c, (ii) b, (iii) c,
(iv) c, (v) a, b, c, (vi) c,
(vii) a, b, c, (viii) c.

Qu. 6. (i) $x - y = 0$, (ii) $x + y = 0$.

Qu. 8. (i) As $x \to 0, 2, 4$, $\dfrac{dy}{dx} \to \infty$.

(ii) When $x = 0$, $\dfrac{dy}{dx} = \pm\sqrt{2}$; as $x \to -2$, $\dfrac{dy}{dx} \to \infty$.

page 140. **Exercise 7c**

9. $r^2 = 1 - \tan^4 \theta$.

page 140. **Exercise 7d**

1. (i) $x < 1$ or $x > 3$, (ii) $x < 1$ or $x > 3$,
(iii) $1 < x < 2$ or $x > 3$.

2. $p - 1 \leqslant x \leqslant p + 1$.

3. $b^2 \geqslant 4ac$, a, $-b$, c all of the same sign.

4. $2 < \lambda < 2\frac{2}{3}$. **5.** $-2 < k < 6$, $0 < k < 6$.

6. $(a + b + c)^2 = 4(bc + ca + ab)$.

7. $\pi/12 \leqslant \alpha \leqslant 7\pi/12$, $13\pi/12 \leqslant \alpha \leqslant 19\pi/12$.

9. $k = 3$, $a = -2$, $b = -2$; $k = -7$, $a = 8$, $b = \frac{1}{2}$.

12. (*a*) $(x+1)(y+1) > 0$, (b) $-1 \leqslant a \leqslant 1\frac{1}{3}$.

13. (a) (i) $\frac{1}{4} < x < 1\frac{1}{2}$, (ii) $x < -1$, $1 < x < 2$.

14. $-1 \cdot 22$. **16.** $k = \frac{1}{2}$.

18. $p = 1$, $q = 2$, $(-4, \frac{1}{9})$.

19. $k \leqslant 1$, $k \geqslant 9$; max. $(1, 1)$, min. $(-\frac{1}{3}, 9)$.

20. Asymptotes $x = \pm 2$, $y = 1$.

21. (a) max. 9. (b) $-2 < k < 8$, $k = -2$, $k = 8$; max. 8, min. -2.

23. $p = -1$, $q = -4$ or $p = -4$, $q = -1$.

24. $0 < p < 1$. **26.** $2 - \log_e 3$, $\pi(\frac{8}{3} - 2\log_e 3)$.

27. $4x + 16y - 9 = 0$, $(-9/16, 45/64)$.

29. $(0, -4)$, $(1, 3)$.

CHAPTER 8

page 147.

Qu. 1. 4. Yes.

Qu. 2. 4.

Qu. 3. 10, *not* 5.

page 150.   **Exercise 8a**

1. $-2, -1, 3, 4$.   **2.** $-7, -1, -1, 5$.   **3.** $\pm \log_e 2$.

4. $0, 2$.   **5.** $\pm 1, \pm 3$.   **6.** $1, 8$.

7. 3.   **8.** 9.   **9.** 4, *not* 1.

10. $5, 1\frac{4}{5}$.   **11.** 5.   **12.** 5, *not* $22\frac{7}{9}$.

13. 14, *not* 6.   **14.** $(4, 1), (-2, -2)$.

15. $(4, 5), (-3, 4)$.   **16.** $(3, 1), (-5, -5)$.

17. $(0, 0), (8, -6)$.   **18.** $(-3, 4), (-6, 3)$.

19. $\pm(\sqrt{5}, \frac{4}{5}\sqrt{5}), \pm(2\sqrt{5}, \frac{2}{5}\sqrt{5})$.   **20.** $2(a+b)/b$.

21. (i) $-(c+d)/c$,   (ii) $-(a+b)/(a-b)$.

22. $t^2 x - y - ct^3 + c/t = 0; \ (-c/t^3, \ -ct^3)$.

23. $(a(1-p)^2, 2a(1-p))$.   **24.** $(-c/t^2, \ -ct^2)$.

25. $(a(a^4-b^4)/(a^4+b^4), \ -2a^2b^3/(a^4+b^4))$.

26. $(3-a)^2 = (6-ab)(b-2)$.

28. $x = \dfrac{Ca-cA}{Ab-aB}, y = \dfrac{Bc-bC}{Ab-aB}; \ (aC-cA)^2 = (bC-cB)(aB-bA)$.

29. $-30, -12, -6, \ 0$.   **31.** $\frac{1}{3}, \frac{1}{2}, 2, 3$.

32. $-\frac{1}{4}, -4$.   **33.** $\frac{1}{5}, 5$.

34. $Ay^2 + By + (C - 2A) = 0$, where $y = x + 1/x$.   **35.** 1, 2.

36. 8 cm, 15 cm, 17 cm.   **37.** 1 cm.

38. $x^2(b^2 + a^2m^2) + 2a^2cmx + a^2(c^2 - b^2) = 0$,

     $y^2(b^2 + a^2m^2) - 2b^2cy + b^2(c^2 - a^2m^2) = 0$,

                 $(-a^2cm/(b^2+a^2m^2), \ b^2c/(b^2+a^2m^2))$.

page 155.   **Exercise 8b**

1. $\dfrac{2a-c}{2b-d}$.   **2.** $\dfrac{3a-4c}{3b-4d}$.   **3.** $\dfrac{2a+3c}{2b+3d}$.

4. $\dfrac{3a-c}{3b-d}$.   **27.** $\frac{1}{4}, \frac{1}{3}, -\frac{1}{2}$.   **28.** $-2, 3, 5$.

29. $1, \frac{1}{2}, \frac{1}{3}$.   **30.** $4, 5, 2$.   **31.** $1, 3, 2$.

32. $\frac{1}{2}, \frac{1}{4}, 1$.   **33.** $-3, 2, -1$.   **34.** $3\frac{25}{71}, \frac{1}{71}, \frac{20}{71}$.

35. $\frac{6}{7}, 8\frac{2}{7}, 6\frac{4}{7}$.

Qu. 10. (i) 3,   (ii) 2,   (iv) -2,   (vi) 4.

Qu. 11. (i) $a^3 + b^3 + c^3$,   (ii) $a(b+c) + b(c+a) + c(a+b)$,

         (iii) $\dfrac{1}{a} + \dfrac{1}{b} + \dfrac{1}{c}$,   (iv) $ab^2c^2 + bc^2a^2 + ca^2b^2$.

Exercise 8c

page 161.

1. $3:2,\ -4:3$.   **2.** $-5:2,\ 6:1$.

3. $-1:1,\ 2:1,\ 2:1$.   **4.** $1:1,\ -2:3,\ -3:2$.

5. $\pm 3:1,\ \pm 1:2$.   **6.** $1:1,\ 1:1,\ 5:1,\ -3:1$.

7. $10:-11:-13$.   **8.** $1:14:11$.

9. $-c:c:a-b$.   **10.** $a:-(a^2+1):a$.

11. $\sin\theta-\cos\theta:\sin\theta+\cos\theta:1$.   **12.** $t_1t_2:t_1+t_2:1$.

13. $x^4+y^4+z^4$.   **14.** $\dfrac{1}{yz}+\dfrac{1}{zx}+\dfrac{1}{xy}$.

15. $x^2(y+z)+y^2(z+x)+z^2(x+y)$.

16. $x^2y+y^2z+z^2x$.   **17.** $xy^2+yz^2+zx^2$.

18. $x^3yz+y^3zx+z^3xy$.   **23.** $(1-t)(1+t+t^2)$.

24. $(4x+y)(16x^2-4xy+y^2)$.   **25.** $(2+3z)(4-6z+9z^2)$.

26. $(5y-z^2)(25y^2+5yz^2+z^4)$.

27. $(a+2b)^3$.   **28.** $(3u-1)^3$.

29. $(b-c)(3a^2+b^2+c^2+bc+3ca+3ab)$.

30. $2x(x^2+3y^2)$.   **31.** $(a+b+c)^2$.

32. $(a+b-c)^2$.

33. $(a+b+c)(a^2+b^2+c^2+2bc-ca-ab)$.

34. $(a+2b+3c)(a^2+4b^2+9c^2-6bc-3ca-2ab)$.

35. $(a-b)(a+b)(a^2+ab+b^2)(a^2-ab+b^2)$.

36. $(x^n-a^n)/(x-a)$.

Qu. 15.   (i) $\frac{4}{3},\ \frac{2}{3},\ -\frac{5}{3}$,   (ii) $0,\ 0,\ 1$,

          (iii) $0,\ \frac{6}{7},\ \frac{5}{7}$,   (iv) $-2,\ -1,\ 3$,

          (v) $5,\ 0,\ -2$,   (vi) $-1,\ 1,\ -1$.

Qu. 16.   (i) $x^3-6x^2+11x-6=0$,   (ii) $x^3-13x+12=0$,

          (iii) $x^3-14x^2+288=0$.

Qu. 17.   (i) $x=\sqrt{y}$,   (ii) $x=y+2$,

          (iii) $x=\frac{1}{2}(y-1)$,   (iv) $x=-7y/3$.

Qu. 20. $3ac=b^2,\ 27a^2d=b^3$.   (Other relations are possible. Those given determine the ratios $c:a,\ d:a$ in terms of $b:a$.)

Exercise 8d

page 167.

1.   (i) $\frac{4}{3},\ -\frac{1}{3},\ -\frac{2}{3}$,   (ii) $0,\ \frac{5}{4},\ \frac{3}{2}$,

     (iii) $-5,\ 4,\ -4$,   (iv) $-\frac{7}{3},\ \frac{13}{3},\ -\frac{1}{3}$.

2.   (i) $x^3-2x^2+5=0$,   (ii) $x^3+3x^2+2x-6=0$.

     (iii) $x^3-x-5=0$.

page 167.

3.   (i) $2y^3 - 3y^2 - 13y + 7 = 0.$ (ii) $7y^3 + 13y^2 - 3y - 2 = 0.$
     (iii) $2y^3 + 15y^2 + 23y - 5 = 0.$ (iv) $4y^3 + 26y^2 - 21y - 49 = 0.$

4.   (i) $y^3 + 3y^2 - 21y + 17 = 0.$ (ii) $y^3 + 3y^2 - 21y - 63 = 0.$
     (iii) $100y^3 + 30y^2 - 3y - 1 = 0.$

5.   (i) $y^3 + 6hy^2 + 9h^2y - g^2 = 0.$ (ii) $g^2y^3 - 9h^2y^2 - 6hy - 1 = 0.$
     (iii) $y^3 + 3gy^2 + 3(g^2 + 9h^3)y + g^3 = 0.$

6.   (i) $a^2y^3 + (2ac - b^2)y^2 + (c^2 - 2bd)y - d^2 = 0.$
     (ii) $d^2y^3 + (2bd - c^2)y^2 + (b^2 - 2ac)y - a^2 = 0.$
     (iii) $a^3y^3 + (3a^2d + b^3 - 3abc)y^2 + (3ad^2 - 3bcd + c^3)y + d^3 = 0.$

7.   (i) $-6h,$ (ii) $9h^2,$ (iii) $18h^2.$

8. $-6h, \ -3g, \ 18h^2.$

9.   (i) $9a^2 - 6b,$ (ii) $-27a^3 + 27ab - 3c,$
     (iii) $81a^4 - 108a^2b + 12ac + 18b^2.$

10.   (i) $2b^3 - 9abc + 27a^2d = 0,$
     (ii) $b^3d - ac^3 = 0, \ dy^3 + cy^2 + by + a = 0, \ 2c^3 - 9bcd + 27ad^2 = 0.$

11. $b^3 - 4abc + 8a^2d = 0.$ 12. $\frac{1}{2}, \ \frac{2}{3}, \ \frac{3}{8}.$

13. $\frac{3}{4}, \ \frac{5}{4}, \ \frac{7}{4}.$ 14. $\frac{5}{2}, \ \frac{5}{2}, \ -\frac{2}{3}.$

15. $\frac{2}{3}, \ \frac{2}{3}, \ -\frac{5}{2}.$ 16. $\frac{3}{2}, \ \frac{3}{2}, \ -\frac{4}{3}.$

17. $g^2 = 4h^3, \ g = h = 0.$ 18. $(\frac{1}{4}t^2, \ -\frac{1}{8}t^3).$

19. $tx + y - at^3 - 2at = 0; \ 27ay^2 = 4(x - 2a)^3.$

20. $2x + t^3y - 3t = 0; \ -1:4, \ 2:1.$

21. $-1, \ 2, \ -3$ and permutations.

22. $2, \ -3, \ 5$ and perms.   23. $1, \ 4, \ -5$ and perms.

24. $1, \ 3, \ -2$ and perms.

25. $-b/a, \ c/a, \ -d/a, \ e/a; \ ey^4 + dy^3 + cy^2 + by + a = 0,$
       $a^2y^4 + (2ac - b^2)y^3 + (c^2 + 2ae - 2bd)y^2 + (2ce - d^2)y + e^2 = 0.$

26. $-0.81, \ 0.39, \ 6.4.$

27. $-0.9397, \ 0.1736, \ 0.7660.$

28. $-1.83, \ 0.226, \ 1.61.$ 29.   Between -4 and $-3; \ -3.$

30. $f = b - a^2, \ g = 2a^3 - 3ab + c.$ Reduce equation to the form
       $x^3 + 3fx + g = 0$ and draw across $y = x^3$ the line $y + 3fx + g = 0.$

31. $(9ad - bc)^2 = 4(b^2 - 3ac)(c^2 - 3bd).$

page 169.   **Exercise 8e**

1.   (i) $\pm 1/27, \ \pm 8,$ (ii) $(-1, 2), \ (-\frac{1}{2}, 1\frac{3}{4}).$

2.   (i) $k, \ -3 - k,$ (ii) $4.$

3.   (i) $-2, 0,$ (ii) $(-3, 2), \ (3, -2), \ (-4, 1\frac{1}{2}), \ (4, -1\frac{1}{2}).$

4.   (i) $A = -1, \ B = -2.$ 5.   (ii) $\frac{1}{2}(3 \pm \sqrt{5}).$

6.   (ii) $t^2 + qt - p^3 = 0, \ 2^{1/3} - 2^{2/3}.$

7. $\frac{1}{2}(1 \pm \sqrt{2}), \ \frac{1}{2}(-1 \pm \sqrt{2}).$

8. $(4, 2), \ (2, 4), \ (12, -6), \ (-6, 12).$ 9.   (i) $(-2, 1), \ (1\frac{1}{3}, 6).$

page 170.

10. (i) $\frac{2}{7}$, $\frac{4}{7}$, $\frac{6}{7}$; $-1\frac{1}{2}$, -3, $-4\frac{1}{2}$;

(ii) $a = 3$, $b = 2$, $x^2 + 2x + 3$, $x^2 - 2x + 3$.

11. (i) $(3, 0)$, $(-1, 4)$,   (ii) -1, 2, -3.

12. (i) -2, 1,   (ii) 3, -1, 2.

13. $b_1c_2 - b_2c_1 : c_1a_2 - c_2a_1 : a_1b_2 - a_2b_1$,

(i) 4, 2, -6,   (ii) $(c_1a_2 - c_2a_1)^2 = (b_1c_2 - b_2c_1)(a_1b_2 - a_2b_1)$.

15. (i) $(3\frac{9}{25}, -25)$,   (ii) $(2, 1)$, $(-1\frac{1}{2}, -2\frac{1}{2})$.

16. (i) -2, $\frac{1}{2}(3 \pm \sqrt{5})$,   (ii) $1 - \frac{1}{2}\sqrt{7}$.

17. (i) 3,   (ii) $-\frac{2}{3}$, $-2 \pm \sqrt{6}$.

18. (i) 3, $-\frac{2}{7}$, $-\frac{1}{3}$.

19. (ii) $x^3 - 2x^2 + 5x - 11 = 0$.   **20.** $-\frac{1}{2}$.

21. (a) $(x - 2)(x^2 + 4x + 9)$,   (b) 3,   (c) $k = -8$, $l = 4$.

22. $(0, 0)$, $(2, 64)$, $(5, -125)$, $0 \leqslant k \leqslant 64$, counting repeated roots twice.

23. (i) $-0\cdot47$, $1\cdot83$,   (ii) $0 \leqslant k < 3$.

24. Max. 32, min. 0, 27; $27 < k < 32$.

25. (a) $y^2 + 31y + 256 = 0$,   (b) $a^3 - 2a^2 + a(3b + 2) + b^2 - 2b = 0$.

26. $c = 2$, $a = 2$, $p = 3$, $b = 3$, $0\cdot855$.

28. (a) $p = -\frac{3}{8}$, $q = \frac{1}{64}$,   (b) $-\frac{3}{4}$.

CHAPTER 9

page 177.

Qu. 1. (i) $y^2 = -4ax$,   (ii) $x^2 = 4by$.

Qu. 4. $y_1x + 2ay - y_1(x_1 + 2a) = 0$.

Qu. 5. $x - ty + at^2 = 0$.

page 180.   **Exercise 9a**

1. $(at_1t_2, a(t_1 + t_2))$.   **2.** $t_1t_2 = -1$.

4. $(0, \frac{1}{2})$.   **5.** $(x - 2)^2 = 4(y + 1)$.

6. $(x + y)^2 + 4(x - y + 1) = 0$.

7. $-t$, $\dfrac{2}{t + t_1}$, $(a(t + 2/t)^2, -2a(t + 2/t))$.

9. (i) $(\frac{1}{2}, -2)$,   (ii) $(8, 8)$.

11. $(a(t_1^2 + t_1t_2 + t_2^2 + 2), -at_1t_2(t_1 + t_2))$.

14. $4a$.

15. $x - y + a = 0$, $x - 16y + 256a = 0$.

Qu. 6. $(\frac{5}{4}, 3)$, $y - 1\frac{3}{4} = 0$.

page 185.   **Exercise 9b**

1. $y + 1 = 0$.   **2.** $(1, \frac{1}{2})$, $\frac{3}{4}$.

3. $(-\frac{3}{4}, -6)$.   **5.** $a^2 = bc$.

page 185.

6. $y = 4 - x^2$, $2x + y - 5 = 0$.

8. $y = 2a/k$.   **9.** $x = a$.

10. $y^2 - 4ax = \frac{1}{4}k^2$.

11. $y^2 = 2a(x - a)$.   **12.** $x(x - a)^2 = ay^2$.

13. $\left(-\dfrac{b}{2a}, \ -\dfrac{b^2 - 4ac}{4a} + \dfrac{1}{4a} \right)$, $y = -\dfrac{b^2 - 4ac}{4a} - \dfrac{1}{4a}$.

14. $yy_1 = 2a(x + x_1)$.   **15.** $(3a, \ \pm 2\sqrt{3}a)$.

16. $2a$.   **18.** $2x(h - x - 2a) + ky = 0$.

19. $3y^2 = 16ax$.   **20.** $y^2 - 2ax - 2ay + 2a^2 = 0$.

21. $(2x + a)y^2 = a(3x + a)^2$.

22. $3x^2 + y^2 - ax = 0$.

Qu. 7. $7 \cdot 2$ cm, 12 cm.

Qu. 10. $\frac{3}{8}$.

Qu. 9. 4, 3.

Qu. 11. $(\pm 3\sqrt{3}/2, \ 0)$

page 192.   **Exercise 9c**

1. (i) $(\pm \sqrt{5}, \ 0)$, $x = \pm 9\sqrt{5}/5$;

  (ii) $(\pm 5\sqrt{15}/4, \ 0)$, $x = \pm 4\sqrt{15}/3$.

2. (i) $2x \cos \theta + 3y \sin \theta - 6 = 0$,   (ii) $9x + 16y - 25 = 0$.

3. (i) $16x - 9y - 7 = 0$,   (ii) $4x + y - 2 = 0$.

4. $3x - 2y - 5 = 0$.   **6.** $\dfrac{\cos \frac{1}{2}(\theta - \phi)}{\cos \frac{1}{2}(\theta + \phi)} = \pm e$.

8. $ex + y - a = 0$.   **10.** $\left(a \dfrac{\cos \frac{1}{2}(\theta + \phi)}{\cos \frac{1}{2}(\theta - \phi)}, \ b \dfrac{\sin \frac{1}{2}(\theta + \phi)}{\cos \frac{1}{2}(\theta - \phi)} \right)$

12. $\dfrac{(2x - ae)^2}{a^2} + \dfrac{4y^2}{b^2} = 1$.

13. $\dfrac{x^2}{a^2} + \dfrac{y^2}{b^2} = \cos^2 \frac{1}{2}k$.   **14.** $4a^2x^2 + 4b^2y^2 = (a^2 - b^2)^2$.

15. $a^2y^2 + b^2x^2 = 4x^2y^2$.   **16.** $b^2x^2 + a^2y^2 = 2a^2b^2$.

page 196.   **Exercise 9d**

1. (i) $2x - y \pm 5 = 0$,   (ii) $x + y \pm 2 = 0$,   (iii) $x - 2y \pm 10 = 0$.

2. (i) $(\frac{9}{13}, \ -\frac{4}{13})$,   (ii) $(-\frac{1}{2}, \ \frac{1}{3})$,   (iii) $(11, \ -6)$.

4. $3x + 2y \pm 2\sqrt{10} = 0$.   **5.** ± 5, $(16/5, \ -9/5)$, $(-16/5, \ 9/5)$.

6. $c^2 < a^2m^2 + b^2$.   **8.** $(x^2 + y^2)^2 = a^2x^2 + b^2y^2$.

9. $a^2y_1x - b^2x_1y - x_1y_1(a^2 - b^2) = 0$.

10. (i) $\left(\dfrac{-a^2mc}{a^2m^2 + b^2}, \ \dfrac{b^2c}{a^2m^2 + b^2} \right)$,   (ii) $\left(\dfrac{2am^2 - nl}{l^2}, \ -\dfrac{2am}{l} \right)$.

11. $8x - 27y = 0$.

12. $mx - y - mae = 0$, $\left(\dfrac{a^3em^2}{b^2 + a^2m^2}, \ -\dfrac{ab^2em}{b^2 + a^2m^2} \right)$, $a^2y^2 + b^2x^2 - ab^2ex = 0$.

page 197.

13. $a^2y^2 + b^2x(x-a) = 0$.
14. $(a^2 + b^2 - x^2 - y^2)^2 = 4(b^2x^2 + a^2y^2 - a^2b^2)$.
15. $b^2hx + a^2ky - (b^2h^2 + a^2k^2) = 0$.
16. $(a^2y^2 + b^2x^2)^2 = a^2(a^4y^2 + b^4x^2)$.
17. $(y^2 - 2ax)^2 = 4a^4 + b^2y^2$.
18. $a^4y^2 + b^4x^2 = 4x^2y^2$.

page 206.   **Exercise 9e**

2. $9b^2x^2 - 9a^2y^2 - 12ab^2x + 12a^2by - a^2b^2 = 0$.
4. $(a^2 + b^2)^2 = 4(a^2x^2 - b^2y^2)$.   **5.** $(-c/t^3, -ct^3)$.
6. $xy = c^2$.   **7.** $2xyc^2 = c^4 - y^4$.   **8.** $y_1x + x_1y - 2x_1y_1 = 0$.
10. $(ct, -ct^3)$.   **11.** $(x^2 + y^2)^2 = 4c^2xy$.
12. $n^2 = 4lmc^2$.   **14.** $x^2 + y^2 = a^2 - b^2$.
15. $(x^2 + y^2)^2 = a^2x^2 - b^2y^2$.
18. $bx \cos \frac{1}{2}(\theta - \phi) - ay \sin \frac{1}{2}(\theta + \phi) - ab \cos \frac{1}{2}(\theta + \phi) = 0$.
20. $\left(\dfrac{x^2}{a^2} - \dfrac{y^2}{b^2}\right)^2 \left(\dfrac{a^6}{x^2} - \dfrac{b^6}{y^2}\right) = (a^2 + b^2)^2$.
22. $c^2(x^2 - y^2)^2 + x^3y^3 = 0$.
24. $4cxy - 2\sqrt{2}xy(x + y) + c(x^2 + y^2) = 0$.

page 208.   **Exercise 9f**

1. $x^2/36 + y^2/20 = 1$.   **2.** $x^2/36 + 4y^2/119 = 1$.
4. $3x \pm 2y = 0$; $3x \pm 2y + 1 = 0$.   **5.** $25x + 20y + 64 = 0$, $4/5$.
6. $x \pm y \pm \frac{1}{6}\sqrt{3} = 0$.   **9.** $(4, -3)$.
10. $(0, 3)$, $3\sqrt{2}$.   **11.** $y + 1 = 0$.
13. $1, (\frac{1}{2}, 2)$; $2x - 2y + 1 = 0$, $2x + 2y - 3 = 0$.
15. $x^2 + y^2 = a^2 + b^2$; $2(a^2 + b^2)$, $4ab$.   **16.** $x + y - \sqrt{2}c = 0$.
18. $y^2 = 2a(x + 2a)$.
19. $\dfrac{b^2x^2}{a^2} + \dfrac{a^2y^2}{b^2} = (a^2 + b^2)\left(\dfrac{x^2}{a^2} + \dfrac{y^2}{b^2}\right)^2$.
20. $64xy = 1$.   **21.** $y^2 = a(x - 3a)$.
22. $\{y^2 + x(x - a)\}^2 = a^2(x - a)^2 + b^2y^2$.
23. $\left(\dfrac{V^2 \sin 2\alpha}{2g}, -\dfrac{V^2 \cos 2\alpha}{2g}\right), y = \dfrac{V^2}{2g}$

CHAPTER 10

page 212.

Qu. 1.   (i) $1 - x + \dfrac{x^2}{2!} - \dfrac{x^3}{3!} + \ldots + (-1)^n \dfrac{x^n}{n!} + \ldots,$

         (ii) $1 + x^2 + \dfrac{x^4}{2!} + \dfrac{x^6}{3!} + \ldots + \dfrac{x^{2n}}{n!} + \ldots,$

page 212.

Qu. 1. (iii) $1 + 3x + \dfrac{9x^2}{2} + \dfrac{9x^3}{2} + \ldots + \dfrac{3^n x^n}{n!} + \ldots,$

(iv) $1 + \dfrac{1}{x} + \dfrac{1}{2x^2} + \dfrac{1}{3! \, x^3} + \ldots + \dfrac{1}{n! x^n} + \ldots,$

(v) $1 - \dfrac{1}{x^2} + \dfrac{1}{2x^4} - \dfrac{1}{3! \, x^6} + \ldots + \dfrac{(-1)^n}{n! \, x^{2n}} + \ldots.$

page 215. **Exercise 10a**

1. (i) $1 \cdot 1052$,   (ii) $0 \cdot 3679$,   (iii) $1 \cdot 6487$.
2. $1 + x^3 + \frac{1}{2}x^6 + \frac{1}{6}x^9 + \ldots + x^{3n}/n! + \ldots.$
3. $1 + \frac{1}{3}x + \frac{1}{18}x^2 + \frac{1}{162}x^3 + \ldots + x^n/(n! 3^n) + \ldots.$
4. $1 - 2x + 2x^2 - \frac{4}{3}x^3 + \ldots + (-1)^n 2^n x^n/n! + \ldots.$
5. $e^2 \{ 1 + x + \frac{1}{2}x^2 + \frac{1}{6}x^3 + \ldots + x^n/n! + \ldots \}.$
6. $1 - \frac{1}{2}x + \frac{1}{8}x^2 - \frac{1}{48}x^3 + \ldots + (-1)^n x^n/(n! 2^n) + \ldots.$
7. $1 + 2x + \frac{3}{2}x^2 + \frac{2}{3}x^3 + \ldots + (n+1)x^n/n! + \ldots.$
8. $1 - 2x^2 + \frac{8}{3}x^3 - 2x^4 + \ldots + (-1)^{n-1}(n-1)2^n x^n/n! + \ldots.$
9. $1 + 4x + 8x^2 + \frac{32}{3}x^3 + \ldots + 4^n x^n/n! + \ldots.$
10. $2 + 3x + \frac{5}{2}x^2 + \frac{3}{2}x^3 + \ldots + (1 + 2^n)x^n/n! + \ldots.$
11. $3 + x + \frac{1}{2}x^2 + \frac{1}{6}x^3 + \ldots + x^n/n! + \ldots.$
12. $4 + 6x + 7x^2 + 6x^3 + \ldots + (1 + 2^n + 3^n)x^n/n! + \ldots.$
13. $10^9/9!$, $10^{10}/10!$
14. $1 + 2x + 3x^2 + \frac{10}{3}x^3.$
15. $e(1 - 3x + \frac{11}{2}x^2 - \frac{15}{2}x^3).$
16. $1 + \frac{1}{2}x^2 - \frac{1}{3}x^3.$ 17. $1 - x + \frac{1}{2}x^2 - \frac{1}{6}x^3.$
18. (i) $\frac{1}{4}$,   (ii) $2\frac{2}{3}$,   (iii) 1.
19. $(1 + x)e^x.$ 20. $(e^{3x} - 1)/(3x).$
21. $\frac{1}{2}(e^x + e^{-x}).$ 22. $\frac{1}{2}(e^x - e^{-x}).$
23. $xe^x - e^x + 1.$ 24. $e^{\frac{1}{2}} - 1.$
25. $3e^{\frac{1}{3}} - 4.$ 26. $3e - 2.$
27. $e + 2.$ 28. $2e^{\frac{1}{2}} - 1.$

Qu. 3. (i) $\dfrac{1}{4}x - \dfrac{1}{32}x^2 + \dfrac{1}{192}x^3 - \ldots + (-1)^{n-1}\dfrac{x^n}{4^n . n} + \ldots,$ $-4 < x \leqslant 4,$

(ii) $\log_e 3 - \dfrac{1}{3}x - \dfrac{1}{18}x^2 - \ldots - \dfrac{x^n}{3^n . n} - \ldots,$ $-3 \leqslant x < 3,$

(iii) $-2x - x^2 - \dfrac{2}{3}x^3 - \ldots - \dfrac{2x^n}{n} - \ldots,$ $-1 \leqslant x < 1.$

Qu. 4. $-1 + x - \frac{1}{2}x^2.$

page 222. **Exercise 10b**

1. (i) $\log_e 3 + \dfrac{1}{3}x - \dfrac{1}{18}x^2 + \dfrac{1}{81}x^3 - \ldots + (-1)^{n-1}\dfrac{x^n}{3^n . n} + \ldots,$

$$-3 < x \leqslant 3.$$

page 222.

1. (ii) $-\frac{1}{2}x - \frac{1}{8}x^2 - \frac{1}{24}x^3 - \frac{1}{64}x^4 - \ldots - \frac{x^n}{2^n \cdot n} - \ldots, \quad -2 \leqslant x < 2.$

(iii) $\log_e 2 - \frac{5}{2}x - \frac{25}{8}x^2 - \frac{125}{24}x^3 - \ldots - \frac{5^n x^n}{2^n \cdot n} - \ldots, \quad -\frac{2}{5} \leqslant x < \frac{2}{5}.$

(iv) $-x^2 - \frac{1}{2}x^4 - \frac{1}{3}x^6 - \frac{1}{4}x^8 - \ldots - \frac{x^{2n}}{n} - \ldots, \quad -1 < x < 1.$

(v) $\frac{2}{3}x + \frac{2}{81}x^3 + \frac{2}{1215}x^5 + \ldots + \frac{2x^{2n-1}}{(2n-1)3^{2n-1}} + \ldots, \quad -3 < x < 3.$

(vi) $-\frac{3x}{2} - \frac{9}{32}x^3 - \frac{243}{2560}x^5 - \ldots - 2 \cdot \frac{3^{2n-1}x^{2n-1}}{(2n-1)4^{2n-1}} - \ldots, \quad -\frac{4}{3} < x < \frac{4}{3}.$

2. $\log_e \frac{2}{3} - \frac{1}{6}x - \frac{5}{72}x^2 - \ldots - \left(\frac{1}{2^n} - \frac{1}{3^n}\right)\frac{x^n}{n} - \ldots, \quad -2 \leqslant x < 2.$

3. $-\log_e 3 + \frac{4}{3}x + \frac{20}{9}x^2 + \ldots + 2^n\{1 + (-1)^n(\frac{1}{3})^n\}x^n/n + \ldots, \quad -\frac{1}{2} \leqslant x < \frac{1}{2}.$

4. $-6x^2 + 28x^3 - 111x^4 + \ldots + (-1)^{n-1}(4^{n-1} - 3^{n-1})12x^n/n + \ldots,$
$$-\tfrac{1}{4} < x \leqslant \tfrac{1}{4}.$$

5. $\frac{1}{2}\log_e 2 + \frac{3}{4}x - \frac{5}{16}x^2 + \ldots + (-1)^{n-1}\{1 + (\frac{1}{2})^n\}x^n/(2n) + \ldots, \quad -1 < x \leqslant 1.$

6. $x + \frac{1}{2}x^2 - \frac{2}{3}x^3 + \ldots - \frac{2}{3n}x^{3n} + \frac{x^{3n+1}}{3n+1} + \frac{x^{3n+2}}{3n+2} - \ldots, \quad -1 \leqslant x < 1.$

7. $1 - \frac{1}{2}x + \frac{1}{3}x^2 - \ldots + (-1)^n x^n/(n+1) + \ldots, \quad -1 < x \leqslant 1.$

8. $-x + \frac{1}{2}x^2 + \frac{2}{3}x^3 + \ldots - (-1)^{3n}\frac{2x^{3n}}{3n} + (-1)^{3n+1}\frac{x^{3n+1}}{3n+1}$
$$+ (-1)^{3n+2}\frac{x^{3n+2}}{3n+2} + \ldots, \quad -1 < x \leqslant 1.$$

9. $x + \frac{1}{2}x^2 + \frac{5}{6}x^3, \quad |x| < 1.$

10. $x + \frac{1}{2}x^2 + \frac{1}{3}x^3, \quad -1 < x \leqslant 1.$

11. $1 + \frac{3}{2}x + \frac{5}{12}x^2, \quad -1 < x \leqslant 1.$

12. $x^2 + x^3 + \frac{11}{12}x^4, \quad -1 \leqslant x < 1.$

13. $0 \cdot 693\ 1, \quad 1 \cdot 099.$

14. $2 \cdot 302\ 6, \quad 0 \cdot 434\ 3.$

15. $1 \cdot 945\ 9.$

16. $1 \cdot 041\ 4.$

17. (i) $-1,$ (ii) $1,$ (iii) $-\frac{2}{3},$ (iv) $0.$

18. $\log_e \frac{4}{3}.$ **19.** $\log_e 2.$ **20.** $\frac{1}{2}\log_e \frac{5}{3}.$

21. $\frac{5}{2}\log_e \frac{7}{5}.$ **22.** $\log_e 3.$ **23.** $\frac{3}{2}\log_e 3 - 1.$

24. $\frac{343}{54}\log_e \frac{5}{2} - \frac{49}{9}.$ **25.** $\frac{1}{3} + \log_e \frac{3}{2}.$

26. $\frac{7}{6} - \log_e \frac{3}{2}.$ **27.** $1\frac{5}{12} - 2\log_e \frac{4}{3}.$

28. $\frac{1}{4} + 2\log_e \frac{3}{4}.$ **29.** $4\frac{2}{3} - 4\log_e 3.$

30. $2\log_e 2 - \frac{1}{2}.$ **31.** $\frac{1}{3} - \frac{1}{3}\log_e \frac{4}{3}.$

32. $\frac{1}{2} - \frac{1}{2}\log_e 2.$ **33.** $5\log_e \frac{5}{4} - 1.$

34. $26\log_e \frac{5}{4} - 5\frac{1}{2}.$

37. s_n tends to a limit between 0 and 1.

page 224. **Exercise 10c**

For standard logarithmic series, see pages 217, 220.

1. $2\left\{\dfrac{m-n}{m+n}+\dfrac{1}{3}\left(\dfrac{m-n}{m+n}\right)^3+\dfrac{1}{5}\left(\dfrac{m-n}{m+n}\right)^5+\ldots+\dfrac{1}{2n-1}\left(\dfrac{m-n}{m+n}\right)^{2n-1}+\ldots\right\},$
$2{\cdot}079\,44.$

2. $1/(1-x^2).$ **4.** $e^y=1+y+\tfrac{1}{2}y^2+\ldots;\ \tfrac{1}{2}(\log_e a)^2-1/a.$

5. (i) $2{\cdot}3979,$ (ii) $(-1)^{n-1}\dfrac{(n-1)3^n}{n!}.$

6. $a(1-r^n)/(1-r),\ 0{\cdot}095\,310.$

7. (ii) $-2/(3n),\ 1/(3n+1),\ 1/(3n+2),$ (iii) $-\tfrac{2}{3}x^3+\tfrac{1}{8}x^4.$

8. (iii) $2e-5.$

9. (i) $2x-\tfrac{1}{3}x^3+\tfrac{2}{5}x^5-\tfrac{1}{2}x^6,\ |x|<1,$ (ii) $e^x=1+x+\dfrac{x^2}{2!}+\dfrac{x^3}{3!}+\ldots+\dfrac{x^n}{n!}+\ldots$

$$e^{-x}=1-x+\dfrac{x^2}{2!}-\dfrac{x^3}{3!}+\ldots+(-1)^n\dfrac{x^n}{n!}+\ldots.$$

10. $p=q=\tfrac{1}{2};\ a=\tfrac{1}{48}.$ **11.** $a=\tfrac{2}{3},\ b=\tfrac{1}{6}.$

13. $\log_e 2+\dfrac{3}{2}x+\dfrac{3}{8}x^2+\dfrac{3}{8}x^3+\ldots+\dfrac{x^n}{n}\left[1-\left(-\dfrac{1}{2}\right)^n\right]+\ldots,\ -1\leqslant x<1.$

15. $\dfrac{1}{2}+\dfrac{1}{4}\left(\dfrac{x^2-1}{x}\right)\log_e\left(\dfrac{1+x}{1-x}\right);\ \dfrac{N}{2N+1}.$

16. $(1+x)^n=1+nx+\dfrac{n(n-1)}{2!}x^2+\dfrac{n(n-1)(n-2)}{3!}x^3+\ldots;\ \log_e 2.$

17. (ii) $10e-4.$

18. (i) $e^x(1+x),$ (ii) $1+\left(\dfrac{1}{x}-1\right)\log_e(1-x),$ (iii) $9\sqrt{3}.$

19. (b) $3e.$ **20.** $5e,\ 15e.$

CHAPTER 11

page 230.

Qu. 1. (i) $2\log_e a+\log_e b,$ (ii) $3\log_e a-3\log_e b,$
(iii) $\tfrac{1}{2}\log_e a+\tfrac{1}{2}\log_e b+\tfrac{1}{2}\log_e c,$ (iv) $\log_e a+\tfrac{1}{3}\log_e b-3\log_e c,$
(v) $-4\log_e c,$ (vi) $b\log_e a.$

Qu. 2. (i) $3,$ (ii) $-2,$ (iii) $4,$
(iv) $2,$ (v) $2x,$ (vi) $3x^2.$

Qu. 3. (i) $1/x,$ (ii) $2/(1+2x),$ (iii) $-1/(1-x),$
(iv) $3/x,$ (v) $\cot x,$
(vi) $\sec x\,\operatorname{cosec} x=2\operatorname{cosec}2x.$

Qu. 4. (i) $6y\dfrac{dy}{dx},$ (ii) $3y^2\dfrac{dy}{dx},$

(iii) $-\sin y\dfrac{dy}{dx},$ (iv) $\dfrac{1}{v}\dfrac{dy}{dx}.$

page 231.

Qu. 4.   (v) $20y^3 \dfrac{\mathrm{d}y}{\mathrm{d}x}$,   (vi) $-\dfrac{6}{y^3}\dfrac{\mathrm{d}y}{\mathrm{d}x}$,

      (vii) $\dfrac{1}{2\sqrt{y}}\dfrac{\mathrm{d}y}{\mathrm{d}x}$,   (viii) $\sec^2 y \dfrac{\mathrm{d}y}{\mathrm{d}x}$.

Qu. 5.   (i) $-\tfrac{2}{3}(x+1)^{-\frac{2}{3}}(x-1)^{-\frac{4}{3}}$,   (ii) $-\dfrac{2x^2+x+4}{(2x-1)^3\sqrt{(x^2+1)}}$,

      (iii) $\dfrac{x\,e^x(x^2-2x-2)}{(x-1)^4}$.

Qu. 6. $\dfrac{10^x}{\log_e 10}$, $\dfrac{10^x}{\log_e 10}+c$.

Qu. 7.   (i) $2^x \log_e 2$,   (ii) $3^x \log_e 3$,

      (iii) $-(\tfrac{1}{2})^x \log_e 2$,   (iv) $5(\log_e 10)10^{5x}$,

      (v) $2x\,10^{x^2}\log_e 10$.

Qu. 8.   (i) $\dfrac{2^x}{\log_e 2}+c$,   (ii) $\dfrac{3^x}{\log_e 3}+c$,

      (iii) $-\dfrac{(\tfrac{1}{2})^x}{\log_e 2}+c$,   (iv) $\dfrac{10^{5x}}{5\log_e 10}+c$.

Qu. 9.   (i) $\tfrac{2}{9}(3x+1)^{\frac{3}{2}}+c$,   (ii) $-\tfrac{1}{6}\cos^6 x+c$,

      (iii) $\tfrac{1}{2}(1+\cos x)^{-2}+c$,   (iv) $5^x/\log_e 5+c$,

      (v) $2^{2x}/(2\log_e 2)+c$,   (vi) $x\log_e x-x+c$.

Qu. 10.   (i) $\tan y = x$,   (ii) $x = \sec y$,   (iii) $p = \cos q$.

Qu. 11.   (i) $2y\dfrac{\mathrm{d}y}{\mathrm{d}x}$,   (ii) $\cos y\dfrac{\mathrm{d}y}{\mathrm{d}x}$,

      (iii) $\sec^2 y\dfrac{\mathrm{d}y}{\mathrm{d}x}$,   (iv) $\sec y \tan y\dfrac{\mathrm{d}y}{\mathrm{d}x}$.

Qu. 12.   (i) $-1/\sqrt{(1-x^2)}$,   (ii) $-1/(1+x^2)$,

      (iii) $1/\sqrt{(-x-x^2)}$.

page 235.   **Exercise 11a**

1.   (i) $3\log_e a+4\log_e b$,   (ii) $\log_e a-\log_e b$,

      (iii) $\tfrac{3}{2}\log_e a-\tfrac{1}{2}\log_e b$,   (iv) $2\log_e a+\log_e b-\tfrac{1}{2}\log_e c$,

      (v) $\tfrac{1}{2}\log_e a+\tfrac{1}{2}\log_e b-\tfrac{1}{2}\log_e c$,

      (vi) $-\tfrac{1}{2}\log_e a-\tfrac{1}{2}\log_e b-\tfrac{1}{2}\log_e c$.

2.   (i) 5,   (ii) 3,   (iii) 4,

      (iv) $\tfrac{1}{2}$,   (v) x^3,   (vi) $-2x$.

3. $2(3-2x)\sqrt{\dfrac{2x+3}{(1-2x)^3}}$.   **4.** $\dfrac{e^{x/2}}{2x^5}(x\sin x-8\sin x+2x\cos x)$.

5. $\dfrac{x-5x^3}{3\sqrt{(x^2+1)^3}\sqrt[3]{(x^2-1)^4}}$.   **6.** $\dfrac{x\tan x-x-1}{x^2 e^x\cos x}$.

7. $\dfrac{3x^2+12x+11}{2\sqrt{\{(x+1)(x+2)(x+3)\}}}$.

page 235.

8. $\dfrac{2(15x^3 - 4x^2 + x - 6)(x^2 + 1)(2x + 1)^2}{(3x - 1)^3}$.

9. $\dfrac{2(2x^2 - 4x + 3)(x - 1)\,e^{2x}}{(2x - 1)^4}$.

10. $(\sin x \cos x + 3x \cos^2 x - 2x \sin^2 x)\sin^2 x \cos x$.

11. $7^x \log_e 7$.

12. $10^{3x}\, 3 \log_e 10$.

13. $-\frac{1}{2}\, 10^{-x/2} \log_e 10$.

14. $-\dfrac{\log_e 10}{10^x}$.

15. $4^{x+1} \log_e 4$.

16. $2x\, 3^{x^2} \log_e 3$.

17. $\dfrac{5^x}{\log_e 5} + c$.

18. $\dfrac{8^x}{\log_e 8} + c$.

19. $-\dfrac{(\frac{1}{3})^x}{\log_e 3} + c$.

20. $\dfrac{3^{2x}}{2 \log_e 3} + c$.

21. $-\dfrac{7^{-x}}{\log_e 7} + c$.

22. $\dfrac{-1}{10^{2x}\, 2 \log_e 10} + c$.

23. $a^x \log_e a$.

24. $\dfrac{a^x}{\log_e a} + c$.

25. $\dfrac{1}{1 + x^2}$.

26. $\dfrac{1}{x \sqrt{(x^2 - 1)}}$.

27. $\dfrac{1}{\sqrt{(-x^2 - 2x)}}$.

28. $-\dfrac{1}{\sqrt{(x - x^2)}}$.

29. $-\dfrac{2x}{x^4 + 1}$.

30. $-\dfrac{2}{(1 + x^2)\sqrt{(2 + x^2)}}$.

31. $\dfrac{1}{1 + 9x^2}$.

32. $\dfrac{-10}{\sqrt{(1 - 25x^2)}}$.

33. $\dfrac{1}{2\sqrt{(-x^2 - x)}}$.

34. (i) 0,   (ii) 0.   The angles are complementary.

36. (i) $\dfrac{2}{\sqrt{(1 - 4x^2)}}$,   (ii) $\dfrac{2x}{\sqrt{(1 - x^4)}}$,   (iii) $\dfrac{1}{\sqrt{(1 - x^2)}}$.

37. $\frac{1}{6}(4x + 3)^{3/2} + c$.

38. $\frac{1}{16}(2x^2 + 1)^4 + c$.

39. $\frac{1}{3}\sec^3 x + c$.

40. $x \log_e x - x + c$.

41. $\frac{1}{3}x^3 \log_e x - \frac{1}{9}x^3 + c$.

42. $-x \cos x + \sin x + c$.

43. $x \sin^{-1} x + \sqrt{(1 - x^2)} + c$.

44. $x \tan^{-1} x - \frac{1}{2}\log(1 + x^2) + c$.

45. $\frac{1}{2}x \sin 2x + \frac{1}{4}\cos 2x + c$.

46. $(x - 1)\,e^x + c$.

47. $-(1 + \log_e x)x^{-x}$.

48. $\left(\cos x \log_e x + \dfrac{1}{x}\sin x\right)x^{\sin x}$.

49. $\dfrac{2 \log_e x}{x}\, x^{\log_e x}$.

50. $\left(\log_e \log_e x + \dfrac{1}{\log_e x}\right)(\log_e x)^x$.

51. $(\log_e \cos x - x \tan x)(\cos x)^x$.

52. $1 + \log_e x$.

page 241 **Exercise 11b**

1. (1, 4) max., (3, 0) min.
2. (2, 3) min.   3. (2, −1) max., (4, 3) min.
4. (1, 1) min.
5. (−2, −$\frac{2}{3}$) max., (2, 10) max., (3, 9$\frac{3}{4}$) min.
6. (0, 108) max., (2, 0) min., (−3, 0) infl.
8. ($\frac{1}{3}\pi$, $\frac{3}{4}\sqrt{3}$) max., (π, 0) infl., ($\frac{5}{3}\pi$, −$\frac{3}{4}\sqrt{3}$) min.
9. ($\frac{1}{2}\pi$, $\frac{1}{2}$) max.   10. (0, −1) max., (2π, −1) max.
11. (0, $\frac{2}{3}$) min., ($\frac{1}{4}\pi$, $\frac{2}{3}\sqrt{2}$) max., ($\frac{3}{4}\pi$, −$\frac{2}{3}\sqrt{2}$) min., (π, −$\frac{2}{3}$) max.,
   ($\frac{5}{4}\pi$, −$\frac{2}{3}\sqrt{2}$) min., ($\frac{7}{4}\pi$, $\frac{2}{3}\sqrt{2}$) max., (2π, $\frac{2}{3}$) min.
12. (1, 1/e) max.   13. (2, 10 tan^{-1} 2−2) max.
15. $\left(2n\pi+\frac{1}{4}\ \pi,\ \dfrac{1}{\sqrt{2}}\ e^{2n\pi+\pi/4}\right)$ max., $\left(2n\pi+\dfrac{5}{4}\ \pi,\ -\dfrac{1}{\sqrt{2}}\ e^{2n\pi+5\pi/4}\right)$ min.
16. $\sqrt{2}$:1.   18. $\sqrt{2}$:1.
20. 1:2.   21. 2 sin^{-1} $\frac{1}{3}$.

page 242 **Exercise 11c**

1. (i) $\dfrac{2x}{(1-x^2)^2}$,   (ii) $\dfrac{-1}{x\sqrt{(x^2-1)}}$,   (iii) e^{2x} (2 cos 3x − 3 sin 3x).

2. (i) $\dfrac{x-3}{2(x-1)^{3/2}}$,   (ii) 3 sin^2 x cos^3 x − 2 sin^4 x cos x,   (iii) log$_e$ x.

3. (i) $-\dfrac{12x-1}{(2x+1)^2(3x-2)^2}$,   (ii) −2 sin 2x e$^{\cos 2x}$,   (iii) 2 cot 2x.

4. (i) $\dfrac{-2}{\sqrt{(2x+1)}\sqrt{(2x-1)^3}}$,   (ii) $\dfrac{1}{1+x^2}$,   (iii) −4 e^{-4x}.

5. (i) x(x^2+1)(x^3+1)^2(13x^3+9x+4),
   (ii) $-\dfrac{\sec^2 x}{\sqrt{(1-\tan^2 x)}}$ = −sec x$\sqrt{(\sec 2x)}$,   (iii) $\dfrac{1-2\ \log_e x}{x^3}$.

6. (i) $\dfrac{x^3+3x-1}{\sqrt{(3x^2-2x)}\sqrt[3]{(x^3+1)^4}}$,   (ii) $\dfrac{\sin^3 x\ (4+\sin^2 x)}{\cos^6 x}$,
   (iii) $\dfrac{1}{1+x^2}$.

7. (i) 2$^{x^2+1}$x log$_e$ 2,   (ii) 2(log$_e$ x + 1)x^{2x},
   (iii) (cot x cos x − sin x log$_e$ sin x)(sin x)$^{\cos x}$.

10. (i) −b/(a^2 sin^3 θ).

11. (i) $\dfrac{3}{4t}+\dfrac{1}{t^3}$,
   (ii) $\frac{1}{3}$ sec^4 t cosec t,
   (iii) $-\dfrac{1}{4a}$ cosec^4 $\frac{1}{2}\theta$.

14. 2$\sqrt{(1-x^2)}$, $\frac{1}{2}$x$\sqrt{(1-x^2)}$+$\frac{1}{2}$ sin^{-1} x+c.

page 244.

15. $\tan^{-1} x$, $x \tan^{-1} x - \frac{1}{2} \log_e (1 + x^2) + c$.

16. 1 max., -1 min.   **17.** $(2^{\frac{1}{3}}, 2^{\frac{2}{3}})$ max., $(0, 0)$ min.

18. $\frac{1}{3} \sqrt{6}$.

19. (i) e^x,   (ii) $n!$ (iii) $(-1)^{n+1}(n-1)! \, x^{-n}$.

20. $\cos (x + \frac{1}{2}\pi)$, $\cos (x + \frac{1}{2}n\pi)$.

23. $\dfrac{-16}{21(x+5)^2} + \dfrac{4}{21(4x-1)^2}$, $\dfrac{32}{21(x+5)^3} - \dfrac{32}{21(4x-1)^3}$.

$(1, \frac{1}{6})$ max., $(-\frac{1}{3}, \frac{9}{49})$ min., $(2, \frac{5}{49})$ infl.

24. $\left(\dfrac{\pi}{4}, \dfrac{2\sqrt{2}-1}{14} \right)$ max., $\left(\dfrac{5\pi}{4}, -\dfrac{2\sqrt{2}+1}{14} \right)$ min.   **25.** $4a (\sqrt{5}-1)$.

26. $5c$.   **32.** $\dfrac{(n+1)!}{(n+1-r)! \, r!} \, u_{n+1-r} \, v_r$.

33.   (i) $e^x(x^3 + 9x^2 + 18x + 6)$,   (ii) $x \cos x + 4 \sin x$,

      (iii) $2^{n-2} e^{2x} \{4x^2 + 4nx + n(n-1)\}$,

      (iv) $(1 - x^2)y_n - 2nxy_{n-1} - n(n-1)y_{n-2}$.

CHAPTER 12

page 249.

Qu. 1.   (i) $n\pi$,   (ii) $(2n+1)\pi$,   (iii) $n\pi + \frac{1}{4}\pi$,

         (iv) $2n\pi + \frac{1}{2}\pi$,   (v) $n\pi + \frac{1}{2}\pi$,

         (vi) $n\pi + (-1)^n \pi/6$,   (vii) $n\pi - \frac{1}{4}\pi$,

        (viii) $n\pi + (-1)^n \frac{1}{4}\pi$,   (ix) $2n\pi \pm \frac{3}{4}\pi$.

Qu. 2.   (i) $n\pi - \frac{1}{4}\pi$,   (ii) $3(4n+1)\pi/8$,

        (iii) $5n\pi \pm 5\pi/6$,   (iv) $2n\pi/3$,

        (v) $(6n+3)\pi/4$,   (vi) $n\pi/5 + (-1)^n \pi/20$.

page 253   **Exercise 12a**

1. $n\pi \pm \frac{1}{6}\pi$.   **2.** $180n° - 32 \cdot 7°$.

3. $360n° \pm 47 \cdot 6°$.   **4.** $n\pi$, $2n\pi \pm \pi/3$.

5. $\frac{1}{6}\pi + 2n\pi/3$.   **6.** $(2n+1)\pi/2$, $n\pi + (-1)^n \pi/6$.

7. $180n° + (-1)^n \, 17 \cdot 6°$.   **8.** $n\pi$.

9. $n\pi/2$.   **10.** $n\pi/2$.

11. $n\pi/3$.   **12.** $n\pi/2$, $n\pi - (-1)^n \pi/6$.

13. $n\pi/3$.   **14.** $\frac{1}{2}\pi + 2n\pi/3$.

15. $2n\pi - \frac{1}{2}\pi$, $n\pi + (-1)^n \pi/6$.

16. $(4n+1)\pi/8$, i.e. $180n° + 22 \cdot 5°$, $180n° - 67 \cdot 5°$.

17. $n\pi$, $n\pi + (-1)^n \pi/6$.

18. $2n\pi$, $2n\pi \pm \pi/3$.   **19.** $n\pi + (-1)^n \pi/6$.

20. $(2n+1)\pi/2$, $(2n+1)\pi/4$.

21. $180n° + 8 \cdot 1°$, $180n° - 12 \cdot 5°$.

page 253.

22. $n\pi - (-1)^n \pi/6.$

23. $2n\pi, 2n\pi + 2\pi/3.$

24. $360n° - 53\cdot1°, 360n° + 36\cdot9°.$

25. $360n° + 163\cdot7°.$

26. $360n° - 119\cdot5°, 360n° + 13\cdot3°.$

27. $(2n+1)90°, 180n° - 63\cdot4°.$

28. $n\pi, 2n\pi \pm \pi/3.$

29. $(2n+1)\pi/2, (4n+1)\pi/4.$

30. $(4n-1)\pi/4.$

31. $(4n+1)\pi/4.$

Qu. 4.   (i) $-\pi/4,$ (ii) $-\pi/6,$ (iii) $\pi,$

        (iv) $\pi/3,$ (v) $\pi/2,$ (vi) $0.$

Qu. 6.   (i) $0\cdot322,$ (ii) $1\cdot824,$ (iii) $0\cdot010.$

        (iv) $-0\cdot201,$ (v) $-1\cdot249,$ (vi) $0\cdot927.$

Qu. 7. $n\pi, (2n+1)\pi/4.$

Qu. 8. $(4n+1)\pi/18, (4n-1)\pi/2.$

page 258   **Exercise 12b**

1.   (i) $\pi/3,$ (ii) $\pi/4,$ (iii) $\pi/6,$

    (iv) $\pi/2,$ (v) $-\pi/3,$ (vi) $-\pi/2.$

2.   (i) $1\cdot107,$ (ii) $0\cdot643,$ (iii) $1\cdot159,$

    (iv) $-0\cdot340,$ (v) $-0\cdot464,$ (vi) $1\cdot318.$

3. $n\pi, 2n\pi \pm \cos^{-1}\frac{1}{6}.$

4. $(2n+1)\pi, 2n\pi \pm \cos^{-1}\frac{2}{3}.$

5. $n\pi, n\pi \pm \sin^{-1}\frac{5}{6}.$

6. $n\pi, n\pi \pm \tan^{-1}(\frac{1}{2}\sqrt{2}).$

7. $2n\pi + \pi/4.$

8. $2n\pi, 2n\pi - 2\pi/3.$

9. $n\pi + \tan^{-1}(3 \tan \alpha).$

10. $n\pi + \alpha - \tan^{-1}\frac{3}{4}.$

11. $2n\pi \pm \alpha, (2n+1)\pi \pm \cos^{-1}(\frac{1}{2}\cos\alpha).$

12. $(2n+1)\pi - \tan^{-1}\frac{4}{3}, (2n+1)\pi - \tan^{-1}\frac{12}{5}.$

13. $(4n+1)\pi/2, (2n+1)\pi - \tan^{-1}\frac{20}{21}.$

14. $n\pi/2.$

15. $2n\pi, (2n+1)\pi/5.$

16. $n\pi/3.$

17. $(4n+1)\pi/10.$

18. $(4n+1)\pi/14.$

19. $n\pi/6.$

20. $(4n-1)\pi/10, (4n+1)\pi/2.$

21. $2n\pi/5; 4\cos^3\theta - 2\cos^2\theta - 3\cos\theta + 1 = 0; -\frac{1}{4}(1+\sqrt{5}).$

28. $\frac{1}{4}\pi.$ 29. $\frac{1}{4}\pi.$ 30. $\frac{1}{4}\pi.$ 31. $0.$

32. $\frac{1}{2}\pi.$

page 260   **Exercise 12c**

8. $\frac{1}{2}R.$ 12. $\frac{1}{2}R.$

18. The reflection in AB of major arc AB.

31. $\frac{1}{2}(B+C), \frac{1}{2}(C+A), \frac{1}{2}(A+B).$

33. I A, $I_2 I_3$ are interior and exterior bisectors of angle A. $2R.$

36. $\sqrt{\dfrac{s(s-b)(s-c)}{s-a}}.$

page 262   **Exercise 12d**

1. (i) $217°, 323°,$ (ii) $60°, 90°, 270°, 300°,$
     (iii) $76·7°, 209·6°.$

2. (i) $0°, 45°, 135°, 180°,$ (ii) $35·3°, 144·7°,$ (iii) $90°.$

3. (i) $18°, 72°,$ (iii) $0°, 30°, 90°, 150°, 180°, 210°, 270°, 330°, 360°.$

4. (i) $(2n+1)\pi/4,\ 2n\pi \pm \tfrac{1}{3}\pi,$
     (ii) $360n° + 29·6°,\ 360n° + 256·7°.$

5. (i) $(2k-\tfrac{1}{2})\pi/5,\ (2k+\tfrac{1}{2})\pi/3,$
     (ii) $(2k \pm \tfrac{1}{3})\pi,$ (iii) $k\pi/6.$

6. (i) $n\pi + (-1)^n \pi/6,\ n\pi - (-1)^n \sin^{-1} \tfrac{3}{4},$
     (ii) $(4n-1)\pi/2,\ (4n+1)\pi/10.$

7. (i) $16c^5 - 20c^3 + 5c,\ \tfrac{1}{4}\sqrt{(10 - 2\sqrt{5})},$ (ii) $k\pi/3.$

8. $2x^2 - x - 1 = 0;\ 1,\ -\tfrac{1}{2};\ \theta = 0,\ \phi = 2\pi/3;\ \theta = 2\pi/3,\ \phi = 0.$

9. (i) $120°, 240°, 300°.$ **10.** $n\pi/5.$

11. (i) $(4t - 4t^3)/(1 - 6t^2 + t^4);\ \tan(4n+1)\pi/16,\ n = 0, 1, 2, 3.$

12. $(4k+1)\pi/10,\ (4k-1)\pi/2;\ \pm 1,\ -\sin(3\pi/10),\ \tfrac{1}{4}(\sqrt{5}+1).$

13. (i) $168·3°, 348·3°.$ **14.** (ii) $30·9°, 71·8°.$

15. $0,\ \sqrt{5}.$ **16.** $98·0°, 149·3°.$

17. (a) $0°, 22\tfrac{1}{2}°, 60°, 67\tfrac{1}{2}°, 112\tfrac{1}{2}°, 120°, 157\tfrac{1}{2}°, 180°.$

18. (i) $360n° - 60·5°,\ 360n° + 166·7°.$

19. $(3n+1)\pi/3 \pm \tfrac{1}{2}\cos^{-1}\tfrac{1}{3}.$ No solution unless α is an odd multiple of $\pi/2$, in which case θ may take any value.

20. (ii) $a^2 + b^2 = 2.$

21. $-c\cos(3\alpha+\beta)/\sin\alpha,\ c\cos(2\alpha+\beta)/\sin\alpha.$

22. $3\cos 2\theta + 4\sin 2\theta + 3,\ A = 3,\ \alpha = \tan^{-1}(4/3),$
     max. $8, \theta = 26·6°;$ min. $-2, \theta = 116·5°.$

24. (i) $(2n+1)\pi/16,\ (2n+1)\pi/4.$

28. $9·81$ cm.   **29.** $44·8$ cm., 567 cm².

33. $(q\sin\alpha + p\cos\beta)/\cos(\alpha-\beta),\ (q\cos\alpha - p\sin\beta)/\cos(\alpha-\beta).$

36. (a) $73·7°, 53·1°.$

CHAPTER 13

page 270.

Qu. 3. (i) $\sin x - x\cos x + c,$ (ii) $\tfrac{1}{2}x\sin 2x + \tfrac{1}{4}\cos 2x + c,$
     (iii) $\tfrac{1}{2}x^2\log_e x - \tfrac{1}{4}x^2 + c,$ (iv) $xe^x - e^x + c.$

Qu. 4. $2xe^{x^2},\ \tfrac{1}{2}e^{x^2}(x^2-1) + c.$ **Qu. 5.** $x\log_e x - x + c.$

Qu. 6. (i) $(1-x^2)^{-\frac{1}{2}},$ (ii) $x\sin^{-1} x + \sqrt{(1-x^2)} + c.$

Qu. 8. (i) $(x^2-2)\sin x + 2x\cos x + c,$ (ii) $e^x(x^2 - 2x + 2) + c.$

1. (i) $2 \sin x - 2x \cos x + c$,   (ii) $\frac{1}{2}(x-1)e^x + c$,

  (iii) $\frac{1}{4} \sin 2x - \frac{1}{2}x \cos 2x + c$,   (iv) $\frac{1}{3}x^3(3 \log_e x - 1) + c$,

  (v) $x \sin (x+2) + \cos (x+2) + c$,   (vi) $\frac{1}{72}(1+x)^8(8x-1) + c$,

  (vii) $\frac{1}{2}xe^{2x} - \frac{1}{4}e^{2x} + c$,   (viii) $\frac{1}{2}e^{x^2} + c$,

  (ix) $-\dfrac{1}{x}(\log_e x + 1) + c$,   (x) $x \tan x + \log_e \cos x + c$,

  (xi) $(n+1)^{-2}x^{n+1}\{(n+1) \log_e x - 1\} + c$,

  (xii) $(\log_e 3)^{-2} . 3^x(x \log_e 3 - 1) + c$.

2. (i) $x \log_e 2x - x + c$,   (ii) $x \sin^{-1} 3x + \frac{1}{3}\sqrt{(1 - 9x^2)} + c$,

  (iii) $2y(\log_e y - 1) + c$,   (iv) $\theta \tan^{-1} \dfrac{\theta}{2} - \log_e (4 + \theta^2) + c$,

  (v) $t \cos^{-1} t - \sqrt{(1 - t^2)} + c$,   (vi) $2e^{\sqrt{x}}(\sqrt{x} - 1) + c$.

3. (i) $\frac{1}{3}e^{x^3}(x^3 - 1) + c$,   (ii) $-\frac{1}{2}e^{-x^2} + c$,

  (iii) $-\frac{1}{2}e^{-x^2}(1 + x^2) + c$,   (iv) $\frac{1}{2}x^2 \sin x^2 + \frac{1}{2} \cos x^2 + c$,

  (v) $\frac{1}{2}x^2 \tan x^2 + \frac{1}{2} \log_e \cos x^2 + c$.

4. (i) $\frac{1}{3}x^2 \sin 3x + \frac{2}{9}x \cos 3x - \frac{2}{27} \sin 3x + c$,

  (ii) $e^x(x^3 - 3x^2 + 6x - 6) + c$ (iii) $\frac{1}{8} \cos 2x(1 - 2x^2) + \frac{1}{4}x \sin 2x + c$,

  (iv) $-e^{-x}(x^2 + 2x + 2) + c$,

  (v) $\frac{1}{6}x^3 + \frac{1}{8}(2x^2 - 1) \sin 2x + \frac{1}{4}x \cos 2x + c$,

  (vi) $\frac{1}{4}x^2\{1 - 2 \log_e x + 2(\log_e x)^2\} + c$.

5. (i) $\frac{1}{8} \sin 2x - \frac{1}{4}x \cos 2x + c$,   (ii) $-e^{-x}(1 + x) + c$,

  (iii) $\frac{1}{168}(1 + 2x)^6(12x - 1) + c$,   (iv) $\frac{1}{2}(\log_e y)^2 + c$,

  (v) $\frac{1}{2}(1 + u^2) \tan^{-1} u - \frac{1}{2}u + c$,   (vi) $-\frac{1}{2}e^{-x^2} + c$,

  (vii) $-e^{-x}(x^3 + 3x^2 + 6x + 6) + c$,   (viii) $-\frac{1}{14}(1 - x^2)^7 + c$,

  (ix) $\frac{1}{4}t^2 - \frac{1}{4}t \sin 2t - \frac{1}{8} \cos 2t + c$,   (x) $\frac{1}{9}e^{3v}(3v - 1) + c$,

  (xi) $y \cot^{-1} y + \frac{1}{2}\log_e(1 + y^2) + c$,   (xii) $\log_e \sin \theta - \theta \cot \theta + c$,

  (xiii) $-\dfrac{1}{4x^2}(1 + 2 \log_e x) + c$.

6. (i) $x \tan x + \log_e \cos x - \frac{1}{2}x^2 + c$.

7. (i) $\frac{1}{2}\pi - 1 \simeq 0 \cdot 571$,   (ii) $e - 2 \simeq 0 \cdot 718$,   (iii) $e^2 + 1 \simeq 8 \cdot 39$,

  (iv) $\frac{1}{2}\pi - 1 \simeq 0 \cdot 571$,   (v) $\dfrac{\pi^2}{4} \simeq 2 \cdot 47$,   (vi) $50 - \dfrac{99}{4 \log_e 10} \simeq 39 \cdot 2$.

Qu. 9. (i) $3(1 - 9x^2)^{-\frac{1}{2}}$,   (ii) $2(1 + 4x^2)^{-1}$,

  (iii) $(9 - x^2)^{-\frac{1}{2}}$,   (iv) $-2(1 - 4x^2)^{-\frac{1}{2}}$,

  (v) $\frac{3}{2}(1 + 9x^2)^{-1}$,   (vi) $6(4 + x^2)^{-1}$,

  (vii) $\frac{1}{2}(2x - x^2)^{-\frac{1}{2}}$,   (viii) $4(5 + 2x + x^2)^{-1}$.

Qu. 10. (i) $\tan^{-1} \dfrac{x}{2} + c$,   (ii) $\frac{3}{2} \tan^{-1} 2x + c$,

  (iii) $4 \sin^{-1} \dfrac{x}{3} + c$,   (iv) $\frac{1}{3} \sin^{-1} 3x + c$.

page 273.

Qu. 10.   (v) $\dfrac{1}{5\sqrt{2}} \tan^{-1} \dfrac{5x}{\sqrt{2}} + c$,   (vi) $\sin^{-1} \dfrac{2x}{\sqrt{3}} + c$,

        (vii) $\dfrac{1}{\sqrt{2}} \tan^{-1} \dfrac{x-1}{\sqrt{2}} + c$,   (viii) $5 \sin^{-1} \dfrac{x+2}{3} + c$.

Qu. 11. $\log_e \tan \tfrac{1}{2} x + c$.   **Qu. 12.** $-2 \log_e \cos \tfrac{1}{2}\theta + c$.

Qu. 14.   (i) $(1+t^2)^{-1}$,   (ii) $(4+4t^2)^{-1}$,   (iii) $2(3+3t^2)^{-1}$.

Qu. 15.   (i) $\tfrac{1}{2} \log_e \tan x + c$,   (ii) $-\tfrac{2}{3}(1+\tan \tfrac{3}{2}\theta)^{-1} + c$,

        (iii) $\log_e \{x + \sqrt{(x^2-1)}\} + c$.

Qu. 16.   (i) $\dfrac{1}{\sqrt{2}} \tan^{-1} \left(\dfrac{1}{\sqrt{2}} \tan x \right) + c$,   (ii) $-\log_e (1 - \tan^2 x) + c$.

Qu. 17.   (i) $\log_e (x^2 + 2x + 10) + \tfrac{1}{3} \tan^{-1} \dfrac{x+1}{3} + c$,

        (ii) $5 \sin^{-1} \dfrac{x+2}{3} + 2\sqrt{(5 - 4x - x^2)} + c$,

        (iii) $\tfrac{1}{2} x - \tfrac{1}{2} \log_e (\sin x + \cos x) + c$,

        (iv) $\tfrac{3}{2} x - \tfrac{5}{2} \log_e (3 \cos x + \sin x) + c$.

Qu. 18.   (i) 1,   (ii) $\tfrac{1}{2}\pi$.

Qu. 19.   (ii) $\tfrac{1}{2}\pi$,   (v) 1,   (vi) $\dfrac{\pi}{3}$.

page 280.   **Exercise 13b**

1.   (i) $2(1-4x^2)^{-\frac{1}{2}}$,   (ii) $3(2+6x+9x^2)^{-1}$,

     (iii) $-\tfrac{2}{3}(1-4x^2)^{-\frac{1}{2}}$,   (iv) $2(8+2x-x^2)^{-\frac{1}{2}}$,

      (v) $(x^2+4)^{-1}$,   (vi) $2(4-9x^2)^{-\frac{1}{2}}$,

     (vii) $-(1+x^2)^{-1}$,   (viii) $\dfrac{1}{x\sqrt{(x^2-1)}}$,

     (ix) $2x^3(1+x^4)^{-1} + 2x \tan^{-1} x^2$,   (x) 0,

     (xi) $\dfrac{1}{x\sqrt{(1-x^2)}} - \dfrac{\sin^{-1} x}{x^2}$,   (xii) $\dfrac{2 \sec^2 x}{1 + 4\tan^2 x}$,

   (xiii) $6(1-9x^2)^{-\frac{1}{2}} \sin^{-1} 3x$,   (xiv) $(4 - \sin^2 x)^{-\frac{1}{2}} \cos x$,

    (xv) 0,   (xvi) $2\sqrt{(1-x^2)}$,   (xvii) $16(x^2+4)^{-2}$.

2.   (i) $\tfrac{1}{3} \tan^{-1} \dfrac{x}{3} + c$,   (ii) $3 \sin^{-1} \dfrac{y}{2} + c$,

     (iii) $\tfrac{2}{3} \tan^{-1} 3u + c$,   (iv) $\tfrac{1}{2} \sin^{-1} 4x + c$,

      (v) $\dfrac{1}{\sqrt{3}} \tan^{-1} \dfrac{2t}{\sqrt{3}} + c$,   (vi) $\tfrac{1}{2} \sin^{-1} \dfrac{2x}{\sqrt{5}} + c$,

     (vii) $\dfrac{1}{\sqrt{6}} \tan^{-1} \dfrac{\sqrt{3}y}{\sqrt{2}} + c$,   (viii) $\dfrac{1}{3\sqrt{6}} \sin^{-1} \sqrt{2}\, x + c$,

     (ix) $\dfrac{1}{3\sqrt{2}} \tan^{-1} \dfrac{(y-2)\sqrt{2}}{3} + c$,   (x) $\dfrac{2}{\sqrt{3}} \sin^{-1} \dfrac{(x-1)\sqrt{3}}{2} + c$.

page 280.

3. (i) $2 \log_e \tan \dfrac{x}{4} + c,$ (ii) $\frac{1}{2} \log_e (\sec 2\theta + \tan 2\theta) + c,$

(iii) $\frac{1}{3} \log_e \tan \frac{3}{2}x + c,$ (iv) $\frac{1}{4} \log_e (\sec 4\phi + \tan 4\phi) + c,$

(v) $\log_e \tan x + c,$ (vi) $\tan \frac{1}{2}y + c,$

(vii) $-(1 + \tan x)^{-1} + c,$ (viii) $\log_e (1 - \cos \theta) + c,$

(ix) $\frac{1}{3} \log_e (3 + \tan \frac{1}{2}x) - \frac{1}{3} \log_e (3 - \tan \frac{1}{2}x) + c,$

(x) $\tan^{-1}(\frac{1}{2} \tan \frac{1}{4}\theta) + c,$ (xi) $\log_e \{x + \surd(x^2 - 9)\} + c.$

4. (i) $\dfrac{1}{\sqrt 3} \tan^{-1}\left(\sqrt 3 \tan x\right) + c,$

(ii) $\frac{1}{4} \log_e (1 + 2 \tan x) - \frac{1}{4} \log_e (1 - 2 \tan x) + c,$

(iii) $\surd 2 \tan^{-1}\left(\dfrac{1}{\sqrt 2} \tan x\right) - x + c,$

(iv) $\frac{1}{6} \log_e (1 + 3 \tan x) - \frac{1}{6} \log_e (1 - 3 \tan x) + c.$

5. (i) $\frac{1}{2} \log_e (x^2 + 3) + \dfrac{5}{\sqrt 3} \tan^{-1} \dfrac{x}{\sqrt 3} + c,$

(ii) $\log_e (y + 3) - (y + 3)^{-1} + c,$

(iii) $\frac{3}{2} \log_e (u^2 + 2u + 5) + \frac{5}{2} \tan^{-1} \dfrac{u + 1}{2} + c,$

(iv) $7 \surd(4x - x^2) - 11 \sin^{-1} \dfrac{x - 2}{2} + c,$

(v) $\frac{1}{2}\theta + \frac{1}{2} \log_e (\sin \theta + \cos \theta) + c,$

(vi) $\frac{1}{2}x + \frac{5}{2} \log_e (\sin x + \cos x) + c,$

(vii) $2x - \log_e (3 \sin x + 2 \cos x) + c.$

6. (i) $1,$ (ii) $\dfrac{\pi}{6}.$

7. (ii) $2,$ (v) $\frac{1}{2},$ (vi) $1,$ (vii) $-\frac{1}{2} (\log_e 2 + 1) \simeq -0{\cdot}847.$

(viii) $-1,$ (ix) $\frac{1}{4}\pi - \frac{1}{2} \sin^{-1} \frac{1}{3} \simeq 0{\cdot}4205,$ (xii) $\frac{1}{20}\pi.$

8. (i) tends to $\pi,$ (ii) tends to infinity.

9. $15\pi.$ 10. $\frac{256}{3}.$

Qu. 20. $\frac{1}{13}e^{2x}(2 \sin 3x - 3 \cos 3x) + c.$

Qu. 21. (i) and (ii) $\frac{1}{5}e^x(\cos 2x + 2 \sin 2x) + c,$ no.

page 283. **Exercise 13c**

1. (i) $\frac{1}{13}e^{3x}(3 \cos 2x + 2 \sin 2x) + c,$ (ii) $\frac{1}{25}e^{4x}(4 \sin 3x - 3 \cos 3x) + c,$

(iii) $\frac{2}{5}e^{-t}(\sin \frac{1}{2}t - 2 \cos \frac{1}{2}t) + c,$

(iv) $\frac{1}{5}e^x\{\sin (2x + 1) - 2 \cos (2x + 1)\} + c,$

(v) $\frac{1}{5}e^{2\theta} (2 + \cos 2\theta + \sin 2\theta) + c,$ (vi) $\frac{1}{17}e^{x/2}(\sin 2x - 4 \cos 2x) + c.$

2. $\frac{1}{2} \tan x \sec x + \frac{1}{2} \log_e (\sec x + \tan x) + c.$

3. (i) $\frac{1}{16} x^4 (4 \log_e x - 1) + c,$ (ii) $y \tan^{-1} 2y - \frac{1}{4} \log_e (1 + 4y^2) + c,$

page 283.

3.   (iii) $-\frac{1}{2}e^{-x^2}+c$,   (iv) $\frac{1}{9}(\sin 3x - 3x \cos 3x)+c$,

     (v) $\frac{1}{4} \cos 2x(1-2x^2)+\frac{1}{2}x \sin 2x+c$,

     (vi) $\frac{1}{13}e^{3x} (3 \sin 2x - 2 \cos 2x)+c$,   (vii) $\frac{1}{4}e^{u^2}(u^2-1)+c$,

   (viii) $\frac{1}{168}(2x-1)^6 (12x+1)+c$,

     (ix) $\frac{1}{4}(x^2-1) \log_e (x-1)-\frac{1}{8}x^2-\frac{1}{4}x+c$,

     (x) $x (\log_e 3x - 1)+c$,   (xi) $\frac{1}{4}e^{2x}(2x^2-2x+1)+c$,

     (xii) $\frac{2}{5}e^{-y}(\sin \frac{1}{2}y - 2 \cos \frac{1}{2}y)+c$,   (xiii) $-\frac{1}{4}x^{-2} (1+2 \log_e x)+c$,

     (xiv) $t \sin^{-1}\dfrac{t}{3} + \sqrt{(9-t^2)}+c$,   (xv) $3x(\log_e x - 1)+c$,

     (xvi) $\frac{1}{6}y^3 + \frac{1}{8} (2y^2-1) \sin 2y + \frac{1}{4}y \cos 2y+c$,

   (xvii) $\frac{1}{2} \sin x^2+c$,   (xviii) $\frac{1}{2}x^2(\log_e x^2-1)+c$,

   (xix) $\frac{1}{2} \sin \theta^2 - \frac{1}{2}\theta^2 \cos \theta^2+c$,

     (xx) $\frac{1}{4}(2x^3-3x) \sin 2x + \frac{3}{8}(2x^2-1) \cos 2x+c$,

     (xxi) $e^{ax} \left\{ \dfrac{1}{2a} - \dfrac{1}{2(a^2+4b^2)} (a \cos 2bx + 2b \sin 2bx) \right\}+c$,

   (xxii) $\frac{3}{8} e^x(\sin x - \cos x) - \frac{1}{40}e^x (\sin 3x - 3 \cos 3x)+c$,

   (xxiii) $4(\sqrt{y}-2)e^{\frac{1}{2}\sqrt{y}}+c$,   (xxiv) $(\log_e 10)^{-2}10^x(x \log_e 10-1)+c$.

4. $C = \dfrac{e^{ax}}{a^2+b^2} (a \cos bx + b \sin bx)$, $S = \dfrac{e^{ax}}{a^2+b^2} (a \sin bx - b \cos bx)$.

6. $\frac{64}{21}$.   7. $\left(\frac{1}{2}\pi - 1, \frac{1}{8}\pi\right)$.   8. $\left(\frac{\pi}{4}-\frac{1}{\pi}, 0\right)$

10. $\frac{1}{10}(1+e^{3\pi})$.

Qu. 22. $\frac{1}{4} \sin x \cos^3 x + \frac{3}{8} \sin x \cos x + \frac{3}{8}x+c$.

Qu. 24.   (i) $\frac{1}{3} \sin x \cos^2 x + \frac{2}{3} \sin x+c$,   (ii) $\sin x - \frac{1}{3} \sin^3 x+c$.

Qu. 25. $\frac{8}{15}$.   Qu. 26. (i) $\dfrac{35\pi}{256}$,   (ii) $\frac{128}{315}$,   (iii) $\dfrac{63\pi}{512}$.

page 287.   **Exercise 13d**

1. $-\frac{1}{4} \cos x \sin^3 x - \frac{3}{8} \cos x \sin x + \frac{3}{8}x+c$, $\dfrac{3\pi}{16}$.

2. $-\frac{1}{5} \cos x \sin^4 x - \frac{4}{15} \cos x \sin^2 x - \frac{8}{15} \cos x+c$, $\frac{8}{15}$.

3.   (i) $\frac{2}{3}$,   (ii) $\dfrac{5\pi}{32}$,   (iii) $\frac{128}{315}$,

     (iv) $\dfrac{3\pi}{16}$,   (v) $\dfrac{63\pi}{512}$,   (vi) $\frac{16}{35}$.

4.   (i) $\frac{8}{35}$,   (ii) $\frac{8}{5}$,   (iii) $\dfrac{5\pi}{96}$.

page 288.

5. (i) $\frac{32}{35}$,   (ii) $\dfrac{3\pi}{8}$,   (iii) $\dfrac{5\pi}{16}$,   (iv) 0,   (v) 0,   (vi) $\frac{256}{315}$,

   (vii) $\dfrac{63\pi}{256}$,   (viii) $\dfrac{35\pi}{128}$.

6. (i) $I_{m-2,n} = \dfrac{m-3}{m+n-2} I_{m-4,n}$, $I_{m,n} = \dfrac{(m-1)(m-3)}{(m+n)(m+n-2)} I_{m-4,n}$.

   (ii) $I_{m-4,n} = \dfrac{(n-1)(n-3)(n-5)}{(m+n-4)(m+n-6)(m+n-8)} I_{m-4,n-6}$,   (iii) $\frac{8}{693}$,

   (iv) (a) $\frac{8}{1287}$,   (b) $\dfrac{5\pi}{4096}$,   (c) $\frac{16}{3003}$,   (d) $\frac{1}{120}$.

7. (i) $\frac{1}{504}$,   (ii) $\dfrac{7\pi}{2048}$.   **8.** (i) $I_n = \dfrac{n}{a} I_{n-1}$,   (ii) $\frac{2835}{8}$.

9. $I_n = \dfrac{1}{n-1} - I_{n-2}$,   (i) $\frac{5}{12} - \frac{1}{2}\log_e 2 \simeq 0.070$,   (ii) $\frac{1}{4}\pi - \frac{76}{105} \simeq 0.062$.

10. (ii) $\dfrac{1328\sqrt{3}}{2835}$.

page 289.   **Exercise 13e**

1.   (i) $\frac{1}{3}\sqrt{(x^2+1)^3} + c$,   (ii) $\frac{2}{3}\sqrt{(x^3+3x-4)} + c$,

   (iii) $\sin u - \frac{2}{3}\sin^3 u + \frac{1}{5}\sin^5 u + c$,   (iv) $\tan\theta + \frac{2}{3}\tan^3\theta + \frac{1}{5}\tan^5\theta + c$,

   (v) $\frac{1}{5}\sec^5 x - \frac{2}{3}\sec^3 x + \sec x + c$,   (vi) $-\frac{1}{2}\cos x^2 + c$,

   (vii) $2\tan\sqrt{x} + c$,   (viii) $\frac{1}{4}\log_e(2x^2+3) + c$,

   (ix) $-\frac{1}{2}e^{-x^2} + c$,   (x) $\frac{1}{2}(\log_e 10)^{-1}10^{y^2} + c$,

   (xi) $\log_e\sec^2\dfrac{\theta}{2} + c$,   (xii) $\frac{1}{2}(\log_e x)^2 + c$,

   (xiii) $\log_e(\sec x + \tan x) + c$.

2.   (i) $\frac{1}{5}(x+1)\sqrt{(2x-3)^3} + c$,   (ii) $\frac{1}{324}(24x+1)(3x-1)^8 + c$,

   (iii) $y + 16(y-4)^{-1} + c$,   (iv) $\dfrac{1}{\sqrt{5}}\sin^{-1}\dfrac{\sqrt{5}\,y}{2} + c$,

   (v) $\dfrac{1}{3\sqrt{3}}\tan^{-1}\sqrt{3}u + c$,   (vi) $\dfrac{1}{2\sqrt{2}}\tan^{-1}\dfrac{u-3}{2\sqrt{2}} + c$,

   (vii) $\dfrac{1}{\sqrt{2}}\sin^{-1}\dfrac{(x-1)\sqrt{2}}{3} + c$,   (viii) $\frac{1}{2}y\sqrt{(4-y^2)} + 2\sin^{-1}\dfrac{y}{2} + c$,

   (ix) $\sec^{-1}3x + c$,   (x) $\frac{2}{3}\tan^{-1}(\frac{1}{3}\tan\frac{1}{2}\theta) + c$,

   (xi) $\frac{1}{3}\log_e\tan\frac{3}{2}u + c$,   (xii) $\frac{1}{2}\tan^{-1}(2\tan x) + c$,

   (xiii) $\log_e\{x + \sqrt{(x^2-16)}\} + c$.

3.   (i) $\frac{1}{3}e^{3x} + c$,   (ii) $(\log_e 10)^{-1}10^y + c$,   (iii) $-\frac{1}{3}e^{-x^3} + c$,

   (iv) $\frac{1}{3}\log_e x + c$,   (v) $\frac{1}{3}\log_e(3x+4) + c$,   (vi) $-\frac{1}{2}\log_e(2x-3) + c$,

   (vii) $\frac{1}{3}\log_e(x+3) + c$,   (viii) $\frac{1}{2}\log_e\dfrac{1+x}{1-x} + c$,   (ix) $x(\log_e x - 1) + c$,

   (x) $2e^{\sqrt{x}}(\sqrt{x}-1) + c$,   (xi) $(x+3)\log_e(x+3) - x + c$.

page 291.

4. (i) $\frac{1}{3}\log_e\frac{3+x}{3-x}+c,$ (ii) $\frac{1}{3}\log_e\frac{y-3}{y}+c,$

   (iii) $\frac{1}{x}+\log_e\frac{x-1}{x}+c,$ (iv) $4(4-x)^{-1}+\log_e(4-x)+c,$

   (v) $-\frac{4x+5}{2(x+1)^2}-\log_e(x+1)+c,$ (vi) $\log_e\frac{(x+1)^3}{x^2-x+1}+c.$

5. (i) $2x\sin\frac{1}{2}x+4\cos\frac{1}{2}x+c,$ (ii) $\frac{1}{2}e^x(x-1)+c,$

   (iii) $\log_e\sin y-y\cot y+c,$ (iv) $-\frac{1}{252}(21y+1)(1-3y)^7+c,$

   (v) $(\log_e 3)^{-2}\,3^x(x\log_e 3-1)+c,$ (vi) $\frac{1}{4}x^2(2\log_e 2x-1)+c,$
   (vii) $t(\log_e t-1)+c,$ (viii) $x\tan^{-1}3x-\frac{1}{6}\log_e(1+9x^2)+c,$
   (ix) $4^x(\log_e 4)^{-1}+c,$ (x) $x(6-x^2)\cos x+3(x^2-2)\sin x+c,$
   (xi) $\frac{1}{2}\sin t^2-\frac{1}{2}t^2\cos t^2+c,$ (xii) $\frac{1}{4}(\theta^2+2\theta\sin\theta+2\cos\theta)+c,$
   (xiii) $\frac{1}{8}(1+4y^2)\tan^{-1}2y-\frac{1}{4}y+c,$
   (xiv) $\frac{1}{4}(2x\tan 2x+\log_e\cos 2x-2x^2)+c,$
   (xv) $\frac{1}{34}e^{3x}(3\cos 5x+5\sin 5x)+c,$
   (xvi) $\frac{1}{4}\sec 2\theta\tan 2\theta+\frac{1}{4}\log_e(\sec 2\theta+\tan 2\theta)+c.$

6. (i) $\frac{1}{4}\log_e(4x^2+3)-\frac{1}{2\sqrt{3}}\tan^{-1}\frac{2x}{\sqrt{3}}+c,$

   (ii) $4\sqrt{(1+2y-y^2)}-3\sin^{-1}\frac{y-1}{\sqrt{2}}+c,$

   (iii) $\frac{2}{5}\theta-\frac{1}{5}\log_e(2\cos\theta-\sin\theta)+c,$
   (iv) $\frac{2}{5}\log_e(4\sin x+3\cos x)-\frac{1}{5}x+c.$

7. (i) $\frac{\pi}{9},$ (ii) $\frac{1}{2\sqrt{2}}\tan^{-1}\frac{18-10\sqrt{2}}{31},$ (iii) $\frac{1}{2},$

   (iv) $\frac{256}{693},$ (v) $\frac{231\pi}{2048},$ (vi) $\frac{5\pi}{128},$

   (vii) $\frac{35\pi}{128},$ (viii) $0,$ (ix) $\frac{128}{230945},$ (x) $\frac{1}{2}\log_e\frac{1}{3}.$

8. (i) $-\frac{1}{5}\cos 5x+c,$ (ii) $3\sin\frac{1}{3}x+c,$ (iii) $\frac{1}{5}\log_e\sec 5x+c,$
   (iv) $2\log_e\sin\frac{1}{2}x+c,$ (v) $\log_e\tan\frac{1}{2}x+c,$
   (vi) $\log_e(\sec x+\tan x)+c$ or $\log_e\tan(\frac{1}{2}x+\frac{1}{4}\pi)+c.$

9. (i) $3\tan\frac{1}{3}x+c,$ (ii) $-\frac{1}{4}\cot 4x+c,$ (iii) $\frac{1}{2}x-\frac{1}{4}\sin 2x+c,$
   (iv) $\frac{1}{2}x+\frac{1}{4}\sin 2x+c,$ (v) $\tan x-x+c,$ (vi) $-\cot x-x+c.$

10. (i) $\frac{1}{3}\cos^3 x-\cos x+c,$ (ii) $\sin x-\frac{1}{3}\sin^3 x+c,$
   (iii) $\frac{1}{2}\tan^2 x+\log_e\cos x+c,$ (iv) $-\frac{1}{2}\cot^2 x-\log_e\sin x+c,$
   (v) $\frac{1}{2}\tan x\sec x+\log_e\sqrt{(\sec x+\tan x)}+c,$
   (vi) $\frac{1}{2}\log_e\tan\frac{1}{2}x-\frac{1}{2}\cot x\operatorname{cosec} x+c.$

page 292.

11.   (i) $\frac{1}{32}(12x - 8\sin 2x + \sin 4x) + c$,

      (ii) $\frac{1}{32}(12x + 8\sin 2x + \sin 4x) + c$,

      (iii) $x - \tan x + \frac{1}{3}\tan^3 x + c$,   (iv) $-\frac{1}{3}\cot^3 x - \cot x + c$,

      (v) $\frac{1}{3}\tan^3 x + \tan x + c$,   (vi) $x + \cot x - \frac{1}{3}\cot^3 x + c$.

12.   (i) $x\sin^{-1}x + \sqrt{(1-x^2)} + c$,   (ii) $x\cos^{-1}x - \sqrt{(1-x^2)} + c$,

      (iii) $x\tan^{-1}x - \frac{1}{2}\log_e(1+x^2) + c$,   (iv) $x\cot^{-1}x + \frac{1}{2}\log_e(1+x^2) + c$,

      (v) $x\sec^{-1}x - \log_e\{x + \sqrt{(x^2-1)}\} + c$,

      (vi) $x\,\mathrm{cosec}^{-1}x + \log_e\{x + \sqrt{(x^2-1)}\} + c$.

13.   (i) $\dfrac{1}{2\sqrt{3}}\tan^{-1}\dfrac{2x}{\sqrt{3}} + c$,   (ii) $\frac{1}{8}\sqrt{(5+8x^2)} + c$,

  †(iii) $\log_e\{x + \sqrt{(1+x^2)}\} + c$,   (iv) $\frac{1}{6}\log_e(2+3x^2) + c$,

      (v) $\frac{1}{3}\sqrt{(3+x^2)^3} + c$,

      (vi) $\frac{1}{4}\log_e(3+2x^2) + \dfrac{1}{\sqrt{6}}\tan^{-1}\dfrac{x\sqrt{2}}{\sqrt{3}} + c$,

      (vii) $\frac{1}{2}\log_e(x^2-4x+7) + c$,

  (viii) $\frac{1}{2}x\sqrt{(x^2+2)} + \log_e\{x + \sqrt{(x^2+2)}\} + c$,

      (ix) $\frac{3}{2}\log_e(x^2-4x+5) - 5\tan^{-1}(x-2) + c$,

      (x) $\frac{2}{135}(9x-4)\sqrt{(2+3x)^3} + c$,   (xi) $\log_e(3+x) + 3(x+3)^{-1} + c$.

14.   (i) $\dfrac{1}{\sqrt{5}}\sin^{-1}\dfrac{\sqrt{5}\,x}{2} + c$,   (ii) $-\frac{2}{27}(3x+2)\sqrt{(1-3x)} + c$,

      (iii) $\frac{1}{3}\log_e\dfrac{3+x}{3-x} + c$,   (iv) $3(16-x)^{-1} + c$,   (v) $-\frac{1}{3}\sqrt{(6-x^2)^3} + c$,

      (vi) $-\frac{3}{2}\log_e(4-x^2) + c$,   (vii) $\frac{1}{2}x\sqrt{(4-x^2)} + 2\sin^{-1}\dfrac{x}{2} + c$,

  (viii) $-\frac{1}{2}\sqrt{(7-2x^2)} + c$,   (ix) $-\frac{1}{4}\sqrt{(3-4x^2)} - \sin^{-1}\dfrac{2x}{\sqrt{3}} + c$,

  †(x) $\log_e\{x + \sqrt{(x^2-9)}\} + c$,   (xi) $-\frac{1}{3}\sqrt{(13-6x-3x^2)} + c$.

15.   (i) $\dfrac{180}{\pi}\sin x° + c$,   (ii) $\frac{1}{32}\sin 4x - \frac{1}{8}x\cos 4x + c$,

      (iii) $2\log_e\tan\frac{1}{2}\theta + c$,   (iv) $-\frac{1}{7}\cos^7 x + \frac{2}{9}\cos^9 x - \frac{1}{11}\cos^{11}x + c$,

      (v) $y\tan y + \log_e\cos y + c$,   (vi) $\sin x - x\cos x + c$,

      (vii) $-\frac{1}{2}\cos x^2 + c$,   (viii) $(u^2-2)\sin u + 2u\cos u + c$,

      (ix) $\frac{1}{8}y - \frac{1}{32}\sin 4y + c$,   (x) $-\frac{1}{42}(3\cos 7x + 7\cos 3x) + c$,

      (xi) $\frac{1}{4}\theta^2 + \frac{1}{4}\theta\sin 2\theta + \frac{1}{8}\cos 2\theta + c$,

      (xii) $\tan^3\frac{1}{3}x - 3\tan\frac{1}{3}x + x + c$,

      (xiii) $-2\cos\frac{1}{2}x + \frac{4}{3}\cos^3\frac{1}{2}x - \frac{2}{5}\cos^5\frac{1}{2}x + c$,

      (xiv) $y\tan y + \log_e\cos y - \frac{1}{2}y^2 + c$,   (xv) $\frac{2}{3}\sqrt{\sec^3\theta} + c$.

16.   (i) $\tan\frac{1}{2}\theta + c$,   (ii) $\frac{1}{4}\log_e\dfrac{1+2\tan\theta}{1-2\tan\theta} + c$,

† See also pp. 443, 445.

page 293.

16. (iii) $\tan x - \sec x + c$,   (iv) $\log_e (\cos \theta + 3 \sin \theta) - \theta + c$,

(v) $-2 \cot \frac{1}{2}x + c$,   (vi) $\frac{4}{3} \tan^{-1} (3 \tan x) + c$,

(vii) $\frac{1}{4} \log_e (\sec 4y + \tan 4y) + c$,   (viii) $\tan x + \sec x + c$,

(ix) $\frac{1}{2} \log_e \dfrac{1 + \tan \frac{1}{2}\theta}{3 - \tan \frac{1}{2}\theta} + c$,   (x) $\frac{1}{2}x + \frac{1}{2} \log_e (\cos x + \sin x) + c$.

17. (i) $-e^{-x}(x^3 + 3x^2 + 6x + 6) + c$,   (ii) $(x+2) \log_e (x+2) - x + c$,

(iii) $2e^{\sqrt{y}} + c$,   (iv) $2\sqrt{\log_e t} + c$,

(v) $\frac{1}{18}(1 + 9x^2) \tan^{-1} 3x - \frac{1}{6}x + c$,

(vi) $\frac{1}{2}(\sin^{-1} x)^2 + c$,   (vii) $(\log_e 4)^{-1} 4^x + c$,

(viii) $x(\log_e 10)^{-1}10^x - (\log_e 10)^{-2} 10^x + c$,

(ix) $\frac{1}{4}x^4 \log_e 2x - \frac{1}{16}x^4 + c$,   (x) $\frac{1}{2}e^x(x^2 - 1) + c$,

(xi) $\frac{1}{2} \log_e (\tan^{-1} 2x) + c$,   (xii) $\frac{1}{3}(\log_e 10)^{-1}10^{t^3} + c$,

(xiii) $\frac{1}{2}\{\log_e (y+2)\}^2 + c$,   (xiv) $\theta \sin^{-1} 2\theta + \frac{1}{2}\sqrt{(1 - 4\theta^2)} + c$,

(xv) $\frac{1}{13}e^{2x}(2 \sin 3x - 3 \cos 3x) + c$,

(xvi) $(2+x) \log_e (2+x) - (2-x) \log_e (2-x) - 2x + c$,

(xvii) $\frac{1}{2}e^y(1-y) \cos y + \frac{1}{2}ye^y \sin y + c$,

(xviii) $\log_e (\log_e x) + c$.

page 294.   **Exercise 13f**

1. (i) $\frac{1}{5}x^5 - 2x^2 - 4x^{-1} + c$,   (ii) $\frac{1}{16}(4 \cos 2x - \cos 8x) + c$,

(iii) $2 \log_e x - \log_e (1+x^2) + c$; $\frac{1}{3}$.

2. (i) $\frac{1}{5}(1+x^2)^{5/2} + c$,   (ii) $\frac{1}{2} \log_e (1+2x) - \log_e (1-2x) + c$,

(iii) $\frac{1}{3}x^3 \log_e x - \frac{1}{9}x^3 + c$, $\frac{1}{16}\pi^2 - \frac{1}{36}$.   **3.** $\dfrac{1}{4(1+x)} + \dfrac{1}{4(1-x)} + \dfrac{1}{2(1+x^2)}$

4. (i) $\log_e \dfrac{x-2}{x-1} + c$,   (ii) $\frac{1}{2}x^2 \tan^{-1} x - \frac{1}{2}x + \frac{1}{2} \tan^{-1} x + c$,

(iii) $x - \log_e (x^2 + 2x + 2) + c$.

5. (i) $2 \log_e (x+2) - \log_e (x+4) + c$,   (ii) $\frac{1}{2}x^2 \log_e x - \frac{1}{4}x^2 + c$,

(iii) $x \sin^{-1} x + \sqrt{(1-x^2)} + c$; $2 + \sqrt{2}$.

6. $\log_e \frac{4}{3} - \frac{1}{6}\frac{4}{3}$.   **8.** (i) $-\frac{1}{2} \mathrm{cosec}^2 x + c$,   (ii) $-\sqrt{(7 - 6x - x^2)} + c$.

9. (i) $\sin^{-1} \dfrac{x+2}{3} + c$,   (ii) $-\frac{1}{2}e^{-x^2}(x^2 + 1) + c$.

10. $\frac{1}{4}\pi + \frac{1}{2}$.   **11.** $\frac{1}{2}\sqrt{19} - 2$.

12. (i) $x + \log_e (x-2) - \log_e (x+2) + c$,

(ii) $-e^{-x^2/2}(x^2 + 2) + c$,   (iii) $\sin^{-1} \dfrac{x-2}{3} + c$; $\dfrac{\pi\sqrt{6}}{12}$.

13. $\frac{1}{13}(3 - 2e^\pi)$.

14. (i) $-(2x-3)^{-\frac{1}{2}} + c$,   (ii) $\sin x - x \cos x + c$,

(iii) $\frac{2}{15}(3x+8)(x-4)^{\frac{3}{2}} + c$; π.

15. (i) Positive,   (ii) zero,   (iii) negative.

16. (i) Positive,   (ii) zero,   (iii) positive.

18. (i) $\frac{1}{4}\pi - \frac{1}{2} \log_e 2$,   (ii) $\frac{1}{4} + \frac{3}{32}\pi$.   **21.** 1.

page 299.   **CHAPTER 14**

Qu. 1. $l \cos \theta$.   **Qu. 7.** ON is negative; $\cos \phi < 0$.

page 303.   **Exercise 14a**

2. 121 m N, 275 m E; 300 m at 066°.
4. 5·80 N, 131·6°.   **6. None**
9. $\frac{1}{2}(b \cos \alpha + a \sin \alpha)$, $a \cos \alpha - \frac{1}{2}b \sin \alpha$.
15. $(a+b) \cos \theta + a\theta \sin \theta$.   18. $(tV \cos \alpha, tV \sin \alpha - \frac{1}{2}gt^2)$.
19. $a \cos \theta$, $b \sin \theta$ m; $-a\omega^2 \cos \theta$, $-a\omega^2 \sin \theta$ m/s².

Qu. 10. PQ.   **Qu. 11.** $l \cos \theta$.

page 310.   **Exercise 14b**

2. (i) When a side is parallel to the axis.
   (ii) When the shorter side is parallel to the axis, and the ratio of the shorter to the longer side equals the cosine of the angle of projection.
3. $ab \cos \theta$.
5. The opposite sides are parallel and equal; the diagonals bisect each other.
8. $\cot^{-1} \sqrt{6}$.   9. 118·5°, 99·1°, 99·1°, 111·6°, 111·6°.
10. Yes.   When corresponding sides are parallel.
11. When a diagonal is parallel to the axis.
12. $\cos^{-1}(b/a)$, $b < a$.
13. When the base, or the altitude to the base, is parallel to the axis.
17. $\tan(\beta - \alpha) = \dfrac{\sec \theta(\tan \beta' - \tan \alpha')}{1 + \sec^2 \theta \, \tan \alpha' \tan \beta'}$.
18. 37·4°.

page 314.   **Exercise 14c**

3. $\dfrac{x^2}{a^2} + \dfrac{y^2}{b^2} = \dfrac{a^2 + b^2}{4a^2}$.
5. An ellipse through the point and the centre of the given ellipse.
8. (i) Always,   (ii) if a side of the square is parallel to the common axis.
17. $2ab$.   22. $\pi/4$.

page 317.   **CHAPTER 15**

Qu. 2. πr^2.
Qu. 7. (i) Cone: $V = \frac{1}{3}\pi R^2 h$, $S = \pi Rl$,
   (ii) cylinder: $V = \pi r^2 h$, $S = 2\pi rh$.
Qu. 9. $\frac{4}{3}\pi a^3$.

page 321. **Exercise 15a**

1. (i) 2:1,   (ii) 3:1,   (iii) 5:1.

2. 1:7:19.   3. 1:3, 1:3:5.

4. $3 \cdot 141\ 03 < \pi < 3 \cdot 142\ 71$.   6. πr^2.

10. $\frac{2}{3}\pi a^3(1 - \cos\alpha)$.   11. $\frac{1}{3}\pi r^2 h$.

14. $x^2 + y^2 - 2ry = 0$.   15. 20:7, 27:5, 25:2.

16. (i) Pair of parallel lines $2\sqrt{(a^2 - b^2)}$ apart,
   (ii) square of side $2\sqrt{(a^2 - h^2)}$,   (iii) $4(a^2 - h^2)$,   (iv) $16a^3/3$.

17. Both $\frac{3}{4}h$ from vertex.   18. $\frac{2}{3}h$ from vertex.

19. $\frac{1}{4}a$, $\frac{1}{3}a$ from centre.   20. $\dfrac{h(4r - h)}{4(3r - h)}$, $\frac{3}{8}r$.

21. $\frac{1}{3}h\left(\dfrac{R + 2r}{R + r}\right)$.   Qu. 10. $\frac{4}{3}Ma^3$.

Qu. 11. Ma^2.   Qu. 12. $\frac{1}{2}Ma^2$.

page 328. **Exercise 15b**

1. (i) $2a^2$,   (ii) $4a^2$.

2. (i) $4a^2$,   (ii) $4a^2$,   (iii) $8a^2$.

3. $\frac{1}{3}Ma^2$.   4. $\frac{1}{3}M(a^2 + b^2)$.   5. $\frac{1}{3}Mh^2$.

6. $\frac{1}{3}M(a^2 + 3h^2)$.   7. $\frac{1}{3}M(a^2 + 3h^2)$.

8. $\frac{1}{6}Ma^2$.   9. $\frac{1}{6}Mh^2$.

10. (i) $2Ma^2$,   (ii) $\frac{2}{6}Ma^2$.   11. $\frac{1}{3}\sqrt{15}a$.

12. $\frac{1}{2}\sqrt{2}a$.   13. $\frac{1}{3}M(a^2 + b^2)$.

14. $\frac{2}{3}Ma^2$.   Qu. 13. Ma^2, $\frac{1}{2}Ma^2$.

page 335. **Exercise 15c**

1. $4mb^2$, $4ma^2$, $4m(a^2 + b^2)$.   2. $4ma^2$, $8ma^2$.

3. (i) $\frac{8}{3}ma^2(2a + 3b)$,   (ii) $\frac{4}{3}m(a + b)^3$.

4. $\frac{4}{3}M(a^2 + b^2)$.   5. $\frac{1}{4}Ma^2$.

6. $\frac{3}{4}Ma^2$.   7. (i) $\frac{16}{3}Ma^2$,   (ii) $\frac{8}{3}Ma^2$.

8. $\frac{2}{3}Ma^2$, $\frac{1}{3}Ma^2$.   9. (i) $\frac{1}{3}Ma^2$,   (ii) $\frac{1}{3}Ma^2$.

10. $\frac{11}{18}Mr^2$.   11. $\frac{1}{2}Mr^2\left(\dfrac{3r + 4h}{r + h}\right)$.

12. $\frac{2}{5}M\left(\dfrac{a^5 - b^5}{a^3 - b^3}\right)$.   13. (i) $\frac{4}{3}Ma^2$,   (ii) $19Ma^2/9$.

14. $\frac{3}{10}Ma^2$.

16. (i) $\pi a^3\sigma\,\Delta x$,
   (ii) $2\pi a\sigma(\frac{1}{2}a^2 + x^2)\,\Delta x$, $M(\frac{1}{2}a^2 + \frac{1}{3}h^2)$, $M(\frac{1}{2}a^2 + \frac{1}{12}h^2)$.

17. (i) $\frac{1}{2}\pi a^4\sigma\,\Delta x$,
   (ii) $\pi a^2\sigma(\frac{1}{4}a^2 + x^2)$, $M(\frac{1}{4}a^2 + \frac{1}{3}h^2)$, $M(\frac{1}{4}a^2 + \frac{1}{12}h^2)$.

18. $M(\frac{1}{4}r^2 + \frac{1}{3}h^2)$,   (i) $M(\frac{1}{4}r^2 + \frac{1}{18}h^2)$,   (ii) $M(\frac{1}{4}r^2 + \frac{1}{6}h^2)$.

page 337.

19. (i), (ii) $M(\frac{1}{3}a^2 + \frac{1}{10}h^2)$.

20. $\frac{5}{4}Mr^2$.

21. $\frac{5}{12}Ma^2$.

23. $\frac{2}{5}Ma^2$.

25. σ.

27. (i) $(\frac{2}{3}a, 0)$, (ii) $\frac{2}{3}\sqrt{3}a$.

28. $\frac{2}{5}Ma^2$.

29. $M(\frac{3}{20}a^2 + \frac{1}{10}h^2)$.

30. $\frac{1}{2}Ma^2 \left(\dfrac{3a+4h}{a+h} \right)$.

31. $\frac{1}{3}M(a^2 + b^2 \sin^2 \theta)$.

32. $Ma^2(1 - \frac{1}{2}\sin^2 \alpha)$.

CHAPTER 16

page 340.

Qu. 1. $lL + mM = 0$, $a + b = 0$. **Qu. 2.** $h^2 \geqslant ab$.

Qu. 4. $h^2 = ab$, $h^2 \geqslant ab$. **Qu. 5.** $\tan^{-1} \left| \dfrac{2\sqrt{(h^2 - ab)}}{a+b} \right|$.

page 342. **Exercise 16a**

1. (i) $2x^2 + 3xy - 2y^2 = 0$,
   (ii) $2x^2 - xy - 3y^2 + 18x - 17y + 28 = 0$,
   (iii) $x^2 + 6xy + 9y^2 + 3x + 9y - 4 = 0$.

2. (i) $3x - 2y = 0$, $x + 7y = 0$, (ii) $x + y - 3 = 0$, $x - y - 2 = 0$,
   (iii) $4x + 5y - 6 = 0$, $2x - 3y + 1 = 0$.

3. (i) $\tan^{-1} \frac{4}{3}\sqrt{19}$, (ii) $\tan^{-1} \frac{3}{2}\sqrt{5}$, (iii) $\tan^{-1} 2\sqrt{11}$.

4. (i) $x^2 - 9xy - y^2 = 0$, (ii) $x^2 - y^2 = 0$; $a + b = 0$.

5. A line-pair joining the origin to the points of intersection of the line and the circle.

6. (i) $n^2 \left(\dfrac{x^2}{a^2} + \dfrac{y^2}{b^2} \right) = (lx + my)^2$, (ii) $n^2 \left(\dfrac{x^2}{a^2} - \dfrac{y^2}{b^2} \right) = (lx + my)^2$,
   (iii) $ny^2 + 4ax(lx + my) = 0$.

7. $n^2(b^2x^2 + a^2y^2) = a^2b^2(lx + my)^2$, (i) $\left| \dfrac{n}{\sqrt{(l^2 + m^2)}} \right|$,
   (ii) $n^2(a^2 + b^2) - a^2b^2(l^2 + m^2) = 0$.

8. $x^2 + y^2 - 4ax = 0$.

9. $(b^2 - a^2)(x^2 + y^2) = a^2b^2$, $b > a$.

10. (i) $ab \sin \theta \cos \theta(x^2 - y^2) + (a^2 \sin^2 \theta - b^2 \cos^2 \theta)xy = 0$,
   (ii) the same.

The tangent and normal at any point on an ellipse bisect the angles between the lines joining the point to the foci.

11. The point dividing (x_1, y_1), (X, Y) in the ratio $\lambda : 1$.
   $(Xx_1 + Yy_1 - a^2)^2 = (X^2 + Y^2 - a^2)(x_1^2 + y_1^2 - a^2)$.

page 344.

12. (i) $\{yy_1 - 2a(x + x_1)\}^2 = (y^2 - 4ax)(y_1^2 - 4ax_1)$,

(ii) $\left(\dfrac{xx_1}{a^2} + \dfrac{yy_1}{b^2} - 1\right)^2 = \left(\dfrac{x^2}{a^2} + \dfrac{y^2}{b^2} - 1\right)\left(\dfrac{x_1^2}{a^2} + \dfrac{y_1^2}{b^2} - 1\right)$,

(iii) $\left(\dfrac{xx_1}{a^2} - \dfrac{yy_1}{b^2} - 1\right)^2 = \left(\dfrac{x^2}{a^2} - \dfrac{y^2}{b^2} - 1\right)\left(\dfrac{x_1^2}{a^2} - \dfrac{y_1^2}{b^2} - 1\right)$.

13. (i) $x + a = 0$,   (ii) $x^2 + y^2 = a^2 + b^2$,   (iii) $x^2 + y^2 = a^2 - b^2$.

Qu. 7. (i) $\sqrt{2}$,   (ii) $\sqrt{34}$,   (iii) $\sqrt{c}$.

Qu. 8. $x - 3y + 1 = 0$, $\tfrac{1}{2}\pi$.

Qu. 9. (i) $7x - 6y = 0$,   (ii) $16x + 8y - 5 = 0$,   (iii) $y = 0$.

Qu. 12. (i), (ii).

page 351.   **Exercise 16b**

2. $x^2 + y^2 - 6x + 4 = 0$.   **3.** $x^2 + y^2 - 5y - 5 = 0$.

4. $x^2 + y^2 - 6x - 2y = 0$.   **5.** (i) $3x - 4 = 0$,   (ii) $x^2 + y^2 + 6x - 8 = 0$.

6. $x^2 + y^2 - 10x - 15y = 0$.   **13.** $x^2 + 2xy\dfrac{dy}{dx} - y^2 - c = 0$.

14. Divides line joining centres in ratio $c^2 + a^2 - b^2$: $b^2 + c^2 - a^2$;

$\dfrac{1}{2c}\sqrt{(a^4 + b^4 + c^4 - 2b^2c^2 - 2c^2a^2 - 2a^2b^2)}$ from radical axis.

page 353.   **Exercise 16c**

2. $\tan^{-1}\left|\dfrac{a_2b_1 - a_1b_2}{a_1a_2 + b_1b_2}\right|$.

3. $(c_2a_1 - c_1a_2)x + (c_2b_1 - c_1b_2)y = 0$,

$(c_2 - c_1)(x^2 + y^2) + 2(c_2g_1 - c_1g_2)x + 2(c_2f_1 - c_1f_2)y = 0$.

4. $y - k = (4h^3 - 1)(x - h)$; ± 1; $(1, 3)$, $(-1, 5)$.

6. $\dfrac{ax + by + c}{\sqrt{(a^2 + b^2)}} = \pm\dfrac{a_1x + b_1y + c_1}{\sqrt{(a_1^2 + b_1^2)}}$, $(2, 1)$, $x^2 + y^2 - 4x - 2y + 4 = 0$.

8. $(-k, 2k)$, $2k$; $y = \tfrac{4}{3}x$.

9. $\sqrt{(g^2 + f^2 - c)}$, $(-g, -f)$, $\tfrac{1}{4}\pi$.   **11.** $2y^2 = x(3 + 4x)^2$.

13. $x - 3y + 13 = 0$, $7x - 3y + 37 = 0$.

14. $(4a, 0)$.   **16.** $x - ty + t^3 = 0$.

17. (i) $gx + fy = 0$,   (ii) $x = 0$,   (iii) $4x + y = 0$,   (iv) $x - y = 0$.

18. (i) $y^2 = 0$,   (ii) $x = 0$, $y = 0$,   (iii) $3x \pm 2y = 0$.

19. $p^3x - py - cp^4 + c = 0$.   **23.** $x^3 = y^2(2a - x)$.

24. $(x^2 + y^2)(x - a)^2 = b^2x^2$.   **25.** $p\dfrac{h\cos\alpha - \sin\alpha}{h\sin 2\alpha + \cos 2\alpha}$.

28. $\tfrac{1}{2}c^2\left(t^2 + \dfrac{1}{t^2}\right)^2$.   **30.** $x + 3y - 4 = 0$.

page 358.

31. $4x^2+2(k-2)xy+(1-4k-k^2)y^2 = 0$; $-5, 1$; $0, -\frac{12}{5}$;
    $x-1 = 0$, $5x-12y-5 = 0$.

32. $x = 0$, $x^2+y^2+2\lambda x+4 = 0$. $(\pm 2, 0)$.
   (i) $x^2+y^2+14x+4 = 0$, $x^2+y^2-6x+4 = 0$,
   (ii) $x^2+y^2-6y-4 = 0$.

33. $(0, 0)$, $(1, 2)$.

CHAPTER 17

page 360.

Qu. 1. $y = Ax+B$, $y = 3x-5$.   **Qu. 2.** $y = x^2+A$.

Qu. 3. $x^2+y^2 = A$.   **Qu. 4.** $s = \frac{1}{2}at^2+At+B$; $s = ut+\frac{1}{2}at^2$.

Qu. 5.   (i) $\dfrac{d^2y}{dx^2} = 0$,   (ii) $\dfrac{dy}{dx} = \dfrac{y}{x}$,   (iii) $\dfrac{dr}{d\theta}+r\tan\theta = 0$,

    (iv) $\dfrac{dy}{dx} = -\dfrac{y}{x}$,   (v) $\dfrac{dy}{dx} = y$,   (vi) $x\dfrac{dy}{dx} = y\log_e y$,

    (vii) $y\dfrac{d^2y}{dx^2} = \left(\dfrac{dy}{dx}\right)^2$,   (viii) $x\dfrac{dy}{dx}\log_e x = y$,

    (ix) $(1+x^2)\dfrac{dy}{dx}\tan^{-1}x = y$.

Qu. 7.   (i) $x^2-y^2+A = 0$,   (ii) $y = Ax$,

    (iii) $x = Ae^{y^2/2}$,   (iv) $x = A\sin y$,

    (v) $\log_e\sqrt{\dfrac{y-1}{y+1}} = e^z+A$,   (vi) $y^2 = 2\sqrt{(x^2+1)}+A$.

Qu. 8. $v^2 = u^2+2as$.

page 363.   **Exercise 17a**

1.   (i) $\dfrac{dy}{dx} = \frac{3}{2}$,   (ii) $\dfrac{dy}{dx} = \dfrac{y+\frac{1}{2}}{x}$,   (iii) $\dfrac{dy}{dx} = \dfrac{y}{x}$,

    (iv) $\dfrac{dy}{dx} = -\dfrac{x}{y}$,   (v) $\dfrac{dy}{dx} = -\dfrac{y}{x}$,   (vi) $\dfrac{dy}{dx} = \dfrac{y}{x-4}$

2.   (i) $\dfrac{d^2y}{dt^2} = -9y$,   (ii) $\dfrac{d^2y}{dt^2} = 3\dfrac{dy}{dt}$,   (iii) $\dfrac{d^2y}{dx^2} = 9y$,

    (iv) $\dfrac{d^2y}{dx^2}-\dfrac{dy}{dx}-6y = 0$,   (v) $\dfrac{d^2y}{dx^2}-8\dfrac{dy}{dx}+16y = 0$.

3. $3x-10y-35 = 0$.   **4.** $y = x^2-3x+1$.

5. $s = A-3t^2$, $s = 12-3t^2$.

6.   (i) $y = e^z-3x\cos x+3\sin x-1$,   (ii) $y = e^z-3x\cos x+3\sin x-e^{\pi/2}$.

7.   (i) $y = Ae^z$,   (ii) $y = \frac{2}{15}(x-1)(3x+2)\sqrt{(x-1)}+A$,

   (iii) $y = A(x+2)$,   (iv) $x = \frac{1}{2}y+\frac{1}{4}\sin 2y+A$,

page 364.

7. (v) $v - 1 = A v e^u$, (vi) $y = x \log_e x - x + A$,

(vii) $\sin y = A e^x$, (viii) $x = y \tan^{-1} y - \log_e \sqrt{(1 + y^2)} + A$,

(ix) $y^2 = x^2 - 2x + A$, (x) $y = A \sqrt{\dfrac{x-1}{x+1}}$,

(xi) $r = \log_e \left(A \tan \dfrac{\theta}{2} \right)$, (xii) $y + 3 = A e^{-1/x}$,

(xiii) $y = A x e^x$, (xiv) $\cos \theta \sin \phi = A$,

(xv) $r = \theta \tan \theta + \log_e (A \cos \theta)$,

(xvi) $2 y^2 = x^2 (\log_e x^2 - 1) + A$,

(xvii) $r = -\theta - \log_e (\cos \theta - \sin \theta) + A$,

(xviii) $2y + 3 = A(x - 2)^2$,

(xix) $x = A - \frac{1}{2} e^{-t} (\cos t + \sin t)$, (xx) $y = 2 \tan (2 e^{-x} + A)$.

8. (i) $y = \tan \theta$, (ii) $(y - 2)^2 = 9 e^{x^2}$, (iii) $y = \dfrac{7x + 1}{7 - x}$,

(iv) $y = \sin (x - \frac{1}{6} \pi)$.

Qu. 9. (i) $x^2 y = x + A$, (ii) $t^2 \log_e x = 3 \sin t + A$,

(iii) $x^2 \sin u = \log_e (kx)$, (iv) $x e^y = 2x + A$.

Qu. 10. (i) $x, x^2 y = \frac{1}{2} e^{x^2} + A$, (ii) $x, x^2 e^y = \frac{1}{3} x^3 + A$,

(iii) $\dfrac{1}{x}, x y^2 = \log_e (kx)$, (iv) $r, r^2 \tan \theta = 2\theta + A$.

Qu. 11. $y = \frac{1}{2} + A e^{-x^2}$; $y = \frac{1}{2} - e^{-x^2}$. **Qu. 12.** x^2.

Qu. 13. (i) $y = 1 + x \tan x + A \sec x$, (ii) $y = x - 4 + A e^{-x}$.

Qu. 14. (i) $\left(\dfrac{y}{x} \right)^2$, (ii) $\dfrac{d^2 y}{dx^2} + xy$, (iii) $x \sqrt{(x^2 + y^2)}$.

Qu. 15. (i), (ii), (iv), (v).

Qu. 16. (i) $x e^{x/y} = A$, (ii) $x^2 - 2xy = A$, (iii) $x - 2y + A xy = 0$.

Qu. 17. $y = x \log_e (A x^2)$.

page 370. **Exercise 17b**

1. (i) $y^2 = \dfrac{A}{x} - \dfrac{1}{x^2}$, (ii) $y^2 = \dfrac{1}{x^2} (\tan 2x + A)$,

(iii) $x \log_e y = \sec x + A$, (iv) $(1 - 2x) e^y = \tan x + A$,

(v) $t^2 e^s = t \sin t + A$, (vi) $r^2 e^u = \cot u + A$.

2. (i) $x^2 \sin y = 3x^2 + A$, (ii) $xy = e^x + A$,

(iii) $x \tan y = e^{x^2} + A$, (iv) $y e^x = \log_e (Ay)$.

3. (i) $y = e^{-2x} (\sin x + A)$, (ii) $s = \frac{1}{2} + A e^{-t^2}$,

(iii) $y = e^{-x^2} (1 + A e^{-x})$, (iv) $r = (\theta + A) \operatorname{cosec}^2 \theta$,

(v) $r = (\theta + A) \cos \theta$, (vi) $y = x^{-2} (\sin x + A)$,

(vii) $y = x \log_e \dfrac{A(x-1)}{x}$, (viii) $y = \dfrac{x}{3} + 3 + A x^{-1/2}$,

page 371.

3. (ix) $y = (x - \sin x + A) \cot \frac{1}{2}x$,   (x) $y = (x-2)^{-1} + A(x-2)^{-3}$,

   (xi) $y = x^2 \cos x + Ax$,   (xii) $y = \frac{1}{2}x \operatorname{cosec} x + \dfrac{A}{x} \operatorname{cosec} x$,

  (xiii) $x = y \log_e (Ay^2)$,   (xiv) $y = x^2 \tan^{-1} x + x^2 \log_e \{k(1+x^2)\}$.

4.   (i) $x^3 = Ae^{y/x}$,   (ii) $x^2(x^2 - 2y^2) = A$,

   (iii) $\tan^{-1} \dfrac{y}{x} = \log_e (Ax)$,   (iv) $3x - y + Axy = 0$,

   (v) $y = Ae^{x^2/2y^2}$,   (vi) $2x = (2x - y) \log_e \{A(2x - y)\}$,

  (vii) $\sin^{-1} \dfrac{y}{x} = \log_e (Ax)$,   (viii) $y = x(Ax - 1)$,

   (ix) $(x + y)(2x - y)^2 = A$,   (x) $\dfrac{2}{\sqrt{3}} \tan^{-1} \left(\dfrac{2y - x}{x\sqrt{3}} \right) = \log_e (Ax)$,

   (xi) $(x^2 + y^2)^{1/2} = x \log_e (Ax)$,   (xii) $y(2x - y) = 2x^2 \log_e \dfrac{A(x + y)^2}{x^3}$.

5. $x^2 - y^2 - 2xy + 4x = A$.

6.   (i) $y - 2 = Ae^{(x-3)/(y-2)}$,   (ii) $(x - y - 3)^2(x + 2y - 3) = A$.

7. $x^2 + y^2 - 2xy - 4x - 8y + A = 0$.

8.   (i) $x - y + A = \log_e (2x + y)$,   (ii) $x + y - 1 = Ae^{x-y}$.

9.   (i) $y = (x + A)(x + 3)^2$,   (ii) $x = (x - y) \log_e (Ax)$,

   (iii) $(y + 3) \sin x = A - \frac{1}{2}e^{-2x}$,   (iv) $\sin y = \dfrac{Ax - 1}{x(x + 3)}$,

   (v) $\tan^{-1} \dfrac{2(y - 1)}{x - 2} = \log_e \{A(x^2 + 4y^2 - 4x - 8y + 8)\}$,

   (vi) $y + 2 = xe^{2x} + Ae^x$,   (vii) $y^{x^2} = A \sin x$,

  (viii) $(2r + 3) \tan \theta = 3\theta + A$,   (ix) $xy = Ae^{y/x}$,

   (x) $y = e^{-x}(x^2 + A) \operatorname{cosec} x$,   (xi) $x^2 + y^2 - 2xy + 2x - 6y + A = 0$.

10.   (i) $y = (x + 1)^4$,   (ii) $u = \sin \theta + 2 \operatorname{cosec} \theta$,

   (iii) $x^2 - 2xy - y^2 = 17$,   (iv) $4y^2 - x^2 = 2y^2 \log_e \dfrac{y}{2}$,

   (v) $e^t(x - 1) = 1 - \dfrac{1}{t}$,   (vi) $\frac{1}{2} \tan^{-1} \dfrac{y}{2x} = \dfrac{\pi}{8} + \log_e \dfrac{8x^3}{4x^2 + y^2}$,

  (vii) $y = \frac{1}{2}x^3 \{(\log_e x)^2 + 1\}$.

Qu. 19.   (i) $y = 2x \log_e x + Ax + B$,   (ii) $y = 2 \sin x - x \cos x + Ax + B$,

         (iii) $y = x^3 + A \log_e x + B$,   (iv) $y^2 = Ax + B - 2 \cos x$.

Qu. 20.   (i) $y = Ae^{Bx}$ or $y = C$,   (ii) $y = Ax(x - 1) + B$.

Qu. 21. $y = \log_e x^2 + Ax^{-1} + B$.

Qu. 22.   (i) $x = a \cos (2t + \epsilon)$,   (ii) $y = a \cos (3x + \epsilon)$,

        (iii) $y = -\frac{8}{3}x^3 + Ax + B$.

Qu. 23. $2 \dfrac{dx}{dt}$.

page 377.

Qu. 24. $x = 2 \cos(\frac{3}{2}t)$,   (i) $\frac{dx}{dt} = \frac{3}{2}\sqrt{(4-x^2)}$,   (ii) $\frac{dx}{dt} = -3\sin(\frac{3}{2}t)$.

Qu. 25. (i) $x = a\cos nt$,   (ii) $x = a\sin nt$.

Qu. 26. (i) $x = a\cos nt$,   (ii) $x = a\sin nt$.

Qu. 27. $\pm\dfrac{\pi}{3}$.

Qu. 28. (i) $y = a\cos(2x+\epsilon)-1$,   (ii) $\theta = a\cos(\sqrt{2}t+\epsilon)+3$,
      (iii) $x = A+Bt-\frac{1}{2}t^2-\frac{3}{8}t^3$.

page 379.   **Exercise 17c**

 1. (i) $y = x\log_e x + Ax + B$,
     (ii) $y = \log_e \sec\theta + A\log_e(\sec\theta + \tan\theta) + B$,
     (iii) $y = 2e^{-x} + Ax + B$,   (iv) $y = A\log_e\{B(2x+1)\}$,
     (v) $y = e^x(x-1) + Ax^2 + B$,   (vi) $x = \log_e\sin t + At + B$.
 2. (i) $y = Ax^4 + B$,   (ii) $A+Bx = e^{-2y}$,
     (iii) $6x = y^3 + Ay + B$ or $y = C$, (iv) $y = A\tan^{-1}x + B$,
     (v) $y = A\sin^{-1}x + B$,   (vi) $y^2(Ax+B)+1 = 0$ or $y = C$,
     (vii) $3y-2 = Ae^{Bx}$ or $y = C$,   (viii) $2y+1 = (Ax+B)^2$ or $y = C$,
     (ix) $Ax+B = (9+Ay^2)^{1/2}$,

     (x) $y = -\dfrac{1}{2\sqrt{A}}\tan^{-1}\dfrac{2x}{\sqrt{A}} + B$, $(A>0)$;

       $y = \dfrac{1}{4\sqrt{(-A)}}\log_e\left\{\dfrac{\sqrt{(-A)}+2x}{\sqrt{(-A)}-2x}\right\} + B$, $(A<0)$,

     (xi) $y = \frac{1}{9}(6x+A)^{3/2} + B$,   (xii) $x = \dfrac{1}{A}\log_e(Ay-2) + B$ or $y = C$.

 3. $s = \log_e\{A(t+B)^{10}\}$.

 4. (i) $y = \log_e\operatorname{cosec} x + A\log_e\tan\dfrac{x}{2} + B$,

     (ii) $\frac{1}{6}x^3 + \frac{1}{6}x^2 - \frac{2}{3}x + A\log_e(x+2) + B$,
     (iii) $36y = 6x^2\log_e x - 5x^2 + Ax^{-1} + B$.

 5. $y+\dfrac{\pi}{4} = \frac{1}{2}\sin^{-1}x + \frac{1}{2}x\sqrt{(1-x^2)}$.

 6. (i) $s = a\cos(5t+\epsilon)$,   (ii) $y = a\cos(\frac{3}{2}x+\epsilon)$,

     (iii) $s = A+B\theta-\frac{1}{27}\theta^3$,   (iv) $y = a\cos\left(\dfrac{\sqrt{3}}{2}t+\epsilon\right)$.

 7. $s = 4\cos\dfrac{3t}{4}$.   **8.** $\dfrac{d^2x}{dt^2} = -9x$, $x = 2\cos\left(3t\pm\dfrac{\pi}{6}\right)$.

 9. At O; 12 s.   **10.** 0, 6 s.

 11. $x = \sqrt{2}\cos\left(2t-\dfrac{\pi}{4}\right)$, $\sqrt{2}$ m, $-\dfrac{\pi}{4}$ s, π s.   **12.** 5 m, 0·927 s, 4π s.

 13. (i) $\dfrac{\pi}{2}$,   (ii) $\dfrac{\pi}{12}$,   (iii) $\dfrac{\pi}{2}$,   (iv) π,   (v) $\dfrac{7\pi}{12}$.

page 381.

14.   (i) $y = a \cos (2x + \epsilon) - 3$,   (ii) $\theta = a \cos \left(\dfrac{3}{\sqrt{2}} \, t + \epsilon \right) + \frac{1}{3}$,

   (iii) $s = A + Bt + \frac{1}{6}t^2 - \frac{2}{3}t^3$,   (iv) $x = \sqrt{2} \cos \left(2t + \dfrac{\pi}{4} \right) - 2$.

15.   (i) $y = A + Be^{2x} - x$,   (ii) $y = A + Be^x$ or $y = C$,
   (iii) $x = a \cos (t + \epsilon)$,   (iv) $y = A \log_e (Bx)$ or $y = C$,
   (v) $(3y - 1)^{\frac{2}{3}} = Ax + B$ or $y = C$,
   (vi) $y = \frac{1}{2}x^2 \log_e x - \frac{3}{4}x^2 + Ax + B$,
   (vii) $x = a \cos (\frac{2}{3}t + \epsilon) + \frac{1}{4}$,   (viii) $x = A + Bu - \frac{3}{4}u^3$,
   (ix) $y = e^x(x - 1) + Ax^2 + B$.

page 381.   **Exercise 17d**

1. $x = A(\sec t + \tan t)$.   **2.** $y = x^2 \log_e x - x^2 + Ax$.

3. $y = \frac{1}{2}e^x \sin x + Ax + B$.   **4.** $r = 2 \tan \dfrac{\theta}{2} - \theta + A$.

5. $(y + 1)^2 = \dfrac{Ax}{x + 1}$.   **6.** $y = a \cos \left(\dfrac{2}{\sqrt{3}} \, x + \epsilon \right)$.

7. $y = (2x - 1)^{-2} + A(2x - 1)^{-4}$.

8. $y + \dfrac{1}{A} \log_e (Ay - 1) = 3x + B$ or $y = C$.

9. $x + 2y = Ax^6(x - 2y)^3$.   **10.** $v = \frac{1}{2}(\log_e u)^2 + A$.
11. $y = e^{-x}(A - x) + B$.   **12.** $y = Ae^x(x - 2) + B$ or $y = C$.
13. $ye^x = x \sin^{-1} x + \sqrt{(1 - x^2)} + A$.   **14.** $y = \frac{1}{2}e^x(1 + \tan x) + A \sec x$.
15. $x^3(x + 2y) = Ae^{2y/x}$.   **16.** $(x + y - 2)^2(2x - y - 10) = A$.
17. $y = Ae^{-x} + \frac{1}{2}x^2 + Bx + C$.   **18.** $2x - 2y + A = \log_e (2x + 2y + 1)$.

19. $y = x \cos A + \sqrt{(1 - x^2)} \sin A$.   **20.** $y = \cot \dfrac{x}{2} \log_e \left(A \sec^2 \dfrac{x}{2} \right)$.

21. $\sqrt{2} \tan^{-1} \dfrac{y}{x\sqrt{2}} = \log_e \{ A(2x^2 + y^2) \}$.

22. $y = 4 \sin \dfrac{x}{2} + A \log_e \sin \dfrac{x}{2} + B$.   **23.** $y\sqrt{(1 - x^2)} = \sin^{-1} x + A$.

24. $x = 4 \cos \left(\frac{2}{3}t + \dfrac{\pi}{3} \right)$.   **25.** $y = \frac{5}{3} \cos 3x$.

26. $s = 2 \cos \left(3t - \dfrac{\pi}{3} \right) + 1$.   **27.** $(x + y)e^{-y/x} + x \log_e (Cx) = 0$.

28. $\dfrac{dx}{dy} = -\dfrac{y - 1}{x - 1}$, $(x - 1)^2 + (y - 1)^2 = k$.

29. $y(x^2 + 1) = (x^2 - 1) \sin x + 2x \cos x + 2$.

30. $y = -x^2 + A \cos x + B \sin x$.   **31.** $\dfrac{dy}{dx} = -\sin x - 2 \sin 2x$.

32. $\sqrt{(x^2 + y^2)} = x(\sqrt{2} + 1) - 1$.   **33.** $xy = (x + 1) \log_e \{ C(x + 1) \}$.
34. $v^2 = 2\mu a(a - x) + ce^{-x/a}$.
36. $xy = -x \cos \frac{1}{2}x + 2 \sin \frac{1}{2}x + c$; 0.   **37.** $(x + 1)y = e^{2x} + Ae^x$.

page 386.   CHAPTER 18

Qu. 2. 0·809.   **Qu. 4.** Between 3·28 and 3·29.

Qu. 5. 0, between 1·89 and 1·90, between −1·90 and −1·89.

page 389.   **Exercise 18a**

3. Between −4 and −3, between 1 and 2, between 2 and 3; between −3·58 and −3·57.

4. Between 5·15 and 5·16.   5. 0, between 1·256 and 1·257.

6. 0·77.   7. Between 4·14 and 4·15.

8. Between −1·945 and −1·944, between −0·209 and −0·208.

Qu. 8. $f(x) \simeq f(a) + f'(a)(x-a) + \dfrac{f''(a)}{2}(x-a)^2$.

Qu. 9. $f(a+h) \simeq f(a) + f'(a)h + \dfrac{f''(a)}{2}h^2$.   **Qu. 11.** $1 + 2h + 2h^2 + \frac{8}{3}h^3$.

Qu. 12. $\frac{3}{5} - \frac{4}{5}(x-\alpha) - \frac{3}{10}(x-\alpha)^2 + \frac{2}{15}(x-\alpha)^3$.

Qu. 13. (i) $1 + x + \dfrac{x^2}{2!} + \dfrac{x^3}{3!} + \dfrac{x^4}{4!} + \dfrac{x^5}{5!}$,   (iii) $1 - \dfrac{x^2}{2!} + \dfrac{x^4}{4!}$.

page 395.   **Exercise 18b**

1. $\log_e 2 - \frac{3}{2} + x - \frac{1}{8}x^2$.

2. $\sin \alpha + (\cos \alpha)(x-\alpha) - \dfrac{\sin \alpha}{2}(x-\alpha)^2$.

3. (i) $1 + \dfrac{1}{e}(x-e) - \dfrac{1}{2e^2}(x-e)^2 + \dfrac{1}{3e^3}(x-e)^3 - \dfrac{1}{4e^4}(x-e)^4$.

   (ii) $1 + \frac{1}{2}\left(x - \dfrac{\pi}{2}\right)^2 + \frac{5}{24}\left(x - \dfrac{\pi}{2}\right)^4$.

4. 0·581.

5. $f(0) + f'(0)x + \dfrac{f''(0)}{2!}x^2 + \dfrac{f''(0)}{3!}x^3 + \dfrac{f'''(0)}{4!}x^4 + \cdots$.

6. (i) $1 + 2x + 2x^2 + \frac{4}{3}x^3 + \cdots$.   (ii) $-x - \dfrac{x^2}{2} - \dfrac{x^3}{3} - \dfrac{x^4}{4} - \cdots$.

   (iii) $1 - \dfrac{x^4}{2!} + \dfrac{x^8}{4!} - \dfrac{x^{12}}{6!} + \cdots$.   (iv) $\dfrac{x}{2} - \dfrac{x^3}{48} + \dfrac{x^5}{3840} - \dfrac{x^7}{645120} + \cdots$.

8. (i) 1·4918,   (ii) 0·1823,   (iii) 0·955,   (iv) 0·199.

9. (i) $x^2 - \frac{1}{3}x^4$,

   (ii) $1 + nx + \dfrac{n(n-1)}{2}x^2 + \dfrac{n(n-1)(n-2)}{6}x^3$,

   (iii) $1 + x \log_e 2 + \dfrac{(x \log_e 2)^2}{2!} + \dfrac{(x \log_e 2)^3}{3!}$,

   (iv) $\dfrac{\pi}{2} - x - \dfrac{x^3}{6}$,   (v) $x + x^2 + \frac{1}{3}x^3 - \frac{1}{30}x^5$,

   (vi) $x - \frac{1}{6}x^3$.

page 397.

10. $x + x^2 + \frac{1}{3}x^3 - \frac{1}{30}x^5 - \frac{1}{90}x^6$.

11. $\log_e 4 + \frac{1}{4}(x-4) - \frac{1}{32}(x-4)^2 + \frac{1}{192}(x-4)^3$; 1·3913.

Qu. 15. (i) $0, 1, 0, -\frac{1}{3}, 0,$ (ii) $x - \frac{1}{3}x^3 + \frac{1}{5}x^5 - \frac{1}{7}x^7$.

Qu. 16. $1 + x + \frac{1}{2}x^2 - \frac{1}{8}x^4$.

Qu. 17. $1 + \frac{1}{2}x - \frac{1}{8}x^2 + \frac{1}{16}x^3 + \frac{49}{384}x^4$.

Qu. 18.   (i) $-2 < x < 2$,   (ii) $-1 \leqslant x < 1$,

     (iii) $-2 \leqslant x \leqslant 0$,   (iv) all values of x,

     (v) $-\frac{1}{2} < x \leqslant \frac{1}{2}$,   (vi) $-1 \leqslant x \leqslant 1$,

    (vii) $-\frac{1}{3} < x < \frac{1}{3}$,   (viii) $-1 < x < 1$,

     (ix) $-1 < x < 1$.

Qu. 19.   (i) $x < -2$ or $x > 2$,   (ii) $x < -1$ or $x \geqslant 1$.

Qu. 20.   (i) $0 < x \leqslant 4$,   (ii) $-1 < x < 1 + \pi$.

page 402.   **Exercise 18c**

1.   (i) $x + \frac{1}{6}x^3 + \frac{3}{40}x^5 + \frac{5}{112}x^7$,   (ii) $x + \frac{1}{6}x^3 + \frac{3}{24}x^5$,

   (iii) $\frac{1}{4}\pi + \frac{1}{2}x - \frac{1}{4}x^2 + \frac{1}{12}x^3 - \frac{1}{40}x^5$,   (iv) $\frac{1}{2}\pi - x - \frac{1}{6}x^3 - \frac{3}{40}x^5$.

2.   (i) $1 + \frac{x^2}{2!} + \frac{x^4}{4!} + \frac{x^6}{6!}$,   (ii) $1 + \frac{x^2}{6} + \frac{7x^4}{360}$,

   (iii) $1 - \frac{3}{2}x^2 + \frac{7}{8}x^4$,   (iv) $x + \frac{1}{3}x^3 + \frac{2}{15}x^5$,

   (v) $-\frac{1}{6}x^2 - \frac{1}{180}x^4$,   (vi) $x - \frac{1}{2}x^2 + \frac{1}{6}x^3 - \frac{1}{12}x^4$,

  (vii) $\log_e 2 + \frac{1}{2}x + \frac{1}{8}x^2 - \frac{1}{192}x^4$.

3.   (ii) $-1 < x < 1$,   (iii) $-1 < x < 1$,

   (iv) $-2 < x < 2$,   (v) $-1 < x \leqslant 1$,   (vi) yes.

6. 0·7494.   **7.** $2a - 1 < x < 1$ if $a < 1$, $1 < x < 2a - 1$ if $a > 1$.

8. $\frac{\pi}{6} - \frac{2}{\sqrt{3}}x + \frac{2}{3\sqrt{3}}x^2 - \frac{8}{9\sqrt{3}}x^3$; $-\frac{1}{2} < x < \frac{3}{2}$.

9. $1 + \sqrt{3}x + x^2 - \frac{1}{3}x^4 - \frac{2\sqrt{3}}{15}x^5 - \frac{4}{45}x^6$.

page 403.   **Exercise 18d**

1. $\frac{-2}{x+3} + \frac{2x+4}{x^2+1}$, $\frac{10}{3} + \frac{20}{9}x - \frac{110}{27}x^2 - \frac{160}{81}x^3$, $-1 < x < 1$.

2. $x + \frac{1}{2}x^2 - \frac{2}{3}x^3 + \frac{1}{4}x^4 + \frac{1}{5}x^5 - \frac{1}{3}x^6$.

6. $\frac{1}{x+2} - \frac{3}{(x+2)^2} + \frac{1}{x-3}$.

8. $\log_e 2 + \frac{1}{2}x + \frac{3}{8}x^2 + \frac{7}{24}x^3, -1 \leqslant x < 1$; $2\log_e \frac{3}{2}$, 1·098.

9. $x - 2x^2 + 2x^3 - \frac{8}{3}x^4$.

11. $(-1)^{r-1}\frac{(r-1)}{r!}3^r$.

13. $-\log_e \left(1 - \frac{1}{x^2}\right)$ expansion valid if $x > 1$ or $x < -1$, but given function not defined for $x < -1$.

page 405.

14. $\frac{1}{4}x^2 + \frac{1}{4}x^3 + \frac{7}{32}x^4$; $\dfrac{2^{n-1}-1}{n2^{n-1}}$; $-1 \leqslant x < 1$.

15. 0·787. 16. 60·3.

17. 4·01. 18. $1 + 4x + \frac{7}{2}x^2 - \frac{22}{3}x^3$.

19. $\dfrac{2}{2x+1} + \dfrac{1-x}{1+x^2}$.

21. $\dfrac{dy}{dx} = 1$, $\dfrac{d^2y}{dx^2} = 3$, $x = (y-1) - \frac{3}{2}(y-1)^2$.

22. $x < -1$ or $x > 0$. 23. $h - \frac{1}{2}h^2 + \frac{1}{4}h^3 - \frac{1}{8}h^4 + \dots$.

25. $x > 0$.

26. $\dfrac{2}{2n+1} + \dfrac{2}{3}\left(\dfrac{1}{2n+1}\right)^3 + \dfrac{2}{5}\left(\dfrac{1}{2n+1}\right)^5 + \dots$, $n < -1$ or $n > 0$.

page 409. **CHAPTER 19**

Qu. 1.

Time (s)	1	3	5	7	9	11	13	15	17	19
Acc. (m/s²)	0·47	0·92	1·06	0·75	0·53	0·44	0·37	0·36	0·35	0·31
Force (N)	35	69	79	56	40	33	28	27	26	23

Qu. 2.

Time (s)	1	3	5	7	9
Vel. (m/s)	15	22·5	30·5	36	30·5

Time (s)	2	4	6	8	10
Acc. (m/s²)	3·8	4·0	2·8	−2·8	−3·5

Time (s)	11	13	15	17	19
Vel. (m/s)	23·5	15·5	9	5·5	2·5

Time (s)	12	14	16	18
Acc. (m/s²)	−4·0	−3·3	−1·8	−1·5

Qu. 3. 0·816, 304/375 $\simeq$ 0·811.

Qu. 4. $\frac{1}{2}d(y_1 + 2y_2 + 2y_3 + 2y_4 + 2y_5 + 2y_6 + 2y_7 + y_8)$,
$\frac{1}{2}d(y_1 + 2y_2 + 2y_3 + 2y_4 + 2y_5 + 2y_6 + 2y_7 + 2y_8 + y_9)$.

Qu. 5. 240, to nearest 10. (First two ordinates are further apart than the others.)

Qu. 6. $a(y_1 + 2y_2 + 2y_3 + 2y_4 + 2y_5 + 2y_6 + 2y_7 + y_8)/14$,
$a(y_1 + 2y_2 + \dots + 2y_{n-1} + y_n)/(2n-2)$.

Qu. 7. 0·6938. **Qu. 8.** 131 m.

Qu. 9. $\frac{1}{3}d(y_1 + 4y_2 + 2y_3 + 4y_4 + y_5)$,
$\frac{1}{3}d(y_1 + 4y_2 + 2y_3 + 4y_4 + 2y_5 + 4y_6 + 2y_7 + 4y_8 + y_9)$.

Qu. 10. 0·6931. **Qu. 11.** 304/375.

page 417.   **Exercise 19a**

1. 1·3 m/s².   **2.** 32 m/s.

3. Time (s) $1\frac{1}{4}$ $3\frac{3}{4}$ $6\frac{1}{4}$ $8\frac{3}{4}$ $11\frac{1}{4}$ $13\frac{3}{4}$ $16\frac{1}{4}$ $18\frac{3}{4}$
   Speed (m/s) 3   10   15   19   21   23   25   26
   1·9 kN, 24 kW.

4. 10·2 cm².   **5.** 310 cm³.

6. 1·86 dm³.   **7.** 22, 38, 42, 50, 53, 48, 36, 15 cm².

8. 157·5 $\simeq$ 160 m/s.   **9.** $\sqrt{24} \simeq 4\cdot9$ m/s.

10. $166\frac{2}{3}$.   **11.** 0·7468.

12. (i) 0·2983,   (ii) 0·2983.   **13.** 3·988, 0·997.

15. $1\frac{1}{3}$.   **17.** $\frac{1}{6}\pi(4+6\sqrt{2}+5\sqrt{6}) \simeq 12\cdot95$.

18. $(15+\sqrt{3})\pi^2/324 \log_e 2 \simeq 0\cdot735$.

page 422   **Exercise 19b**

1. 1·325.   **2.** 0·26, 2·54.   **3.** $-0\cdot475$, 1·395.

4. 3·104.   **5.** 1·857, 4·536.   **6.** 1·32.

7. 3·521.   **8.** 0·159, 3·146.   **9.** $-0\cdot8267$, 10.

10. 2·34, 4·68, 9·13 cm.   **11.** 0·653r.   **12.** 4·1%.

13. 1·16r.

Qu. 14. The fourth difference is zero.

Qu. 15. 40, 21, 8, 1, 0, 5, 16.

Qu. 16. 4, 13, 10, 1, -8, -11, -2.   **Qu. 17.** $3x^2+\frac{1}{4}$.

Qu. 18. Corrected values of $\Delta^4(e^x)$: 3, 2, 5, 3, 6.

Qu. 19. 33; 82, 29, 13, 11, 17.

Qu. 21. $y = y_0+\frac{1}{2}\,\Delta y_0-\frac{1}{8}\,\Delta^2 y_0+\frac{1}{16}\,\Delta^3 y_0$.

Qu. 22. 1392.   **Qu. 23.** 1396·8.

page 434.   **Exercise 19c**

1. See tables of squares.

2. 0, 1, 3, 6, 10, 15, 21, 28, 36, 45; 0, 0, 1, 4, 10, 20, 35, 56, 84, 120.

3. 42 should be 40.   **4.** -25, -8, 0, 10, 27.

5. 101, 183, 288.   **6.** 11·18.

7. 13·9.   **8.** 8727 should be 8772.

10. 12, 1·0012 (1·0013 to 4 places).   **11.** (i) 0·0043,   (ii) 0·0294.

12. (i) 598,   (ii) 740.   **13.** 0·1736, 0·3420.

CHAPTER 20

page 438.   **Exercise 20a**

3. $-1 < \tanh x < 1$.

14. $\dfrac{\tanh A+\tanh B}{1+\tanh A \tanh B}$, $\dfrac{\tanh A-\tanh B}{1-\tanh A \tanh B}$.

page 439.

15. $\log_e 1\cdot 8$.   **16.** $a^2 = b^2 + c^2$.

Qu. 1.   (i) $\sinh A + \sinh B = 2 \sinh \frac{1}{2}(A+B) \cosh \frac{1}{2}(A-B)$,

         (ii) $\cosh A + \cosh B = 2 \cosh \frac{1}{2}(A+B) \cosh \frac{1}{2}(A-B)$,

       (iii) $\cosh A - \cosh B = 2 \sinh \frac{1}{2}(A+B) \sinh \frac{1}{2}(A-B)$,

       (iv) $\operatorname{sech}^2 \theta = 1 - \tanh^2 \theta$,

       (v) $\operatorname{cosech}^2 \theta = \coth^2 \theta - 1$,

       (vi) $\cosh 3\theta = 4 \cosh^3 \theta - 3 \cosh \theta$,

      (vii) $\tanh 3\theta = (3 \tanh \theta + \tanh^3 \theta)/(1 + 3 \tanh^2 \theta)$.

Qu. 2.   (i) $2 \sinh 2x$,   (ii) $\frac{1}{2} \cosh \frac{1}{2}x$,   (iii) $\sinh \frac{1}{2}x$,

      (iv) $2 \cosh 4x$,   (v) $2 \sinh \cosh x = \sinh 2x$,

      (vi) $6 \cosh^2 2x \sinh 2x$.

page 442.   **Exercise 20b**

2. $\sinh x + c$, $\cosh x + c$.

3.   (i) $3 \sinh 3x$,   (ii) $2 \cosh 2x$,   (iii) $\sinh 2x$,

      (iv) $6 \sinh^2 x \cosh x$,   (v) $6 \operatorname{sech}^2 2x$,   (vi) $-\operatorname{sech}^2 x \tanh x$,

     (vii) $3 \sinh 6x$,   (viii) $-\frac{1}{2} \operatorname{cosech}^2 x \sqrt{(\tanh x)}$,

     (ix) $2 \tanh \frac{1}{2}x \operatorname{sech}^2 \frac{1}{2}x$.

4. (i) $\frac{1}{6} \cosh 3x + c$.   (ii) $6 \sinh \frac{1}{3}x + c$.

5. (i) $2 \operatorname{cosech} 2x$,   (ii) e^{2x},   (iii) $\frac{1}{2} \operatorname{sech}^2 \frac{1}{2}x$.

6. (i) $\frac{1}{2} \tanh 2x + c$,   (ii) $-\operatorname{sech} x + c$.

7. 4.   **8.** $1/\sqrt{(1+x^2)}$.   **10.** 0.

11. (i) $\frac{1}{10} \cosh 5x + \frac{1}{2} \cosh x + c$,   (ii) $\frac{1}{8} \sinh 4x + \frac{1}{4} \sinh 2x + c$.

12. $2/(e-1)^2$.

13. $bx \cosh \theta - ay \sinh \theta - ab = 0$,

     $ax \sinh \theta + by \cosh \theta - (a^2 + b^2) \sinh \theta \cosh \theta = 0$,

     $$\frac{4a^2x^2}{(a^2+b^2)^2} - \frac{b^2}{4y^2} = 1.$$

14. $\left(\dfrac{d^2y}{dx^2} - 4y\right)(3 \coth 3x \coth 2x - 2) = 5\left(\coth 2x \dfrac{dy}{dx} - 2y\right)$.

15. $x - \dfrac{x^3}{3} + \dfrac{2x^5}{15}$.

16. $x + \dfrac{x^3}{3} + \dfrac{x^5}{5} + \ldots + \dfrac{x^{2n+1}}{2n+1} + \ldots$, $\log_e \sqrt{\dfrac{1+x}{1-x}}$.   **17.** $\log_e \sqrt{\dfrac{1+x}{1-x}}$.

18. $(\pm \sinh^{-1} \frac{1}{2}, -5\sqrt{5})$ min., $(0, -11)$ max.

Qu. 3. (i) $x = a \cosh \theta$,   (ii) $x = a \sinh \theta$.

Qu. 5.   (i) $x = a \tan \theta$ (or $x = a \sinh \theta$),

       (ii) $x = a \tanh \theta$ (or $x = a \sin \theta$ or $x = a \cos \theta$).

Qu. 7. (i) L^0,   (ii) L^{-1}.

Qu. 8. The two expressions differ by a constant (possibly zero).

Qu. 10. (i) $0\cdot 8813$,   (ii) $1\cdot 3169$,   (iii) $0\cdot 5515$.

Qu. 11. $\frac{1}{12} \cosh 3\theta - \frac{3}{4} \cosh \theta + c$.

page 447. **Exercise 20c**

1. $\sinh^{-1}\frac{1}{3}x + c$.

2. $3\sin^{-1}(x-2) + c$.

3. $\cosh^{-1}(2x) + c$.

4. $\cosh^{-1}\dfrac{x+2}{2} + c$.

5. $\frac{2}{3}\sqrt{3}\tan^{-1}\dfrac{2x+1}{\sqrt{3}} + c$.

6. $\sinh^{-1}(x-3) + c$.

7. $\frac{1}{2}\cosh^{-1}(8x+1) + c$.

8. $\frac{1}{2}\sin^{-1}\dfrac{8x-3}{3} + c$.

9. $\sinh^{-1}\frac{1}{2} \simeq 0.481$.

10. $\cosh^{-1}\frac{4}{3} \simeq 0.795$.

11. $\sinh^{-1}4 - \sinh^{-1}3 \simeq 0.2763$.

12. $\frac{1}{2}\sqrt{2}\sinh^{-1}\sqrt{2} \simeq 1.146$.

13. $\frac{1}{2}x + \frac{1}{4}\sinh 2x + c$.

14. $\frac{1}{3}\sinh^3 x + \sinh x + c = \frac{1}{12}\sinh 3x + \frac{3}{4}\sinh x + c$.

15. $(12x - 8\sinh 2x + \sinh 4x)/32 + c$.

16. $x - \tanh x + c$.

17. $\log_e \cosh x + c$.

18. $\log_e \sinh x - \frac{1}{2}\operatorname{cosech}^2 x + c$.

19. $x - \tanh x - \frac{1}{3}\tanh^3 x + c$.

20. $\log_e\{(e^x - 1)/(e^x + 1)\} + c = \log_e \tanh \frac{1}{2}x + c$.

21. $\frac{1}{5}\sinh^5 x + \frac{1}{3}\sinh^3 x + c$.

22. $-\frac{1}{3}\operatorname{sech}^3 x + c$.

23. $\frac{1}{8}\cosh 8x - \frac{1}{2}\cosh 2x + c$.

24. $\frac{1}{8}\sinh 4x + \frac{1}{4}\sinh 2x + c$.

25. $x\sinh^{-1}x - \sqrt{(1+x^2)} + c$.

26. $x\tanh^{-1}x + \frac{1}{2}\log_e(1-x^2) + c$.

27. $\frac{1}{2}x\sinh 2x - \frac{1}{4}\cosh 2x + c$.

28. $x^2\cosh x - 2x\sinh x + 2\cosh x + c$.

29. $\frac{1}{4}e^{2x} + \frac{1}{2}x + c$.

30. $-\cosh\dfrac{1}{x} + c$.

31. $\frac{3}{2}x - \frac{1}{4}e^{-2x} + c$.

32. $\frac{2}{5}x + \frac{3}{5}\log_e(3\sinh x - 2\cosh x) + c$.

33. $\sqrt{(x^2 - 1)} + \cosh^{-1}x + c$.

34. $\sqrt{(x^2 + 1)} - \sinh^{-1}x + c$.

35. $\sqrt{(x^2 - 1)} - \cosh^{-1}x + c$.

36. $\frac{1}{2}x\sqrt{(x^2 - 4)} - 2\cosh^{-1}\frac{1}{2}x + c$.

37. $\frac{1}{2}(\sqrt{2} + \sinh^{-1}1) \simeq 1.15$.

38. $\frac{1}{4}\sqrt{3} - \frac{1}{2}\cosh^{-1}2 \simeq -0.225$.

39. $4\sinh 4 - \cosh 4 - 3\sinh 3 + \cosh 3 \simeq 61.9$.

40. $\log_e 2 - \frac{3}{4}\pi$.

41. $\dfrac{1}{\sqrt{a}}\cosh^{-1}\dfrac{2ax+b}{\sqrt{(b^2 - 4ac)}} + k$.

42. $\log_e\{x + \sqrt{(x^2 - 1)}\}$.

43. $\frac{1}{2}x\sqrt{(x^2 + a^2)} + \frac{1}{2}a^2\sinh^{-1}\dfrac{x}{a} + c$;

$\frac{1}{2}x\sqrt{(x^2 - a^2)} - \frac{1}{2}a^2\cosh^{-1}\dfrac{x}{a} + c$;

$p^2 < q$, $\frac{1}{2}(x+p)\sqrt{(x^2 + 2px + q)} + \frac{1}{2}(q - p^2)\sinh^{-1}\dfrac{x+p}{\sqrt{(q - p^2)}} + c$;

$p^2 > q$, $\frac{1}{2}(x+p)\sqrt{(x^2 + 2px + q)} - \frac{1}{2}(p^2 - q)\cosh^{-1}\dfrac{x+p}{\sqrt{(p^2 - q)}} + c$;

$p^2 = q$, $\frac{1}{2}(x+p)^2 + c$ or $\frac{1}{2}x^2 + px + k$.

page 449. **Exercise 20d**

2. (i) 0, $-\log_e 3$,   (ii) $\log_e(2 + \sqrt{3}) \simeq 1.32$, $\log_e(\frac{3}{2} + \frac{1}{2}\sqrt{5}) \simeq 0.96$.

page 449.

3. *b.*
          4. $\dfrac{d^4y}{dx^4} - (p^2+q^2)\dfrac{d^2y}{dx^2} + p^2q^2y = 0.$

5. (i) (a) $\dfrac{2}{1-4x^2}$,   (b) $\dfrac{1}{\sqrt{(1+x^2)}}$;   (ii) $\dfrac{-b(3\sinh^2\theta+1)}{a^2\sinh^3\theta\cosh^3\theta}.$

6. $a^2\{\sqrt{2}-\log_e(1+\sqrt{2})\}.$

7. $ab(\tfrac{1}{2}\sinh 2\theta - \theta),\ \left(\dfrac{4a\sinh^3\theta}{3(\sinh 2\theta - 2\theta)},\ 0\right).$

8. $(1{\cdot}20,\ 1{\cdot}81).$
          9. $4\sqrt{3}\pi a^2 b.$

10. (i) (a) $2x + \dfrac{8x^3}{3!} + \ldots + \dfrac{(2x)^{2n-1}}{(2n-1)!} + \ldots,$

      (b) $\dfrac{2x^2}{2!} + \dfrac{8x^4}{4!} + \ldots + \dfrac{2^{2n-1}x^{2n}}{(2n)!} + \ldots,$

   (ii) $1 + \dfrac{x^2}{2!} + \dfrac{x^4}{4!} + \ldots + \dfrac{x^{2n}}{(2n)!} + \ldots,$ $1 - \dfrac{x^2}{2} + \dfrac{5x^4}{24}.$

11. (i) $(\sinh x - x)/x^2,$ (ii) $\cosh 2 - 1.$

13. (i) $2a^2\log_e(1+\sqrt{2}),$ (ii) $\left(a\left(\dfrac{\sqrt{2}-1}{\log_e(1+\sqrt{2})}\right),\ 0\right),$

  (iii) $\pi^2 a^3/4.$

14. (i) $\tfrac{1}{2}\tanh^2 x + c,$ (ii) $\log_e(1+\tanh x)+c,$

  (iii) $\sqrt{(x^2+a^2)} + a\sinh^{-1}\dfrac{x}{a} + c.$

15. (i) $\dfrac{2\sqrt{3}}{3}\tan^{-1}\dfrac{2x-1}{\sqrt{3}} + c,$ (ii) $\tfrac{1}{32}(12x + 8\sinh 2x + \sinh 4x) + c,$

    $\cosh^{-1}4 - \cosh^{-1}2 \simeq 0{\cdot}747,\ \ 1-\dfrac{1}{e}.$

16. (i) $\tfrac{1}{2}x\sqrt{(x^2-a^2)} - \tfrac{1}{2}a^2\cosh^{-1}\dfrac{x}{a} + c,\ \ |x|\geqslant a,$

  (ii) $\tfrac{1}{2}x\sqrt{(x^2-a^2)} + \tfrac{1}{2}a^2\cosh^{-1}\dfrac{x}{a} + c,\ \ |x|\geqslant a.$

  (iii) $\tfrac{1}{8}x\sqrt{(x^2-a^2)}(2x^2-a^2) - \tfrac{1}{8}a^4\cosh^{-1}\dfrac{x}{a},\ \ |x|\geqslant a.$

17. $P = R = \dfrac{2a}{4ac-b^2},\ \ Q = \dfrac{b}{4ac-b^2}.$

  $\dfrac{1}{4} - \dfrac{1}{12\sqrt{3}}\log_e(2+\sqrt{3}),\ \ \tfrac{19}{648}.$

18. $\sinh\{x + \tfrac{1}{2}(n-1)y\}\sinh\tfrac{1}{2}ny\ \operatorname{cosech}\tfrac{1}{2}y,$
    $\cosh\{x + \tfrac{1}{2}(n-1)y\}\sinh\tfrac{1}{2}ny\ \operatorname{cosech}\tfrac{1}{2}y.$

19. $\tfrac{1}{2}\pi,\ \tfrac{5}{32}\pi.$

20. $\tfrac{1}{6}\cosh^5 x\sinh x + \tfrac{5}{24}\cosh^3 x\sinh x + \tfrac{5}{16}\cosh x\sinh x + \tfrac{5}{16}x + c.$

page 453. CHAPTER 21

Qu. 2. $\frac{3}{2}\pi a^2$.

Qu. 3. $\dfrac{a^2}{2k}\sinh 2k\pi$.

Qu. 4. $\frac{1}{12}\pi a^2$.

Qu. 5. $2\pi + \frac{3}{2}\sqrt{3}$.

Qu. 8. πab.

Qu. 9. $17\frac{1}{15}$.

Qu. 10. $\frac{4}{3}$.

Qu. 11. $3\pi a^2$.

Qu. 12. (i) $\dfrac{dy}{dx}$, (ii) $\dfrac{ds}{dx}$, (iii) $\dfrac{ds}{dx} = \sqrt{\left\{1 + \left(\dfrac{dy}{dx}\right)^2\right\}}$.

Qu. 13. $\frac{14}{27}$.

Qu. 14. $\log_e 3$.

Qu. 15. $\frac{1}{2}\sinh^{-1}(2\sqrt{2}) + 3\sqrt{2}$.

Qu. 16. $a\alpha$.

Qu. 17. $\dfrac{a}{27}\{(9t^2+4)^{3/2} - 8\}$.

Qu. 18. $6a$.

Qu. 19. $8a$.

Qu. 20. $\varDelta r,\ r\,\varDelta\theta$.

Qu. 21. $a\sqrt{(1+k^2)}(e^{2k\pi}-1)/k$.

Qu. 22. $\frac{1}{2}a\{\sinh^{-1}\pi + \pi\sqrt{(1+\pi^2)}\}$.

Qu. 23. $8a$.

Qu. 24. $28\sqrt{5}\pi$.

Qu. 25. (i) $4\pi a^2$, (ii) $2\pi ah$.

Qu. 26. $\dfrac{8\pi a^2}{3}(2\sqrt{2}-1)$.

Qu. 27. $\frac{12}{5}\pi a^2$.

Qu. 28. $2\pi b\sqrt{(2b^2+a^2)} + \sqrt{2}\pi a^2\sinh^{-1}\left(\dfrac{\sqrt{2}b}{a}\right)$.

Qu. 29. $\frac{64}{5}\pi a^2$.

Qu. 30. 15π cm³.

Qu. 31. $\frac{1}{3}\pi r^2 h,\ \pi rl$.

Qu. 32. $\dfrac{r(r+h)}{2r+h}$.

Qu. 33. $\dfrac{2}{\pi}$ (radius) from centre along axis of symmetry.

Qu. 34. (i) $2\pi^2 a^2 b$, (ii) $4\pi^2 ab$.

Qu. 35. (i) $\pi(a+b)l$, (ii) $\frac{1}{3}(a^2+ab+b^2)/(a+b)$.

Qu. 36. $4\pi a^2\sin\alpha,\ \dfrac{a\sin\alpha}{\alpha}$.

Qu. 38. $\dfrac{1}{4a}$.

Qu. 39. (i) $2\sqrt{5}/25$, (ii) 2.

Qu. 41. $-1/\{2a(t^2+1)^{3/2}\}$.

Qu. 42. $-1/(4a)$.

Qu. 43. $\varDelta\psi,\ \dfrac{ds}{d\psi}$.

Qu. 44. $x^2+y^2-4cx-4cy+6c^2 = 0$.

Qu. 45. The centre of curvature.

Qu. 46. kx^2.

Qu. 47. (i) $\frac{1}{2}$, (ii) $\frac{1}{2}$, (iii) 0, (iv) $\frac{1}{2}a^2/b$.

page 466. **Exercise 21**

1. $9\pi/2$.

2. a^2.

3. $2c^2\log_e a$.

4. $2\pi a^2(\cos\alpha + 2\sin\alpha),\ \frac{1}{2}a(\cos\alpha + 2\sin\alpha)/(1+\alpha)$.

6. $\frac{3}{8}\pi a^2$.

7. $\frac{4}{3}\pi(a^2-r^2)^{3/2}$.

8. $\left(\dfrac{a^2-b^2}{a}, 0\right)$.

9. $c\sinh\dfrac{x}{c},\ 2\pi c\left(x\sinh\dfrac{x}{c} - c\cosh\dfrac{x}{c}\right) + 2\pi c^2$.

10. $\frac{1}{2}\sinh 2\pi$.

11. $\frac{176}{105}\pi$.

page 467.

12. (i) $\frac{3}{2}\pi a^2$, (ii) $\dfrac{16a}{9\pi}$, (iii) $\frac{8}{3}\pi a^3$.

13. $24a$, $12\pi a^2$. **14.** (i) $\dfrac{1}{8a}$, (ii) $16a$, (iii) $\dfrac{216\sqrt{3}\pi a^2}{35}$.

15. $1\frac{1}{2}$, $x^3 + y^3 - 3xy = 0$. **16.** $2\cdot58$.

18. $ct^4 - t^3x + ty - c = 0$, $\left(\dfrac{c(3t^4 + 1)}{2t^3}, \dfrac{c(t^4 + 3)}{2t}\right)$.

19. $\sqrt{(x^2 + 1)} + \frac{1}{2}\log_e \dfrac{\sqrt{(x^2 + 1)} - 1}{\sqrt{(x^2 + 1)} + 1} - \sqrt{2} - \frac{1}{2}\log_e (3 - 2\sqrt{2})$,

$\pi\{\sinh^{-1} x + x\sqrt{(1 + x^2)} - \log_e (1 + \sqrt{2}) - \sqrt{2}\}$.

20. $\frac{1}{2}\sqrt{2}\pi a^2\{\sqrt{2}\cosh\theta\sqrt{(2\cosh^2\theta - 1)} - \cosh^{-1}(\sqrt{2}\cosh\theta)$
$- \sqrt{2} + \cosh^{-1}\sqrt{2}\}$.

page 473. **CHAPTER 22**

Qu. 2. $\frac{5}{2} + \frac{1}{2}i$.

page 473. **Exercise 22a**

1. (i) $-i$, (ii) 1, (iii) i, (iv) -1, (v) -1,
(vi) $-i$, (vii) i.

2. (i) $4 + 3i$, (ii) 9, (iii) $1 - 5i$, (iv) $2i$.

3. (i) $-7 + 22i$, (ii) $8 + i$, (iii) 2, (iv) 25,
(v) $u^2 + v^2$, (vi) $2x^2 - 2y^2 + 5ixy$,
(vii) $-3q + 2ip$, (viii) $p^2 + 4q^2$.

4. (i) $-i$, (ii) $\dfrac{2 + 3i}{13}$, (iii) $\dfrac{4 + 7i}{5}$, (iv) $\dfrac{9 + 40i}{41}$,
(v) $\dfrac{x - iy}{x^2 + y^2}$, (vi) $\dfrac{x + iy}{x^2 + y^2}$, (vii) $\frac{4}{13}$.

5. (i) $-5 + 12i$, (ii) $-9 - 40i$, (iii) $x^2 - y^2 + 2ixy$.

6. (i) $-2 + 2i$, (ii) $-2 - 2i$, (iii) $-\frac{1}{4}(1 + i)$.

7. (i) 1, (ii) $\cos 2\theta + i \sin 2\theta$, (iii) $\cos(\theta + \phi) + i \sin(\theta + \phi)$.

8. (i) $\cos\theta + i\sin\theta$, (ii) $\dfrac{1 - \cos\theta + i\sin\theta}{2 - 2\cos\theta} = \frac{1}{2}(1 + i\cot\frac{1}{2}\theta)$,

(iii) $\dfrac{1 + \cos\theta + i\sin\theta}{2 + 2\cos\theta} = \frac{1}{2}(1 + i\tan\frac{1}{2}\theta)$, (iv) $\cos(\theta + \phi) + i\sin(\theta + \phi)$.

9. $\cos 3\theta - i\sin 3\theta$.

11. (i) $4iab(3a^4 - 10a^2b^2 + 3b^4)$,
(ii) $2(a^6 - 15a^4b^2 + 15a^2b^4 - b^6)$, (iii) $(a^2 + b^2)^6$.

12. (i) $3 \pm i$, (ii) $-6 \pm 2i$, (iii) $\frac{1}{4}(-3 \pm \sqrt{7}\,i)$,
(iv) $\frac{1}{6}(5 \pm \sqrt{59}\,i)$.

15. (i) $\frac{1}{2}(1 + i)(1 - i^{n+1})$, (ii) $\frac{1}{5}(1 - 2i)\{1 - (-2i)^{n+1}\}$.

page 474.

16. $\dfrac{1-(ix)^n}{1-ix} - n(ix)^n, \dfrac{\{1-(n+1)(ix)^n + n(ix)^{n+1}\}(1-x^2+2ix)}{(1+x^2)^2}.$

17. $\dfrac{1}{x-ia} + \dfrac{1}{x+ia}, \dfrac{i}{x+ia} - \dfrac{i}{x-ia}.$

18. $\dfrac{1}{3(x-1)} + \dfrac{-1+\sqrt{3}\,i}{3(2x+1-\sqrt{3}\,i)} - \dfrac{1+\sqrt{3}\,i}{3(2x+1+\sqrt{3}\,i)}.$

19. $\log_e(x^2+a^2)+c, \ i\log_e\left(\dfrac{x+ia}{x-ia}\right)+c.$ **20.** $2l^2 = n^2.$

Qu. 4. (i) $[a+c, 0],$ (ii) $[ac, 0],$ (iii) $[a-c, 0],$ (iv) $[a/c, 0].$

Qu. 6. $-y+ix, \ -x-iy, \ y-ix.$

Qu. 7. (i) 5, (ii) 1, (iii) 1, (iv) 1, (v) 3, (vi) $\sqrt{2}.$

Qu. 8. (i) $\frac{1}{4}\pi,$ (ii) 0, (iii) $\pi,$ (iv) $-\frac{1}{4}\pi,$
(v) $\frac{1}{3}\pi,$ (vi) $\frac{1}{3}\pi,$ (vii) $-\frac{2}{3}\pi.$

page 480. **Exercise 22b**

1. (i) $\sqrt{2}, \ \frac{1}{4}\pi,$ (ii) $\sqrt{13}, \ \tan^{-1}(-\frac{2}{3})+\pi,$
(iii) $\sqrt{13}, \ \tan^{-1}(\frac{2}{3})-\pi,$ (iv) $5, \ \tan^{-1}(-\frac{4}{3}),$
(v) $5, \ \tan^{-1}(-\frac{3}{4})+\pi,$ (vi) $1, \ \frac{1}{3}\pi,$
(vii) $1, \ \frac{5}{6}\pi,$ (viii) $1, \ \pi.$

2. $2i, \ -2+2i, \ -4; \ \frac{1}{4}\pi, \frac{1}{2}\pi, \frac{3}{4}\pi, \ \pi.$

3. (i) $\frac{1}{2}+\frac{1}{2}\sqrt{3}\,i, \ i, \ -\frac{1}{2}+\frac{1}{2}\sqrt{3}\,i; \ \frac{1}{6}\pi, \frac{1}{3}\pi, \frac{1}{2}\pi, \frac{2}{3}\pi.$
(ii) $2+2\sqrt{3}\,i, \ 8i, \ -8+8\sqrt{3}\,i; \ \frac{1}{6}\pi, \frac{1}{3}\pi, \frac{1}{2}\pi, \frac{2}{3}\pi.$

6. No.

7. (i) $\cos\frac{5}{12}\pi+i\sin\frac{5}{12}\pi,$ (ii) $\cos\pi+i\sin\pi = -1,$
(iii) $\cos\frac{1}{2}\pi+i\sin\frac{1}{2}\pi = i,$ (iv) $2(\cos\frac{1}{2}\pi+i\sin\frac{1}{2}\pi) = 2i,$
(v) $2\sqrt{2}\{\cos(-\frac{1}{12}\pi)+i\sin(-1\frac{1}{12}\pi)\},$ (vi) $4(\cos\frac{4}{3}\pi+i\sin\frac{4}{3}\pi).$

8. (i) $\cos(-\frac{1}{2}\pi)+i\sin(-\frac{1}{2}\pi),$ (ii) $\cos\frac{2}{3}\pi+i\sin\frac{2}{3}\pi,$
(iii) $2\sqrt{2}(\cos\frac{5}{12}\pi+i\sin\frac{5}{12}\pi),$ (iv) $4\{\cos(-\frac{5}{6}\pi)+i\sin(-\frac{5}{6}\pi)\}.$

11. $x^2+y^2, \ 1, \ 0.$ **12.** When $\arg z_1 = \arg z_2.$

14. $|z_1| = |z_1-z_2| < |z_2| < |z_1+z_2| < |z_1z_2|.$

15. $|z_1+z_2| < |z_1z_2| < |z_1| < |z_2| < |z_1-z_2|.$

16. These complex numbers correspond to:
(i) the mid-point of $P_1P_2,$ (ii) the centroid of $\triangle P_1P_2P_3,$
(iii) the point dividing P_1P_2 in the ratio $m_2:m_1.$

17. (i) $|z_1| = |z_2|,$ (ii) $\arg z_2 - \arg z_1 = \pm\frac{1}{2}\pi$ (or $\pm\frac{3}{2}\pi$).

18. $2x-y-1 = 0, \ \frac{1}{5}\sqrt{5}.$ **19.** The straight line $P_1P_2.$

23. $(x-a)^2+(y-b)^2 = r^2.$ A circle, centre $(a, b),$ radius $r.$

26. (i) The parabola $y^2 = 4ax,$ (ii) The ellipse $\dfrac{x^2}{a^2} + \dfrac{y^2}{a^2(1-e^2)} = 1.$

27. (i) $|z| = 2,$ (ii) $|z-1| = 1,$ (iii) $|z+a| = b.$

28. (i) $|z-3| = \sqrt{2},$ (ii) $|z+3-2i| = 5.$

page 483.

29. $z = (\ddot{r}\cos\theta - r\dot\theta\sin\theta) + i(\ddot{r}\sin\theta + r\dot\theta\cos\theta)$, $(\dot{r}, r\dot\theta)$, $(\ddot{r} - r\dot\theta^2, 2\dot{r}\dot\theta + r\ddot\theta)$.

Qu. 9. (i) $\cos 5\theta + i \sin 5\theta$,   (ii) $\cos 2\theta - i \sin 2\theta$,

  (iii) $\cos 3\theta + i \sin 3\theta$,   (iv) $\cos 5\theta + i \sin 5\theta$,

  (v) $\cos(\theta - \phi) + i \sin(\theta - \phi)$,   (vi) $\cos(\theta + \phi) + i \sin(\theta + \phi)$.

Qu. 10. (i) $\cos\theta + i \sin\theta$,   (ii) $\cos\frac{2}{3}\pi + i \sin\frac{2}{3}\pi = -\frac{1}{2} + \frac{1}{2}\sqrt{3}\,i$,

  (iii) $\cos\frac{3}{2}\theta + i \sin\frac{3}{2}\theta$.

Qu. 11. $1, \; -\frac{1}{2} \pm \frac{1}{2}\sqrt{3}\,i$.   **Qu. 13.** $-1, \; \frac{1}{2} \pm \frac{1}{2}\sqrt{3}\,i$.

page 487.   **Exercise 22c**

1. $\cos 5\theta + i \sin 5\theta$.   **2.** $\cos 4\theta + i \sin 4\theta$.

3. $\cos\theta + i \sin\theta$.   **4.** $\cos\theta + i \sin\theta$.

5. $\cos 8\theta + i \sin 8\theta$.   **6.** $\cos 3\theta + i \sin 3\theta$.

7. $\cos(2\phi + 3\theta) + i \sin(2\phi + 3\theta)$.   **8.** $\cos(6\theta - 6\phi) + i \sin(6\theta - 6\phi)$.

9. $1, \; -1, \; i, \; -i; \quad \frac{1}{2}\sqrt{2}(1 \pm i), \; \frac{1}{2}\sqrt{2}(-1 \pm i)$.

10. $1, \; \cos\frac{2}{3}\pi \pm i \sin\frac{2}{3}\pi, \; \cos\frac{4}{5}\pi \pm i \sin\frac{4}{5}\pi$.

  $(x^5 - 1) = (x - 1)(x^2 - 2x\cos\frac{2}{5}\pi + 1)(x^2 - 2x\cos\frac{4}{5}\pi + 1)$.

11. $1, \; -1, \; \frac{1}{2}(1 \pm \sqrt{3}\,i), \; \frac{1}{2}(-1 \pm \sqrt{3}\,i), \; (x^2 + x + 1)(x^2 - x + 1)$.

12. Rotate the radius vectors through an angle of $\pi/2n$.

13. When n is a prime number.   If n is odd but not prime, the first property will hold for some roots but not for others.   The second holds for all n.

14.   (i) $(x - 1)(x^2 - 2x\cos\frac{2}{7}\pi + 1)(x^2 - 2x\cos\frac{4}{7}\pi + 1)(x^2 - 2x\cos\frac{6}{7}\pi + 1)$,

  (ii) $(x + 1)(x^2 - 2x\cos\frac{1}{3}\pi + 1)(x^2 - 2x\cos\frac{3}{5}\pi + 1)$,

  (iii) $(x - 1)(x + 1)\left(x^2 - 2x\cos\frac{\pi}{n} + 1\right)\dots\left(x^2 - 2x\cos\frac{(n-1)\pi}{n} - 1\right)$.

15. $\cos 3\theta = 4\cos^3\theta - 3\cos\theta$.   **16.** $\sin 3\theta = 3\sin\theta - 4\sin^3\theta$.

17. $\cos 4\theta = 8\cos^4\theta - 8\cos^2\theta + 1$.

21. $\tan 6\theta = \dfrac{6t - 20t^3 + 6t^5}{1 - 15t^2 + 15t^4 - t^6}$,

$$\tan 2n\theta = \frac{\binom{2n}{1}t - \binom{2n}{3}t^3 + \dots + (-1)^{n-1}\binom{2n}{2n-1}t^{2n-1}}{1 - \binom{2n}{2}t^2 + \dots + (-1)^n\,t^{2n}},$$

$$\tan(2n+1)\theta = \frac{\binom{2n+1}{1}t - \binom{2n+1}{3}t^3 + \dots + (-1)^n t^{2n+1}}{1 - \binom{2n+1}{2}t^2 + \dots + (-1)^n\binom{2n+1}{2n}t^{2n}},$$

where $t = \tan\theta$.

22. $-5{\cdot}03, \; -0{\cdot}668, \; 0{\cdot}199, \; 1{\cdot}50$.   **23.** $-1{\cdot}69, \; -0{\cdot}431, \; 0{\cdot}225, \; 1{\cdot}14, \; 10{\cdot}8$.

24. $u = x^3 - 3xy^2, \; v = 3x^2y - y^3$.   **25.** $\dfrac{ac + bd + i(bc - ad)}{c^2 + d^2}$.

26. (i) $\pm(2 + 3i)$,   (ii) $\pm\dfrac{1}{\sqrt{2}}(1 + i)$.